INDUSTRIAL ENERGY MANAGEMENT AND UTILIZATION

INDUSTRIAL ENERGY MANAGEMENT AND UTILIZATION

Larry C. Witte
University of Houston
Houston, Texas

Philip S. Schmidt
University of Texas
Austin, Texas

David R. Brown
David R. Brown, Inc.
Austin, Texas

⦿ **HEMISPHERE PUBLISHING CORPORATION**
A member of the Taylor & Francis Group

New York   Washington   Philadelphia   London

INDUSTRIAL ENERGY MANAGEMENT AND UTILIZATION

    3 4 5 6 7 8 9 0   B C B C   8 9 8

BookCrafters, Inc. was printer and binder.

Library of Congress Cataloging-in-Publication Data

Witte, L. C.
  Industrial energy management and utilization.

  Bibliography: p.
  Includes index.
  1. Industry—Energy conservation. 2. Industry—
Power supply. I. Schmidt, Philip S. II. Brown,
David R. (David Robert), date. III. Title.
TJ163.W57   1987   333.79   86-22776
ISBN 0-89116-322-0

CONTENTS

Index   659

PREFACE

Effective energy utilization is desirable for conserving the earth's natural fuel resources as well as for reducing the cost of energy consumption. Industry is by far the biggest user of energy and, although much has been done to use energy more effectively in the past few years, many industries can profit by implementing some of the strategies described in this text.

The material presented here is an outgrowth of the authors' involvement in industrial energy conservation and auditing. Some of the material was first prepared for a set of short courses offered at the University of Houston and the University of Texas at Austin. It has been altered and augmented to make it suitable for a text that can be used for a senior-level elective or first-year graduate course. In the fall of 1983 the text material was used in a senior elective course at the University of Houston. The practicality of the material has been retained so that practicing engineers might also use the text effectively without the need for formal instruction in the topic areas.

Energy utilization and management involve a unique blend of several disciplines in which most engineers have some knowledge. Economics, management, thermodynamics, heat transfer, fluid mechanics, electricity, and control theory are all important aspects of intelligent energy utilization. However, bringing all these disciplines to bear on the energy utilization field has been accomplished only in a fragmentary fashion up to now. This text brings these subjects together in the context of energy utilization and management.

Chapters 1 through 4 address the socioeconomic aspects of energy utilization. Some of this material might be omitted in a course, depending on the background of the students and the intent of the instructor. The material contained in these chapters will, however, prove useful to anyone who is involved in developing an energy management program.

Chapter 5 is a cursory review of topics crucial to analyzing thermal-fluid systems. Although the chapter might be redundant to some readers, it should serve as a refresher for those who have training in the thermal sciences but who have become unfamiliar with the field. The chapter is not directed to readers unschooled in the thermal sciences; they must consult the basic texts in the field to gain the proper background.

Chapters 6 through 11 address specific areas where conservation practices have proved successful. These represent the heart of energy management techniques and, as such, most students will find them especially interesting. The chapters are written such that they need not be covered in serial fashion.

Chapter 12 describes cogeneration, which involves the simultaneous generation of electricity and steam for processes. Analysis of cogeneration systems involves technical and economic concepts covered in the preceding chapters; hence, this chapter is best covered after preceding chapters are thoroughly understood.

Many mechanical, chemical, or industrial engineers involved in energy management overlook opportunities for conservation in electrical systems. Chapters 13 and 14 provide a review of the principles of electricity and ways to conserve energy in electrical systems, respectively.

Chapter 15 shows how six of the most energy intensive industries in the United States use energy. Sources of energy, intermediate and end uses of energy, and opportunities for energy conservation are all detailed for these industries. The chapter is intended to be a reference for readers and instructors to show where the techniques and strategies described in the text might be applied.

Thanks are due the many people who have participated directly and indirectly in the development of this text. Constructive criticism from short-course participants and from the University of Houston students to whom the text material was presented is appreciated. Thanks are also due the Texas Industrial Commission for encouraging the development of the text through direct sponsorship of short courses and other conservation activities.

We appreciate the efficiency and courtesy of Barbara Bodling of Hemisphere in working to bring the text to completion. Special thanks go to M. A. Williams for the knowledge that he has conveyed to us about energy audits and the management and organization of energy conservation programs. Finally, we must say thanks to our respective families for their sacrifices as we worked early and late to complete this project.

Larry C. Witte
Philip S. Schmidt
David R. Brown

INDUSTRIAL ENERGY MANAGEMENT
AND UTILIZATION

ENERGY: CONSUMPTION, CONSERVATION, AND RESOURCES

If the evolution of the earth as a livable planet were compressed to the time scale of a year, human history would occupy little more than a small part of December 31. In fact, the industrial era would last only a few seconds. However, in that very short period of time, humans have undone what it took natural processes millions of years to accomplish. We have converted a very significant fraction of the stored hydrocarbon in the earth's crust to heat and to materials that we consider useful.

It is becoming more and more difficult and costly both to locate and to produce additional hydrocarbon-based fuel for the production of heat. We have somewhere between a 10- and a 50-year supply of readily extractable oil and gas. Even solid fuels (coal, shale, etc.) will not last indefinitely. Uranium to fuel nuclear reactors also would prove to be in short supply if we were to implement the use of reactors on a grand scale in the industrialized countries of the world.

Several questions immediately arise. What can be done about this situation? Must we of necessity return to a "harder," less energy-intensive standard of living? Is increased exploration for and production of fossil-based fuels the answer? (For the long term, it obviously is not.) Can nuclear fission reactors in the short term, and fusion reactors in the long term, provide the energy needed to offset the loss of fossil fuel production? Can solar energy satisfy our long-term energy needs? What about the other so-

called alternative energy resources—wind, waves, geothermal, and others
— can they have a significant impact on our requirements?

These are some of the most difficult questions facing humans today. The
answers will have severe economic effects on many countries throughout the
world. Even so-called underdeveloped countries may be relegated to
undeveloped status because of this energy crisis.

We do not know the answers to all of these questions. However, we do
know the history of energy consumption in the United States, Europe,
Japan, and other highly industrialized societies. We also know with some
degree of certainty the proven reserves of natural fuels in the world. And, in
recent times, a large scientific effort has been devoted to assessing alternative ways of converting energy.

Lately we have seen an increased willingness on the part of the general
populace to conserve energy, perhaps not from patriotic motivation or for
the welfare of the earth. More likely this willingness is due to economic
necessity as the price of energy has risen more rapidly than the general level
of prices in our economy. In the United States we are just beginning to adopt
conservation techniques that have been used for many years in energy-
starved Europe.

Our prospects are not all bleak. It appears that with sound conservation
measures, with increased production of previously noneconomic energy
resources, with the addition of alternative energy supplies, and with sound
legislative and fiscal policies we can survive the short-term crisis without
serious degradation of our standard of living. In the long term, there appear
to be feasible solutions to the energy problem.

This book emphasizes the conservation aspects of energy management.
Industrial energy conservation is the main theme. Industry consumes vast
quantities of energy, and the potential for savings is correspondingly vast.
Industry usually has stable capital and cash flows and can undertake energy
conservation projects. This is not always the case in residential, commercial,
and institutional buildings or in transportation. Mass transit systems, for
example, may need transitory government subsidies and taxing authority.

1.1 HISTORICAL PATTERNS

History can reveal past patterns of energy consumption and can perhaps
suggest new courses of action that could be beneficial. The pattern of world
energy consumption for the twentieth century is shown in Fig. 1.1. The drop
in consumption between 1930 and 1935 reflects the seriousness of the worldwide depression that occured then. The advent of liquid and gaseous forms
of fuel and the relative decline of solid fuels such as wood and coal are also
evident in recent years. Figure 1.2 shows a similar plot for the United States.
The trends are similar, with an even greater move away from solids toward

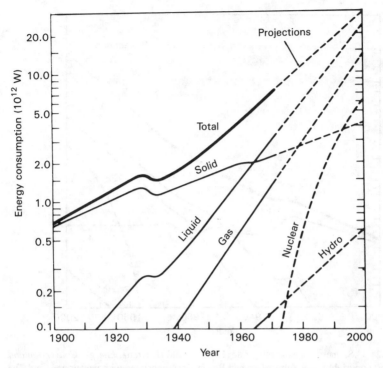

Figure 1.1 World energy consumption rate. From Jerrold H. Krenz, *Energy: Conversion and Utilization*, copyright © 1976 by Allyn and Bacon, Inc., reprinted with permission, and *The Energy Decade 1970–1980*, Ballinger, 1982.

fluid fuels. This reflects the move away from coal toward oil and gas between the world wars.

Another problem in the United States and Canada, and to an extent in Europe, is that far more energy per capita is being consumed than in most other countries. Figure 1.3 illustrates this quite graphically. The United States consumes about 10 times as much energy per capita as the world average. In contrast, India, which is making some strides toward modernization, consumes only about one-fiftieth as much energy per capita as the United States. This has led the United States to import greater and greater amounts of fuel, as shown in Fig. 1.4. Japan and Europe, with little domestic fuel production, are even more dependent on fuel imports than is the United States. This is clearly shown in Fig. 1.5, which illustrates the more recent history of world oil production. Figure 1.6 shows the major exporters of oil and dramatizes the western world's dependence on the OPEC[1] cartel.

[1] The Organisation of Petroleum Exporting Countries includes Iraq, Iran, Kuwait, Qatar, Saudi Arabia, United Arab Emirates, Venezuela, Equador, Algeria, Libya, Ghana, Nigeria, and Indonesia.

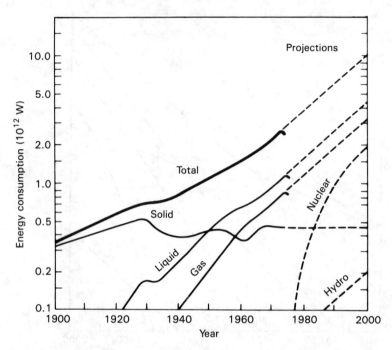

Figure 1.2 U.S. energy consumption rate. From Jerrold H. Krenz, *Energy: Conversion and Utilization,* copyright © 1976 by Allyn and Bacon, Inc., reprinted with permission, and *The Energy Decade 1970–1980,* Ballinger, 1982.

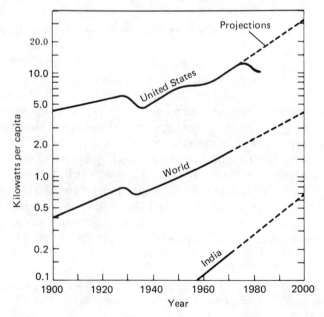

Figure 1.3 Per capita energy consumption. From Jerrold H. Krenz, *Energy: Conversion and Utilization,* copyright © 1976 by Allyn and Bacon, Inc., reprinted with permission.

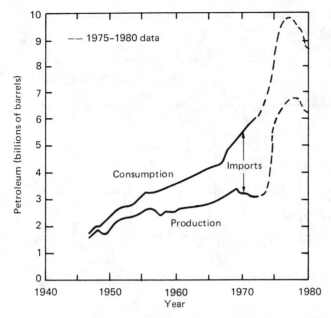

Figure 1.4 U.S. petroleum production and consumption. From Krenz [1]. The 1975–1980 data are from the *World Almanac*, 1983, pp. 141–142.

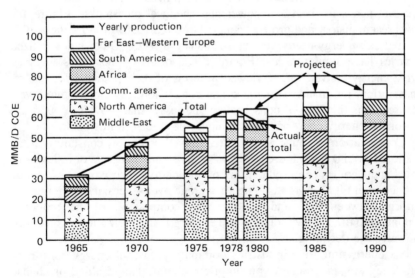

Figure 1.5 World oil production. MMB/D COE refers to 1 million barrels per day crude oil equivalent. This would be approximately equal to 5.8×10^{12} Btu or 1.7×10^{9} kWh. From [2].

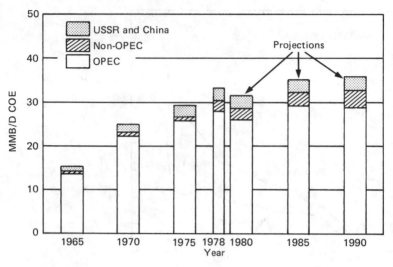

Figure 1.6 World oil exports by producer. From [2].

1.2 POTENTIAL FOR ENERGY SAVINGS

A very substantial fraction of energy demand in the United States fills our industrial needs for heat and power and chemical feedstocks.[2] Figure 1.7 shows recent history and projects demand through 1990. In 1978 about 30% of America's energy needs were in the industrial sector. Other industrialized countries have similar needs. Figure 1.7 also shows the losses incurred during energy conversion. If viewed in terms of their potential for energy conservation, the industrial demand and conversion losses constitute a vast resource. It is this area of consumption and losses that provides opportunities for the energy conservation techniques described in this book.

Conservation can have a substantial impact on energy consumption, as illustrated in Fig. 1.8. Figure 1.8*d* shows the steady progress of industry in reducing the energy required per unit of production since 1965. (The unit of production in Fig. 1.8*d* is based on an average over several industries and is uniform between 1965 and 1990.) Other components of the economy responded to the 1973 Middle East oil embargo, and their energy consumption rates are declining.

Six energy-intensive industries dominate industrial energy consumption. They are, in order of consumption, chemicals and allied products; primary metals; petroleum and coal products; stone, clay, and glass products; paper and allied products; and food and kindred products. Table

---

[2] Fuel used to produce material goods rather than heat or power.

1.1 shows consumption by these industries in selected years between 1947 and 1980. All of these industries show substantial increases in total energy use, but, as seen in Fig. 1.8*d*, the rate of consumption per unit of production has steadily declined. We conclude that industry has not been wasteful of energy, even though until recently energy was considered "cheap." But continued progress in conservation is needed, and even small savings are more and more difficult to achieve.

1.2.1 Industrial Processes

It is possible to categorize energy consumption patterns in industry by the kind of process involved. These categories cut across many industries and allow a treatment of conservation potential and practice common to those industries. Arranged in approximate order of energy consumption, these categories are:

- Boiler fuel: The energy required to produce steam for process energy and space conditioning.
- Direct process heat: The energy supplied to dryers, kilns, ovens, process heaters, reheat furnaces, etc. Boiler fuel is excluded.
- Feedstock: The equivalent energy of materials consumed as feedstock to produce material goods.
- Mechanical drive: The energy supplied primarily to electric motors for liquid pumping, conveyors, grinders, sorters, etc.
- Space conditioning: The energy expended directly to heat, cool, or ventilate work space.
- Lighting.

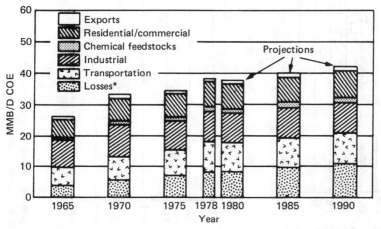

*Incurred in converting from one energy form to another, e.g., electricity generation.

Figure 1.7 Total U.S. energy demand by market. From [2].

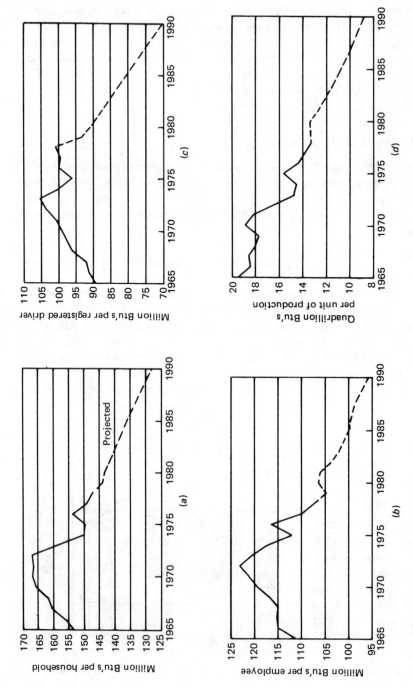

Figure 1.8 Conservation efficiency in several sectors of the U.S. economy. (a) Residential; (b) commercial; (c) motor gasoline; (d) industrial heat and power. From [2].

8

Table 1.1 Gross Energy Consumed by High-Energy-Using Manufacturing Groups, Selected Years 1947–1980 (trillion Btu's)

	1947	1954	1958	1962	1967	1971	1975	1980
Purchased by all manufacturing	8,738	9,766	10,696	12,485	15,463	17,060	20,919	24,383
By six high-energy-using groups								
Food and kindred products	857	883	951	992	1,098	1,284	1,284	1,453
Paper and allied products	635	801	932	1,068	1,367	1,560	1,725	1,771
Chemicals and allied products	1,023	1,753	2,282	2,592	3,257	3,473	4,997	6,020
Petroleum and coal products	550	695	1,018	1,252	1,543	1,820	2,083	2,302
Stone, clay, and glass products	929	1,032	1,058	1,178	1,341	1,444	1,617	1,864
Primary metal industries	2,547	2,499	2,182	2,833	3,340	3,364	4,031	4,655
Sum of six groups	6,541	7,663	8,423	9,915	11,964	12,947	15,737	18,065
By all other manufacturing	2,197	2,103	2,273	2,570	3,517	4,113	5,182	6,318

Source: Ford Foundation, *Energy Consumption in Manufacturing,* 1974 [3].

Table 1.2 shows a breakdown of the primary categories between the various energy sources in manufacturing.

It should be pointed out that the energy consumption per unit of production can be strongly affected by nontechnical factors. For example, when plant capacity utilization is low, the energy cost per output will increase, since there are certain threshold energy consumption rates required to keep equipment operational. When capacity utilization is very high, the energy cost per output will also be high if equipment is run above its optimum efficiency level. These factors must be taken into account when the potential for energy conservation is being considered.

1.2.2 Efficiency Factors

An industrial plant can be viewed simply as a system to which raw materials, energy, and labor are provided and from which material goods, waste materials, and, always, waste energy are obtained. The waste energy is generally heat energy that is rejected to the surroundings because of the cyclic nature of many plant processes. It is worthwhile from an energy and cost savings viewpoint to minimize the heat actually rejected to the atmosphere. However, according to the second law of thermodynamics, a certain amount of heat must be returned to the environment for cyclic devices[3]; hence, total reclamation of the waste heat may not be possible. In many cases, waste materials can be recycled as fuels, for instance, the paper industry uses wood by-products as fuel.

Overall efficiencies of industrial plants are strongly influenced by the thermal efficiencies of their components. Figure 1.9 shows some estimated

---

[3]This limitation resulting from second law considerations is discussed further in Chapter 9.

Table 1.2 1974 Btu Consumption in the Manufacturing Sector by Application and Fuel Type (10^{12} Btu)

FUEL	APPLICATION								
	Process steam	Direct heat	Indirect heat (nonsteam)	Mechanical drive	Internal electrical generation	Electrolytic	Feedstock	Other	Total
Coal	544.8	237.6	0	0	288.0	0	0	263.9	3,642.1
Oil	706.6	425.6	710.7	46.3	114.0	0	1,882.5	517.3	4,220.0
Natural gas	1,974.1	1,727.9	2,008.7	137.4	429.7	0	475.4	1,582.9	8,109.0
Electricity	0	63.7	39.0	809.8	−292.5	373.8	0	1,013.0	2,091.9
Other	1,005.7	65.8	25.3	11.7	22.4	13.0	1,518.1	935.3	1,842.9
Total	4,231.1	2,520.6	2,783.8	1,005.1	501.5	386.9	3,876.0	4,312.5	20,105.9

Source: Federal Energy Administration, *Energy Consumption in the Manufacturing Sector,* 1977 [4].

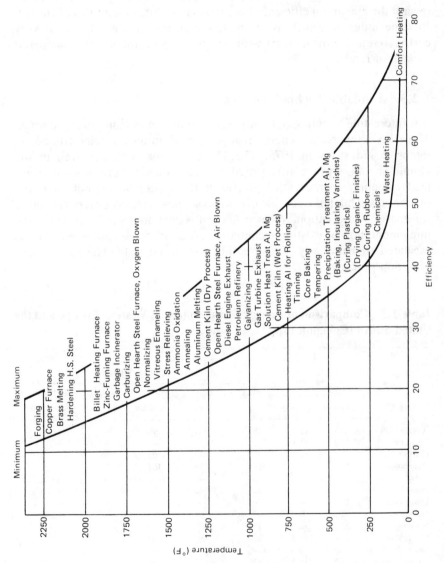

Figure 1.9 Approximate efficiencies for industrial processes. Reprinted from Chiogioji [6] by courtesy of Marcel Dekker, Inc.

limits of the thermal efficiences for a host of processing devices in terms of the process temperature. Processes involving higher temperatures (forging, metal working, etc.) have very high thermal losses. Yet with new equipment designed with energy conservation in mind, it will be possible to move toward the maximum efficiency line, thus saving large quantities of energy. In some industries, such as the metals industry, new equipment is very costly, so refitting the industry with energy-efficient equipment is hampered by lack of funds.

1.2.3 Available Technology

It is interesting to compare U.S. practice with that of European countries. Table 1.3 compares the energy cost per ton of product in selected energy-intensive industries in 1975. The United States compares well in the aluminum industry (a relatively modern industry and long considered to be a leader in energy conservation practice in the United States) but not in the other areas. Thus it is clear that in many cases new technology is not required for energy conservation. What is needed is economic justification for the implementation of existing technology. Since 1975 the cost of energy has escalated tremendously, thus creating a more favorable climate for conservation practices.

Table 1.3 Comparison of Energy Consumption in Western Europe and the United States (Chiogioji [6])

| Country | Sectorial energy consumption, 1975 (% TPE)[a] | | Specific efficiencies (10^4 kcal/ton of product) | | | | |
	Industry	Transport	Crude steel	Pulp and paper	Cement	Petroleum products	Aluminum
Austria	30.7	19.2	450	344	90	—	1,346
Belgium	41.2	11.8	—	—	—	—	—
Canada	26.0	19.1	555	673	95	180	—
Denmark	16.8	17.9	236	353	164	53	—
Italy	34.4	15.4	334	340	96	45	—
Japan	47.3	—	513	512	120	46	1,385
Luxembourg	71.0	—	701	—	—	—	—
Netherlands	32.5	10.0	470	—	131	—	1,290
Norway	33.9	12.2	189[b]	659	115	—	1,591
Sweden	31.6	12.3	398	489	140	49	1,648
Switzerland	18.3	17.8	—	557	—	—	—
United Kingdom	28.5	14.3	478	627	138	73	2,107
United States	27.7	25.0	543	579	161	90	947
West Germany	36.1	13.4	326	438	91	80	1,481–1,503

[a] TPE, total primary energy.
[b] Electricity only.
Source: IGT Highlights, Supplement, April 11, 1977.

Opportunities abound for the application of current technology to waste heat recovery. Using waste heat from a high-temperature process as input to a lower-temperature process can be a highly efficient approach. Using heat exchange devices to recover waste heat for space heating, process heating, or perhaps direct recycling as input energy is a common practice that has an enhanced potential as energy becomes more costly. The fine-tuning of combustion with the aid of in situ gas analyzers and on-line microprocessor computing can save huge amounts of energy and can be helpful in environmental pollution control as well. These techniques and others are described in later chapters of this text.

1.3   ALTERNATIVE ENERGY SOURCES

The words "alternative energy sources" are rather vague and may mean different things to people living in different parts of the world. Geothermal energy in the United States represents a minute fraction of total energy conversion, while in Iceland it is a major source of energy. An American would view geothermal energy as an alternative energy source, but an Icelander would not.

In industrialized countries the production of heat energy is dominated by conventional hydrocarbon combustion. Collection of heat directly from the sun, the energy of wind, waves, hydroelectric power, geothermal energy, and even nuclear generation are of less importance.

Nuclear electrical generation is an established technology. It can have a substantial impact on meeting the projected demands for electrical energy in the next several decades. The industry is currently beset with licensing problems related to safety and radioactive waste disposal. Solutions to these problems that can meet with public and governmental acceptance will allow the industry to move forward. (In this text we omit descriptions of the technology of nuclear power generation. The reader is directed to the works of Foster and Wright [7] and Sesonske [8] for the details of various aspects of reactor operation.)

In this section we discuss the potential for energy conversion for solar energy and for water, wind, and geothermal power. Some details of the technologies also are presented.

1.3.1   Solar Energy

An often-stated philosophical point is that all terrestrial energy ultimately comes from the sun and thus should be called solar energy. However, solar energy is commonly understood to mean the instantaneous collection of solar flux for immediate usage or for short-term storage for later use.

The amount of solar energy that reaches the earth is immense but very diffuse. The average worldwide flux at the earth's surface is estimated to be about 180 W/m^2. The average energy incident per day (24 hours) is 4.37 kWh/m^2. This of course occurs during the daylight hours. It must also be realized that no type of collection device is able to convert all of the incident solar flux into useful energy.

Collecting the sun's radiant energy and converting it into heat is a reasonably well developed technology. The key constraint in effective utilization lies in matching the temperature level and heat flow to a specific application. Generally, for industrial process heat or for solar electric power generation, large quantities of heat at relatively high temperatures are required.

Solar Power Generation Solar power generation on a large scale is an emerging technology. Two approaches are being considered. The first is photovoltaic conversion. The application of direct photovoltaic conversion is limited by efficiency and cost factors. Although improvements have been made, the efficiency of a single-crystal silicon cell is on the order of 15%. Other materials such as polycrystalline silicon and cadmium sulfide are considerably less efficient, so their potentially lower cost is offset by lower efficiency. Capital investment per kilowatt-hour is on the order of 10 times that of a conventional steam-electric power plant. Photovoltaics are presently limited to applications where cost is not a concern.

The other approach is called solar thermal power generation. The name refers to the fact that radiant energy is collected and converted to electric output at a central power plant, such as that shown in Fig. 1.10. Heat input to the boiler can be accomplished either through a field of distributed collectors or by reflecting the sun's rays with a heliostat[4] field onto a central receiver. The distributed system involves individual collectors, which are used to heat a working fluid. The collectors are connected by a piping network so that the hot fluid can be delivered to a central conversion plant. The major disadvantage of this system is the cost of the piping network and thermal and mechanical losses that occur between the collectors and the central station.

The heliostat system, shown schematically in Fig. 1.11, consists of a field of many heliostats that reflect solar radiation onto a central receiver. This system minimizes heat loss, but since heliostats must be individually steered, the cost is driven up. Hildebrandt and Vant-Hull [9] give an excellent description of this type of system. A 50-mW system is currently being tested at Barstow, California, under the sponsorship of the U.S. government and Southern California utilities.

Solar power generation will not directly affect energy conservation. It

---

[4] Heliostat is a mirrored reflecting surface arranged to reflect in a controlled manner.

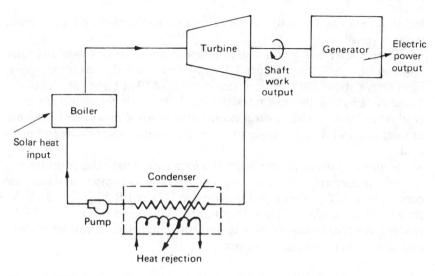

Figure 1.10 Typical central energy conservation plant for solar collection system.

can ease the dependence on fossil fuels and clearly is important if we are to meet our energy needs.

Solar Industrial Process Heat The use of solar energy for low-grade process heat can save other scarce energy resources. In 1972 industrial process heat was estimated to consume over 15 quadrillion (10^{15}) Btu [10]. In 1974, approximately 9.5×10^{15} Btu was consumed as process heat in the United States alone [4]. Fraser [11] estimates that by the year 2000 about 20% of industrial process heat could be furnished by solar technology. About 25% of industrial process heat is used at temperatures below 350°F (177°C). These temperatures can be achieved with moderately concentrating solar collectors. Also, about 5% of industrial heat occurs at temperatures less than 212°F (100°C). This application could be met by flat-plate collectors. The economics of solar process heat is promising since the col-

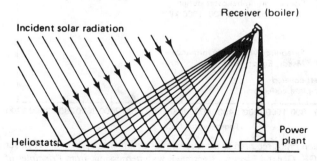

Figure 1.11 Heliostat field for concentration of solar insolation.

lector system can be designed for a specific application and year-round usage is possible.

To use solar energy for higher-temperature processes would require concentrating collectors. Figure 1.12 shows qualitatively that for temperatures above about 300°C concentration ratios of 10 or higher are required. These systems usually must track the sun to maintain focus and are more costly than devices with lower concentration ratios. Consequently, the use of solar energy for high-temperature process heat applications is less practical.

Figure 1.13 shows the configuration for a typical flat-plate collector and for two moderately concentrating collectors, the compound parabolic concentrator (CPC) developed by Winston [12] and Rabl [13] and the V-trough or trapezoidal collector described by Howell and Bannerot [14]. The reader is referred to Kreith and Kreider [10] for detailed discussions of these and many other types of collectors.

1.3.2 Other Alternative Energy Sources

Several other sources are currently being used to varying extents to fulfill the world's needs for energy. Ocean power conversion, wind conversion, and geothermal energy conversion are all viable options in today's energy mix. For brevity, these sources will not be discussed in detail here.

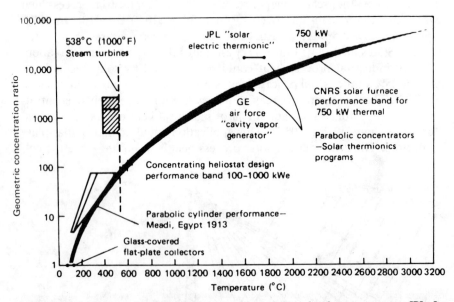

Figure 1.12 Concentration ratios required to achieve various absorber temperatures. JPL, Jet Propulsion Laboratory; GE, General Electric. Reprinted, with permission, from *Principles of Solar Engineering* by F. Kreith and J. Kreider, Hemisphere, Washington, D.C., 1978 [10].

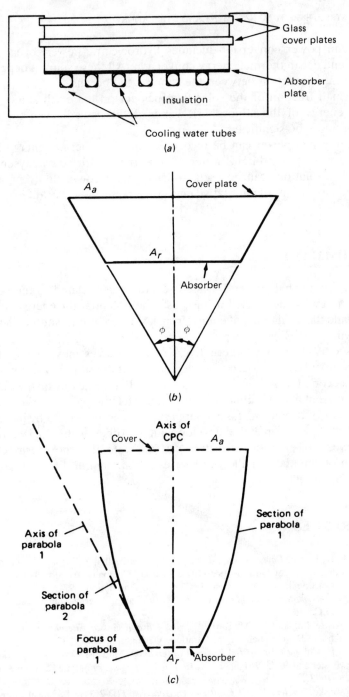

Figure 1.13 Solar collector geometries. (a) Flat-plate collector, two cover plates; (b) moderately concentrating, trapezoidal groove collector geometry; (c) moderately concentrating, compound parabolic collector geometry.

Ocean power conversion includes hydroelectric power, ocean wave power, and ocean thermal energy conversion. All these sources depend on the natural interaction between the sun and the oceans.

A small fraction of the solar flux incident on the earth shows up as kinetic energy of the atmospheric winds. If harnessed, even this small fraction might be significant.

Geothermal power can be recovered from the thermal energy in the outer crust of the earth. Hot water, such as that produced by geysers, and hot rock formations can be used to drive power conversion equipment. Detailed discussions of alternative energy sources for the world are given in [15–22].

1.4 SUMMARY

In this opening chapter we discussed several issues related to the way in which energy has been used in the past, the potential for energy conservation, and the availability of energy from less conventional sources, such as solar, wind, and wave power.

It seems clear that "alternative" energy sources must eventually be used, as fossil-based resources are depleted. But the situation is clouded by complex economic, political, and legal issues that create drastic fluctuations in the apparent need to implement alternative technologies. As time passes, the urgency of decreasing fossil energy supplies accompanied by increasing energy costs will force the development of other technologies. In the meantime, conservation measures can alleviate the short-term energy supply problem and can allow industries to maintain profits in the face of increasing energy costs.

REFERENCES

1. Jerrold H. Krenz, *Energy Conversion and Utilization*, Allyn & Bacon, Boston, 1976.
2. The National Energy Outlook 1980–1990, Shell Oil Company, August 1980.
3. *Energy Consumption in Manufacturing*, The Ford Foundation, 1974.
4. *Energy Consumption in the Manufacturing Sector,* Federal Energy Administration, Washington, D.C., 1977.
5. John A. Belding and W. M. Burnett, *From Oil and Gas to Alternative Fuels*, Energy Research and Development Administration, Washington, D.C., 1976.
6. Melvin H. Chiogioji, *Industrial Energy Conservation*, Marcel-Dekker, New York, 1979.
7. A. R. Foster and R. L. Wright, Jr., *Basic Nuclear Engineering,* 2d ed., Allyn & Bacon, Boston, 1973.
8. A. Sesonske, *Nuclear Power Plant Design Analysis*, TID-26241, Technical Information Center, Office of Information Services, U.S. Atomic Energy Commission, Washington, D.C., 1973.

9. A. Hildebrandt and L. Vant-Hull, Solar Thermal Power System Based on Optical Transmission, *Solar Energy* 18:31–39, 1976.
10. F. Kreith and J. Kreider, *Principles of Solar Engineering*, Hemisphere, Washington, D.C., 1978.
11. M. D. Fraser, Intertechnology Corporation Assesses Industrial Heat Potential, *Solar Engineering* 1:11–12, September 1976.
12. R. Winston, Solar Concentrators of a Novel Design, *Solar Energy* 16:89, 1974.
13. A. Rabl, Optical and Thermal Properties of Compound Concentrators, *Solar Energy* 18: 497, 1976.
14. J. R. Howell and R. Bannerot, Moderately Concentrating Flat-Plate Solar Energy Collectors, ASME Paper 75-HT-54, 1974.
15. H. H. Landsberg and Sam H. Schurr, *Energy in the United States*, Random House, New York, 1968.
16. N. Rosenberg, *Technology and American Economic Growth*, Harper & Row, New York, 1972.
17. M. King Hubbert, *Energy Resources*, National Academy of Sciences, Washington, D.C., 1962.
18. M. King Hubbert, *World Power Data*, U.S. Government Printing Office, Washington, D.C., 1972.
19. M. King Hubbert, Energy Resources, in *Resources and Man*, National Academy of Sciences/National Research Council, W. H. Freeman, San Francisco, 1969.
20. J. W. Reed, *Wind Power Climatology of the United States*, SAND 74-0348, Sandia Laboratories, Albuquerque, N.M., 1975.
21. R. E. Wilson and P. S. B. Lissaman, *Applied Aerodynamics of Wind Power Machines*, NSF/RANN GI 41840, Oregon State University, Corvallis, 1974.
22. G. J. Nunz, Hot Dry Rock Geothermal Energy, *Mech. Eng.* 102(12):26, 1980.

ENERGY AUDITS AND ENERGY CONSERVATION

2.1 INTRODUCTION: THE ENERGY AUDIT CONCEPT

The dictionary variously defines the word *audit* as "a formal examination and verification of financial accounts," "a methodological examination and review," or "an examination with intent to verify." While historically audits have concentrated on a determination of the financial conditions of an organization or business, it is wholly appropriate, in an era in which energy has become a major cost factor in many manufacturing processes, to extrapolate these definitions to a consideration of the energy utilization in industrial operations. In many industrial processes, in fact, energy and profit are so closely related that financial audits and energy audits are close kin.

Since the formal declaration of the energy crisis in 1974, various meanings have been attached to the term *energy audit*, often tied to programs of government assistance in energy conservation. These audits may range from a cursory listing of obvious points of energy waste in a plant to lengthy and complex technical and economic analyses of individual process units. Here we will use the term energy audit to denote an in-depth study of a facility (1) to determine how and where energy is being used or converted from one form to another, (2) to identify opportunities to reduce energy usage, (3) to evaluate the economics and technical practicability of implementing these reductions, and (4) to formulate prioritized recommendations for implementing process improvements to save energy.

2.2 ELEMENTS OF ENERGY AUDITS

While detailed procedures for energy audits may vary somewhat from industry to industry and even from plant to plant within an industry, certain basic elements are common to all comprehensive energy audits, regardless of the nature and size of the operation. Briefly, these may be defined as follows:

1. Historical review of energy-related records to establish a baseline against which progress can be measured.
2. Preplanning walk-through of the plant to identify major energy-using components, to familiarize the audit team with the general energy and material flows through the process, and to identify obvious sources of energy waste, such as leaks and uninsulated equipment.
3. Detailed definition of data requirements.
4. Computation of mass and energy flows and estimation of energy losses.
5. Enumeration of energy conservation opportunities (ECO).
6. Estimation of energy savings potential for each ECO.
7. Determination of cost and profitability potential for the implementation of ECOs.
8. Establishment of priority recommendations for ECO implementation.
9. Establishment of a continuous monitoring effort for major energy-using systems.

Some of these elements will be discussed in detail below.

2.2.1 Historical Review

An examination of records of energy consumption within a plant can provide valuable insight into the nature of energy usage and can be used to establish a baseline from which future consumption can be gauged. Fuel and electricity billings for at least 12, preferably 24, months prior to the audit should be used. In examining these records several important points should be considered.

Billing records must be analyzed carefully to ensure that energy consumption is actually the quantity being determined. Direct dollar expenditures for fuel or electricity are usually a poor measure of energy consumption, since utility bills normally include several cost components in addition to energy. For example, electricity bills include a basic service charge, and generally kilowatt-hours are not all charged equally. In some cases, additional charges are levied for peak demand and reactive power (i.e., power factor), neither of which is directly related to gross kilowatt-hours. The heating value of natural gas may vary substantially from month to month, so the fuel volume shown on a gas bill (e.g., thousand cubic feet) may not consistently reflect energy consumption. Thus, the figures appearing on utility bills must be interpreted appropriately to determine actual energy consumption.

Another factor to take into account in examining historical records is the normalization of energy consumption data to reflect changes in production. Energy consumption is typically keyed to production rate and possibly other variables, such as seasonal weather conditions. Thus, it is desirable to interpret past energy consumption on a "per unit" basis. The variables used to normalize energy data will differ from one situation to another. For example, suppose that in a particular installation a single product is produced and that most of the energy consumption for the plant goes into process equipment for that product. A reasonable figure to use for tracking energy consumption would, therefore, be Btu's per pound of product. In another situation the primary energy use might be for space heating and cooling. In this case a more appropriate measure of energy use would be Btu's or kilowatt-hours per degree-day. This would take into account variations in energy consumption with changes in outside temperature.

Normalizing energy consumption data becomes more complicated when energy use in the process is diversified and more than a single product is produced. Suppose that a plant produces two products, A and B, and that the relative production rates of these products vary from month to month. If A requires 50% more energy per pound than B, it is clear that a given total production of A plus B may result in quite different total energy requirements, depending on the relative proportion of each product produced. It would be desirable to weight the production figures for the products on the basis of their relative energy requirements to produce a composite normalizing variable on which to base energy consumption. This naturally leads to the question of where the weighting factors come from, since it is tacitly assumed that some idea of the relative energy requirement for each product is known. These numbers must come from the audit itself or at least from an estimate of the energy requirements of each unit process derived from other studies of similar operations. Indeed, in large complex facilities, such as those typical of the chemical industry, one of the primary objectives of the audit may be to establish these tracking parameters for use as management controls.

In much of the literature describing energy consumption for industrial processes, electricity and direct fuel use are lumped into a single energy figure using an average conversion rate of, say 10,000 Btu/kWh. While this procedure may be useful for some purposes, such as consideration of co-generation, for most applications it tends to obscure information of primary interest, that is, the energy and dollar savings realizable by implementing an energy conservation action for a specific piece of equipment. Therefore, it is recommended that fuel and electricity use be tracked separately. In many cases this will require that different forms of energy be normalized differently, based on the way in which each form is used in the particular plant. Example 2.1 illustrates some of the ideas discussed above.

Example 2.1

A plant making prepared animal feeds from grain produces two products, A and B. Product A is a meal produced by grinding the grain and then drying it to a specified moisture content. Product B is pelletized and is made by further drying the meal and pressing it into pellets in an extruder. The process is shown schematically below. The energy intensiveness of each step is indicated in kilowatt-hours per pound of throughput in that step.

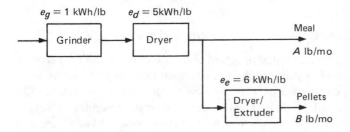

Production figures and total electricity consumption for a 3-month period are as follows:

Month	A (lb)	B (lb)	kWh
January	20,000	0	122,000
February	10,000	10,000	178,000
March	6,000	12,000	182,000

It is desired to normalize the energy consumption to accurately reflect plant energy use trends.

Solution: Designating the energy consumption per pound of each product by e_A and e_B, respectively, where $e_A = 1 + 5 = 6\,\text{kWh/lb}$ and $e_B = 1 + 5 + 6 = 12\,\text{kWh/lb}$, the total monthly energy E is

$$E = Ae_A + Be_B$$

$$= e_A\left(A + \frac{e_B}{e_A}B\right)$$

or

$$e_A = \frac{E}{A + (e_B/e_A)B}$$

Thus, in terms of energy consumption, each unit of B produced is equivalent to e_B/e_A units of A. For the case presented above, each pound of B is "worth" 2 pounds of A in energy consumption. The monthly energy can thus be normalized in terms of energy per equivalent pound of A by dividing E by the quantity $(A + 2B)$. The table below shows the energy normalized to simple total production and to equivalent A production.

Month	E (kWh)	$\dfrac{kWh}{(A + B)}$	$\dfrac{kWh}{(A + 2B)}$
January	122,000	6.10	6.10
February	178,000	8.93	5.93
March	182,000	10.11	6.07

Note that raw kilowatt-hours shows a sharp increase from January to February. If this energy consumption is normalized simply on the basis of total pounds of production $(A + B)$, a continuous upward trend is indicated. However, when the differing energy requirements of the two products are taken into account, it is clear that actual per unit energy consumption remained approximately constant during the period.

2.2.2 Walk-Through and Process Familiarization

Before embarking on a detailed program of data acquisition and analysis, it is important to develop a thorough familiarity with the process and its associated mass and energy flows. Process flow sheets and plant engineering drawings may be available to help in this task, although, invariably, changes will have been made that do not appear on the drawings. In any case it is important that the audit team see the equipment in operation to develop a firsthand sense of its size, location, complexity, and operating "habits." A first-pass awareness of the overall energy flows in the system can be obtained by relating flow sheets to equipment in the field. The locations of temperature, pressure, and flow sensors and their typical readings at the time of the walk-through should be noted. Conversations with equipment operators can be very revealing. It will be helpful to carry along some simple instrumentation, such as a hand-held electronic thermometer and a wand-type airflow meter. Rough readings can help in defining requirements for more precise and detailed measurements to be made later.

2.2.3 Definition of Data Requirements

Credibility is the key to the effectiveness of an energy audit, and this means that the audit must be based on actual, not hypothetical, operating data. Process measurements will be required, and the definition of data is thus a matter of critical importance. The auditor must approach this problem with several questions in mind. What data are already available from existing in-place instrumentation, and are these data reliable? At what points in the system can instruments be installed without interfering with process operation? What equipment and how much lead time will be necessary?

It will be readily apparent that defining these data requirements is, in fact, essentially the process of strategic and tactical planning of the audit itself. Required energy and mass flow calculations must be anticipated at this stage, necessitating careful forethought concerning the overall effort. For example, during the walk-through it might be observed that the boiler in a plant is the primary direct fuel consumer and that the stack exhaust temperature is sufficiently high (say 500 °F) to indicate the possibility of a stack heat recovery system. To evaluate this project, the magnitude of the stack heat loss must be known. This will necessitate an energy balance for the boiler, which will require certain data, such as exhaust temperature, stack gas composition, fuel flow rate, and fuel heating value. Furthermore, a typical load profile of the boiler is needed to estimate the overall heat loss on an annual basis, and this implies that tests should be run at several firing rates. Instrumentation and personnel requirements for the tests can be planned with these considerations in mind, and arrangements can be made for installation of measuring devices during temporary shutdowns. In addition, during the walk-through, observations can include the availability of space for a heat recovery unit, existing structural limitations, and any interference problems that may arise with existing ducts and piping.

To further delineate what is required in the data definition stage, a more extensive discussion of several items of special interest will be presented.

Process Flowsheets As mentioned earlier, accurate flow sheets for the process are necessary as a guide to the flows of mass and energy. These form the basis for planning the audit and must either be available or be created by the audit team after the walk-through. Even if sheets are already available, they should be checked to ensure that all modifications have been incorporated. Flow sheets should indicate locations of existing instrumentation and, if available, nominal flows, temperatures, and pressures. It should be recognized that although an instrument may be indicated on paper, there is no assurance that it is functioning properly. Also, it should be recognized that nominal process design conditions are just that, and that audit calculations should be based on measured conditions whenever possible.

Historical Energy Use and Production Data The need for past records of fuel and electricity usage and related production data for normalization has been discussed earlier. In defining exactly what information is required, it will be necessary for the audit team to anticipate the way in which the energy-use baseline is to be established. While most information should be available from company utility and production records, it may be necessary to contact electricity or fuel suppliers directly to obtain certain information, such as electrical demand charts.

Batch-Based Variables Certain quantities unique to batch processes may be required. These include capacities of batch heaters and cycle times, which can be obtained from equipment specifications, direct measurement, or discussions with equipment operators.

Nameplate Specifications of Major Energy-Using Equipment All major pieces of energy-using equipment include a number of pertinent energy-related characteristics in their basic manufacturers' specifications. For a boiler, for example, these might include full-load firing rates, fuel pressure ratings, and draft airflow requirements. For electric motors, they would include full-load horsepower rating, current, and speed. While nameplate specifications do not generally reflect the actual operating condition of the device, they can be quite useful in conjunction with measured quantities. For example, in a gas-fired boiler, the fuel pressure at the burner is closely related to firing rate. While the actual fuel input rate may be difficult to determine at all firing rates, the fuel pressure is quite easy to measure, and an indication of the actual loading of the boiler can be obtained by observing the pressure compared with its nominal full-load value. This will be discussed in more detail in Chapter 6.

Similarly, the revolutions per minute of an induction motor can be easily measured with a tachometer or strobe. The actual value, when compared with the nameplate value, determines the percentage slip of the motor. This parameter may be related to the motor load, which in turn can be used with data available from the manufacturer to determine the operating efficiency.

Material Properties Thermodynamic, physical, and chemical properties are frequently of importance in carrying out energy balance calculations and technical and economic feasibility analyses. These include such quantities as specific heats and densities of fluids, thermal conductivities, and corrosion characteristics (e.g., fuel sulfur content).

Dimensional Data A retractable tape measure is a valuable tool in carrying out an energy audit. Lengths and diameters of piping runs, thickness of insulation, and available space for retrofitting equipment are all important in estimating losses and evaluating the feasibility of implementing energy conservation measures. For example, the diameter of a boiler stack is important in determining the cost of a stack-mounted heat recovery unit. The audit team should anticipate such needs in examining drawings and manufacturers' records.

Production Data As already mentioned, production data are valuable in establishing a baseline for plant energy consumption. They may also be quite important in computing process energy requirements. Quantities such as the number of batches processed per day or the monthly throughput of a product through specific unit operations are required in extending energy balance calculations to a total energy consumption per month or year.

Equipment Operating Profiles In industrial operations, even those of a continuous nature, equipment rarely operates on a truly steady basis. Boilers, heaters, compressors, and the like undergo load fluctuations in time. In estimating monthly or annual energy use for such equipment, and the potential for energy recovery by retrofitting or replacement, knowledge of the load profile is needed. Direct measurement is, of course, the most reliable source of data. It may be quite difficult to obtain such information, however, particularly over an extended period of time; thus, it may be necessary to rely on other sources.

For example, in a steel foundry operation in which electric arc furnaces are used for melting, monthly demand charts obtainable from the electric company might provide a reliable indication of cycle times and loads. These can usually be read to a resolution of about 15 minutes.

Descriptive input from equipment operators should not be underrated as a source of such data. An experienced operator can usually describe, with a fair degree of accuracy, typical instrument readings during a production cycle and the length of time over which these readings occur. This information may then be related to energy consumption profiles during a cycle and over an extended period of time. Plant production records can provide long-term information, such as periods of equipment downtime for vacation, maintenance, or shift-change operations.

Fuel and Electricity Costs In estimating the potential savings to be realized from the implementation of energy conservation initiatives, energy costs are a critical factor. Cost savings estimates must consider not only current fuel costs but also those projected over the lifetime of the investment. This is especially important in a period of rapidly rising fuel and electricity prices.

While current energy costs are easily obtained from utility records, projections are much less certain. Projections can be obtained from such organizations as the U.S. Department of Energy, the Electric Power Research Institute, the Gas Research Institute, energy companies, and trade associations and are frequently published in trade and professional literature. It is up to the auditor to evaluate the credibility of these projections, and it is often desirable to carry out cost savings estimates based on low-, medium-, and high-energy cost escalation rates to "bracket" the predictions.

Guesstimates of Incomplete Information It is disconcerting, but always seems to be true, that in conducting an energy audit not all of the data deemed necessary or desirable for comprehensive evaluations of energy conservation opportunities can be obtained. Recognizing and accepting this fact in advance will help alleviate some of the frustration that is typical in the audit process.

In many cases the audit team will have to rely on the judgment of its members to fill in missing information. Again, the value of plant operating and maintenance personnel should be recognized. An operator's statement that a storage tank is too hot to touch can be quite meaningful if no other temperature information is available. Similarly, the appearance of a jet of steam issuing from a leak can be a reasonable indicator of the pressure in the line, particularly to an experienced maintenance worker who has seen thousands of such jets.

Defining the requirements for data and how the data are to be obtained is one of the most challenging tasks in the performance of an in-depth energy audit. The audit team functions somewhat as a team of detectives, constructing a complete picture from indirect and often incomplete sources of information. The viability of the result rests on the reliability of the data acquired. For this reason, the need for careful attention to this phase of the audit cannot be overstressed.

2.2.4   Measurements in Energy Audits

Instrumentation plays a crucial role in an energy audit, providing operating data in place of assumptions and guesses. Measurements specific to various types of tests will be discussed in more detail in subsequent chapters, but it is appropriate here to discuss briefly the application of several general-purpose instruments common to most audits.

Temperature General-purpose temperature measuring devices are invaluable in providing information for energy balances and heat loss estimates. Hand-held temperature indicators such as the one shown in Fig. 2.1 can be used with surface or immersion probes to determine important temperatures, for example, for the surfaces of storage tanks or for stack gases.

For surfaces that are inaccessible, the radiation-type thermometer shown in Fig. 2.2 is especially useful. Some care must be exercised in the use of radiation thermometers, however, because the indicated temperature depends on the radiant emissivity of the surface, which may vary widely with surface condition and composition. Radiation thermometers also tend to be less accurate at relatively low temperatures. One way to compensate for emissivity variations is to measure the temperature of an easily accessible surface, similar to the surface of interest, with both the radiation device and a hand-held surface probe. The radiation thermometer emissivity adjustment can be calibrated in place in this way. This method can be used, for example, to scan the surface of a large storage tank for hot spots, which indicate a breakdown of insulation.

Pressure Pressure in steam and compressed air lines is an important determinant of their energy content. The most common means of measuring

Figure 2.1 Hand-held digital thermometer. (Courtesy of Wahl Instruments, Inc.)

these pressures is the Bourdon gauge, illustrated in Fig. 2.3. Inexpensive gauges of this type are available for pressures ranging from low vacuum to high positive pressure. Appropriate mounting ports must, of course, be provided in the lines to be measured; this must be anticipated in the data definition phase of the audit, and adequate lead time must be provided to allow for installation.

For measuring low pressures, such as boiler draft, a simple manometer such as the one shown in Fig. 2.4 is convenient. These can also be used for air or flue gas velocity measurement, as discussed below.

Fluid Flow   Measurement of fluid flow rate and fluid velocity is one of the more difficult challenges in auditing and monitoring energy consumption, because most fluid flow meters are relatively expensive and require rather elaborate installation. For measurement of steam, compressed air, or process liquid and gas flows, obstruction-type flow meters, including orifices, calibrated nozzles, and venturis, are the most common. These are shown schematically in Fig. 2.5. The fluid passing through the obstruction accelerates, producing a drop in static pressure. The static pressure change can be measured and related to flow rate by using appropriate calibration factors.

Low-pressure flows, such as air in ducts or flue gas, can be measured with a Pitot tube, which relates the fluid velocity to the difference between

Figure 2.3 Bourdon pressure gauge. (Courtesy of Dresser Industries Instrument Division)

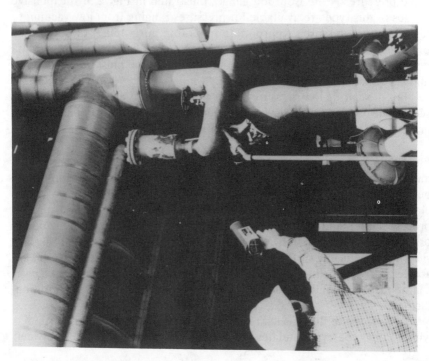

Figure 2.2 Radiation thermometer. (Courtesy of Wahl Instruments, Inc.)

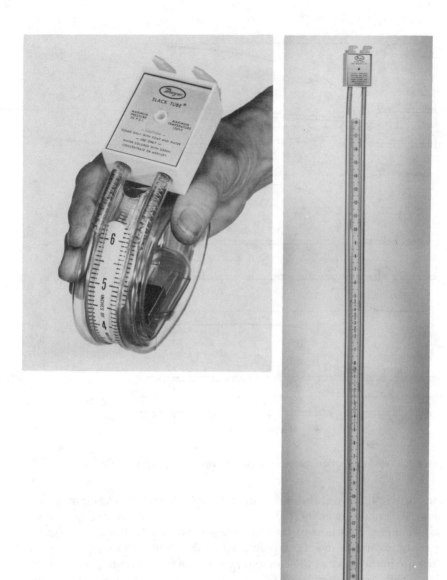

Figure 2.4 U-tube manometer. (Courtesy of Dwyer Instruments, Inc.)

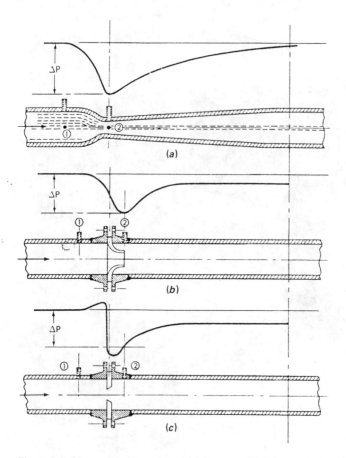

Figure 2.5 Obstruction flow meters. (*a*) Venturi; (*b*) flow nozzle; (*c*) orifice flow meter.

the impact pressure at the end of the tube and the static pressure measured along its length, as illustrated in Fig. 2.6. The annular averaging element shown in Fig. 2.7 works on basically the same principle, but it averages the impact pressure from several points across the stream and senses the static pressure in the wake of the element rather than from a point parallel to the flow. This device can be used for gross flow measurement at elevated pressures and can be "hot-tapped" into live steam lines.

Fuel Flow Metering of gaseous and liquid fuels represents a special case of flow metering. Because of the hazardous nature of these fluids, meters must be constructed and installed with particular attention to safety. High accuracy is also a primary consideration, since in this case the measurand is the energy source itself. For this reason, positive displacement meters, such as the lobe-type meter shown in Fig. 2.8, are most commonly used. At each

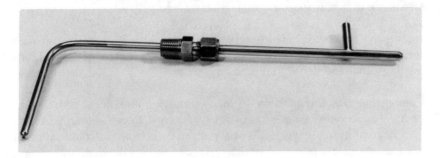

Figure 2.6 Pitot tube. (Courtesy of United Sensor and Control Corp.)

Figure 2.7 Annular averaging flow meter. (Courtesy of Dietrich Standard Corp.)

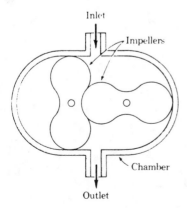

Figure 2.8 Lobe-type gas meter. From J. P. Holman, *Experimental Methods for Engineers,* 4th ed., McGraw-Hill, New York, 1984. Reprinted with permission.

rotation of the lobe or vane a fixed volume of liquid or gas is passed through the meter. This volume can be related to the energy content of the fuel if its heating value and density are known.

Combustion Gas Composition In many industrial situations, combustion equipment accounts for the largest use of energy. In evaluating the efficiency of combustion devices, it is necessary to determine the composition of the exhaust gases, particularly their oxygen, carbon dioxide, and carbon monoxide content. The Orsat analyzer, shown in Fig. 2.9, has traditionally been used for determining these constituents. A gas sample is drawn into the sample chamber of the analyzer and is sequentially bubbled through several chemical solutions, each of which absorbs specific gas components. By measuring the volume change of the sample after each absorption, the components of interest can be determined. A variation of the Orsat analyzer, which has been packaged in a more convenient form, is shown in Fig. 2.10. In essence, this is a single-component Orsat analyzer; the gas sample is drawn into the instrument and bubbled through its solution by inverting it several times. The rise in the height of the column of absorbing solution in the sample tube indicates directly the percentage of carbon dioxide or oxygen. A separate sample must be drawn for each constituent, since a different analyzer must be used for each component gas.

Measurements of carbon monoxide can be obtained with the CO

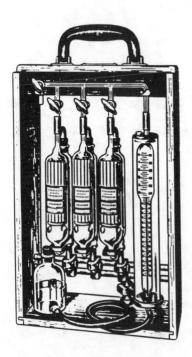

Figure 2.9 Orsat analyzer.

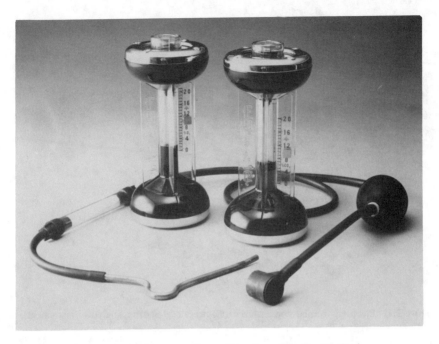

Figure 2.10 Single-element flue gas analyzer. (Courtesy of Bacharach, Inc.)

analyzer shown in Fig. 2.11. A small sample of fixed volume is drawn through a glass tube containing crystals that change color upon absorbing carbon monoxide. The length of the dark streak in the tube indicates the percentage of CO.

All of the devices described above can provide reliable gas composition data, but the measurements are rather time consuming and to some extent

Figure 2.11 Carbon monoxide analyzer. (Courtesy of Bacharach, Inc.)

Figure 2.12 Electronic combustion analyzer. (Courtesy of Teledyne Analytical Instruments)

depend on the skill and experience of the user. The electronic combustion analyzer shown in Fig. 2.12 is much more convenient and provides a direct meter reading of the percentage of oxygen and combustibles (usually calibrated in terms of carbon monoxide) in the gas being sampled. This instrument is more expensive than the "wet" analyzers described above, but it has the advantage of continuous sampling and essentially instantaneous indication of these two critical gas components. This can be especially helpful in certain boiler and heater tests and adjustments described in Chapter 6.

Electrical Measurements Electrical measurements in an energy audit can be valuable not only with respect to energy consumption in electrical devices but also as a means of determining the operating characteristics of other energy-using equipment. For example, a record of the amount of time a solenoid valve is activated can provide information on the energy consumption of a steam-heated storage tank. A simple measurement of the armature current in a motor, in conjunction with information from the nameplate, can be used to determine the horsepower loading. Two electrical instruments in particular are most useful in energy audits. These are the clamp-on hand-held AC ammeter and the recording ammeter.

Clamp-on ammeters, such as the one shown in Fig. 2.13, are inexpensive and easy to use. The jaws of the ammeter clamp around one of the current-carrying leads without having to break any connections; this is especially important when it is impossible or inconvenient to interfere with normal equipment operation. The jaws act as the secondary loop of a trans-

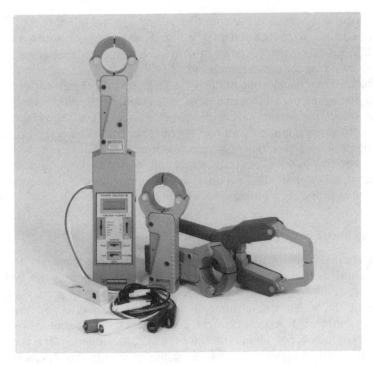

Figure 2.13 Clamp-on hald-held ammeter. (Courtesy of Esterline-Angus Instrument Corp.)

former, and the meter provides a direct indication of the current flowing in the line. In three-phase equipment, currents in the three legs can be measured independently and combined mathematically to give the total average current to the device.

The recording ammeter, shown in Fig. 2.14, similarly clamps around current-carrying lines without breaking the circuit. The advantage of a recording device is that it provides continuous information over an extended period of time to indicate the characteristic operating cycle as well as the load on the equipment. This is useful, for example, when evaluating an air compressor that operates periodically to recharge a storage tank.

Recording instruments are also available for measuring voltage, real power, and reactive power in AC circuits. Reactive power can be determined indirectly with a recording wattmeter by using a phase-shiftng transformer. Transformers containing integral timers that periodically switch the transformer in and out of the circuit can be purchased. These transformers permit the dual use of the wattmeter to determine real and reactive power over a period of, say, 7–10 minutes. These quantities determine the power factor of the device — a parameter of considerable interest in evaluating potential improvements in electrical equipment operations. Three-phase wattmeters require a substantial amount of wiring into the circuit, representing both an inconvenience and a safety hazard when working with live lines. In some cases, however, they can provide information that is not obtainable from a current measurement alone.

Instrumentation is usually one of the most difficult kinds of expenditure to justify to management. Unlike maintenance expenses or capital investments for increased production, measurement devices, especially those not used directly for process control, are hard to relate to profit. Nonetheless, better information on how efficiently a process is actually functioning can lead to great rewards in terms of reduced fuel and electricity consumption, and most companies that have invested in instrumentation for energy monitoring have found it to be highly profitable.

2.2.5 Mass and Energy Balances

A financial audit is based on the fundamental premise that (contrary to popular opinion) money does not vanish into thin air but can always be accounted for. Similarly, the basic principles of conservation of mass and energy provide the basis for determining where energy is used in a system and where losses occur. Energy, of course, is never actually lost, but in any process not all of the energy put in is used for the purpose intended. It is the primary goal of an energy audit and its resulting energy conservation program to increase process efficiency and thereby reduce the amount of energy lost from productive use.

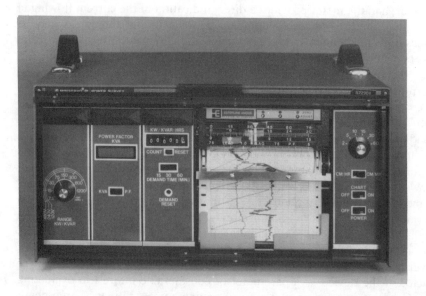

Figure 2.14 Recording ammeter. (Courtesy of Esterline-Angus Instrument Corp.)

The subject of mass and energy balances is treated in detail in fundamental texts on thermodynamics, such as that by Van Wylen and Sonntag [1]. It is beyond the scope of this book to cover these subjects in depth; however, two examples are given to illustrate the application of mass and energy balance principles to energy audits.

Example 2.2

A starch conditioner in a corrugated-box plant mixes starch, water, and steam to produce glue. In one plant the conditioner feeds two glueing lines, requiring 200 and 250 lb/hr, respectively. The amounts of starch and water required are 300 lb/hr and 100 lb/hr, respectively. What is the steam consumption?

Solution: The basic principle of conservation of mass states that all mass that enters the system must either leave or be stored inside. Assuming that the level of the starch mixture in the tank is constant (i.e., there is no change in the amount of mass stored), what goes in must come out. Hence,

$$m_{str} + m_{wat} + m_{stm} = m_{g1} + m_{g2}$$

$$m_{stm} = (m_{g1} + m_{g2}) - (m_{str} + m_{wat})$$

$$= (200 + 250) - (300 + 100)$$

or

$$m_{stm} = 50 \text{ lb/hr}$$

Example 2.3

In the preceding example, what is the energy content of the glue? To answer this question, we need information on the physical properties of the glue constituents, and then we can use the law of conservation of energy to determine the energy content of the mixture. The following information was determined in the audit:

Starch temperature: 85°F
Specific heat: 0.45 Btu/lb °F
Water temperature: 150°F
Specific heat: 1.0 Btu/lb °F
Steam: Saturated at 15 psig

To determine the energy inflow to the system, it is necessary to choose a common baseline condition for all constituents. A typical baseline is atmospheric temperature and pressure, about 70°F and 0 psig. Starting from this baseline, the energy content of each of the constituent streams can be calculated:

Starch:

$$E_{str} = m_{str}\, c_{str}\, (85 - 70)$$

$$= 300 \text{ lb/hr} \times 0.45\, (85 - 70)$$

$$= 2025 \text{ Btu/hr}$$

Water:

$$E_{\text{wat}} = m_{\text{wat}} \, c_{\text{wat}} \, (150 - 70)$$

$$= 100 \times 1.0 \times 80$$

$$= 8000 \text{ Btu/hr}$$

Steam:

From the steam tables (Appendix B), the enthalpy of saturated steam at 15 psig is 1164.3 Btu/lb. For water, at atmospheric pressure and 70°F, the enthalpy is about 38 Btu/lb. Thus,

$$E = m_{\text{stm}} \, (h_{\text{sat,15psi}} - h_{\text{wat,70F}})$$

$$= 50 \text{ lb/hr} \times (1164.3 - 38)$$

$$= 56,315 \text{ Btu/hr}$$

Applying the principle of conservation of energy,

$$E_{\text{str}} + E_{\text{stm}} + E_{\text{wat}} = E_g = m_g \, e_g$$

where e_g is the energy content of the glue (Btu/lb):

$$e = \frac{2025 + 56,315 + 8000}{450}$$

$$e = 147.4 \text{ Btu/lb}$$

As illustrated in the examples above, mass and energy balances serve two vital purposes in an energy audit. First, they convert basic thermodynamic information, such as temperatures, pressures, and flow rates, into energy terms. Second, they provide additional resources to allow us to infer important information from incomplete data. Throughout the remainder of this book we will see many other examples of the application of these important principles.

Shortcut Aids for Estimating Energy Losses As discussed earlier, it is often difficult to obtain complete enough data on a process to permit a direct accounting for all energy flows. Thus, it is necessary to make indirect estimates of some quantities. Much of this book is devoted to the subject of estimating techniques for various types of equipment. In this section we will cite several sources of useful information that can be especially helpful in energy audits.

One of the most valuable sources of information is the *Energy Conservation Program Guide for Industry and Commerce* (EPIC), developed by the National Bureau of Standards [2]. EPIC contains sections on organizing an industrial energy conservation program and estimating energy losses in various types of equipment, as well as other resource material, such as a review of financial analysis methods and a listing of pertinent trade and professional organizations.

Figure 2.15 illustrates the kind of technical information to be found in EPIC. This figure shows the annual rate of heat loss from uninsulated pipes containing steam at various pressures. One can quickly estimate from the figure that a 6-in. line, 200 ft long, carrying steam at 400 psig will lose about 4900 million Btu per year. Calculation of this loss from basic heat transfer equations would require a tedious and time-consuming procedure involving combined radiation and convection, and an iterative solution of the external pipe temperature. Having this procedure done for us in advance and presented in an easily interpreted format represents a tremendous convenience.

A word of caution is in order, however, concerning the use of graphic and tabular shortcut methods. Many assumptions are inherent in any presentation of complex calculations in a condensed handbook form, and it is virtually impossible to list all of these assumptions. Therefore, the user must be sure that he or she understands the basis of the charts and tables before using them. For example, in Fig. 2.15, the ambient air temperature and velocity on which the curves are based are not shown. The curves should thus be used with care, and the user should at least be prepared to spot-check the data with detailed calculations.

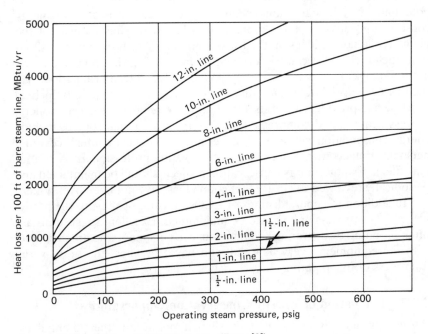

Figure 2.15 Extract from EPIC Handbook. (From [2])

Most of the figures and tables presented in EPIC are drawn from engineering handbooks, manufacturers' literature, and technical periodicals. Several resources of this type that are of special value in energy auditing are listed in references [3–6].

The amount of time that can be saved by using these calculation aids cannot be overestimated, provided they are used with appropriate care. At various places throughout this book, figures and tables drawn from manufacturers' literature are presented. This type of information can be especially valuable. For example, insulation manufacturers provide extensive tables of heat losses from surfaces, both insulated and uninsulated, and some manufacturers also provide computer calculation services to assist in energy audits. Manufacturers should be contacted to determine the input data required for these calculations, and this information should be incorporated in planning the data acquisition phase of the audit.

2.2.6 Enumerating Energy Conservation Opportunities

The procedures discussed in the preceding sections serve to identify the most significant sources of energy loss in a plant or process unit. The next important phase of the audit is to identify ways of decreasing this loss and to evaluate the energy savings and profitability potential in implementing changes. The EPIC handbook noted earlier lists a number of energy conservation opportunities and illustrates how to estimate their energy savings potential. An audit team should begin early to develop a checklist of energy conservation ideas for the various types of systems characteristic of the plant. While not all ideas on the list will prove practical in the given situation, such a checklist can be very helpful in ensuring that no stone is left unturned in seeking possible energy savings. Several technical periodicals, such as *Chemical Engineering* and *Plant Engineering*, frequently include articles that provide checklists for common equipment such as boilers, process heaters, and compressed air systems. Additional sources include energy conservation newsletters, government manuals, conferences, and personal contacts with other energy conservation professionals. A continuing file of possible energy conservation projects should be maintained and periodically reviewed to ensure that good ideas have not been overlooked.

In addition to general listings of ECOs, a listing specific to the particular plant or process unit should be maintained. Excellent suggestions are often proposed by operating and maintenance personnel who are in contact with the equipment every day and who have, perhaps, the best overall awareness of its operating characteristics.

2.2.7 Evaluation of Energy Conservation Opportunities

An energy audit is of value only when it points out the potential for improved profitability achievable by implementing energy conservation projects. To accomplish this, each energy conservation opportunity must be reviewed to determine whether it is applicable to the process and, if so, what the associated energy savings will be in the particular operation being evaluated. Obviously, it would not be productive to carry out a detailed evaluation of every possibility, since there may be thousands. The energy balances carried out earlier in the audit can help rank these so that the most attractive choices can be made for detailed evaluation. For example, in the course of the calculations it may become clear that boilers account for most of the energy use in the plant, while electric motors are relatively small energy users. Thus, emphasis in the evaluation phase of the audit would be placed first on improving boiler efficiency, and secondary consideration would be given to motor replacement.

The evaluation procedure consists of recalculating energy and mass balances for each item, assuming that an energy conservation modification has been made. The cost of the modification must be considered, and a lifetime economic evaluation must be carried out to determine the profitability of the measure.

Several points should be noted. First, the energy savings should be cited separately from the economic consequences of an action, because it may be desirable to consider this factor in itself. The aggregate total of energy savings accomplished in an overall program may be of interest to management because of its implications for production capability in times of fuel curtailment. For example, if a 20% reduction in overall fuel use can be effected, the plant might still be able to maintain full production during a temporary fuel cutback to predetermined fuel allocation levels. Second, it is necessary to consider fuel costs over the lifetime of the project, not simply at present levels. Third, various measures of economic performance may be used, each of which has a somewhat different interpretation. The appropriate measure will depend on the accounting practices in use in the particular operation, the capital intensiveness of the project, and its economic life. These considerations will be discussed further in Chapter 3.

2.3 PRESENTATION OF ENERGY AUDIT RESULTS

No matter how accurate the data and how careful and comprehensive the calculations, an energy audit is only as good as the quality of the final report. An energy audit report is essentially a sales document. Its purpose is to "sell" management on the idea of investing money in energy conservation measures in competition with other investment alternatives. Like all good sales material, it must be concise, direct, and convincing.

Table 2.1 shows a suggested outline for a typical energy audit report. The report should begin with an executive summary, which briefly describes the procedures that were carried out and concentrates primarily on the final

Table 2.1 Suggested Outline for an Energy Audit Report

Executive Summary
Chapter 1: Introduction
Chapter 2: Energy Audit Procedures
Chapter 3: Plant Energy Distribution
Chapter 4: Evaluation of Energy Conservation Opportunities
Chapter 5: Recommendations for Project Implementation
Appendixes:
   Data Compilation
   Example Calculations
   Equipment Cost Estimates and Quotations

results. The executive summary should contain a table of energy conservation project recommendations listed in order of priority as in the example shown in Table 2.2.

The introductory chapter should describe the concept of the audit and what it is intended to produce and should familiarize the reader with the plant's energy systems. It should be remembered that many persons reading the report will not have the intimate knowledge of the details of the process acquired by the members of the audit team, and thus the introduction should include a listing of the major points of energy use within the process. This will lead into a discussion in the next chapter of the points of concentration of the audit.

Table 2.2 Project Summary Table

Project priority	Description	Potential energy savings ($/yr)	Estimated cost ($)	Payback period (yr)
1	Reduce boiler excess air from 60 to 25%	8,750	1,000	0.2
2	Pressurize condensate return system	3,740	1,400	0.4
3	Utilize flash steam in starch tank	2,500	2,000	0.8
4	Utilize flash steam in paper conditioner	3,500	4,000	1.1
5	Insulate hot-plate backing	3,100	1,100	0.4
6	Relamp fluorescent fixtures	1,500	200	0.2
7	Retrofit HPS lamps on present mercury vapor fixtures	2,150	4,750	2.2

Table 2.3 Tabulated Presentation of Energy Distribution

Steam use	Steam consumption (lb/hr)	Energy value (Btu/hr)	Percent of total
Machinery			
C-flute	812	0.74×10^6	10
B-flute	512	0.44×10^6	6
Double backer	694	0.59×10^6	8
Hot plate (1)	715	0.61×10^6	8
Hot plate (2)	1,248	1.06×10^6	15
Starch tank	84	0.09×10^6	1
Paper conditioning	200	0.20×10^6	2
(Subtotal)			(50)
Miscellaneous			
Piping losses		0.20×10^6	3
Boiler radiation and convection losses		0.17×10^6	2
Stack losses		1.70×10^6	23
Flash steam losses	387	0.44×10^6	6
Miscellaneous and unaccounted		1.24×10^6	16
Total[a]		7.48×10^6	100

[a]Estimated fuel supplied to boiler.

Following the introduction, the general procedures carried out in the audit should be described. It is not necessary to include detailed data here, but the procedures for obtaining the data should be outlined in brief. For example, the measurements and methods used to conduct boiler tests should be summarized, but the tabulations of actual test data should be relegated to an appendix.

The next chapter should summarize the distribution of energy use in the plant. Calculation procedures need not be covered at great length, although they might be described briefly; example calculations can instead be presented in an appendix. The primary objective of this chapter should be to justify quantitatively the major points of interest. Energy distributions can be presented either in tabular form as shown in Table 2.3 or graphically as illustrated in Fig. 2.16.

The following chapter should cite the energy conservation opportunities considered in the study and summarize the results of the energy savings calculations and economic evaluations for each. The results should be presented primarily in tabular form as shown in Table 2.4.

The last chapter of the report should contain the priorities and recommendations for implementation of energy conservation opportunities. The

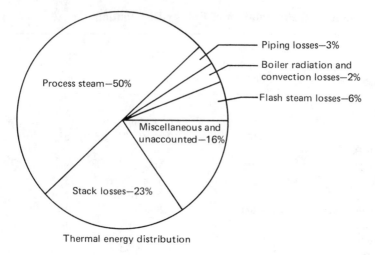

Thermal energy distribution

Figure 2.16 Pie chart illustrating plant energy distribution.

priority assigned to each item should be explained. A short payback period will not always be the sole criterion in setting priorities. An ECO that produces large energy savings, for example, might take precedence over another that offers relatively little savings, even though the latter might have a shorter payback period. Since this chapter is really the high point of the report, it is expected that much of the executive summary will be drawn from it, and hence a certain amount of redundancy is acceptable.

Finally, the report should include appendixes that contain most of the detailed data, calculations, manufacturers' quotes on equipment, and other material valuable for future reference but not essential to a general under-

Table 2.4 Energy Savings Estimates

Item	Existing heat loss (Btu/yr)	Proposed fiber-glass insulation thickness (in.)	Heat loss with proposed insulation (Btu/yr)	Value of potential energy savings
Underside of hot plate	310×10^6	2	21×10^6	$1,350/yr
Steam header	50×10^6	2	4×10^6	$200/yr
Steam header at boilers	127×10^6	2	6×10^6	$530/yr
Steam to equipment	33×10^6	2	2×10^6	$136/yr
Total	520×10^6		33×10^6	$2,216/yr

standing of the audit and its results. If the report is too lengthy, these appendixes might be incorporated in a separate volume.

The importance of good reporting should be recognized early in the planning of the audit, and a substantial amount of time and effort must be devoted to it. All too often, preparation of the report is viewed as a necessary evil by engineers who would really rather spend more time testing and calculating. A realistic assessment of the time and effort necessary to do a first class job of documentation can spell the difference between a report that produces results and one that gathers dust on a shelf.

2.4 CONCLUSIONS

Energy audits can take many forms, from superficial inspections to comprehensive engineering studies. This chapter has concentrated on the latter type of audit and has outlined some of the procedures used in carrying out such a project. The importance of careful planning and attention to detail has been stressed, as has the necessity of basing the audit on measured data whenever possible.

A thorough evaluation of energy use in a plant can provide the "master plan" for an ongoing and effective program of energy conservation. The results will usually justify the effort.

REFERENCES

1. G. J. Van Wylen and R. E. Sonntag, *Fundamentals of Classical Thermodynamics*, 2d ed., Wiley, New York, 1973.
2. National Bureau of Standards, *Energy Conservation Program Guide for Industry and Commerce* (EPIC), NBS Report 118, Washington, D.C., 1976.
3. Robert H. Perry and Cecil H. Chilton, *Chemical Engineers' Handbook*, 5th Ed., McGraw-Hill, New York, 1973.
4. Marshall Sittig, *Practical Techniques for Saving Energy in the Chemical, Petroleum, and Metals Industries*, Noyes Data Corp., Park Ridge, N.J., 1977.
5. Fred S. Dublin et al., *How to Save Energy and Cut Costs in Existing Industrial and Commercial Buildings*, Noyes Data Corp., Park Ridge, N.J., 1976.
6. Richard J. Reed, *Combustion Handbook*, 2d ed., North American Manufacturing Corp., Cleveland, 1978.

PROBLEMS

2.1 A beet sugar plant converts raw sugar beets to granulated sugar, with molasses and animal feed (produced from the pulp) as saleable by-products. The proportions of these three products will vary from month to month as the composition of the beet crop varies. The steam and electricity requirements of the process for each of the products are shown below, along with monthly production figures for the peak summer quarter. Gas cost to the plant for steam production is $4.20/10^3$ ft^3 (heating value of approximately 10^6 Btu/10^3 ft^3) and electricity cost is $0.07/kWh.

1. Assuming a boiler efficiency of 80% and an electrical distribution system efficiency of 99%, estimate the fuel and electric bills for each month.

2. Allocate the monthly usage of gas and electricity to each product on a percentage basis.

3. Normalize total fuel and total electricity use per month to a per-ton-of-sugar basis and determine the range of variation in this parameter over the 3-month period. Alternatively, try normalizing energy use on a per-ton-of-total product basis and see which parameter gives the most consistent measure of plant energy use for the period.

Energy and Production Records

Product	Steam energy (10^6 Btu/ton)	Electricity (kWh/ton)	Production (tons)		
			June	July	August
Sugar	13.2	110	13,500	10.800	12,000
Molasses	8.1	85	2,700	2,500	2,430
Feed	5.8	78	6,300	5,000	5,600

2.2 In a food processing plant, dried pears are rehydrated for baby food by mixing with hot water and steam. One particular rehydrator produces 200 lb$_m$/hr of pears at 140°F with an input of 120 lb$_m$/hr of dried pears at 70°F, hot water at 140°F, and saturated steam at 1 atm. Using 70°F as a baseline, the energy content of dried pears is zero and rehydrated pears is 49 Btu/lb$_m$. The specific heat of water is about 1 Btu/lb$_m$ °F and at 1 atm the latent heat of saturated steam is 970.3 Btu/lb$_m$. How much hot water and steam are required for the process?

2.3 In Chapter 15, Section 15.3, the energy consumption for a blast furnace/basic oxygen furnace steel plant is outlined. On the basis of this profile, prepare a work plan for an energy audit of a steel mill. The work plan should include the following elements:

1. A listing, in order of priority, of the major pieces of energy-using equipment.
2. A listing of measurements and calculations to be made on each major equipment item to evaluate the magnitude of energy losses in that unit.
3. A listing of instrumentation needed for the audit.
4. A listing of energy conservation measures possible for each unit.
5. A statement of how the cost reduction associated with each energy conservation measure can be estimated.

2.4 Repeat Problem 2.3 for a writing-paper plant as detailed in Chapter 15, Section 15.4.

2.5 Repeat Problem 2.3 for a low-density polyethylene plant as detailed in Chapter 15, Section 15.5.

THREE

ECONOMIC ANALYSIS

Most energy conservation projects require an investment of capital funds, either for retrofit or for new installations. The intent is to reduce operating or production costs by lowering energy consumption. As in any outlay of capital, the investment must be made now, while the benefits will be reaped in the future. The energy auditor or manager must be able to quantify the payback of a potential project. He or she should have a reasonably accurate estimate of how long it will take to recoup the investment and of the profit the project will realize over its lifetime. The management of industrial firms requires this type of analysis in order to aid in decisions to implement an energy conservation project. We call this project investment analysis, which implies a concern with the purchase of new equipment or the replacement of old equipment or auxiliaries.

The following sections contain rudimentary treatments of some aspects of economic analysis related to energy conservation. The reader is directed to Riggs [1] for a more sophisticated treatment of the general principles of engineering economy.

3.1 ECONOMIC INVESTMENT ANALYSIS

Most energy conservation projects deal with the investment of relatively large amounts of money. The key word is, of course, relatively, and here it means the size of the investment compared to the yearly net income. Depending on the size of the industrial concern, this could range from a few hundred to millions of dollars. The management of a small foundry, for example, might want to invest in a control system for a small fired heater, requiring a few thousand dollars. In contrast, hundreds of thousands of dollars might be required for a large paper mill to adapt its boilers to burn waste wood by-products. The required capital outlays for these two extreme cases might be relatively equal on the basis of the percentage of net income.

Capital investments are recovered over a period of years. The benefits of the investments must be realized over the time when the energy conservation equipment operates. It is difficult to foresee the exact nature of future benefits in all cases. Furthermore, an investment is generally irrevocable, that is, once the investment is made, it is very hard to retract. This can be the result of several factors. Installation costs can be a major portion of the investment. Equipment may become an integral part of a process, and millions of dollars can be lost if the entire process line is shut down. This problem is typical of the refining and petrochemical industries. Careful attention should be given to the economics of a potential energy conservation project.

3.1.1 Elements of a Capital Project

An investment project can be separated into elements so that a quantitative analysis can be conveniently performed. These elements are *initial investment,*[1] *returns on investment,* and *economic life.* A knowledge of these three elements generally allows us to anticipate the history of a project.

Initial Investment (I_o) The initial investment includes not only the price of the equipment and material but also all of the ancillary costs of the finished project. These ancillary costs include transportation, installation, licensing fees, and in some cases the "working" capital required to start up the equipment. Application of insulation to steam pipes is an example of a project where the installation costs might be of the same order as the material costs.

A classical example of the working capital required to initiate a project is given by the Alaska pipeline. To fill the 900-mile pipeline so that oil could be received at Valdez, approximately $21 million worth of oil at $8 per barrel had to be supplied. This is a part of the required working capital that will not be recovered before the pipeline is finally shut down.

Returns on Investments After an initial investment is made, returns will be realized over the lifetime of the project. We hope to gain a net return of cash from the project and use this money to amortize the investment. Net returns are desired; a return can be either discrete or continuous. Returns are typically credited at discrete times (monthly, yearly, etc.) rather than on a continuous day-to-day basis.

---

[1]A glossary of terms is provided at the end of the chapter for those unfamiliar with the jargon of economics. Words and phrases that are italicized are defined in the glossary.

Salvage Value The economic worth of the equipment at the end of its economic life must be included in the returns on investment. Taxes, operating costs, and maintenance costs must be subtracted from the gross project income.

Economic Life The economic return on an investment will not last indefinitely; thus, some estimate of the expected economic life is required. Equipment used for energy conservation has a physical lifetime. Pumps, blowers, and heat exchangers do not last forever. Economic life, which is used in analysis of investment projects, is usually the best estimate of the length of time that equipment can be economically used. In addition to projected physical lifetimes for equipment, there are other limits on economic life. These limits have been taken into account by the Internal Revenue Service in stipulating the economic life for certain investments, as shown in Table 3.1.

Economic life may also be arbitrarily agreed on by contract—in a lease or a franchise, for example. Regardless of how it is determined, economic life reflects a balance of earned income, or savings, against operating and investment costs.

Cash Flow Diagrams The initial investment and the returns on an investment, taken along with the economic life, allow a graphic projection of the details of an investment project. Figure 3.1a shows such a graph for a simple project. Starting at the present, the initial investment, I_o, and the funds accrued over each year of the economic life are plotted at discrete yearly points. The initial investment is considered to be a negative *cash flow*. Figure 3.1a shows a project for which there is the same cash flow for each year of the economic life and no salvage value at the end of the project. Cash flow is one of the most important elements of an investment project. It is the net intake of funds (or savings of funds for energy conservation projects) for each time period of the economic life.

The timing of cash flow is also important, since money may take on a different value as time passes. This is called the time value of money. It will

Table 3.1 IRS Guidelines for Asset Economic Life

Conventional steam power plant	28 yr
Nuclear power plant	20 yr
Office equipment	10 yr
Heavy trucks	6 yr
Automobiles	3 yr

Source: Depreciation Guidelines and Rules, U.S. Government Printing Office, Washington, D.C. (published periodically).

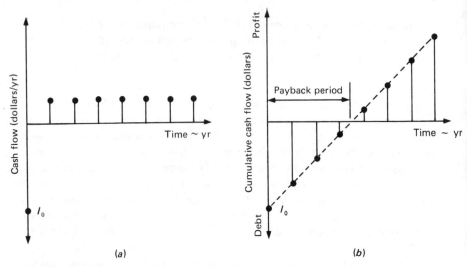

Figure 3.1 (a) Cash flow diagram for uniform, discrete returns, no salvage value; (b) cumulative cash flow corresponding to (a).

be discussed in detail later in terms of rate of return and discounted cash flows.

There are many variations of cash flow diagrams. A cumulative cash flow diagram can be constructed for each cash flow diagram. Such graphs are instructive because they clearly show when an initial investment will be recovered. The integrated effect of cash flow is shown; at each time period the actual debt or profit is plotted. Figure 3.1b shows a cumulative cash flow plot corresponding to the cash flow plot of Fig. 3.1a. When the cumulative cash flow crosses the axis at zero, the capital has been recovered. This is the payback period. Many industrial firms use the payback period as one of the major indicators of the attractiveness of an investment.

Figures 3.2 and 3.3 show other cash flow diagrams. We see that cash flows need not be uniform; salvage value can be an important part of the cash flow; and cash flows need not be tabulated at discrete intervals. Figure 3.3 illustrates a project for which the cash flow is plotted continuously. Savings and loan institutions generally use this technique for calculating interest on savings accounts in order to maximize the return to the investor.

The ability to construct a cash flow diagram cannot be overemphasized. If you can draw the picture, you have the data required to make the decision to implement the project. The picture may be mental rather than graphic, but its elements must be understood.

3.1.2 Sources of Funds

An engineer preparing a proposal for an energy conservation project might not be concerned with the source of funds for the project. However, management must have access to appropriate funds to implement the project.

The principal source of funds for many large firms is *retained earnings*. Retained earnings are the net funds made up of after-tax earnings less the dividends paid to stockholders. Contrary to what some people might think, retained earnings are not funds that simply lie idle. They are normally invested by the company in capital projects.

Another method of obtaining the funds necessary for a project is to borrow them from an outside institution. Companies may use this method if retained earnings are insufficient to cover needed investments. There is a definite and specified cost for the use of borrowed money: *interest*. This is the premium paid to a lender for the use of its money. Interest becomes a negative part of the cash flow of a project.

The third common source of funds is *equity financing*, the selling of additional stock in the company to raise capital. Equity financing often indicates that the company does not have adequate retained earnings or that it has exhausted its line of credit for borrowing. Thus, this is usually the last resort for financing a project.

3.1.3 Interest and Depreciation

Interest and *depreciation* are integral parts of an investment project. Depreciation is the allowable *amortization* of an initial investment over its economic lifetime as a non-cash flow operating expense. Depreciation for energy conservation devices or material might also be described as the loss

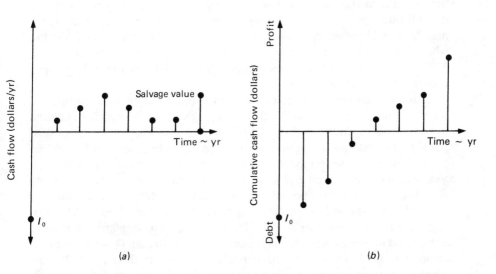

Figure 3.2 (*a*) Cash flow for discrete, nonuniform returns when the investment has salvage value; (*b*) cumulative cash flow for Fig. 3.2*a*.

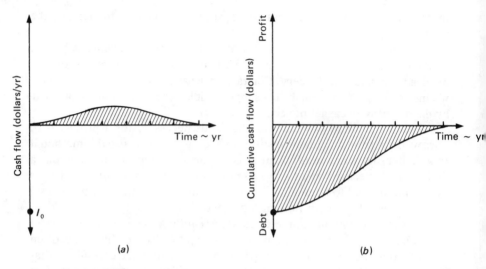

Figure 3.3 (a) Continuous cash flow with no salvage value; (b) cumulative cash flow for continuous returns of (a).

caused by deterioration of function, or obsolescence. Actually, depreciation is an accounting technique that is not strictly related to the physical lifetime of equipment. This is described very well in reference [1]. Interest is the cost of using borrowed money.

Interest The interest rate is the ratio of the interest payable at the end of a time period to the money owed or the principal at the beginning of the period. Interest is generally assumed to be based on a year even though it may be payable at shorter times. Take as an example a debt of $100 for which interest of $12 is payable after 1 year. The interest rate per annum is 12%. This is equivalent to a stated interest rate of 1% per month, 3% per quarter, or 6% semiannually.

Interest may be calculated in two ways, either as simple or as compound interest. Simple interest implies that the repayment of interest is calculated over the life of the loan. Compound interest involves the calculation of payable interest based on the amount owed at the beginning of the appropriate time period for repayment. It is common practice in business to use a yearly repayment system and to apply compound interest. In the following discussion we mean compound interest when we refer to interest rates.

Table 3.2 shows four possible repayment plans for a loan of $100,000 over a 10-year period at a 12% interest rate. These plans all involve compound interest since interest is paid each year. In plan D, for example, a lump sum of $310,584.82 for principal and interest at the end of 10 years would represent a compound interest of 12% over 10 years — $210,584.82 plus the principal, $100,000.

Equivalence The payments in Table 3.2 differ in magnitude but are made on different dates. Each repayment plan gives the borrower access to $100,000 at the outset of the loan. *Equivalence* is a concept used in economics that implies that different repayment plans are equivalent if they give the borrower the same amount of money at the start of the loan. The four plans in Table 3.2 are equivalent only for an interest rate of 12%. The type of repayment plans that would repay $100,000 at 12% interest are certainly not limited to the plans shown in Table 3.2. Any number of equivalent plans could be devised.

Present Worth The lender views a loan differently than a borrower does. The lender makes an investment (loan) in return for a contractual repayment plan that includes the loan value plus interest. The investment needed to secure the repayment contract is the *present worth* of the repayments. To the borrower, on the other hand, the present worth represents the present

Table 3.2   Four Plans for Repayment of $100,000 in 10 Yr at 12% Interest

End of year	Interest due (12% of money owed at start of year)	Total money owed before year-end payment	Year-end payment	Money owed after year-end payment
		Plan A		
0				$100,000
1	$12,000	$112,000	$ 12,000	100,000
2	12,000	112,000	12,000	100,000
3	12,000	112,000	12,000	100,000
4	12,000	112,000	12,000	100,000
5	12,000	112,000	12,000	100,000
6	12,000	112,000	12,000	100,000
7	12,000	112,000	12,000	100,000
8	12,000	112,000	12,000	100,000
9	12,000	112,000	12,000	100,000
10	12,000	112,000	112,000	0
		Plan B		
0				$100,000
1	$12,000	$112,000	$22,000	90,000
2	10,800	100,800	20,800	80,000
3	9,600	89,600	19,600	70,000
4	8,400	78,400	18,400	60,000
5	7,200	67,200	17,200	50,000
6	6,000	56,000	16,000	40,000
7	4,800	44,800	14,800	30,000
8	3,600	33,600	13,600	20,000
9	2,400	22,400	12,400	10,000
10	1,200	11,200	11,200	0

Table 3.2 Four Plans for Repayment of $100,000 in 10 Yr at 12% Interest (*Continued*)

End of year	Interest due (12% of money owed at start of year)	Total money owed before year-end payment	Year-end payment	Money owed after year-end payment
		Plan C		
0				$100,000
1	$12,000	$112,000	$17,698.41	94,301.59
2	11,316.19	105,617.78	17,698.41	87,919.37
3	10,550.22	98,469.69	17,698.41	80,771.28
4	9,642.55	90,463.53	17,698.41	72,765.42
5	8,731.85	81,497.27	17,698.41	63,798.86
6	7,655.86	71,454.72	17,698.41	53,756.31
7	6,450.76	60,207.07	17,698.41	42,508.66
8	5,101.04	47,604.70	17,698.41	29,411.29
9	3,589.35	33,500.65	17,698.41	15,802.33
10	1,896.27	17,698.49	17,698.41	0.00
		Plan D		
0				$100,000
1	$12,000	$112,000	$ 0.00	112,000
2	13,440	125,440	0.00	125,440
3	15,052.80	140,442.80	0.00	140,492.80
4	16,859.14	157,351.44	0.00	157,351.94
5	18,882.23	176,234.17	0.00	176,234.17
6	21,148.10	197,382.27	0.00	197,382.27
7	23,685.87	221,068.14	0.00	221,068.14
8	26,528.18	247,596.32	0.00	247,596.32
9	29,711.56	277,307.80	0.00	277,307.88
10	33,276.95	310,584.82	310,584.82	0.00

sum that can be secured with the promise of future payment. A repayment plan is equivalent to the present worth regardless of whether it is viewed by the borrower or lender. The present worth of all four plans in Table 3.2 is $100,000 and the plans are equivalent.

The concept of present worth provides a convenient method for comparing repayment plans even though they might involve vastly different repayment sums. The four payment series in Table 3.2, for example, require total repayments of $220,000, $166,000, $176,984.10, and $310,584.82.

Equivalence depends on the interest rate. For example, consider a $10,000 loan at an interest rate of 6% for 10 years. Table 3.3 shows four cash flow repayment plans for the terms of the loan. The disparity in total repayment sums is large. The plans are equivalent, however, with a present worth of $10,000. Table 3.4 shows the present worth of the four plans for different interest rates. The present worths of the plans increase as the interest rate is lowered, and, conversely, a higher interest rate lowers the present worth. The comparison in Table 3.4 also shows that equivalence of the plans exists only for a 6% interest rate.

A present sum is always equivalent to a larger sum of future payments at some interest rate. In economic considerations we sometimes want to determine the interest rate that will make two time series of repayments equivalent. This interest rate for different time series of repayment is called the *rate of return* on an investment.

Interest Formulas Mathematical formulas are available for computing the required repayment of money at a given rate of interest and present worth over some period of time. We use the following symbols:

i, interest rate per interest period
n, number of interest periods
P, present worth
F, sum that is equivalent to P at an interest rate i after n periods
A, end-of-period payment

Table 3.5 summarizes several variations on the basic interest formulas, $F = P (1 + i)^n$. The factor $(1 + i)^n$ is called the single-payment compound amount factor. When multiplied by the present worth, it gives the equivalent sum after n periods. We adopt the standard symbology $(F/P, i, n)$, which stands for the factor $(1 + i)^n$. As an example, calculate the single payment, F, required to repay a loan of $10,000 at 20% at the end of 10 years. The relation is $F = P (F/P, i, n) = 10,000 (1 + 0.20)^{10} = \$61,917$.

Table 3.3   Four Equivalent Series of Payments ($10,000 at 6% for 10 Years)

Year	Investment	Plan A Annual interest payment plus lump-sum final payment	Plan B Annual payment plus interest due	Plan C Uniform annual payments	Plan D Single repayment
0	$10,000				
1		$ 600	$1,600	$1,359	
2		600	1,540	1,359	
3		600	1,480	1,359	
4		600	1,420	1,359	
5		600	1,360	1,359	
6		600	1,300	1,359	
7		600	1,240	1,359	
8		600	1,180	1,359	
9		600	1,120	1,359	
10		10,600	1,060	1,359	$17,910

Table 3.4 Present Worth at Various Interest Rates of Payment Series in Table 3.3

Interest rate (%)	Plan A	Plan B	Plan C	Plan D
0	$16,000	$13,300	$13,590	$17,910
2	13,590	12,030	12,210	14,690
4	11,620	10,940	11,020	12,100
6	10,000	10,000	10,000	10,000
8	8,660	9,180	9,120	8,300
10	7,540	8,460	8,350	6,900

Inversion of the basic formula gives $P = F/(1 + i)^n$, where $1/(1 + i)^n$ is called the single-payment present worth factor, $(P/F, i, n)$. Application of the equation $P = F(P/F, i, n)$ gives the present worth of a future sum.

A fund designed to produce a desired amount of money at the end of a specified period by means of a series of uniform end-of-period payments is called a *sinking fund*. The factor $i/[(1 + i)]^n - 1) = (A/F, i, n)$ is called the sinking fund factor. To illustrate, we calculate the annual payment required to produce $100,000 at the end of 20 years at an interest rate of 15%. The sinking fund factor is $(A/F, 15, 20) = 0.15/[(1.15)^{20} - 1] = 0.00976$. Thus, the required annual payment is $A = 100,000 (0.00976) = \$976.00$.

Table 3.5 Variations of the Basic Interest Formula

Basic interest formula:

$$F = P (1 + i)^n \qquad (1 + i)^n = (F/P, i, n)$$

Single-payment compound amount factor

Variations:

$$P = F \left[\frac{1}{(1 + i)^n} \right] \qquad \frac{1}{(1 + i)^n} = (P/F, i, n)$$

Single-payment present worth factor

$$A = F \left[\frac{1}{(1 + i)^n - 1} \right] \qquad \frac{i}{(1 + i)^n - 1} = (A/F, i, n)$$

Sinking fund factor

$$A = P \left[\frac{i (1 + i)^n}{(1 + i)^n - 1} \right] \qquad \frac{i (1 + i)^n}{(1 + i)^n - 1} = (A/P, i, n)$$

Capital recovery factor

$$F = A \left[\frac{(1 + i)^n - 1}{i} \right] \qquad \frac{(1 + i)^n - 1}{i} = (F/A, i, n)$$

Series compound amount factor

$$P = A \left[\frac{(1 + i)^n - 1}{i (1 + i)^n} \right] \qquad \frac{(1 + i)^n - 1}{i (1 + i)^n} = (P/A, i, n)$$

Series present worth factor

Source: Adapted from Grant et al. [2].

On the other hand, if we want to know the uniform end-of-the-year payment that can be secured for n years for a present investment P we use the relation $A = P[i(1 + i)^n/(1 + i)^n - 1]$, where the bracketed term is called the capital recovery factor, $(A/P, i, n)$. To demonstrate its use, let us find the uniform annual payment that we could secure for an investment of $100,000 at an interest rate of 18% over a 20-year period. The capital recovery factor $(A/P, 18, 20) = [0.18 (1.18)^{20}/(1.18)^{20} - 1)] = 0.1868$. Then, the annual payment is $100,000 (0.1868) = \$18,680$.

In a similar vein, the future worth of a series of uniform payments uses the series compound amount factor, which is $(F/A, i, n) = [(1 + i)^n - 1]/i$. Take the prior example of a set of 20 annual payments of $18,680 at 18% interest. What is the worth of this payment series at 20 years? The result is $F = A(F/A, 18, 20) = 18,680 (146.63) = \$2,739,048$.

We might also need to know the present worth of a uniform series of payments. To get this, we use $P = A[(1 + i)^n - 1]/i(1 + i)^n = A(P/A, i, n)$, the series present worth factor. In accordance with the previous example, we will calculate the present worth of a series of payments of $18,680 for an interest rate of 18% for 20 years. The series present worth factor, $(P/A, i, n)$ is $[(1.18)^{20} - 1] / 0.18 (1.18)^{20} = 5.353$. The present worth is $P = 18,680 (5.353) = \$99,994$. The value rounded off is \$100,000, which was the present worth for the series of examples that we just completed.

Table 3.5 summarizes five of the variations of the basic interest formula that are frequently used in economic investment analysis.

Continuous Compounding of Interest A sum P invested over n years at a nominal interest rate[2] r with m compounding periods per year gives a compound factor:

$$F = P\left(1 + \frac{r}{m}\right)^{mn} \tag{3.1}$$

Let $m/r = k$. Then $m = rk$ and

$$F = P\left(1 + \frac{1}{k}\right)^{rkn} = P\left[\left(1 + \frac{1}{k}\right)^k\right]^{rn} \tag{3.2}$$

Continuous compounding corresponds to allowing m to increase without limit or

$$\lim_{k \to \infty} F = Pe^{rn} \tag{3.3}$$

since

$$\lim_{k \to \infty} \left[1 + \frac{1}{k}\right]^k = e \tag{3.4}$$

---

[2]Nominal interest is interest compounded m times per year at the rate r/m per period.

Unknown Interest Rates In some cases the investment sum is known and a prospective repayment plan is envisioned. The effective interest rate is desired.

It is a simple matter to calculate the interest rate for a single-investment, single-repayment plan. We can solve the equation $F = P(1 + i)^n$ for the interest rate and get

$$i = \left(\frac{F}{P}\right)^{1/n} - 1 \tag{3.5}$$

However, it is more complicated to calculate an unknown interest if a repayment series is involved. This is especially true when nonuniform annual payments are projected. An interpolation process is usually required for these cases. The following examples illustrate techniques for finding unknown interest rates.

Example 3.1 Single Repayment Plan — Unknown Interest

Calculate the interest rate for a loan of $12,100 that requires a single repayment of $17,910 after 10 years.

$$i = \left(\frac{17,910}{12,100}\right)^{1/10} - 1$$

$$i = 0.040 = 4\%$$

This interest rate agrees with that listed in Table 3.4 for plan d for 4% and a present worth of $12,100.

Example 3.2 Rate of Return on Infrared Thermograph

Calvin Caloric is considering the purchase of an infrared thermography system for use in his energy auditing business. The system will cost $12,000. Caloric's accountant estimates that the device will have a 10% salvage value at the end of 8 years. The accountant also estimates the cash flows for the system as shown in the following table.

Cash Flow Estimate for Infrared Thermograph Ownership

Year	Receipts	Disbursements	Net cash flow
0		− $12,000	− $12,000
1	+ $4,200	− 500	+ 3,700
2	+ 3,900	− 900	+ 3,000
3	+ 3,600	− 1,200	+ 2,400
4	+ 3,500	− 1,400	+ 2,100
5	+ 3,100	− 1,400	+ 1,700
6	+ 3,000	− 1,500	+ 1,500
7	+ 2,800	− 1,500	+ 1,300
8	+ 2,650	− 1,500	+ 1,150
8[a]	+ 1,200		+ 1,200
Total	+$27,950	− $21,900	+ $ 6,050

[a]Salvage.

The receipts result from clients' fees and occasional rentals. Disbursements result mainly from maintenance and repair, but storage, taxes, and insurance premiums are also included. What is the rate of return or the "effective interest" for this investment?

To solve this, use present worth calculations at different interest rates (rates of return) to bracket the actual projected cash flow history. The correct rate of return would yield a present worth of zero at the end of 8 years.

At 11% interest, at the end of the first year the present worth of the $3,700 cash flow is

$$P = F(P/F, i, n) \qquad P = \$3,700 \left(\frac{1}{1.11} \right) = \$3,333$$

For the second year

$$P = \$3,000 \left[\frac{1}{(1.11)^2} \right] = \$2,435$$

For the third year

$$P = \$2,400 \left[\frac{1}{(1.11)^3} \right] = \$1,755$$

The present worth of the estimated cash flow can be calculated in this fashion for each of the three interest rates shown in the following table.

Present-Worth Calculations to Determine Prospective Rate of Return

Year	Estimated cash flow	Present worth		
		At 11%	At 12%	At 13%
0	− $12,000	− $12,000	− $12,000	− $12,000
1	+ 3,700	+ 3,333	3,304	3,274
2	+ 3,000	+ 2,435	+ 2,392	+ 2,349
3	+ 2,400	+ 1,755	+ 1,708	+ 1,663
4	+ 2,100	+ 1,383	+ 1,335	+ 1,288
5	+ 1,700	+ 1,009	+ 965	+ 923
6	+ 1,500	+ 802	+ 760	+ 720
7	+ 1,300	+ 626	+ 588	+ 553
8	+ 2,350	+ 1,020	+ 949	+ 884
Total	+ $6,050	+ $363 − 336	+ $1	+ $346

A rate of return of about 12% is the closest match. This represents the cash flows before income taxes are paid. Caloric must now consider this rate of return in comparison to other options such as equipment rental.

Example 3.2 is an illustration of a technique also known as discounted cash flow. This technique is frequently used to find the rate of return on invested capital. It requires calculating the present worth of year-by-year net cash flows at a given discount rate, as illustrated in the above example.

Depreciation We now return to the concept of depreciation, which was touched on previously. It is a concept used primarily in accounting. When depreciation is allowable, it is possible to include its effects in cash flow diagrams. This is called *amortization*. The amortized cost of an asset is a prepaid operating expense that can be apportioned over the years of the asset's economic life. The actual value of the asset is not being amortized; rather, the cost is being spread out over its economic life. The common term *book value* describes the difference between the value of the asset and the total depreciated costs at a given time. This difference is more accurately called unamortized cost.

In depreciation accounting the cost of an asset, less salvage value, is written off the books. Several ways of performing this write-off exist. Some of the common methods are discussed in the following sections.

Straight-Line Depreciation The straight-line method is the most straightforward system of depreciation. The straight-line rate is found by subtracting the estimated salvage value as a percentage of first cost from unity and dividing by the service life. We write

$$f_{s-l} = \frac{1 - (\text{salvage value/first cost})}{n}$$

where n is the economic lifetime in years. The rate f_{s-l} then yields a uniform write-off for each year of the service life. At the end of its economic life, the asset's costs have been amortized, and only the salvage value remains as the book value.

Declining-Balance Depreciation The production of income by an asset is sometimes higher during the early years of its service life. In 1954 the Internal Revenue Service approved several methods of allowing for a greater depreciation write-off in the early years of life. The declining-balance method is one of the approved methods. These accelerated write-off methods are restricted to assets with a life of 3 years or longer.

Declining-balance depreciation involves applying the depreciation rate, f_{d-b}, to the remaining book value. A $10,000 asset whose depreciation rate is 10% would give a depreciation charge of 0.10 ($10,000) = $1,000 for the first year. For the second year, a depreciation charge of 0.10 ($9,000) = $900 is allowed. For the third year, 0.10 ($8,100) = $810 is the charge. The procedure is applied over the lifetime of the asset.

The most frequently used version of this method is the double-rate declining-balance method. In this case the rate is 200%/service life. An asset with a service life of 10 years would have a double-rate depreciation, f_{d-r}, of 20% per year. This is double the straight-line rate of an asset with a service life of n years with zero salvage value, hence the name. The salvage value is always taken as zero when computing f_{d-r}.

ECONOMIC ANALYSIS **63**

The actual f_{d-r} that should be used is usually based on the intent of the 1954 tax law — that is, to allow for about two-thirds of the asset to be written off in the first half of its estimated life.

Sum-of-the-Year's Digits Depreciation The digits of the number of years of estimated service life are added together. The depreciation rate for the first year is the ratio of the last digit to the sum of the digits. For a 10-year service life, the sum-of-the-year's digits are $1 + 2 + 3 + \ldots + 10 = 55$. The first year's rate is $10/55 = 18.2\%$. In the second year the rate is $(n - 1)/\Sigma n = 9/55 = 16.4\%$, and so forth, down to a last year's rate of $1/55 = 1.82\%$. This method is allowable by the 1954 tax laws. It allows for a write-off of about three-fourths of the depreciable cost in the first half of the estimated life.

Example 3.3 Comparison of Depreciation Methods

A new burner control system is to be purchased by Modern Energy Systems for use in its steam generator. The cost is $24,000. Compare the straight-line, double-rate declining balance, and sum-of-the-year digits methods for a service life of 6 years and a salvage of 10% of the system's initial cost.

With the straight-line method,

$$f_{s-l} = \frac{1.00 - 0.1}{6} = 15\%$$

Thus for each year the depreciation charge is $0.15 (24,000) = \$3,600$.
With the double-rate method

$$f_{d-r} = \frac{2.00}{6} = 33.3\%$$

In the first year the write-off is $0.333 (24,000) = \$8,000$. In the second year it is $0.333 (24,000 - 8,000) = \$5,333$. The following table gives the remaining values.

Comparison of Depreciation Methods

	Depreciation charge per year			End-of-year book value		
Year	Straight line	Declining balance	Sum/digits	Straight line	Declining balance	Sum/digits
0				24,000	24,000	24,000
1	3,600	8,000	6,178	20,400	16,000	17,822
2	3,600	5,333	5,143	16,800	10,667	12,679
3	3,600	3,556	4,114	13,200	7,111	8,565
4	3,600	2,370	3,086	9,600	4,741	5,479
5	3,600	1,580	2,057	6,000	3,161	3,422
6	3,600	1,054	1,028	2,400	2,107	2,394
Total	21,600	21,893	21,606			

The sum-of-the-year's digits method would give

$$f_{s-d} = \frac{n + 1 - i}{\Sigma n}$$

for each year, where i is the appropriate year up to n. For the first year, f_{s-d} = 6/21 = 28.6%. The write-off is 0.286 (24,000 − 2,400) = $6,178. For the second year, f_{s-d} = 5/21 = 23.8%. The write-off is $5,141. The remaining values are shown in the table.

3.1.4 Basic Income

Returns on investment are the net after-tax earnings of a project. The gross income is the difference between revenue and costs. Revenue and costs are clearly cash flows because money flows into or out of a project account. The taxable income is the gross income minus allowances. An example of an allowance is a depletion allowance that is permitted by the Internal Revenue Service for certain natural resources that cannot be renewed in a foreseeable period.

Taxes are found by applying the appropriate tax rate to the taxable income. The corporate tax rate for firms with taxable income in excess of about $75,000 per year is 48%. The tax collector shares in the rate of return for an investment, so it is important to factor taxes into the basic income equation as a negative cash flow. (Actual calculations of taxes should be approached very carefully with full attention given to year-by-year changes in tax rates. A competent tax accountant should be consulted if questions arise.)

After-tax earnings are the basic income information required for investment decisions. After-tax earnings are also known as after-tax net cash flow.

In many energy conservation projects, it may not be possible to increase revenues, but energy conservation can reduce costs. The net effect is the same: an increase in after-tax earnings.

Let us look at a simple example of a calculation of after-tax earnings.

Example 3.4   After-Tax Earnings

A $120,000 heat recovery unit is being considered by Modern Energy Systems. The investment should result in a reduction in net annual operating expenses of $30,000 per year for 8 years. The investment will be depreciated for income tax purposes by the straight-line method with zero salvage value. The effective tax rate is 48%. What are the earnings before and after taxes? Also, what are the after-tax earnings for sum-of-the-year's digits depreciation and a 10% investment tax credit taken at the time of investment?

The calculations for straight-line depreciation are simple and are shown in the following table.

After-Tax Earnings for Straight-Line Depreciation

Year	Revenue	Depreciation	Taxable earnings	Taxes	After-tax earnings
0	−120,000				
1	30,000	15,000	15,000	7,200	22,800
2					
3					
4					
5					
6					
7					
8					

With the sum-of-the-year's digits method, the calculations are somewhat more complicated. The results are shown in the following table.

After-Tax Earnings for Sum-of-the-Year's Digits Depreciation with 10% Investment Tax Credit

Year	Revenue	Depreciation	Taxable earnings	Taxes	After-tax earnings
0	− 108,000[a]				
1	30,000	26,667	3,333	1,600	28,400
2		18,148	11,852	5,689	24,311
3		12,530	17,470	8,386	21,614
4		8,702	21,298	10,223	19,777
5		5,995	24,005	11,522	18,478
6		3,997	26,003	12,431	17,519
7		2,442	27,558	13,228	16,772
8		1,153	28,847	13,846	16,154

[a] $120,000 investment less the $12,000 tax credit.

Note that the investment tax credit does not affect after-tax earnings. However, it does affect the effective rate of return after taxes since it reduces the initial negative cash flow by 10%.

3.1.5 Life-Cycle Costing

Life-cycle costing is a type of economic investment analysis in which proper attention is given to economic factors other than initial cost. Energy conservation installations reduce costs rather than produce additional revenue. Life-cycle costing reflects those savings in cost over the projected lifetime of the installation.

Table 3.6 Life-Cycle Cost Parameters

Initial cost, including hardware, transportation, installation, system
  designer's fee, special building features, and value of space used by
  system
Down payment and mortgage interest rate
Life-cycle time
Depreciation rate and final salvage value
Repair and replacement costs
Maintenance costs
Fuel costs
Inflation rate of cost of fuel
General inflation rate
Local, state, and federal taxes

Life-cycle costing uses the traditional accounting concepts of interest, depreciation, present worth, discounted rate of return, and after-tax earnings. The details of a life-cycle cost are represented by the discounted cash flow type of analysis. The preferred method of life-cycle costing is the annualized present worth calculation, which was illustrated in Example 3.2. Table 3.6 summarizes the parameters that commonly affect the life-cycle costs of energy systems.

3.1.6 Evaluation of Projects

The goal of an economic investment analysis is to decide whether to implement an energy conservation project. The decision is usually made by management, based on company-approved policies. Several techniques are in general use for evaluation. The payback (or payout) method, the minimum attractive rate of return (MARR) method, the annualized costs method, and the internal rate of return method are used for project evaluation.

In the following sections we demonstrate the use of these techniques and point out their merits and demerits.

Payback The payback period is the number of years required to recover an initial investment from net cash flows. It is defined as

$$\text{Payback} = \frac{\text{initial investment}}{\text{net annual cash flow}}$$

For example, consider a project with an initial investment of $25,000; a net cash flow per year of $8,000; and, at the end of 5 years, a salvage value of $5,000. The payback period without salvage is 3.125 years, and with salvage is 2.5 years. If management considers 2.5 years to be an acceptable payback period, the project could be implemented.

The advantage of the payback scheme is its simplicity. However, it does not always measure profitability because the cash flows after the payback period are ignored. Furthermore, the time value of money also is not directly considered. The payback period method of evaluation provides an indication of risk, however, because it quickly tells how invested capital can be recovered. *does not tell you the profitability of project*

Minimum Attractive Rate of Return The MARR method is also called the net present worth method because it is based on the difference between present worth of the revenue and costs for the project. If the net present worth is positive, the project is economically attractive. The time value of money, a specified minimum acceptable rate of return, and the cash flows are used in the analysis.

Alternative investment plans are treated like interest-paying projects. The equation

$$\text{Present worth} = \sum_{t=1}^{t=n} \frac{\text{returns } (t)}{(1 + i)^t} - \text{investment} \qquad (3.6)$$

can be used to reduce the project to a net present worth. Here, t is time and i is the MARR. When the present worth is negative, the project is incapable of providing enough net revenue to return the funds corresponding to the minimum allowable rate of return. Conversely, a present worth greater than zero indicates an attractive project.

Choices between alternative projects can be based on the highest net present worth of the projects. The following example illustrates the concept of MARR.

Example 3.5   MARR Project Evaluation

The table below shows four alternative projects. All projects involve uniform cash flows. The MARR is 15%. Calculate the net present worth for the alternative projects and decide which is most attractive.

End-of-year	Project 1	Project 2	Project 3	Project 4 (do nothing)
0 (initial investment)	– $5,000	– $8,000	– $10,000	0
1–10 net revenues	$1,400	$1,900	$2,500	0

Solution: The net present worth (P) is calculated by using Eq. (3.6).

Project 1:

$$P = A(P/A, i, n) - I_0$$
$$= 1,400(5.0184) - 5,000 = 2,026.32$$

Project 2:

$$P = 1,900(5.0184) - 8,000 = 1,535.72$$

Project 3:

$$P = 2,500(5.0184) - 10,000 = 2,547.00$$

Project 4:

$$P = 0$$

Projects 1, 2, and 3 are all attractive. Project 3 would be the best alternative.

The MARR method works with any cash flow pattern, and the calculations are relatively easy. In some cases the calculations are more difficult than those in the annualized costs method, which will be described next.

Annualized Costs This method is useful when investment projects with different economic lifetimes are being considered. The annual cost of an investment can be found from

Annual cost = (operating + maintenance costs) + depreciation
  + [(investment – salvage value) × $(A/P, i, n)$]
  – salvage value × (i) $\qquad$ (3.7)

This equation allows projects with different economic lifetimes to be placed on the same footing — costs per year. The project that gives the lowest annualized costs of a set of alternatives should be selected.

The annualized costs method can be extended to include net revenues if the revenues can be tied to the investment alternatives. Then we have

Annual worth = net annual revenue – annual costs

We would pick the alternative that maximizes the annual worth. Furthermore, if the annual worth for a single project is calculated to be less than zero, the project will not earn enough to pay off the investment. If the annual worth is greater than zero, the project may be attractive.

The following example illustrates the method of annualized costs for economic evaluations.

Example 3.6 Purchase of Pumps

A pump is being considered for a service period of 60 years. Pump 1 costs $10,000, has a life of 30 years, and has a salvage value of $600. It also has a resale value of $2,500 at the end of 20 years. Operation costs are $5,000 per year for the first 20 years; after that, they increase at the rate of $100 per year up to the 30-year life. Maintenance costs are $600 per year.

Pump 2 costs $6,000, has a life of 20 years, and has a salvage value of $100. Operation costs are \$6,000 per year and maintenance is \$800 per year. On the basis of annualized costs, which pump would you select, if the minimum rate of return is 20%?

Solution: Neglect depreciation. The annual cost (AC) of pump 1 for 20 years is, according to Eq. (3.7),

$$AC = 5,000 + 600 + (10,000 - 2,500) \, (A/P, \, 20, \, 20) - 2,500 \, (0.20)$$

$$(A/P, 20, 20) = \frac{i(1 + i)^n}{(1 + i)^n - 1} = \frac{(0.20)(1.20)^{20}}{(1.20)^{20} - 1}$$

$$= 0.2054$$

$$AC = \$6,640.50$$

The annual cost of pump 1 for 30 years is

$$AC = 5,000 + 600 + \underbrace{\frac{10}{30} \, (100) \sum_{n=1}^{10} n}_{\text{uniform gradient}} + \underbrace{(10,000 - 600)(0.2009)}_{(A/P, \, 20, \, 30)} - 600 \, (0.20)$$

$$AC = \$9,441.79 \; (7556.80)$$

The annual cost of pump 2 for 20 years is

$$AC = 6,000 + 800 + (6,000 - 100)(0.2054) - 100 \, (0.2)$$

$$= \$8,031.86$$

The choice would be to purchase pump 1, use it for 20 years, and sell it for \$2,500.

The annualized costs method involves easy computations, especially for simple cash flow patterns. Furthermore, most managers are familiar with the concept of annualized costs. The method is especially useful when projects with unequal economic lifetimes are being considered.

Internal Rate of Return If the minimum attractive rate of return is difficult for managers to estimate, the internal rate of return may be used as an economic evaluation tool. This technique corresponds to finding the unknown interest rate for the projected revenues and costs of the project. This is also called the discounted cash flow method and was covered in Section 3.1.3. Example 3.2 illustrated an application of the technique.

The internal rate of return for a project can be compared to the company's acceptable rate of return to determine whether a project is economically feasible. Most managers are familiar with this concept. The method provides an indication of the relative merits of projects.

Variations of these evaluation techniques are used in some circumstances. Lesso [3] gives an excellent treatment of these variations as applied to energy conservation projects.

The reader is also directed to a discussion in reference [3] of the "risk" associated with investment projects and of the effects of inflation on investment analysis. The risk is a measure of the uncertainty of projections of costs and revenues into the future. Estimates of future inflation levels are clearly one of the major uncertainty factors. An adequate treatment of risk and inflation effects is beyond the scope of this text.

3.2 UTILITY RATE STRUCTURES

Energy markets for natural gas and electricity are subject to energy-pricing structures that basically follow a declining block rate. The central idea of a declining block rate is that the more energy used, the lower the price per unit. Declining block rate structures are artifacts from the era of cheap energy. Energy utilities for decades campaigned for higher consumption of their product. One means of doing this was through lower block rates for higher consumption.

The cost of electricity generation declined at a moderate rate for several decades prior to the 1970s. This reflected higher efficiency for power plants brought about by technological improvements. Fuel costs also declined during that period. Natural gas, for example, was considered almost a nuisance in the 1940s and 1950s and was quite cheap. However, as the price of coal increased, as environmental pollution standards were imposed, and as natural gas pipelines were constructed, natural gas became an important and inexpensive boiler fuel. Many industries in the northeastern United States converted from coal to gas under the influence of these factors.

Since the 1974 oil embargo, concern over energy costs has increased dramatically. The cost of natural gas has increased tenfold over the past 5 years and will rise even more after total deregulation. Coal is plentiful, but mining it efficiently is difficult. Thus, the present cost of burning coal is also much higher than it was prior to the embargo. These increased fuel costs are passed along to consumers, whether they are residential, commercial, or industrial. It is well understood in the United States that a cost incurred by industry becomes part of the price of the marketed product.

Fuels other than natural gas are usually not provided by a franchised utility. Consumers generally make their own arrangements for the purchase of coal or fuel oil. Electricity generation is almost monopolized by utilities with the exception of some industries that generate electricity onsite. Rate structures for electricity and natural gas can be treated with some generality, and consequently are discussed in the following sections.

3.2.1 Electrical Rates

Rates that determine a consumer's utility bill are set by regional and state agencies and depend on a multifaceted scheme for determining the costs of power plants and the distribution system, the cost of the meter, the cost of fuel, and, finally, administrative costs.

System Costs   Utilities must be able to meet the peak demands of their customers. This means that systems must grow to meet the added load or, in some nongrowth cases, they must be maintained to meet the needs of customers. The costs of building and maintaining the system are passed along to customers via their utility bills.

Power plants have become large in recent times because scaling up to larger sizes results in greater efficiency. Plants ranging in size from 500 to 1,000 MW are common, and some nuclear installations have units with capacities greater than 1,000 MW. The required structures and the erection costs are correspondingly large. The return toward coal as a prime boiler fuel has drive costs even higher because of the added requirements for flue gas scrubbers, stack precipitators, coal yards, and coal-handling equipment.

Energy generated at a power plant is fed into a system of step-up transformers, transmission lines, and step-down transformers, which permits distribution of the electricity over the utility district. The costs of these components are passed along to consumers. There are electrical resistance losses in all of these distribution system components, which represent a loss of product by the utility.

The cost of metering also can add significantly to energy costs. A meter for a large industrial firm can cost $1,000 or more. In some industrial plants, several meters may be required.

System costs can be likened to capital costs in that they are incurred by the utility at the "front end" of an investment period. These costs are recovered from sales over the system lifetime.

Cost of Fuel   Fuel was cheap and abundant for many utilities as recently as 1970, but now the availability of fuels is a major operating problem and the cost of fuel is a major component of energy costs, as most consumers know.

Administrative Costs   The utility must maintain a staff of personnel to provide engineering, marketing, and distribution services. Taxes, insurance, and interest on borrowed money must be paid. Compliance with government regulations also costs money. Public utilities are guaranteed a certain return on stockholders' investments. This currently runs between 12 and 16%. All of these costs are passed on to consumers.

Rate Schedules   Plant energy conservation managers must understand the basis on which their electric bills are calculated. Then they can devise appropriate strategies for reducing both their energy consumption and their utility bill.

A typical rate schedule is shown in Table 3.7. This schedule represents a two-part billing system made up of the energy charge and the demand

Table 3.7 Typical Rate Schedule for Large Power Users in the Midwest

Rate:

Demand charge:

First 25 kW of billing demand	$2.15 per kW per month
Next 475 kW	$1.90 per kW per month
Next 1,500 kW	$1.75 per kW per month
Excess over 2,000 kW	$1.47 per kW per month

Energy charge:

First 2,000 kWh per month	$0.02 per kWh
Next 18,000 kWh	$0.015 per kWh
Next 180,000 kWh	$0.011 per kWh
Next 550,000 kWh	$0.009 per kWh
Next 250,000 kWh	$0.0085 per kWh
Excess over 1,000,000 kWh	$0.0082 per kWh

Determination of billing demand:
The billing demand for the month shall be computed as the highest average load in kilowatts occurring during any 15-minute period.

Load factor discount:
A discount of $0.0015 per kilowatt-hour is allowed on that portion of a customer's monthly kilowatt-hour consumption in excess of 360 hr use of his billing demand for the month.

Power factor provision:
Monthly average power factor shall be determined by means of a reactive component meter matched to record only lagging reactive kilovolt ampere hours, used in conjunction with a standard watt-hour meter. Monthly average power factor is defined as the quotient obtained by dividing the kilowatt-hours used during the month by the square root of the sum of the squares of the kilowatt-hours used and the lagging reactive kilovolt amperes supplied during the same period.
In the case of customers with maximum demands of 150 kW or more, the monthly rate shall be decreased 0.2% for each whole 1% by which the monthly average power factor exceeds 80% lagging, and shall be increased 0.3% for each whole 1% by which the monthly average power factor is less than 80% lagging.

Fuel clause:
The monthly amount computed shall be increased or decreased, in accordance with the Fuel Clause set forth in Section 10 of this tariff.

Source: Adapted from Thumann [4].

charge. The energy charge is basically the cost of the electricity generated and reflects in part the capital costs and basic fuel costs. Most utilities have added what is called a fuel adjustment to the energy charge. This is a direct charge passed on to the customer that reflects the escalation of fuel costs not included in the initial fixing of the energy charge. The energy and fuel adjustment charges are based on kilowatt-hours of electrical usage.

The demand charge reflects the rate at which energy is used. Demand charges are usually reserved for industrial and commercial users rather than residential users. They are based on the maximum rate of electrical usage over a specified period of time, usually 15 minutes. Many utilities have moving 15-minute "windows." The customer does not necessarily know when the demand is being sampled. The demand charge is for the capacity that the utility must provide for peak requirements. Various techniques for "demand limiting" are discussed in Chapter 14.

For customers who do not have demand meters, an estimate of demand is made based on total consumption during the billing period.

An additional charge made by some utilities to industrial users is the power factor charge. This accounts for lost power caused by a mismatch between line and load impedance. Power factor correction is a very common practice in industrial plants. Details of this practice are given in Chapter 14.

Table 3.8 shows typical usage patterns for several different types of enterprises for a recent 1-year period. It is instructive to calculate a monthly bill for one of these cases.

Example 3.7 Calculating an Electric Bill

Calculate the electric bill for the manufacturing plant represented in Table 3.8 for October, using the rate schedule of Table 3.7.

The calculation breaks down as follows:

Energy charge: total kWh = 408,000

First	2,000 kWh at 2¢	$ 40.00
Next	18,000 kWh at 1.5¢	$ 270.00
Next	180,000 kWh at 1.1¢	$1,980.00
Remaining	208,000 kWh at 0.9¢	$1,872.00
Total	408,000 kWh	$4,162.00

Demand charge: total kW = 1,650

First	25 kW at $2.15	$ 53.75
Next	475 kW at $1.90	$ 902.50
Remaining	1,150 kW at $1.75	$2,012.50
Total	1,650 kW	$2,968.75

Load discount factor:

360 hr × 1,650 kW = 594,000 kWh

Therefore no LDF is allowed in accordance with Table 3.10.

Total bill $7,130.75

The result in Example 3.7 is the basic electric bill for 1 month. Fuel adjustment costs would be added to this. Power factor measurements were not available and were not included in the bill calculation.

Table 3.8 Electrical Usage Patterns for Typical Commercial Enterprises: Demand (kW) and Energy Use (kWh)

Month	Office building		Shopping center		Grocery store		Apartment complex		Convenience store		Manufacturing plant	
	kW	kWh	kW	kWh	kW	kWh	kW	kWh	kW	kWh	kW	kWh
May	345	101,000	1,425	571,000	300	151,840	234	91,400	30	14,698	1,545	398,000
June	293	95,000	1,455	680,000	312	165,280	231	100,000	30	17,588	1,635	398,000
July	375	118,000	1,440	661,000	310	177,920	240	124,000	32	17,739	1,650	408,000
August	420	156,500	1,395	667,000	322	164,320	240	118,200	35	17,437	1,755	488,000
September	555	130,000	1,455	733,000	317	162,880	273	123,200	31	18,963	1,695	448,000
October	540	207,000	1,455	646,000	319	153,600	270	109,800	32	16,003	1,650	408,000
November	570	169,500	1,545	675,000	286	153,800	234	95,600	30	17,490	1,605	410,000
December	600	172,500	1,740	768,000	250	136,800	234	76,800	29	12,684	1,515	363,000
January	660	194,500	1,935	892,000	250	139,520	192	56,600	28	12,049	1,485	340,000
February	668	215,500	1,905	795,000	247	129,120	153	62,800	30	13,097	1,470	406,000
March	660	223,500	1,740	719,000	262	136,480	240	71,200	31	15,001	1,515	383,000
April	600	177,500	1,515	633,000	290	134,400	237	76,600	32	16,102	1,590	363,000

Source: W. M. Lesso, Notes for Industrial Energy Auditing Short Course, University of Houston, Texas, August 1980.

Fuel charges are determined by a utility company according to a prescribed formula. A typical formula [5] is as follows:

$$\text{Fuel cost/kWh} = \frac{FC}{S} + \frac{E - AC}{S}$$

where FC = estimated cost of fuels, including refunds and the cost or revenues of purchased or sold power for the calendar month

S = estimated sales in kilowatt-hours for the month

E = actual cost of fuels, including refunds and the cost or revenues of purchased or sold electrical energy for the second preceding calendar month

AC = actual cost recovered in the second preceding calendar month

The energy manager should be aware that different rate schedules are used by some utilities, depending on the type of service provided. Tables 3.9, 3.10, and 3.11 show typical schedules used by a moderate-sized utility system in the southwestern United States. Table 3.10 applies to industrial users, and Table 3.9 applies to residential customers. Table 3.11 applies to the specialized category of commercial space conditioning.

The difference in utility bills calculated from different schedules is exemplified below.

Example 3.8   Effect of Rate Schedule on Utility Bill

Calculate the utility bill for a shopping center (Table 3.8) for the month of January using (1) the general schedule shown in Table 3.10 and (2) the specific schedule, commercial space conditioning (CSC) in Table 3.11. The fuel adjustment cost in both cases in 3.112¢/kWh.

Solution:

General schedule: Total energy: 892,000 kWh
    Maximum power: 1,935 kW

Fuel adjustment charge:	
892,000 kWh at $.03112/kWh	$27,759.04
Customer charge:	121.00
Energy charge:	
892,000 kWh at $.00152/kWh	10,275.84
Capacity (demand) charge:	
1935 kW at $3.22/kW	6,230.70
Total bill	**$44,386.58**

CSC schedule:

Fuel adjustment charge:	$27,759.04
Energy charge:	
First 50 kWh at $0.05/kWh	2.50

Table 3.9   Electric Rate Schedule 1: General Service, Multiple Fuels

Applications:

This rate is applicable to all electric service required by any customer to whom no other specific rate applies and when the electricity provided by the City is used in conjunction with other forms of energy.

Electric service of one standard character will be delivered to one point of service on the customer's premises and is measured through one meter.

Character of Service:

Alternating current, 60 cycles, single phase or three phase.

Rate:

Applicable to a customer whose electric service meets or exceeds 30 kilowatts per month for any two months within the most recent six summer billing months or as determined by the City. This rate will be applied for a term of not less than one year (twelve months).

	Billing months of November through April	Billing months of May through October
Customer charge	$4.50	$4.50
Energy charge	1.2¢ per kWh all kWh	1.2¢ per kWh all kWh
Capacity charge	$3.22 per kW all kW	$5.78 per kW first 30kW
		$3.98 per kW all additional kW
Fuel charge	As prescribed	As prescribed

The kilowatt (kW) of the current billing month shall be the maximum indicated or recorded by metering equipment installed by the City.

When the power factor is less than 85%, kilowatt (kW) shall be determined by multiplying the indicated kW by 85%, and dividing by such lower peak power factor.

Rate:

Applicable to a customer whose electric service does not meet or exceed 30 kilowatts per month for any two months within the most recent six summer billing months or as determined by the City.

Source: Abridged from City of Austin (Texas) ordinances.

Table 3.10   Electric Rate Schedule II: General Service, Multiple Fuels

Application:
  This rate is applicable to all electric service required by any customer who receives at 12,500 volts or higher and whose electric service does not exceed 3,000 kilowatts for any two months within the previous twelve months.

Character of service:
  Alternating current, 60 cycles, three phase 12,500 volts or higher

Rate:

	Billing months of November through April	Billing months of May through October
Customer charge	$121.00	$121.00
Energy charge	1.152¢ per kWh all kWh	1.152¢ per kWh all kWh
Capacity charge	$3.22 per kW all kW	$5.78 per kW first 30 kW
		$3.95 per kW all additional kW
Fuel Charge	As prescribed	As prescribed

The kilowatt (kW) for the current billing month shall be the maximum indicated or recorded by metering equipment installed by the City. When the power factor is less than 85%, kilowatt (kW) shall be determined by multiplying the indicated kW by 85% and dividing such lower peak power factor.

Source: Abridged from City of Austin (Texas) ordinances.

Table 3.11   CSC — Commercial Space Conditioning Rate

Net monthly rate: 5.0¢ for the first 50 kWh
                  3.3¢/kWh for the next 550 kWh

May–October (inclusive) billings	November–April (inclusive) billings
2.6¢/kWh for the next 2400 kWh*	1.3¢/kWh for the next 2400 kWh*
1.2¢/kWh for the next 5000 kWh*	1.2¢/kWh for the next 5000 kWh*
0.6¢/kWh for all additional kWh	0.6¢/kWh for all additional KWh

*Add 100 kWh for each kW of billing demand from 20 to 100 kW, and 70 kWh for each kW of billing demand from 100 kW to 500 kW and 50 kWh for each kW of billing demand in excess of 500 kW.

Minimum charge—$1.20 plus $1.50 per kW of billing demand.

Fuel cost adjustment—The above energy charges shall be increased or decreased by 0.004¢/kWh for each whole one quarter of a cent by which the average delivered cost of the fuel during the next preceding six calendar months exceeds 18¢ or is less than 16¢ per million Btu.

Source: Abridged from the City of Austin (Texas) ordinances.

Next 500 kWh at $0.033/kWh	18.15
Next 2,400 kWh at $0.013/kWh	31.20
Next 5,000 kWh at $0.012/kWh	60.00
Remaining 884,000 kWh at $0.006/kWh	5,304.00
Total energy charge	$5,415.85

Demand charge:
    Add 50 kWh × (1935–500) = 71,750 kWh
        to the categories "2,400" and "5,000"

First 2,400:	
71,750 kWh at $0.013/kWh	932.95
Second 5,000:	
71,750 kWh at $0.012/kWh	861.00
Total demand charge	1,793.75
Total charge	$34,968.64

Clearly the customers benefit from using the CSC schedule if they qualify for it.

3.2.2 Gas Rates

Gas rate schedules are similar to electricity schedules. A declining block rate with fuel adjustment charges is most common. In many states the industrial customer may choose different rates depending on the customer's willingness to experience curtailment of service. Table 3.12 shows a typical rate schedule along with the curtailment priorities.

It is clear that industrial rate 3 would be the choice of an industry whose usage is high enough. The problem lies in the curtailment priorities. In the winter of 1978, for example, many industries in Ohio and the surrounding states were without gas service for weeks because of curtailment caused by the severity of the weather. The customer must abide by the selected rate for a period of 1 year. It is not possible to use one rate in summer and another in winter.

The choice of the appropriate rate is usually based on a simple economic study that includes the probable number of days of curtailment per year for each rate, along with the cost of replacing the equivalent lost energy with a backup supply such as fuel oil. Average fuel consumption figures per month are used to calculate the cost of the backup energy supply. Example 3.9 illustrates this procedure.

Example 3.9 Selecting Gas Rates

The heat value of natural gas is approximately 1,000,000 Btu/Mcf (thousand cubic feet), and for No. 2 fuel oil it is 140,000 Btu/gal. Thus, it takes 7.14 gallons of fuel oil to replace 1 Mcf of natural gas. Pygmy Industries requires a monthly average of 800 Mcf. Mammoth Industries, on

Table 3.12 Typical Gas Rate Schedule[a] with Curtailment Clauses

Monthly rates:
  Subject to Company's limitations on the availability of each rate. Customer shall receive service under its choice of one of the following rates in accordance with the rate selected by Customer as provided in the contract:

	Ind. rate 1	
First	125 Mcf or less	$356.25 (2.85/Mcf)
All over	125 Mcf at	2.75/Mcf

	Ind. rate 2	
First	600 Mcf or less	$1,560.00 (2.60/Mcf)
All over	600 Mcf at	2.50/Mcf

	Ind. rate 3	
First	1,250 Mcf or less	$3,062.50 (2.45/Mcf)
All over	1,250 Mcf at	2.40/Mcf

Curtailment:
  Subject to governmental regulation, gas service under this schedule of industrial rates shall be subject to curtailment, interruption or discontinuance in a particular service area when necessary in the judgment of the Company for it to maintain residential and commercial rate service and industrial service having a higher priority. Services shall be furnished by Company and received by Customers in accordance with the following order of priority.
  (1) Residential and commercial rate service
  (1) Military rate service
  (2) Industrial rate 1 service
      The following priorities in descending order shall be observed as subpriorities for rate 1.
      1. Small commercial (less than 100 Mcf on a peak day).
      2. Large commercial (100 Mcf or more on a peak day) and industrial requirements for pilot lights and plant protection gas.
      3. Small industrial (less than 3,000 Mcf on an average day) requirements for feedstock and process gas needs.
      4. Large industrial (3,000 Mcf or more on an average day) requirements for feedstock and process gas needs.
      5. Industrial requirements not specified in priorities 3, 4, or 6.
      6. (1) (a) Boiler and other indirect flame applications (300 Mcf or less on an average day) with alternating fuel capabilities.
             (b) Boiler and fuel and other indirect flame applications (more than 300 Mcf on an average day and less than 3,000 Mcf on an average day) with alternate fuel capabilities.
             (c) Boiler fuel and other indirect flame applications (3,000 Mcf or more on an average day) with alternate fuel capabilities.
  (3) Public free school rate service
  (4) Industrial rate 2 service
  (5) Industrial rate 3 service
  (5) Special electric generation service
  (6) Dump interruptible sales made subject to interruption or curtailment at Seller's sole discretion under contracts or tariffs which provide in effect for the sale of such gas as Seller may be agreeable to selling and Buyer may be agreeable to buying from time to time.

[a] Typical of midwestern U.S. rates. Rates have been arbitrarily adjusted upward.

the other hand, needs 8,000 Mcf per month on the average. Based on economics, which rates should these two firms pick if the fuel oil is valued at $44.00 per 42-gallon barrel? The probable curtailment for each rate is:

Rate 1:   0.5 hr/month
Rate 2:   5 hr/month
Rate 3:   15 hr/month

Pygmy:

Rate 1: First 125 Mcf $356.25
         Remaining 675 Mcf $1,856.25

         Total $2,212.50

½ hour per month would account for

$$\frac{0.5}{24\,(30)} = \frac{0.5}{720} = 0.0694\% \text{ of } 800 \text{ Mcf} = 0.55 \text{ Mcf}$$

This would require only 3.9 gallons of fuel oil, which would cost only $4.10, which is negligible.

Rate 2: First 600 $1,560.00
         Remaining 1,945   486.25
                  194.5

5 hours per month gives

$$\frac{5}{24\,(30)} \times 800 = 5.55 \text{ Mcf} = \$41.00 \text{ of fuel oil}$$

         Total $ 2,087.25

Rate 3: Clearly out of the question. Pygmy does not use enough gas.

This moderate-sized industry should choose rate 2.

Mammoth:

Rate 2: First 600 Mcf $1,560.00

         Fuel oil equivalent $\dfrac{5\,(8,000)}{720}$ = 55.5 Mcf = $410.00 of fuel oil

         Remaining gas 7,344.5 Mcf $18,361.25

         Total $20,331.25

Rate 3: First 1,250 Mcf $3,062.50
         Fuel oil: three times rate 2
                  (166.5 Mcf) $1,230.00
         Remaining 6,583.3 Mcf $15,800.40

         Total $20,092.90

The large industry would probably want to select rate 2 even though rate 3 gives a slightly lower bill. The added risk of curtailment is not offset by the slightly lower bill.

3.3 SUMMARY

This chapter presents a rudimentary treatment of economics for those without such a background. The technical aspects of energy conservation projects must be accompanied by an economic analysis so that a company or a client can intelligently assess the potential for dollar as well as energy savings.

Utility rate structures also were discussed because many industries purchase much of their energy from utilities. In some cases the rate structures are quite complex, requiring an economic analysis in order to determine the appropriate utility and rate structure.

GLOSSARY

Amortization: The gradual payoff of a debt by contribution to a sinking fund at the time of each periodic interest payment. In depreciation it is the gradual write-off of the book value of an asset at discrete time intervals over the asset's economic life.

Book value: The value of an asset that has been amortized. The book value depends on the depreciation method selected and the estimated salvage value; it decreases over the asset's economic life.

Cash flow: The net annual flow of funds during the economic life of an investment. Positive cash flows result from revenues; negative cash flows result from costs.

Depreciation: An accounting technique in which a prepaid cost, such as that of a piece of equipment, is apportioned over the investment's economic life. Depreciation becomes a part of the cash flow for a project.

Economic life: The length of time over which equipment is considered to function economically. It can be determined by lease, franchise, contract, and/or Internal Revenue Service guidelines.

Equity financing: Method by which investment funds are raised by selling additional stock in a company.

Equivalence: The concept in which different time-series of repayments are equated, if they secure access to the same amount of money at the beginning of the project. Equivalence clearly depends on the rate of return.

Initial investment: The funds required at the outset of a project to secure the needed equipment, working capital, and/or services.

Interest: The cost of using borrowed money or the income from lent money. The interest rate is the ratio of interest payable at the end of a period to the money owed at the beginning of the period.

Rate of return: The interest rate that renders different time series of repayment for an investment equal to one another.

Retained earnings: Net funds available when dividends are taken from after-tax earnings. Retained earnings are usually invested in projects requiring an initial outlay of funds.

Return on investment: The net revenues from a project that are available to pay back the investment and to realize a profit.

Present worth: The investment required to secure a given time series of repayments. For a loan, the present worth is called the principal. Present worth can be evaluated at any time after project initiation.

Salvage value: The value of a piece of equipment at the end of its economic life.

REFERENCES

1. J. L. Riggs, *Engineering Economy,* McGraw-Hill, New York, 1977.
2. E. L. Grant, W. G. Ireson, and R. S. Leavenworth, *Principles of Engineering Economy,* 6th ed., Ronald Press, New York, 1976.
3. W. G. Lesso, Economic Analysis, in *Energy Management Handbook,* Wiley, New York, 1982.
4. A. Thumann, *Plant Engineers and Managers Guide to Energy Conservation,* Van Nostrand Reinhold, New York, 1977.
5. Y. Y. Haimes, *Energy Auditing and Conservation: Methods, Measurements, Management, and Case Studies,* Hemisphere, Washington, D.C., 1980.

PROBLEMS

3.1 Construct cash flow and cumulative cash flow diagrams for the economic project history given in the table. Salvage value is $3,000. Estimate the payback period and the total profit for the project period.

End of year	Costs	Revenue
0	15,000	
1	400	3,500
2	400	4,000
3	500	4,500
4	600	4,500
5	700	4,000
6	800	3,500
7	900	3,500
8	1,000	2,500

3.2 A heat recovery unit was purchased in 1974 by Modern Energy Systems for $20,000. The net savings per year achieved by the unit up to 1981, when the unit was sold for salvage for $1,500, are shown in the table. What was the rate of return obtained for the heat recovery unit?

Year	1974	1975	1976	1977	1978	1979	1980	1981
Savings	3,800	4,500	5,000	5,500	6,500	6,500	6,900	6,500

3.3 Plot the cash flow and cumulative cash flow diagrams for Problem 3.2. What were the payback period and the total profit for the venture?

3.4 Prepare a table similar to Table 3.2, showing four repayment plans for $10,000 in 5 years at an interest rate of 14%.

3.5 What effective yearly interest rate corresponds to a nominal rate of 15% compounded semiannually? Compounded quarterly? Compounded monthly?

3.6 Find the rate of interest for which the two payment plans below are equivalent.

Plan 1	Plan 2
Uniform payments of $1,000 for 8 years	Lump-sum payment of $15,325 at end of 8 years

3.7 An investment of $50,000 is expected to yield receipts of $7,500 per year for 15 years. What is the rate of return on this investment?

3.8 Calvin Caloric purchased an infrared thermography system for his energy auditing firm. The first cost was $12,000, the estimated service life is 10 years, and the trade-in value (salvage) is estimated to be $3,000. Calculate the book value of the system at the end of 6 years by the straight-line, double-declining balance, and sum-of-the-year's digits methods. Which method gives the greater write-off in the first year of ownership?

3.9 The heat loss from a bare steam pipe costs $405 per year. Insulation that will reduce the heat loss by 90% can be installed for $275. Alternatively, insulation that will reduce the loss by 86% can be installed for $165. Both types of insulation require no additional expense, and have no salvage value. Determine which type of insulation should be purchased if the pipe's estimated life is 10 years and the expected rate of return is 20%.

3.10 Carnot Chemical Company is considering two alternative recuperators for air preheating. The process off-gas used for air preheating is mildly corrosive, so that conventional metal-tube heat exchangers have shortened lifetimes. You, as the process engineer, are asked to recommend either a ceramic-tube unit, which costs more but has an extended life, or a conventional unit. Both units would save the same amount of fuel per year. Details of the two alternatives are given below. Use the straight-line depreciation method, and an effective interest rate of 20%. Which unit would you recommend?

	Ceramic-tube unit	Metal-tube unit
First cost	$75,000	$45,000
Operating cost/yr	1,000	1,000
Maintenance cost/yr	500	500 first year, increases 100/yr thereafter
Service life	20 yr	10 yr
Salvage value	10,000	5,000

3.11 Clear-Cut Glass Company is considering the purchase of a 10-hp electric motor, which will run an average of 7 hours per day for 270 days per year. It is anticipated that the annual costs for taxes and insurance average 2.5% of first cost. The company must make 25% on invested capital before income taxes and it must recover capital invested in machinery within 5 years.

Motor A costs $700 and has a guaranteed efficiency of 0.85% at the indicated operating load. Motor B costs $600 and has a guaranteed efficiency of 80% at indicated load. Electricity costs 4.2¢/kWh. Calculate the annual cost of each motor and indicate which motor should be purchased.

3.12 Modern Energy Systems has bought a new $6,000 adhesive applicating machine for installing thermal insulation. It has an expected life of 7 years with no salvage value. Taxes, insurance, maintenance, and electricity are estimated at $1,500 for the first year, $1,700 the second year, $1,900 the third year, and increasing by $250 each year thereafter. What is the equivalent uniform annual cost of this adhesive applicator if the interest rate is 15%?

3.13 Calvin Caloric is designing a new industrial facility for a client. He presents two alternatives to the client: building A, which costs less but has a shorter life, and building B, which costs considerably more but has a longer lifetime and lower operating and maintenance costs because it has better materials and more thermal insulation. Data on the buildings are summarized below.

	Building A	Building B
First cost	$50,000	$100,000
Life	20 years	40 years
Salvage value	$10,000	$30,000
Annual operating and maintenance costs	$10,000	$5,000

Compare these two plans, using a minimum attractive rate of return of 15%. Which building should the client choose?

3.14 Gibbs Refining Company needs to install piping for a pumping service. It is proposed to use either 5-in. or 6-in. pipe, which have first costs of $4,500 and $6,000, respectively. The pumping cost for the 5-in. service is $900 and for the 6-in. service is $550. System lifetime should be 15 years with no salvage value. Annual property taxes are 2% of first cost and annual income taxes are 3% of first cost. Compare the present worth of the costs of 15 years' service for an interest rate of 14%.

3.15 Manufacturers' data indicate that the lowest available cost for various sizes of sheathed electrical cable are as listed below:

Size	Ohms/1,000 ft	Cost/1,000 ft
1/0	0.098	$1,110
2/0	0.078	$1,140
3/0	0.062	$1,150
4/0	0.049	$1,155

The cost of electricity is 4.5¢/kWh. In your plant a cable is to deliver 200 amps for 2,400 hr/year. Your company bases capital recovery costs on 23% interest, a 25-year life, and no salvage value. Taxes are 1.5% of first cost per year. Which cable size should be selected to minimize annual costs?

3.16 Three Rivers Steel Company is considering an energy conservation project in which blast furnace gas will be recovered and supplied to a boiler used to provide steam for electrical generation. Blast furnace gas has a heating value of 100 Btu/ft^3 at standard conditions. The steam-electric plant cycle efficiency is 33%. The current cost of purchased electricity is 6¢/kWh (1 kWh = 3412 Btu), but it is expected to escalate in cost at 5% per year.

The installation cost of the project is $5 million, but the 10% investment tax credit lowers the net investment to $4.5 million. The estimated economic lifetime of the project is 10 years. Assuming that recovered gas can replace purchased electricity, calculate the required volume of gas in cubic feet that must be recovered from the blast furnace per year to satisfy a minimum attractive rate of return (MARR) of 15%. Maintenance and other costs are 10% of projected yearly savings. Salvage value is zero, and the company uses the sum-of-the-year's digits depreciation technique.

3.17 Southern Pine Paper Company is contemplating the installation of automatic combustion controls linked to electronic flue gas analyzers for their boilers. It is estimated that 50,000 Btu per ton of paper produced can be saved with this system. The plant produces 70,000 tons of paper per year. Fuel is valued at $2.60/Mcf and its heating value is 10^6 Btu/Mcf. The installed cost of the equipment is $55,000. The expected lifetime is 10 years, the MARR is 17%, and the salvage value is $4,000. Maintenance and other costs are 5% of projected annual savings. Use the present-worth method to determine whether the project is cost effective.

3.18 Calculate the electric utility bill for the manufacturing plant in Table 3.8 for the month of December using the rate schedule of Table 3.10. Use 3.15¢/kWh for the fuel adjustment charge.

3.19 Assume that the average power factor of the manufacturing plant represented in Table 3.8 is 0.75 (lagging) for the month of December. Calculate the electric bill for December including the effect of the power factor billing provision of Table 3.10. If measures were taken to increase the power factor of 0.85, how many dollars could be saved for the month of December? Use 3.15¢/kWh for the fuel adjustment charge.

3.20 Huge Tool Company uses 10,500 Mcf of natural gas per month on the average. Calculate the monthly bill for gas for each of the three rates shown in Table 3.12. Pick the best rate for the company if the curtailment probabilities are: rate 1, 3 hr/month; rate 2, 7 hr/month; rate 3, 20 hr/month. The price of alternative fuel oil is $46.00 per 42-gallon barrel. The heating value of natural gas is 1,000,000 Btu/Mcf and for fuel oil it is 140,000 Btu/gal.

3.21 Spindletop Oil Company used 1,500,000 kWh of electricity during a month when its production was high to meet customers' needs. The maximum demand during that month was 4,000 kW as measured by the power utility. Calculate the electricity bill for Spindletop for an 83% lagging power factor, using the typical rate schedule of Table 3.7.

3.22 Fourier Heat Systems is considering retrofitting process heaters with new, more efficient burners. Currently, the gas usage is 1500 Mcf per month. It is estimated that the new burners will cut gas consumption by 15%. The installed cost of the burners is $40,000 and their expected life is 10 years. The MARR for the company is 18%. Using the rate schedule of Table 3.12, determine whether the retrofit should be done. Operating, maintenance, and other costs are about 1% of first cost per year.

3.23 A small electric utility company bills its customers for basic service plus a fuel charge. The table below gives a recent history of sales and fuel purchases. Fuel costs fluctuate because of supply and demand. Calculate the customer's fuel charge for the month of April based on these data.

	Kilowatt-hours sales (kWh × 10^8)	Fuel consumed (Mcf × 10^6)	Fuel cost ($/Mcf)	Recovered fuel costs ($ × 10^6)
January	4.55	1.55	3.00	4.11
February	4.48	1.53	2.45	4.21
March	4.39	1.50	2.75	4.65
April (est.)	4.32	1.47	2.65	—

MANAGEMENT AND ORGANIZATION OF ENERGY CONSERVATION PROGRAMS

4.1 THE HUMAN ASPECT OF ENERGY CONSERVATION

4.1.1 Energy Conservation and People

While much of this book is devoted to the technical and "hardware" aspects of energy conservation, it must be remembered that without people machines do not run. To a large degree, the successful implementation of an energy conservation effort is a "people problem."

Automation may control some of the equipment in an industrial plant, but most production operations require human operators. One of the primary objectives of management in an energy conservation program is optimizing human efforts to produce more efficient operation. This has several aspects.

First, standard operating procedures that might have been suitable in a period of low energy costs must be revised to fit the times, and operating personnel must be trained in the new standards and convinced to use them. For example, a boiler operator accustomed to setting firing rates on the basis of a "good-looking" flame must be educated in the meaning and implications of excess combustion air and given incentives to operate the equipment so as to optimize this parameter, even though it may not agree with the operator's previous notions of ideal operating practice.

Second, a system must be devised to maintain continuity in the energy conservation effort. Unfortunately, energy conservation is not a one-time business; if improved operating practices are not sustained on a continuing basis, gains may quickly be lost. Thus, the term "maintenance" in energy conservation must be applied to people as well as equipment.

Third, people must be educated to accept not only new procedures but also eventually new and better machines and processes for production. Fear of replacement by a machine is as old as the industrial revolution itself. As older machines are replaced by more efficient ones, the advantages of the new machines must be made clear to those charged with their operation, and they must be participants in the process of change rather than uninterested (and possibly hostile) bystanders.

In meeting all of these objectives, the key word is "involvement." A good manager will involve everyone in the organization, from chief executive to hourly worker, in all phases of the energy conservation program. This chapter is largely devoted to the concept of involvement and ways to achieve widespread and active involvement in various types of industrial operations.

4.1.2 The Involvement Tree

Involvement of people in the effort of energy conservation must be undertaken at several levels to achieve results. The process of involvement in the energy conservation effort may be visualized in terms of a tree. In this analogy, involvement is rooted in awareness of the importance of energy conservation and its meaning to each individual. Growing out of this awareness is the main body of the effort (the trunk of the tree), the program plan. The plan leads to actions (branches) which eventually bear fruit in the form of reduced energy expenditures and more productive operation.

The energy management task and the process of involvement may be illustrated by the process of selling cars. The public must first be made generally aware of the virtues of owning a shiny new car; this is the purpose of those eye-catching commercials on prime-time television. The commercials do not sell cars, but they do create interest and arouse in the viewer a desire to look further. The local dealer's advertisement in the Sunday paper motivates a plan — to go down and have a look. Again, the advertisement does not sell cars, but it does bridge the gap between an abstract idea and a concrete activity. Once the potential customers are in the showroom, the salesperson must convert the plan into action by getting them to sign up to buy a car.

In a very real sense, energy management is a selling job. Management must be sold on the need to expend precious resources on energy conservation, and operating personnel must be sold on the desirability of changing old habits. The energy manager sells to everyone in the company, a task that would be impossible without an effort to involve the sellees themselves in the management process.

4.2 ELEMENTS OF AN ENERGY MANAGEMENT PROGRAM

Energy managers must wear many hats to produce an effective overall program. Their functions include promoting energy conservation, planning and implementing projects, measuring and monitoring results, and reporting progress to others inside and outside the organization. Although the complexity of the program will depend on the size of the company and the importance of energy as a cost element, the basic ingredients of an organized energy conservation effort are essentially the same in most situations.

4.2.1 Promoting Energy Conservation

As discussed in Section 4.1, the roots of any energy conservation program are set in the awareness of people in the organization of the importance of energy conservation and of their personal role in achieving it. The program should promote awareness at all levels: national, regional, corporate, and personal. Educating employees about the nature of the overall energy problem and how it affects the nation and the region in which they live helps to give them a broader sense of the importance of their own activities and their contribution to solving a major national problem. Similarly, the place of energy conservation in meeting the overall business goals of the company should be made clear. Each employee must be made to feel that his or her own efforts can directly influence business performance.

A key element in promotion is the visible commitment of top management to energy conservation. This can be demonstrated in a number of ways. Some large companies have produced films and videotapes with an introduction by the chief executive officer stressing the importance of energy conservation to the company's profitability. Some company presidents have prepared letters to plant managers that emphasize the same message and are suitable for inclusion in newsletters. Of course, money speaks loudest in the business world. Allocations of funds for energy conservation projects should be publicized as evidence of management commitment.

Figure 4.1 illustrates one effective and widely used approach to increasing energy conservation awareness in industry: the use of posters and signs. These devices, if not overused, keep the subject matter in the forefront of employees' consciousness. The examples shown in Fig. 4.1 illustrate some interesting points. The phrase "Saving Energy for Them" places energy conservation in a larger context meaningful to most employees — the welfare of their children. That is, it stresses the societal importance of the problem. The other examples address specific activities: turning off

Figure 4.1 Energy conservation awareness posters [2].

machines and closing doors. Posted in the appropriate places, such posters serve as reminders to implement specific here-and-now actions.

Another effective means of promoting employee awareness is the publication of energy conservation checklists related to particular activities within their jobs. These might take the form of bulletins such as the "Energy-Gram" shown in Fig. 4.2. This bulletin discusses pump operation and illustrates for a typical case how much energy a specific job action (shutting down a backup pump) can save. It then recommends a direct course of action employees can take to determine the feasibility of applying this action in their own situation.

In formulating a promotional effort, it is important that quantitative information be presented in a way that is easily understood. For example, the fact that a particular action saves 5 million Btu's per hour is much more meaningful to the average individual when expressed in terms of equivalent gallons of gasoline that could be saved per year or the number of homes that could be heated for the winter.

A number of other techniques can be used to maintain a sense of awareness among employees. A few that have been used successfully are:

- Continuous posting of energy conservation progress on billboards or in the company newsletter.
- Publication in a memorandum or newsletter of budgeted energy expenditures and how they compare to the expenditures that would be required if energy conservation efforts were not effectively implemented.
- Direct recognition of employees who make special contributions through their ideas or actions by rewards, publicity, and congratulatory letters or calls from company management.

4.2.2 Program Planning

The second key element in energy program management is planning. This is the crucial step that converts a general idea into a procedure for action. The plan also sets the framework for the overall effort, making the program "hang together" as a unified whole.

The energy audit, as discussed in Chapter 2, forms a basis for the program plan. It highlights areas to be emphasized and provides a quantitative framework for allocating resources. By providing estimates of the amount of energy that can be conserved in various ways, it also serves as the key to setting program goals.

OPERATING TWO PUMPS IN PARALLEL

When two pumps are taking suction from the same location and are pumping into a common discharge line, they are said to be operating in parallel. While it is sometimes necessary to operate pumps in parallel in order to maintain a desired flow rate, in most cases it is an inefficient way to operate and energy is wasted.

Here is an example of the cost of operating two pumps in parallel when only one is required. A 300-hp turbine drive spare pump was being run in parallel with a motor driven pump. Investigation showed that the spare pump only needed to be run during the summer months. Savings that resulted from shutting down the spare pump when it was not needed amounted to about 78,000 MMBtu/yr.

Before operating two pumps in parallel, ask yourself the following questions to determine if it is really necessary to have another pump on the line.

1. Can the flow be increased by speeding up a turbine, or by some other means?
2. Is there mechanical trouble in one pump? Try running the spare pump by itself.
3. Is the condition requiring two pumps a temporary one? If so, be sure to shut down the second pump when it is no longer needed.
4. Should this situation be investigated by the energy conservation group in the utilities division? If you are in doubt, assume that it should. They are glad to follow up on all leads for conserving energy.

Figure 4.2 Energy-gram bulletin [3].

Setting Goals Setting goals is fundamental to the planning process. Energy conservation goals, based on estimates in the audit, should be challenging, yet realistic. This is an excellent opportunity to involve people from all of the various plant operating functions in the planning process. In fact, it is crucial that inputs be obtained from those responsible for meeting the goals. People do not like to have goals dictated to them without consultation, and such goals are not likely to meet with a successful response.

Setting Priorities Another important element in program planning is setting priorities for projects. The energy audit provides the necessary quantitative background, but it almost always uncovers more opportunities than can be reasonably covered in the initial effort. Even if availability of capital is not a limiting factor (a rare situation indeed), manpower usually is, and priorities define what must be done first. Economic return on investment will, of course, be a principal consideration, but other concerns must be taken into account as well. Figure 4.3 shows a suggested format, extracted from Ref. 1, for evaluating projects on the basis of their influence on product quality and yield, production rate, safety, pollution, and other factors in addition to economics. Another approach is illustrated in Fig. 4.4. Each candidate project is assigned a "priority index" on the basis of its economics, the state of the art of the technology involved, and the degree of risk with regard to possible "unanticipated problems." The projects are summarized as shown in the figures, and the indices are used by several members of a project evaluation team to "grade" projects. The final project priorities are then assigned by consensus judgment of the team.

Allocation of Resources The resources of a company are finite, in terms of both dollars and people. In assessing the requirements for these resources and their allocation, capital is often the easier of the two to set limits on. The allocation of human time is relatively difficult to establish quantitatively.

The best approach to time allocation is to outline in as much detail as possible the tasks involved in carrying out a particular project (i.e., develop a work statement for the project) and then assign time estimates to each task for the various job classifications required. An example is given in Fig. 4.5a, which shows the work statement for a detailed survey of plant electricity use. Figure 4.5b shows the estimate of man-hours required to perform each task, including the several types of technical and support personnel required. If this particular project is given high priority, the total man-power requirements must be matched to available personnel or, if outside assistance is needed, to the funds available to acquire it. This leads to the next consideration in the planning cycle — scheduling.

Scheduling In energy conservation, as in most pursuits, the principle that "work expands to fit the time allotted" usually applies. Like goals, schedules should be challenging, but not so challenging as to be unrealizable. Gross underestimation of deadlines is counterproductive; once it is recognized that a schedule is not realistically achievable, it becomes meaningless. Schedules can be simple or complex as the situation demands. Figure 4.6 shows a simple checklist-type schedule for a small plant in the housekeeping

ENERGY CONSERVATION PROJECT
EVALUATION SUMMARY

Calculated

Return on investment _____ %

Pay back period _____ months

Other _____ _____

Btu/unit of production: Now _____ After project implemented _____

Benefits/Problems

Product quality _____

Product yield _____

Production rate _____

Safety _____

Pollution _____

Maintenance-manpower/materials _____

Utilities _____

Working conditions _____

Employee attitude _____

Community _____

Other benefits/problems connected with implementation:

_____

_____

_____

Comments: _____

_____

Project rating: _____

_____

Planned authorization request date: _____

Figure 4.3 Project evaluation summary [1].

ENERGY CONSERVATION PROJECT
EVALUATION SUMMARY

Capital _____ or Expense _____

Department _____

Date _____

Project No. _____ Person Responsible _____

Project Title: _____

Description of Project: _____

_____

_____

_____

_____

Location: _____

Financial Evaluation

  Estimated

    Energy saving (electric power kWh/yr   steam lb/yr   etc.)

    Utility or Raw Material   Saving

    _____   _____ /yr

    _____   _____ /yr

    _____   _____ /yr

    Total energy saving   _____MBtu/yr

    Total energy cost saving   _____$/yr

    Other cost saving due to:

    _____   _____ $/yr

    Additional cost due to:

    _____   _____ $/yr

    Net cost saving   _____ $/yr

    Cost of project   _____ $

Figure 4.3 Project evaluation summary [1] (*Continued*).

phase of its energy conservation program. As the program advances and grows to encompass large capital projects, scheduling may become more complicated. For example, a project to retrofit several boilers with heat recovery units will involve careful planning to mesh with normal plant production and maintenance requirements. In very large projects, sophisticated management tools such as critical path methods may be useful. In

PRIORITY INDICES
DEFINITIONS

Priority indices	Economics	Technology	Risk/feasibility
"A" (Good)	Well defined and attractive	Present "T" adequate and confirmed by corporate experience	Negligible risk, very feasible
"B" (Maybe)	Well defined but only marginally acceptable	Only minimal advance or existing "T" needed: or we simply lack any confirming experience	Minor operational risks and/or feasibility doubts
"C" (Hold)	Either poorly defined or marginally unacceptable	Present "T" inadequate but may be easily adapted	Poorly defined risks and/or doubtful feasibility
"D" (No!)	Clearly unattractive	Requires major "T" breakthrough	High risks or serious feasibility problems

EXECUTIVE SUMMARY FORMAT

Energy Definition

Title   (Few words: include any "project number" etc.)
Scope/objective   (Simple, concise statement of changes to be made to equipment or procedures: 3 lines. . .)

Energy impact   Reduces . . . . by . . . . per hr
(at full rates)   Increases . . . by . . . . per hr
Net reduction in energy usage factor . . . . MMBtu/Mlb
Net reduction in plant energy usage . . . . MMBtu/hr

(. . . MMMBtu/yr at 1980 plan)

Economics

Annual savings	Plant basis	Departmental basis
1980	$ — — — M	$ — — — M
1985	— — —	— — —
Payback period (1980)	— — — yr	— — — yr
Return on investment (before tax)	— — — %	— — — %
Estimated installed cost $ . . . . . . M (±50%), 1980		

Overall priority

Priority indices
_____ Economic
_____ Technology
_____ Risk/feasibility
_____ % Probability of achieving estimated savings**
**Overall team judgment

Figure 4.4 Priority ranking of projects [4].

any case, the schedule becomes one scale by which program progress can be measured.

4.2.3 Program Implementation

The implementation of an energy conservation program is as difficult or as easy as the level of effort expended in the planning stage makes it. A review of the planning elements discussed above will show that the plan is, in effect, a checklist to be used in putting the program into practice. Dollars must be

ELECTRICAL ENERGY SYSTEM

1. General data collection

    A. Determine previous year's usage and develop electrical energy intensiveness plot.
    B. Determine billing basis, including demand and power factor provisions.
    C. Develop electricity cost projections.
    D. Draw up general electric distribution schematic.
    E. List potential schedule loads.

2. Specific data collection

    A. Obtain complete nameplate inventory of all major motors and count of all smaller motors, with horsepower ratings.
    B. Obtain typical load profile on largest motors.
    C. Obtain manufacturers' load efficiency curves on major motors (if available).
    D. Survey existing lighting, obtain type, wattage, and fixture manufacturer.
    E. Obtain load profile for air compressors and inventory air uses.
    F. Measure electrical heating loads for curtain coater.
    G. Obtain typical operating profile (duty cycle) for various process equipment daily, monthly, yearly. Run during lunch, break, night, etc.

3. Efficiency and power consumption computations

    A. Estimate energy consumption in kilowatt-hours per year for office, lighting, ventilation, process equipment.
    B. Estimate typical efficiencies for motors and check proper sizing.
    C. Correlate estimates to monthly billings in kilowatt-hours and kilowatts.

4. Evaluation of energy conservation opportunities

    A. Replacement of motors with high-efficiency motors.
    B. Applicability of demand control (peak load) management system.
    C. Replacement of mercury vapor lamps with high-pressure sodium, and conventional fluorescent lamps with efficient fluorescent lamps.
    D. Light switch control and possible natural illumination.

5. Reporting

    A. Prepare priority list of energy conservation opportunities.
    B. Prepare report.

(a)

Figure 4.5 (a) Detailed project work statement [5].

	Man-hour estimate			
Electrical system estimate	Project engineer	Engineering assistant	Senior technical associate	Maintenance technician
1. General data collection	3	3	0	5
2. Specific data collection	11	19	4	11
3. Efficiency and power distribution calculations	9	12	2	3
4. Evaluation of energy conservation opportunities	16	10	2	8
5. Reporting and preparation of recommendations to management	22	15	7	8

(b)

Figure 4.5 (*Continued*). (*b*) project man-hour estimate summary [5].

committed to the program by management, and commitments of manpower must be obtained from all levels of plant activity. If the planning effort has been thorough these requirements will fall naturally into place. Again, involvement is the key. By involving in the planning process all parties responsible for actually carrying out the program, its execution becomes everyone's responsibility and not the task of the energy manager alone.

To put an energy conservation program plan into effect, it is usually helpful to develop a set of "action plans" for each project. An example is shown in Fig. 4.7. The action plan, which may be based on the work statement prepared earlier, designates the person who is responsible for each action, what is expected of him, and when it is due. Copies are sent to the members of the team involved in the particular project as a reminder of their roles. Shortly before the due date, the program manager should tactfully follow up with each individual listed on the action plan to ensure that work is progressing on schedule. If necessary, of course, a revised plan can be issued.

4.2.4 Measuring, Monitoring, and Reporting

Measuring, monitoring, and reporting mean tracking the progress of an energy conservation program and communicating pertinent information to those who must act on it. This may have different meanings to different people in a business organization. Those concerned directly with hardware are primarily interested in the performance of specific pieces of equipment. The information they require is quite different from that needed by corporate level management, whose interest is in overall performance of entire plants or even multiplant divisions. Measuring, monitoring, and reporting will be considered in turn here, although they are inseparable elements of a single integral management activity.

Measuring   In an energy management sense, measuring means much more than reading instruments. It involves setting up a complete system for accurately determining energy consumption. The system usually does involve hard data, but it also includes detailed procedures by which relevant data are

---

PLANNED ACTIVITIES IN ENERGY CONSERVATION

The following activities will take place each month and will not be repeated in the plans for individual months:

1. Meeting of energy conservation committee
2. Meeting of each committee member with the key supervisors in area
3. Updating of energy saving project lists
4. Communications of progress with updated plot of Btu's per unit of production
5. Weekend audit by department supervision

April	1. Distribute booklet of selected ECOs
	2. Review status of corrective actions regarding first energy saving survey
	3. Publish bulletin on energy saving tips for driving
	4. Give technical talk on steam traps
May	1. Give energy conservation course in supervisor training
	2. Departments develop energy balances
	3. Distribute "savEnergy" decals
	4. Give technical talk on economics of insulation
	5. Send letter to employees at home with energy saving tips for the home
June	1. Trained foremen start holding monthly meetings with their people on energy conservation
	2. Publish bulletin on air conditioning tune-up
	3. As a result of energy balances, committee conducts energy surplus survey
	4. Coordinator gives talk at high school
July	1. Hold poster contest
	2. Publish bulletin on energy saving during vacations
	3. Survey steam pressure reducing stations
	4. Give technical talk on lighting
August	1. Publicize poster contest winner in plant and community newspaper
	2. Distribute posters
	3. Survey compressed air requirements
	4. Publish bulletin—"Don't use compressed air for cooling"
September	1. Publish bulletin—"Tune up space heating systems"
	2. Survey steam condensate system
	3. Coordinator prepares talk with slides for local technical society meeting
	4. Committee rides through plant at night and suggests corrective action on lighting
October	1. Demonstrate infrared survey to detect excessive heat losses
	2. Conduct steam trap survey
	3. Publish bulletin on steam tracing systems
	4. Give technical talk on combustion

Figure 4.6   Checklist-type program schedule [1].

ENERGY CONSERVATION ACTION PLAN FORM

Date _____

Project phase _____ Originator _____

_____

Objective _____

_____

Action plan title _____

Monitoring method _____

Step	Person responsible
1.	
2.	
3.	
4.	
5.	

Figure 4.7 Project action plan [5].

to be acquired and processed and a plan for obtaining the information on a regular basis. The quantities to be measured must be clearly defined, and the people responsible for obtaining data designated. Fig. 4.8 illustrates a measurement procedure for steam losses through steam traps. The required instrumentation and its placement are clearly defined, along with a step-by-step test and calculation procedure. Such a test definition, and an action plan for carrying out the test on a regular basis, constitute the the measuring system.

When aggregating measured results on a plantwide basis, it is important to take into consideration possible interactions between energy-using systems (Fig. 4.9). In a complex manufacturing facility, energy is often exchanged not only with the centralized utility system (e.g., the boiler-house) but also between operating units. If the system for energy accounting is not carefully constructed, it is possible for favorable changes in operation of one unit to induce changes in other units that reflect negatively on their operation. This might be the case, for example, in a steam system where high-pressure condensate from one operating unit is flashed to produce lower-pressure steam for another. A reduction in steam usage in the high-pressure unit would result in less steam being available to the low-pressure unit, requiring more low-pressure steam from the utility steam system. At best, an oversimplified energy accounting system inaccurately reflects the real performance of systems in the plant. At worst, it can result in counter-productive changes in operating practice for individual units, which may actually increase overall plant energy consumption.

Monitoring Monitoring means putting the measuring system into the "production mode," i.e., making measurements on a regular basis and tracking important indicators over time to determine energy conservation progress in comparison with established goals. An extensive discussion of how to devise meaningful energy tracking indices in a given plant situation was presented in Chapter 2. As pointed out there, it is usually unsatisfactory to monitor total energy consumption only, since this parameter also reflects variations in operating conditions and production rate. Thus, a normalized energy use parameter is used. In monitoring progress in an energy conservation program, it is often useful to track certain other indicators as well. Some indicators of interest are:

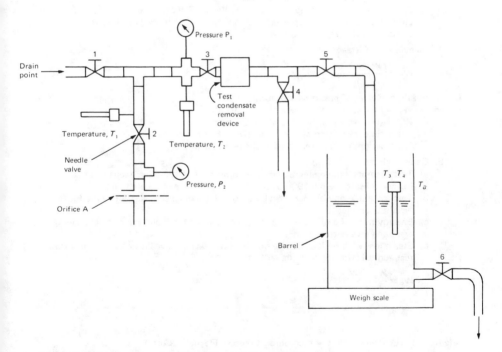

Figure 4.8 (*a*) Test setup definition. (Courtesy Flexitallic Corp.)

STEAM LOSS TESTING IN FIELD UNDER LOAD CONDITIONS

1. Test Arrangement. The recommended test arrangement for steam loss testing is shown in Figure A. The minimum water capacity of the barrel shall be 8 ft^3. Orifice A shall be of sufficient size to pass condensate: expected at drain point times 2 (calculate condensate load from Flexitallic Design Handbook #474). Gauges and thermometers appropriate for conditions.

2. Preliminary Test Procedures.
   a) Start with all valves closed and barrel empty.
   b) Open valve 1 fully and valve 2 slightly more than half. Allow system to warm up to temperature T_1 corresponding to pressure P_1.
   c) Adjust valve 2 to minimum opening to maintain $T1$ at temperature corresponding to P_1. Record P_2.
   d) Calculate condensate capacity of Orifice A at inlet pressure (P_2) from Flexitallic Design Handbook #474. Record as (W_{wn}).
   e) Open valves 3–4 wide. Close valve 2. Allow system to come to normal operating conditions.
   f) During warm-up, fill barrel approximately half full with water whose temperature (T_3) is 30° F *below* ambient temperature (T_a). Weigh and record weight of barrels and initial contents as (W_1).

3. Test Procedure.
   a) Quickly close valve 4 while opening valve 5. Start stopwatch simultaneously.
   b) Discharge into barrel until barrel is approximately ¾ full or until temperature (T_4) is 30° F *above* ambient temperature (T_a). Gently agitate water with non-absorbent peddle while so discharging.
   c) When barrel is ¾ full or temperature (T_4) is 30° F *above* ambient (T_a), rapidly close valve 5 and stop stopwatch. Open valve 4. Record weight of barrel and contents as W_2. Record time as t.

4. Calculation of Steam Loss.

$$W_s = \left(\frac{W_2 - W_1}{t}\right) - W_{wm} \tag{A-1}$$

where: W_s = weight of steam passed, lb/hr
   W_2 = final weight of barrel and contents, lb
   W_1 = start weight of barrel and contents, lb
   t = time, expressed as % of hour (i.e., 15 min = 0.25)
   W_{wn} = weight of condensate normal (load), lb/hr

5. General Notes.
   a) Test accuracy may be increased by employment of a cooling exchanger discharging condensate (below 190° F), continuously to a weight scale tank.
   b) Test accuracy may be increased by employing a balance scale, with time for fix weight drop method.
   c) Employment of a suitable hose attached to valve 4 will eliminate the changeover error and give better results.
   d) Steam loss with no load can be obtained by leaving valve 2 open while performing test and substituting formula (A-2) for (A-1).

$$W_s = \frac{W_2 - W_1}{t} \tag{A-2}$$

(b)

Figure 4.8 (*Continued*). (b) test procedure. (Courtesy Flexitallic Corp.)

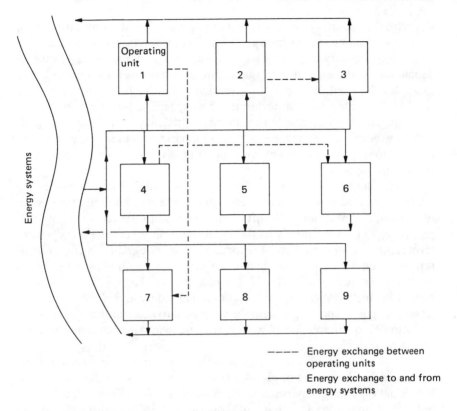

Figure 4.9   Interactions between plant operating units [6].

1. Actual energy consumption compared with the energy that would have been used a year ago under the same production conditions. This might also be shown in terms of comparative dollar expenditures for energy, i.e., what the energy bill would have been in the current year had the energy use per unit production of a year ago been in effect.
2. Cumulative dollars saved by energy conservation compared with cumulative dollars spent. This could also be represented in percentage terms as the ratio of savings to expenditures.
3. Number of energy conservation projects completed and total dollars expended on them as a function of time.
4. Number of people completing plant energy conservation training programs.
5. Number of people receiving awards or other direct recognition for energy conservation contributions.

Reporting Reporting means transmitting meaningful information in a concise way, along with its interpretation, to those responsible for acting on it. Energy conservation reports contain various types of information depending on the area of the organization to which they are directed. For example, the periodic report to a particular unit's operations supervisor would be oriented toward performance of specific equipment and would highlight direct action needed to improve efficiency. On the other hand, reports to corporate management might emphasize trends in energy cost and overall progress of entire plants and operating divisions toward their energy conservation goals.

In devising an overall reporting system, it is desirable to consider the needs of management at various levels and attempt to integrate the system to minimize duplication of effort. The output information for plant-level reports can be used directly as the input data for generating a corporate-level report. Another aspect of reporting system integration is government reporting requirements. Certain data are generally required by government regulatory agencies such as the U.S. Department of Energy, U.S. Environmental Protection Agency, and Occupational Safety and Health Administration. A good reporting system should take into account these data needs and attempt to minimize cost by integrating them into the internal corporate reporting system.

Examples of corporate energy conservation reports are shown in Figs. 4.10a and b. The report in Fig. 4.10a is used for tracking the progress of individual operating departments in a plant. Each operating unit is broken out and the annual goal for that unit shown in terms of both Btu's and dollars saved. The savings for each quarter are then listed and totaled to show the extent of progress toward the goal. This convenient summary provides a clear unit-by-unit picture of energy conservation progress for use by plant management.

Figure 4.10b illustrates a different type of report, intended to display progress in the completion of budgeted energy conservation projects. The report has two parts, the first showing projects completed and the second showing open projects. It is assumed that at the beginning of the reporting year all projects would appear in the second category, and as the year progressed open projects would move into the completed category. Concurrently, expected savings would translate into actual savings.

In large organizations computers are indispensable for energy conservation reporting. Development of the software for a comprehensive corporatewide reporting system may cost hundreds of thousands of dollars, yet most corporations have found that the potential savings more than justify the investment. The advent of low-cost general-purpose microcomputers has now brought computer-based reporting within reach of even small operations.

Figures 4.11a and b illustrate the structure of a typical computer-

ENERGY CONSERVATION—PERFORMANCE SUMMARY
QUARTERLY REPORT

Division _____ Location _____ Coordinator _____ Qtr. _____ Year _____

Department	Oper. Area	Goals		Completed Savings								Year-to-Date Total		Year-to-Date % of Goal		Total Cost	
				1st Qtr.		2nd Qtr.		3rd Qtr.		4th Qtr.							
		10^9 Btu	10^3 \$	10^9 Btu	10^3 \$	10^9 Btu	10^3 \$	10^9 Btu	10^3 \$	10^9 Btu	10^3 \$	10^9 Btu	10^3 \$	10^9 Btu	10^3 \$	C	NC

(a)

Figure 4.10 Energy conservation progress reports [7].

103

Figure 4.10 Energy conservation progress reports [7] (*Continued*).

(b)

ENERGY CONSERVATION—PROJECT COMPLETIONS
QUARTERLY REPORT

Division _____ Location _____ Coordinator _____ Qtr. _____ Year _____

Dept.	Identification (Project)	Suggestor (Name)	Completion Date—Qtr.	Annualized Savings (10^3 $)	Total Cost (10^3 $)	Type

---

ENERGY CONSERVATION—OPEN PROJECT STATUS
QUARTERLY REPORT

Division _____ Location _____ Coordinator _____ Qtr. _____ Year _____

Dept.	Identification (Project)	Suggestor (Name)	Estimated Completion Date—Qtr.	Status	Expected Annual Savings (10^3 $)	Total Cost (10^3 $)	Type

based energy reporting system in a large multiplant corporation. Figure 4.11a shows the organization of the basic computer model, which includes several elements. A plant profile for each plant is maintained as a semipermanent file in the computer memory. The plant profile includes basic information about plant energy-using systems such as HVAC, lighting, various process units, and climatic control of production areas. These data stay relatively fixed as they pertain to operating characteristics of permanent equipment. In addition to the plant profile, there is a file called the transaction history in which variable data for the reporting interval, such as production rates for various products, energy-related process measurements, and weather data, are entered. The central computational block of the system draws data from both files and, using appropriate computational algorithms, generates the necessary results and prepares reports for use by various entities within the corporate organization.

The overall flow of information in this comprehensive reporting system is shown in Fig. 4.11b. Input data from individual plants are put in a standardized format and transmitted to a central data processing facility. Here the data are checked for completeness, accuracy, and any revisions of basic plant systems (i.e., changes in elements of the plant-profile data base) that may have occurred during the preceding reporting period. If revision is necessary, the input reports are referred back to the plant or to the engineering department for correction. Finally, input data are processed to produce the output reports. These include a management summary report, which aggregates overall data for each plant, and detailed engineering reports for transmission to the engineering department and respective plant operating personnel.

4.3 ORGANIZATION OF ENERGY CONSERVATION PROGRAMS

Even in a relatively small plant, energy management involves many people. The details of an energy conservation organization will depend on the size and nature of the manufacturing operation; generally, the more dollars spent on energy, the more elaborate the organization that is justified to reduce this expenditure. It is clearly desirable to organize in such a way as to involve in the decision-making process all segments of the production organization concerned with energy. In this section the general structure of a large corporate energy management program will be discussed, starting at the plant-level organization. Of course, not all elements of this typical organizational structure will be pertinent to every situation. However, certain aspects can be adapted to energy management even in a small company.

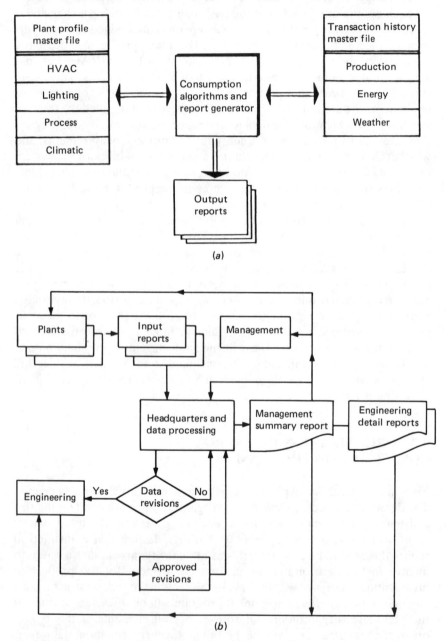

Figure 4.11 Computerized energy reporting system [8].

4.3.1 Plant-Level Organization

A typical plant-level energy conservation organization is shown in Fig. 4.12. This organizational structure will be discussed in some detail as it should have broad applicability to both large and small plants. In general, the plant-level organization has a relatively short-term and technical focus. It is concerned primarily with directly affecting production operations on a day-to-day basis. The plant energy conservation coordinator is the key individual in the organization. The coordinator is primarily responsible to the plant manager and, in the case of a multiplant corporation, may report on energy conservation progress to divisional or corporate management as well. The plant coordinator chairs the plant energy management committee, which, in a small plant, may be composed of representatives such as production supervisors from individual operating units. This committee structure might become cumbersome in a large plant because of the large number of operating departments. In this case, the committee might be limited to perhaps five or six members by selecting representatives from such key areas as engineering, maintenance, production, labor relations, and public relations. The committee assists the plant coordinator in developing and implementing the overall plant program and provides a channel of communications to major areas of plant operation. In addition to the plant energy management committee, each operating department should have a designated energy conservation representative who has primary responsibility for progress in that department and who reports to the plant coordinator. In turn, in a two- or three-shift operation, a representative from each operating shift should be designated to report to the department representative. It should be emphasized that in most plants these various assignments are handled on a part-time basis by regular operating personnel, such as shift maintenance foremen or unit operating engineers. Even the position of plant coordinator is a part-time responsibility in all but the largest plants. The amount of time devoted to these responsibilities is largely determined by the importance of energy as an element of overall manufacturing costs.

In addition to the various operating unit representatives, a special maintenance department representative may be designated, since this department generally plays a pivotal role in the energy conservation program.

4.3.2 Division-Level Organization

The division-level organization (Fig. 4.13) is concerned primarily with monitoring the energy conservation progress of the several plants in a division and planning the overall division program in the medium term (e.g., on a quarterly basis). It facilitates the transfer of useful information between geographically dispersed but operationally similar plants and can bring to

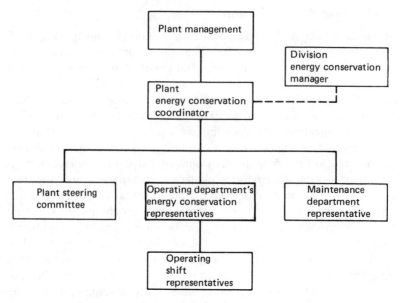

Figure 4.12 Plant-level energy management organization [7].

bear divisionwide resources, such as central engineering and research capabilities, in support of plant-level efforts.

Heading the divisional effort is the division energy conservation manager, to whom the individual plant coordinators report. The division manager chairs the divisional steering committee, which, like the plant-level committee, assists in involving all elements of the division in energy conservation program planning and implementation. The steering committee generates ideas for employee awareness, technical projects, and improved operating procedures applicable on a broad basis throughout the division. The committee also acts as an impartial review body to assess the annual energy conservation program plans of the various plants and ensure uniformity of effort.

Since engineering is often handled on a centralized basis in a large corporation, a suborganization within the engineering department, headed by an engineering department energy conservation coordinator reporting directly to the division manager, may be desirable. Also, since product distribution is a major energy-consuming function and is normally handled on a divisionwide basis, a similar organization might exist within that department, concentrating primarily on energy conservation in the transportation area.

4.3.3 Corporate-Level Organization

In a large multidivisional corporation, a corporate-level energy management organization is established (Fig. 4.14), reporting directly to top-level corporate management. This organization usually has responsibility for all aspects of energy supply as well as energy conservation and is concerned with integration of the divisional energy conservation programs into the overall business plans of the corporation. It also has primary responsibility for energy-related interactions with outside entities such as government, trade associations, and energy suppliers. The focus is relatively long-range, e.g., forecasting future energy prices and planning for long-term energy supplies.

Again, the organization is based around a steering committee, which at this level comprises representatives from the major business planning elements of the corporate structure. In addition, there may be a technical committee primarily oriented toward research, development, engineering, and operations, since energy conservation, in the final analysis, requires careful technical and economic evaluation. The division program managers report directly to the corporate energy conservation director.

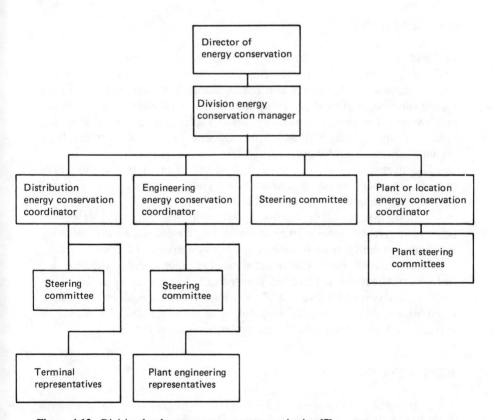

Figure 4.13 Division-level energy management organization [7].

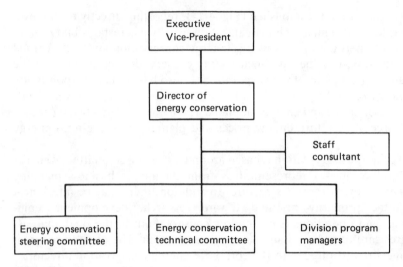

Figure 4.14 Corporate energy management organization [7].

4.4 SUMMARY

In this chapter we have concentrated primarily on the human side of industrial energy conservation, recognizing that people are a key element in the overall success of a program and that without them the technical effort cannot succeed. A great deal of emphasis has been placed on the concept of involvement. Involving all levels of plant and corporate personnel in energy conservation is a major undertaking and one that requires innovation, organization, and hard work.

The particular importance of careful planning in the success of a program has been stressed. The value of energy audits as a vehicle for program planning has been pointed out.

Various aspects of measuring, monitoring, and reporting of energy conservation results have been considered. These elements are critical to the continuing viability of an organized energy conservation effort.

Finally, some basic elements of energy management organization have been discussed. It has been pointed out that managers at the plant, division, and corporate levels, while sharing a common goal, have differing orientations. A common denominator at all of these levels, however, is broad-based involvement in the energy conservation process.

REFERENCES

1. U.S. Department of Commerce, Energy Conservation Program Guide for Industry and Commerce, NBS Handbook 115, Washington, D.C., 1974.

2. M. Weisenthal, Achieving Energy Goals through Employee Participation, Industrial Energy Conservation Technology Conference (IECTC), Houston, April 1981.
3. B. G. Davidson and F. J. Kanewski, People Power Saves Plant Energy, IECTC, Houston, April 1981.
4. D. Corwin, Energy Audits in Process Units, IECTC, Houston, April 1981.
5. P. S. Schmidt, and M. A. Williams, Energy Audits in Industrial Plants, seminar presented to Tenneco Corp. Energy Management Group, March 1980.
6. N. A. Wyhs, and J. E. Logsdon, Development of an Energy Consumption Model for a Multi-Product Chemical Plant, IECTC, Houston, April 1980.
7. M. A. Williams, Initiating, Organizing, and Managing Energy Management Programs, in *The Energy Management Handbook*, W. Turner et al., eds., Wiley, New York, 1982.
8. H. W. Kympton, and B. Bowman, Implementation of a Corporate Energy Accounting and Forecasting Model, IECTC, Houston, April 1981.

ANALYSIS OF THERMAL–FLUID SYSTEMS

The analysis of energy conservation systems requires the use of relationships from the thermal-fluid sciences: classical thermodynamics, heat transfer, and fluid mechanics. The reader should have a background in the thermal-fluid sciences for the most effective use of the material presented here. However, for convenience, some aspects of the thermal-fluid sciences are briefly reviewed in this chapter.

The review is not comprehensive, because some aspects of thermo-dynamics and heat transfer are best explained at the point in the text where they are most useful. For example, combustion is treated in Chapter 6 (Energy Conservation in Combustion Systems), combined conduction and convection are included in Chapter 8 (Heat Exchangers), and some aspects of applied thermodynamics are covered in Chapter 9 (Heat Recovery).

The material is in compact form, with many of the relationships simply tabulated for reference in later chapters. The reader is urged to consult elementary texts in thermodynamics [1], heat transfer [2], and fluid mechanics [3] for detailed expository treatments of the fundamentals.

5.1 THERMODYNAMICS

Thermodynamics deals with the transformation of energy and the accompanying changes in the state of matter. There are two ways to describe the global or macroscopic processes that are observed "thermodynamically":

1. Apply the laws of averaging to predict the macroscopic behavior of a collection of microscopic atoms and molecules.

2. Develop a set of general physical laws of macroscopic behavior without regard to the microscopic structure of matter.

Classical thermodynamics is based on the second strategy, although the first method can be used to develop laws and relationships that are identical to the laws of classical thermodynamics. The first method is called statistical thermodynamics.

5.1.1 First Law of Thermodynamics

The first law is a "conservation" law, which states that energy is conserved during various types of changes in the state of matter. To use the first law, we need to define an appropriate quantity of space and/or matter to which we can apply the law.

Systems and Surroundings   A system is a defined portion of matter and/or space set aside for study. Everything outside the system constitutes its surroundings. It is helpful to break this broad definition of a system into particular types of systems:

Isolated system: A system that exchanges neither mass nor energy with its surroundings. Mass and energy are constant in an isolated system.

Closed system: A system that can exchange energy but not mass with its surroundings.

Open system: A system that can exchange either energy or mass with its surroundings.

Two conservation laws are based on these definitions. The first is the conservation of mass: *the mass of an isolated system is constant.* We can also include a closed system here, if relativistic effects in the system are ignored. The second is a restrictive form of the first law of thermodynamics: *the energy of an isolated system is constant.*

Energy   Energy is a concept that must be considered in some detail. A system can contain energy in various forms, including:

Internal thermal energy: Energy stored in the molecules without reference to relativistic effects or external fields. The symbol U is used to represent this form of energy.

Kinetic energy: Energy tied up in motion of the system (or parts of it). Kinetic energy for a specified mass, m, in a system is

$$KE = \frac{1}{2} mV^2$$

Potential energy: Energy contained in the system because of elevation above a reference plane. It is

$$PE = \varrho g z$$

where ϱ is material density, g is gravitational acceleration, and z is the elevation measured above some reference plane.

A system can contain energy in many other forms, such as electrical, magnetic, chemical, and surface tension energy. A system can exchange energy with its surroundings in two ways — as *work, W*, or *heat, Q*. Work and heat are energy forms that can be identified only as they cross a system boundary. They must never be confused with E, the energy contained within the system.

Work Transfer Consider the closed system made up of the gas contained in a piston cylinder (see Fig. 5.1). When the piston moves, the work done on the piston is

$$W = \int_1^2 F \, dx = \int_1^2 p \, dV \tag{5.1}$$

where F is the force exerted by the gas, $F = pA$, and x is linear distance. In differential form we write

$$dW = F \, dx = p \, dV \tag{5.2}$$

If we adopt the convention that work done on a system is positive, the gas does work $dW = -p \, dV$ as it expands. Work is a so-called inexact differential function, which means that its value depends on the p-V path taken during a change of the system between a beginning and a final state. We indicate this by the symbol dW.

Other forms of work are rotating-shaft work, electrical work, and work required to overcome surface tension when liquid-vapor-solid interfacial surfaces are created.

Heat Transfer Heat is another inexact differential function, which is similar to work in that it is identifiable only as it crosses a system boundary. Heat can be expressed in terms of system properties; the most common method is to use the absolute temperature and a property called entropy, S, in the following way:

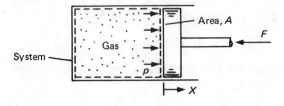

Figure 5.1 Piston-cylinder system capable of transferring work.

$$đQ = T \, dS \qquad (5.3)$$

We are somewhat premature in using this definition, because the property entropy is a result of the second law of thermodynamics, which will be discussed later.

The effects of heat can be explained by using the results of Joule's famous experiments in the 1840s. He observed that, for a given amount of work input to water contained in a closed, insulated system, a predictable temperature rise in the water was measured. Joule knew that the measured ΔT was related to a change in the internal energy of the water, $W \propto \Delta T$. He related the two in the following way:

$$W = J \, (mc)_{H_2O} \, \Delta T \qquad (5.4)$$

where m is the mass of water, c is the heat capacity of the water, and J is the "mechanical" equivalent of heat. Joule measured J to within a few percent of its currently accepted value of 778 ft lb$_f$/Btu.

Joule also observed that if the system sat for a long time following the input of work, it would return to its initial temperature. This occurred because heat was slowly transferred out of the system to exactly offset the input of work to the system.

First Law The foregoing discussion of heat, work, and energy leads to a statement of the first law of thermodynamics for a closed system:

$$dU = đQ + đW \qquad (5.5)$$

When we include other forms of energy storage, we get

$$dE = đQ + đW \qquad (5.6)$$

where Q and W are positive when transferred into the system. Another convention is the assumption that Q into a system is positive and W from a system is positive. Then we can write

$$Q = \Delta E + W \qquad (5.7)$$

for a closed system. We shall use Eq. (5.6) as the basic form of the first law.

5.1.2 Properties and States

The dictionary definitions of property and state are applicable to thermodynamic systems. A *property* is "an inherent or essential quality or peculiarity," while a *state* is defined as "a condition of person or thing." We use the commonsense meanings; the state of a system is the exact condition that a system is in, and a property is an inherent quality typical of the state of the system. Furthermore, the state of the system is the sum of its properties.

Temperature, Pressure, and Equilibrium Equilibrium is defined as "a state of balance or equality between opposing forces." Thermodynamic equilibrium is used in exactly the same way. A system in equilibrium has no spontaneous changes occurring in it, and all of its properties are constant with time.

Various forms of equilibrium can be specified. Mechanical equilibrium means that there is no motion resulting from force imbalances within the system. Thermal equilibrium means that no heat is transferred because of temperature differences within a system or between systems. Chemical equilibrium implies that no reactions occur because of differences in concentrations of different mass species in a system. Classical thermodynamics is based on the concept of equilibrium states, even when changes in the system are occurring.

We know that the property called temperature exists because of its observed effects. Reynolds and Perkins [1] give an excellent discussion of the existence of temperature and its effects. Pressure results in a measurable force, so we readily accept its existence.

Intensive, Extensive, and Specific Properties We have introduced the thermodynamic properties E, U, T, p, V, and S, which can be classified in the following ways:

Extensive properties	Intensive properties	Specific properties
U	p	u
E	T	e
V		v
S		s
KE		c
PE		etc.

An extensive property varies in direct proportion to the extent of the system. Intensive properties are independent of the extent of the system. Doubling the size of a system does not necessarily double its pressure or temperature, but doubling the volume of a closed system containing a gas while keeping its p and T constant will double the internal energy and entropy. Specific properties are those that are basically extensive but that become intensive when written in terms of a unit mass or unit volume of the system.

The State Principle An important concept in thermodynamics is the state principle: The number of independently variable thermodynamic properties for a specified system is equal to the number of relevant reversible work modes plus one.

There are several modes by which systems can do work. Each involves the product of an intensive property acting through an extensive property. In general, these modes take the form $W = \int F\,dx$, where F is a generalized force and x is a generalized displacement. We assume that the work occurs during a process composed of equilibrium states.

We know from observation that a simple, pure, single-phase system, which can exchange work with its surroundings only in the ($-p\,dV$) mode, can have only two properties varied independently. For example, for a gas, p = fn (T, v). Once we know T and v, we have specified the state of the system. This clearly complies with the state principle. In fact, observations of this type actually led to the concept of the state principle.

5.1.3 Simple Compressible Substances

Ideal Gases   A simple substance is one with only one relevant reversible[1] work mode. A simple compressible substance is one in which the work mode is $dW = -p\,dV$. An ideal gas is one important example of such a substance and is completely defined by a combination of Boyle's and Charles's laws. The result is the ideal gas equation of state,

$$pv = RT \tag{5.8}$$

where R is a constant for a particular gas. The ideal gas equation of state also can be written in terms of the number of moles in a system as:

$$p\bar{v} = \bar{R}T \tag{5.9}$$

where $\bar{v}$ is the volume per mole of substance and $\bar{R}$ is the universal gas constant; $\bar{R} = 1{,}545.3$ ft lb$_f$/lb-mole°R, which is independent of the particular gas in the system. Furthermore, R is simply $\bar{R}/M$, where M is the molecular weight of the substance.

For ideal gases, according to the state principle, there are two possible independent properties. So we could write, in addition to the ideal gas equation of state, functions such as fn $(s, p, T) = 0$ or fn $(p, u, v) = 0$. We deal with three state variables for ideal gases, two of which can be independently varied.

Real Substances   Suppose we must deal with a real substance — one that might exhibit ideal gas behavior over certain temperature and pressure ranges but not over the entire spectrum of thermodynamic states accessible to the substance. The property diagram of Fig. 5.2 shows possible states for a substance typical of water in terms of pressure and temperature. Only in the gaseous region would the ideal gas equation suffice. Even there, the p–T state must be sufficiently far removed from the phase transition lines before the ideal gas equation of state can be used.

Reversible means that the process occurs through a sequence of equilibrium states.

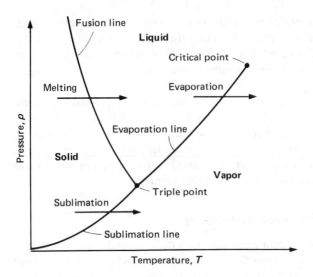

Figure 5.2 Pressure-temperature diagram for a substance typical of water.

Many processes of interest in energy conservation involve the liquid-to-vapor phase change. A plot of p–v behavior along the vaporization line of Fig. 5.2, which allows us to visualize better how the substance behaves, is shown in Fig. 5.3.

Over the years, reliable properties of substances in their subcooled liquid, saturated liquid and saturated vapor, and superheated vapor states have been compiled through a combination of careful experimentation and accurate correlations. The so-called steam tables (Appendix B) are typical of the results of these efforts; steam properties are discussed at greater length in Chapter 7.

To obtain properties of saturated systems — for example, point a on the constant-temperature line between f and g on Fig. 5.3—we define the *quality, x* of a saturated system:

$$x = \frac{v_a - v_f}{v_g - v_f} = \frac{v_a - v_f}{v_{fg}} \tag{5.10}$$

where we use $v_{fg} = v_g - v_f$ for simplicity. For any other specific intensive property, we write, for example

$$u = u_f + x u_{fg} \tag{5.11}$$

$$s = s_f + x s_{fg}$$

Appendix B provides tabular and graphic property data for water, Freon-11, Freon-12, Freon-22, ammonia, and nitrogen.

Specific Heats   The relationship between variables that define the state of a substance can be used to define other useful thermodynamic properties. Specific heat at constant volume is one such material property. It is defined as

$$c_v \equiv \left. \frac{\partial u}{\partial T} \right)_v \qquad (5.12)$$

The specific heat for a constant-pressure process is defined as

$$c_p \equiv \left. \frac{\partial h}{\partial T} \right)_p \qquad (5.13)$$

where $h = u + pv$ is the enthalpy.

For ideal gases, u and h are functions only of temperature, so that integration of Eqs. (5.12) and (5.13) gives

$$u - u_{\text{ref}} = \int_{T_{\text{ref}}}^{T} c_v \, dT \qquad (5.14)$$

and

$$h - h_{\text{ref}} = \int_{T_{\text{ref}}}^{T} c_p \, dT \qquad (5.15)$$

Furthermore, Eqs. (5.14) and (5.15) are exactly true for constant-volume and constant-pressure paths, respectively, regardless of the gas. They are approximately true for incompressible liquids.

Compressibility Factor   For gases at high pressure or vapors near saturation, the compressibility factor can be used to account for real behavior. It is defined as

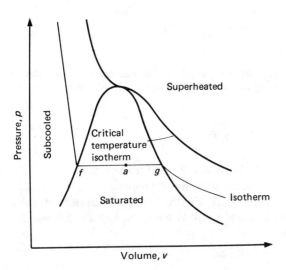

Figure 5.3 Property diagram for liquid-vapor behavir of a substance typical of water.

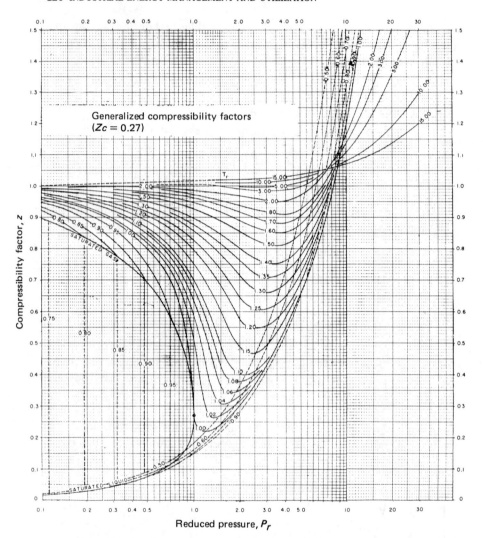

Figure 5.4 Generalized compressibility chart. Reprinted with permission from G. Van Wylen and R. Sonntag, *Fundamentals of Classical Thermodynamics*, 2d ed., Wiley, New York, 1973.

$$Z \equiv \frac{pv}{RT}$$

where $Z = 0$ for incompressible fluids and $Z = 1$ for ideal gases. Figure 5.4 shows a plot of the Z factor in terms of reduced properties, $T_r = T/T_{cr}$ and $P_r = p/p_{cr}$, for the case of $Z_{cr} = 0.27$. Figure 5.4 can be used to develop generalized charts for enthalpy and entropy changes. Figures 5.5 and 5.6

show these graphs. The starred quantities on these graphs are evaluated at low pressure, where $P_r \rightarrow 0$. The unstarred quantities are evaluated at the same temperature as the starred quantities but for the actual existing P_r. These charts can be used to find the changes in enthalpy and entropy for any given change in state.

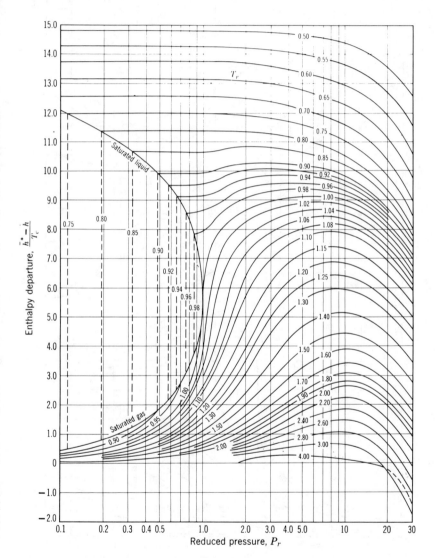

Figure 5.5 Generalized enthalpy correction chart (based on Fig. 5.4). Reprinted with permission from G. Van Wylen and R. Sonntag, *Fundamentals of Classical Thermodynamics*, 2d ed., Wiley, New York, 1973.

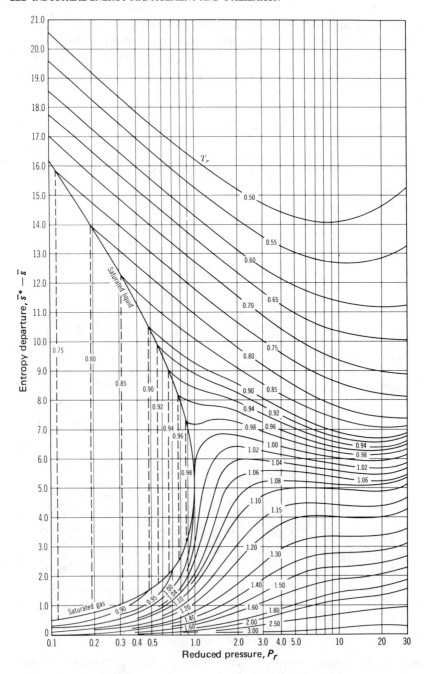

Figure 5.6 Generalized entropy correction chart (based on Fig. 5.4). Reprinted with permission from G. Van Wylen and R. Sonntag, *Fundamentals of Classical Thermodynamics*, 2d ed., Wiley, New York, 1973.

5.1.4 First-Law Forms for Thermal-Fluid Systems

Closed System We have developed the basic form of the first law for a closed system; see Eq. (5.6). For a cycle—i.e., a sequence of processes through which a system passes, eventually returning to its initial state—the first law is

$$\oint dQ = - \oint dW \qquad (5.16)$$

This simply means that heat must be balanced by work if the initial and final states of the system are to be the same.

Open System Energy conversion devices often consist of components through which mass flows. The first law for an open system represents a balance of energy just as it does for a closed system. But because energy is circulated through the system by virtue of the mass flow itself, the character of the first law is different. Figure 5.7 shows a diagram of a control volume that we might use to analyze an open system. The control volume represents the system to be analyzed.

Two approaches to analysis of open systems can be taken, depending on the nature of the process. For steady processes, the steady-state, steady-flow approach, in which it is assumed that the state of material is constant at any point in the system, is applicable. Omitting the development of the equation, we write the first law in rate form

$$\dot{Q}_{cv} + \dot{W}_{cv} + \sum_{in} \dot{m} \left(h + \frac{V^2}{2} + gz \right) = \sum_{out} \dot{m} \left(h + \frac{V^2}{2} + gz \right) \,(5.17)$$

for a system that is stationary. The summations allow for the possibility of multiple inlet and outlet ports.

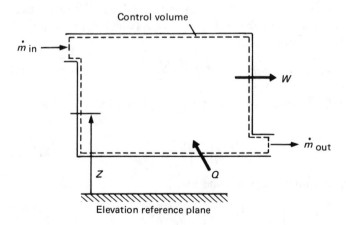

Figure 5.7 Control volume for first-law analysis.

The other approach uses a uniform-flow, uniform-state assumption. This is based on the condition that, at points where mass crosses the system boundary, material states are constant with time. The state of the mass in the system can vary with time but is spatially uniform at any time. The first law for this type of stationary system is

$$Q_{cv} + W_{cv} + \sum_{in} m\left(h + \frac{V^2}{2} + gz\right) = \sum_{out} m\left(h + \frac{V^2}{2} + gz\right)$$

$$+ \{m_2 u_2\} - \{m_1 u_1\}$$

(5.18)

where 1 and 2 refer to the initial and final states, respectively.

The steady-state, steady-flow approach is applicable to flow-through turbines, pumps, compressors, nozzles, and other similar devices. The uniform-flow, uniform-state approach is useful in analyzing the filling of reservoirs, tanks, pressure vessels, etc. Equation (5.17) is just a special case of Eq. (5.18).

Useful Work in Open Systems For a steady-flow system, in which no Q is transferred (adiabatic), we can write

$$đw = dh = (du + p\,dv) + v\,dp \tag{5.19}$$

if we neglect kinetic and potential energy. If we analyze the behavior of a unit mass of material moving through the system, we get

$$đq = du + p\,dv = 0$$

or

$$du = -p\,dv$$

Equation (5.19) then gives the useful work

$$w = \int v\,dp \tag{5.20}$$

in contrast to a closed system, when $w = \int p\,dv$. Equation (5.20) accounts for the energy required to push the fluid in and out of the control volume.

For an open system through which an ideal gas (which behaves according to $pv^n = $ const.) flows

$$W_{open} = \int_1^2 v\,dp = p_1^{1/n} v \int_1^2 \frac{dp}{p^{1/n}} \tag{5.21}$$

where 1 and 2 refer to initial and final states. Equation (5.21) can be integrated to

$$W_{open} = n\frac{(pv)_2 - (pv)_1}{n-1} \tag{5.22}$$

For a closed system involving the same gas

$$W_{closed} = \int p\,dv = p_1 v_1^n \int_1^2 \frac{dv}{v^n} \tag{5.23}$$

which simplifies to

$$w_{closed} = \frac{p_2 v_2 - p_1 v_1}{n - 1} \qquad (5.24)$$

We see that the work for an open system is n times that for a closed system between the same two end states.

Various values of n represent different types of ideal gas behavior:

$n = 0$	$p = $ const.	isobaric
$n = 1$	$T = $ const.	isothermal
$n = k$	$S = $ const.	isentropic ($k = c_p/c_v$)
$n \rightarrow \infty$	$V = $ const.	isometric

Equations (5.22) and (5.24) can be used to calculate useful work for different values of n for both open and closed systems.

5.1.5   Efficiency and Coefficient of Performance

The efficiency of a thermodynamic system is

$$\eta = \frac{\text{energy delivered as work}}{\text{energy absorbed by the system}} \qquad (5.25)$$

For example, consider an adiabatic turbine (Fig. 5.8). The energy delivered as work, according to Eq. (5.17), is

$$w = h_2 - h_1 \qquad (5.26)$$

However, there are losses caused by irreversibilities, such as friction and flow imperfections, that cause more energy to be consumed in the process. It is convenient to represent the process on h - s coordinates, which are called Mollier coordinates. Figure 5.9 is a Mollier diagram representing the turbine expansion process. The dashed line represents a reversible adiabatic expansion. A reversible adiabatic process is an isentropic process. We now define the efficiency of this process as

$$\eta_{turb} = \frac{\text{actual work delivered by the turbine}}{\text{maximum possible work from the turbine}}$$

and $\qquad (5.27)$

$$\eta_{turb} = \frac{h_1 - h_2}{h_1 - h_{2s}}$$

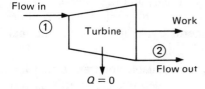

Figure 5.8   Adiabatic turbine.

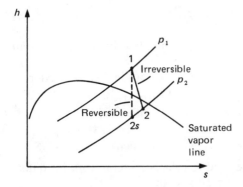

Figure 5.9 Mollier diagram showing turbine expansion.

Figure B.1 in Appendix B is a Mollier diagram for steam that can be used to quickly find the work output of a steam turbine if its efficiency is known.

Application of Eq. (5.25) to a device that consumes work, such as a pump or compressor, results in

$$\eta_{\text{pump}} = \frac{(\Delta h)_s}{(\Delta h)_{\text{actual}}} \tag{5.28}$$

Thermodynamic Cycle Efficiency We can also characterize the overall efficiency for a complete cycle of thermodynamic processes. A good illustrative example is a steam power plant that converts chemical energy to mechanical energy to electrical energy. The steam behaves according to what is called the Rankine cycle, shown in Fig. 5.10. The "thermal" efficiency of this cycle is

$$\eta_{\text{th}} = \frac{\text{net work delivered by the steam}}{\text{net heat input to the steam}} \tag{5.29}$$

The enthalpy rise across the pump, line $1 \rightarrow 2$, is usually small compared to the enthalpy drop across the turbine, line $3 \rightarrow 4$. Consequently, we write for the Rankine cycle

$$\eta_{\text{th}} = \frac{h_3 - h_4}{h_3 - h_2} \tag{5.30}$$

Some energy conversion cycles involve only gaseous behavior. They are analyzed as so-called air-standard cycles, using the assumptions that (1) air is the working fluid and behaves as an ideal gas, and (2) combustion and exhaust processes are replaced by heat exchangers. Table 5.1 lists various power cycles with their corresponding thermodynamic cycles and thermal efficiencies.

Refrigeration Cycles Several methods are available for transferring heat from a cold system to hotter surroundings. We describe the effectiveness of

Table 5.1 Air-Standard Cycles

Cycles	P-V cycle	Efficiency

Carnot cycle

Process	Description
3–4	isothermal heat addition
4–1	isentropic compression
1–2	isothermal heat rejection
2–3	isentropic expansion

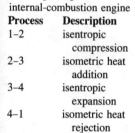

$$\eta_{th} = 1 - \frac{T_L}{T_H} = 1 - \frac{T_4}{T_1} = 1 - \frac{T_3}{T_2}$$

$$\eta_{th} = 1 - r_{ps}^{(1-k)/k} = 1 - r_{vs}^{1-k}$$

$$r_{ps} = \frac{P_1}{P_4} = \frac{P_2}{P_3} = \left(\frac{T_3}{T_2}\right)^{k/(1-k)}$$

$$r_{vs} = \frac{V_4}{V_1} = \frac{V_3}{V_2} = \left(\frac{T_3}{T_2}\right)^{1/(1-k)}$$

Otto cycle
spark-ignition
internal-combustion engine

Process	Description
1–2	isentropic compression
2–3	isometric heat addition
3–4	isentropic expansion
4–1	isometric heat rejection

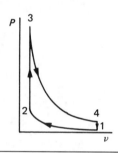

$$\eta_{th} = 1 - \frac{T_1}{T_2} = 1 - (r_r)^{1-k}$$

$$r_r = \frac{V_1}{V_2} = \frac{V_4}{V_3}$$

Diesel cycle
compression-ignition
engine

Process	Description
1–2	isentropic compression
2–3	isobaric heat addition
3–4	isentropic expansion
4–1	isometric heat rejection

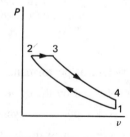

$$\eta_{th} = 1 - \frac{T_1(T_4/T_1 - 1)}{kT_2(T_3/T_2 - 1)}$$

Brayton cycle
gas turbine

Process	Description
1–2	isentropic compression
2–3	isobaric heat addition
3–4	isentropic expansion
4–1	isobaric heat rejection

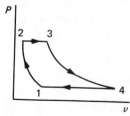

$$\eta_{th} = 1 - \frac{T_1}{T_2} = 1 - \frac{1}{(P_2/P_1)^{(k-1)/k}}$$

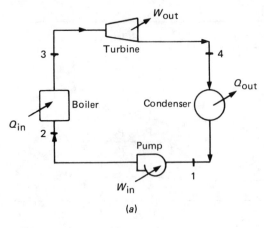

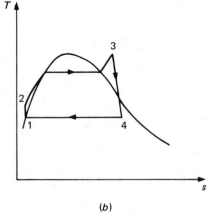

Figure 5.10 Steam-electric power plant. (*a*) Cycle components; (*b*) Rankine cycle.

these systems in terms of the coefficient of performance (COP), where

$$\text{COP} = \frac{\text{energy delivered by the cycle}}{\text{required work input to the cycle}} \tag{5.31}$$

Let us take a vapor compression cycle as an example; see Fig. 5.11. The vapor compression cycle is almost a reversed Rankine cycle, with the exception being the use of an expansion valve to reduce the refrigerant pressure from condenser to evaporator. The expansion valve is called a throttling valve; a throttling process is one of constant enthalpy.

The COP of the system depicted in Fig. 5.11 is

$$\text{COP} = \frac{h_4 - h_3}{h_1 - h_4} \tag{5.32}$$

when it is used as a refrigeration system.

Heat Pump   If we use the vapor compression cycle to deliver heat to a hot region, we call the system a heat pump. We apply the same definition of COP, but in this case

$$\text{COP} = \frac{h_1 - h_2}{h_1 - h_4} \tag{5.33}$$

Heat pumps are taking on added importance for energy conservation. They can be used for heating from an electrical source (compressor) in situations where direct combustion is not practical. Furthermore, heat pumps can be used to upgrade the temperature level of waste heat[2] recovered at a lower temperature.

[2] See Section 9.3.1 for details of the application of the heat pump to heat recovery.

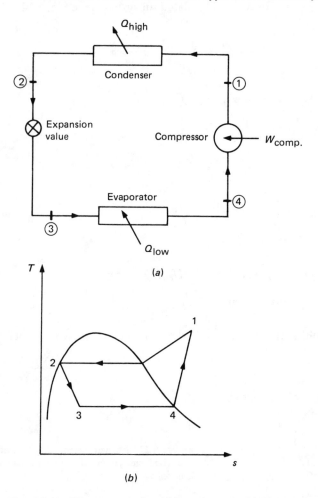

Figure 5.11   Vapor compression refrigeration cycle. (*a*) Cycle components; (*b*) cycle diagram.

5.1.5 Second Law of Thermodynamics

We begin our discussion of the second law of thermodynamics by stating a physical law: The entropy of an isolated system can never increase. This law cannot be proved within the framework of classical thermodynamics. However, we know it is true because we can express it in terms of things that we know to be true and have seen happen over and over.

Statistical thermodynamics can be used to rigorously demonstrate the validity of the second law. Reynolds and Perkins [1] give the axioms and derivations required to prove the law. They demonstrate that entropy is a thermodynamic property and that, accordingly, it must be a function of two other thermodynamic properties, say $S = S(U,V)$.

We can gain some insight into the function $S(U,V)$ by examining a rigid isolated system, C, composed of two rigid isolated subsystems; see Fig. 5.12. For this system,

$$S_C = S_A + S_B$$

or

$$S_C = S_A(U_A,V_A) + S_B(U_B,V_B)$$

For the isolated system C,

$$U_C = U_A + U_B$$

Define

$$r = \frac{U_A}{U_C} \quad (1-r) = \frac{U_B}{U_C}$$

Then

$$S_C = S_A(rU_C,V_A) + S_B[(1-r)U_C,V_B] = \text{fn}(r)$$

only because U_C and V_A do not vary.

Equilibrium would occur at the point where a change in r would not

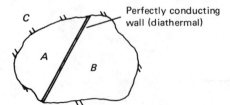
Perfectly conducting wall (diathermal)

Figure 5.12 Isolated system composed of two rigid subsystems that can exchange heat only.

precipitate a change in S_C. Mathematically, we write

$$\frac{dS_C}{dr} = 0 = \frac{\partial S_A}{\partial U_A}\bigg)_{V_A} \frac{dU_A}{dr} + \frac{\partial S_B}{\partial U_B}\bigg)_{V_B} \frac{dU_B}{dr} \qquad (5.34)$$

or

$$\frac{\partial S_A}{\partial U_A}\bigg)_{V_A} U_C + \frac{\partial S_B}{\partial U_B}\bigg)_{V_B} (-U_C) = 0 \qquad (5.35)$$

or, finally,

$$\frac{\partial S_A}{\partial U_A}\bigg)_{V_A} = \frac{\partial S_B}{\partial U_B}\bigg)_{V_B} \qquad (5.36)$$

We now introduce the notion of temperature. We know that for A and B to be in equilibrium, their temperature must be the same. Let us define

$$\frac{\partial S}{\partial U}\bigg)_V = \text{fn}(T) \equiv \frac{1}{T} \qquad (5.37)$$

The inverse temperature relationship is chosen for the consistency with the ideal gas equation of state. From Eq. (5.34) we get

$$dS_C = U_C \left(\frac{1}{T_A} - \frac{1}{T_B} \right) dr \qquad (5.38)$$

which applies to the isolated system C.

According to Eq. (5.38), the following must be true

$$\begin{aligned}
&\text{if } T_A > T_B \qquad dr < 0 \quad \text{or} \quad dU_A < 0 \\
&\text{if } T_A < T_B \qquad dr > 0 \quad \text{or} \quad dU_A > 0
\end{aligned} \qquad (5.39)$$

in order that dS_C will be greater than zero to ensure that the entropy of the isolated system can never decrease. Thus we see that energy must flow from the hotter to the colder body.

The above development leads to the Clausius statement of the second law: *It is impossible to manufacture any device which, operating in a cycle, has no effect other than cooling of one thermal reservoir and heating a second reservoir at a higher temperature.*

An alternative is the Kelvin-Planck statement: *It is impossible for any device to operate in a cycle and produce work while exchanging heat only with reservoirs at a single fixed temperature.* The statements are equivalent; the proof is given in all basic thermodynamic texts.

Now, use $s = s(u,v)$, which is the intensive form of the entropy equation. Then

$$ds = \frac{\partial s}{\partial u}\bigg)_V du + \frac{\partial s}{\partial v}\bigg)_u du \tag{5.40}$$

Referring back to Fig. 5.12, if we allow the wall separating A and B to be diathermal *and* flexible, then Q and W could be transferred internally in C. This will lead to

$$\frac{\partial S}{\partial V}\bigg)_u \equiv \frac{p}{T} \tag{5.41}$$

Using Eqs. (5.40) and (5.41), we can now write

$$ds = \frac{1}{T} du + \frac{p}{T} dv$$

or

$$T \, ds = du + p \, dv = dh \tag{5.42}$$

which allows entropy to be described in terms of the measurable variables T, u, p, and v.

We can use the second law for an isolated system, dS (isolated system) $\geq$ 0, to show that

$$dS \geq \frac{dQ}{T} \tag{5.43}$$

for any system. The equality sign would apply if a reversible heat transfer is assumed to occur, while the $>$ sign applies to irreversible heat transfer. We use Eq. (5.43) in the form

$$dS = \frac{dQ}{T}\bigg)_{rev} \tag{5.44}$$

to evaluate the entropy of a given substance measured above a specific datum state; thus

$$S - S_{ref} = \int_{rev} \frac{dQ}{T}\bigg)_{rev} \tag{5.45}$$

Entropy values evaluated in this way are listed in the property tables of Appendix B.

Second Law for Thermal-Fluid Systems We can now state the second law for various types of thermal-fluid systems:

Closed system, cyclic process:

$$\oint \frac{dQ}{T} \leq 0$$

Closed system, process 1 to 2:

$$\int_1^2 \frac{dQ}{T} \leq S_2 - S_1$$

Open system, SSSF:

$$\sum_{\text{out}} \dot{m}s \geq \sum_{\text{in}} \dot{m}s + \frac{\dot{Q}}{T}$$

Open system, USUF:

$$(ms)_2 - (ms)_1 + \sum_{\text{out}} ms > \sum_{\text{in}} ms + \frac{Q}{T}$$

5.1.6 Mixtures of Gases

When two or more species of gases are mixed, a convenient way to represent the properties of the mixture is needed. We adopt the idea of "independent" gases to accomplish this. A mixture of independent gases is formed when the molecules of the constituent gases behave independently of one another. This means primarily that the molecules exert no long-range forces on one another.

Consider a box containing molecules A and B, as in Fig. 5.13. For independent gases A and B, we find that:

1. The pressure exerted by the gas A, p_A, is the net momentum exchanged by the A molecules hitting the wall. If the B molecules were absent, this momentum exchange would be unaltered, and vice versa. The total pressure then becomes

$$p = p_A + p_B \tag{5.47}$$

If there are more than two species of gas present, we would generalize this to

$$p = \sum_i p_i \tag{5.48}$$

which is Dalton's law.

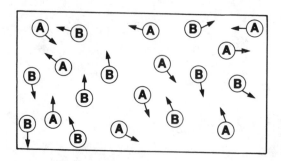

Figure 5.13 Idealized illustration of box containing molecules A and B of different species.

2. The removal of species B molecules at any given instant would not alter the energy of the A molecules. We can describe this behavior for multiple species as

$$U = \sum_i U_i \qquad (5.49)$$

Other extensive properties can be represented in the same way:

$$S = \sum_i S_i$$
$$H = \sum_i H_i \qquad (5.50)$$

and so forth. For thermal equilibrium, the temperature must be uniform throughout the system.

We now define the mole fraction

$$y_i \equiv \frac{n_i}{\sum_i n_i} = \frac{n_i}{n}$$

and the mass fraction

$$x_i \equiv \frac{m_i}{\sum_i m_i} = \frac{m_i}{m}$$

It logically follows that

$$\sum_i y_i = 1$$

and

$$\sum_i x_i = 1$$

Dalton's law, Eq. (5.48), can now take on different forms for an ideal gas. First, we define a partial volume,

$$V_i \equiv \frac{n_i \bar{R} T}{p}$$

which is the volume that the ith component would occupy if it existed alone at p and T. Then for an ideal gas

$$y_i = \frac{n_i}{n} = \frac{V_i p / \bar{R} T}{V p / \bar{R} T} \tag{5.51}$$

so that

$$y_{i_{ideal}} = \frac{n_i}{n} = \frac{V_i}{V}$$

We can now write the Amagat-Leduc law

$$V = \sum_i V_i \tag{5.52}$$

We can also relate mole and mass fractions by introducing the molecular weight, M,

$$x_i = \frac{n_i M_i}{\sum_i n_i M_i} = \frac{y_i M_i}{M_{mix}} \tag{5.53}$$

where $M_{mix} = \Sigma_i y_i M_i$ is the mean value of the molecular weight for the mixture. Alternatively, we write

$$y_i = \frac{x_i / M_i}{\sum_i x_i / M_i} \tag{5.54}$$

Finally, we can write for an independent mixture

$$u = \sum_i x_i u_i \qquad \bar{u} = \sum_i y_i \bar{u}_i$$

$$h = \sum_i x_i h_i \qquad \bar{h} = \sum_i y_i \bar{h}_i$$

$$s = \sum_i x_i s_i \qquad \bar{s} = \sum_i y_i \bar{s}_i$$

$$c_p = \sum_i x_i c_{pi} \qquad \bar{c}_p = \sum_i y_i \bar{c}_{pi}$$

5.1.7 Psychrometry

Psychrometry is the science of air-water vapor mixtures. An air-water vapor mixture does not differ from any other ideal gas mixture in its behavior, except that the water vapor will condense if its partial pressure is high enough.

Air and water vapor mixed together at a total pressure of 1 atm is called "atmospheric" air. Relative humidity, ϕ, is defined as the ratio of the water vapor pressure to the saturated vapor pressure at the temperature of the mixture. Figure 5.14 shows the relationship between points on the $T\text{-}s$ diagram that yields relative humidity.

Humidity ratio, ω, is defined as the ratio of the mass of water vapor to the mass of air in atmospheric air, $\omega = m_v/m_a$. This can be shown to be $\omega = v_a/v_v$, where v_a and v_v are the specific volumes of air and water vapor, respectively. Humidity ratio and relative humidity are related by $\omega = (v_a/v_g)\,\phi$, where v_g is the specific volume of saturated water vapor at the temperature of the mixture.

A convenient way to describe atmospheric air is to base the condition of the air-water vapor mixture on three temperatures: the dry-bulb, wet-bulb, and dew-point temperatures. Dry-bulb temperature is the temperature that is measured by an ordinary thermometer placed in atmospheric air.

Wet-bulb temperature is the temperature that is measured when the bulb of a liquid-in-glass thermometer is exposed to air at a relative humidity of 100%. Wet-bulb temperature is usually measured with a psychrometer, in which a water-saturated wick around the thermometer bulb promotes cooling to the saturation point. The difference between the dry-bulb and wet-bulb temperatures is a measure of the amount of water vapor in atmospheric air.

The dew-point (point 2 in Fig. 5.14) is the saturation temperature of the water vapor at its existing partial pressure. It is the temperature at which water vapor would begin to condense if the mixture were cooled at constant pressure. The dew-point and dry-bulb temperatures are identical for a relative humidity of 100%.

The state diagram for a mixture of air and water vapor is called a psychro-

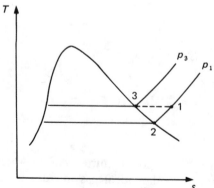

Figure 5.14 Thermodynamic behavior of water in air: $\phi = p_1 p_3 / T_2$, dew point.

metric chart; it is based on the concepts and definitions described in this section. Appendix B contains such a chart for atmospheric air.

5.2   HEAT TRANSFER

Heat transfer is the engineering science that deals with the prediction of energy transport caused by temperature differences. The field is broken down into three basic categories: conduction, convection, and radiation heat transfer.

Conduction is characterized by energy transfer by internal microscopic motion such as lattice vibration and electron movement. Conduction will occur in any region where mass is contained and across which a temperature difference exists.

Convection is characterized by motion of a fluid region. In general, the effect of the convective motion is to augment the conductive effect caused by the existing temperature difference.

Radiation is an electromagnetic wave transport phenomenon and requires no medium for transport.

5.2.1   Conduction Heat Transfer

The basic law of conduction is Fourier's law:

$$\dot{Q} = -kA\frac{dT}{dx} \tag{5.55}$$

The heat flux, $\dot{Q}$, depends on the area A across which energy flows and the temperature gradient dT/dx at that plane. The coefficient of proportionality, k, is a material property called thermal conductivity. Fourier's law applies for both steady and transient cases. If the gradient can be found at any given point and time, the heat flux density, $\dot{Q}/A$, can be calculated.

Conduction Equation   The control volume method of thermodynamics applied to a solid region gives an energy balance that is called the conduction equation. For brevity, the details of this development are omitted; see [2] for this derivation. The conduction equation is

$$G + k\nabla^2 T = \varrho c\,\frac{\partial T}{\partial \tau} \tag{5.56}$$

where ϱ is the material density, c is the specific heat capacity, and τ is time. This equation gives the temperature distribution in space and time. The heat generation term, G, might be caused by either chemical, electrical, or nuclear effects in the control volume. Equation (5.56) can be written alternatively as

$$\nabla^2 T + \frac{G}{k} = \frac{\varrho c}{k} \frac{\partial T}{\partial \tau}$$

The ratio $k/\varrho c$ is also a material property, called thermal diffusivity and denoted by α. Appendix D gives the thermophysical properties of many engineering materials commonly encountered in industrial applications.

Many conduction heat transfer problems of interest in energy conservation studies are steady and one-dimensional. These cases are treated as the need arises in other chapters; for example, one-dimensional conduction across plane and cylindrical layers is treated in Chapters 8 and 10.

5.2.2 Convective Heat Transfer

Convective heat transfer is considerably more complicated than conduction because motion of the medium is involved. In contrast to conduction, where many geometric configurations can be treated analytically, theory alone will give convective heat transfer relationships for only a few cases. Consequently, convection is largely a semiempirical science. That is, correlating equations for heat transfer are based on the results of experimentation.

Convection Modes Convection can be split into several subcategories. For example, *forced* convection refers to the case where the velocity of the fluid is completely independent of the temperature of the fluid. On the other hand, *free* (or natural) convection occurs when the temperature field causes fluid motion through buoyancy effects.

Convection can be further separated according to geometry into external and internal flows. Internal refers to channel, duct, and pipe flow, and external refers to unbounded fluid flow cases. There are other specialized forms of convection, for example, the change-of-phase phenomena — boiling, condensation, melting, freezing, etc. Change-of-phase heat transfer is difficult to predict analytically. Tong [4] gives many correlations for boiling and two-phase flow. Lienhard [2] gives a good up-to-date treatment of the fundamentals of boiling and condensation. Some considerations related to overall heat transfer coefficients that involve boiling and condensation in and around heat exchanger tubes are presented in Chapters 7 and 8.

Analyses and experimentation in convective heat transfer aim to obtain values of h, the heat transfer coefficient, which is defined by

$$\dot{Q} = hA(T_w - T_f) \tag{5.57}$$

where T_w is the surface temperature and T_f is the adjoining fluid temperature. The coefficient, h, is the ratio of heat flux density and the temperature difference between the wall and the fluid.

Dimensionless Heat Transfer Parameters Because experimentation is required to develop correlations for convective heat transfer, the use of generalized dimensionless quantities in these correlations is preferred. In this way, experimental data are applicable to a wider range of conditions and fluids. Some of these parameters, which are called "numbers," are defined as follows:

Nusselt number:

$$Nu = \frac{hL}{k}$$

where k is the fluid conductivity and L is measured along the boundary between liquid and solid. Nu is a nondimensional heat transfer coefficient.

Reynolds number:

$$Re = \frac{LV}{v}$$

where V is fluid velocity and v is kinematic viscosity. Re is discussed further in the section on fluid mechanics. It describes the character of the flow, i.e., whether it is laminar or turbulent.

Prandtl number:

$$Pr = \frac{c\mu}{k}$$

where c is specific heat and μ is viscosity. Pr is the ratio of momentum transport to heat transport characteristics of a fluid. It is important in all convective cases and is a material property.

Grashof number:

$$Gr = \frac{g\beta(T - T_f)L^3}{v^2}$$

where g is gravitational acceleration, β is the coefficient of expansion, and T is the fluid temperature. Gr plays the same role in natural convection as Re does in forced convection, i.e., it describes the character of the flow.

Stanton number:

$$St = \frac{h}{\varrho Vc}$$

St is also a nondimensional heat transfer coefficient. The Stanton number is $St = Nu/(RePr)$.

In general, data are correlated by using relations between dimensionless numbers; for example, in forced convection cases, Nu = fn(Re,Pr) is the functional relation. Then it is possible, from analysis, experimentation, or both, to write a correlation equation that can be used for design calculations. These correlations are called working formulas.

Forced Convection Past Plane Surfaces The average heat transfer coefficient for a plate of length L may be calculated from

$$Nu_L = 0.664(Re_L)^{1/2}(Pr)^{1/3} \qquad (5.58)$$

when the flow is laminar, i.e., if $Re_L \leq 400,000$. The fluid properties should be evaluated at the mean film temperature, T_m, which is the arithmetic average of the fluid and the surface temperature.

For turbulent flow, the most useful correlation includes both laminar leading-edge effects and turbulent effects. It is

$$\overline{Nu} = 0.0036(Pr)^{1/3}[(Re_L)^{0.8} - 18,700] \qquad (5.59)$$

where the transition Re is 400,000.

Forced Convection inside Cylindrical Tubes Fluids flowing through pipes, tubes, and ducts occur often in industry, in both laminar and turbulent flow. Most heat exchangers involve the cooling or heating of fluids in tubes. Single pipes and/or tubes are also used to transport hot or cold liquids in industrial processes. Most of the formulas listed here are for the $0.5 \leq Pr \leq 100$ range.

Laminar Flow For the case $Re_D \leq 2300$, Nusselt showed that $Nu_D = 3.66$ for long tubes at a constant tube wall temperature. The fluid properties are evaluated at the bulk temperature, T_b, which is the flow-averaged temperature at a given cross section.

Sieder and Tate developed the following convenient formula for short tubes:

$$\overline{Nu}_D = 1.86(Re_D)^{1/3}(Pr)^{1/3}\left(\frac{D}{L}\right)^{1/3}\left(\frac{\mu}{\mu_w}\right)^{0.14} \qquad (5.60)$$

A short tube in laminar flow is one where $L/D \leq 0.05$ RePr. The fluid properties are evaluated at T_b except for the quantity μ_w, which is the dynamic viscosity evaluated at the temperature of the wall.

Turbulent Flow McAdams suggested the following empirical relation for cases in which Re > 2,300,

$$\overline{Nu}_D = 0.023(Re_D)^{0.8}(Pr)^n \qquad (5.61)$$

where $n = 0.4$ for heating and $n = 0.3$ for cooling.

Equation (5.61) applies as long as the difference between the pipe surface temperature and the bulk fluid temperature is not greater than 10°F for liquids or 100°F for gases.

For temperature differences greater than the limits specified for Eq. (5.61), or for fluids more viscous than water, the following expression from Sieder and Tate will give better results:

$$\overline{Nu}_D = 0.027(Re_D)^{0.8}(Pr)^{1/3}\left(\frac{\mu}{\mu_w}\right)^{0.14} \tag{5.62}$$

The McAdams equation requires only a knowledge of the bulk temperature, whereas the Sieder-Tate expression also requires the wall temperature. Many people prefer Eq. (5.61) for that reason.

Nusselt found that turbulent heat transfer in short tubes could be represented by the expression

$$\overline{Nu}_D = 0.0036(Re_D)^{0.8}(Pr)^{1/3}\left(\frac{\mu}{\mu_w}\right)^{0.14}\left(\frac{D}{L}\right)^{0.18} \tag{5.63}$$

In turbulent flow, for $Pr > 0.7$, a tube may be considered short if $L/D \leq 40$.

Petukhov [5] recommends the following equation for turbulent fully developed flow, where all properties but μ_w are evaluated at $(T_b + T_w)/2$,

$$\overline{Nu}_D = \frac{(f/8)\, Re_D\, Pr}{1.07 + 12.7\sqrt{f/8}\,(Pr^{2/3} - 1)}\left(\frac{\mu}{\mu_w}\right)^n \tag{5.64}$$

where $0 \leq \dfrac{\mu}{\mu_w} < 40$

$10^4 \leq Re_D < 5 \times 10^6$
$0.5 \leq Pr < 200$ for 6% accuracy
$200 \leq Pr < 2000$ for 10% accuracy
$n = 0.11$ for uniform $T_w > T_b$
$n = 0.25$ for uniform $T_w < T_b$
$n = 0$ for uniform q_w and/or for gases

The friction factor for smooth pipes is

$$f = \frac{1}{(1.82 \log_{10} Re_D - 1.64)^2} \tag{5.65}$$

Alternatively, the friction factor can be found from a so-called Moody plot of f against Re for pipes and ducts; see Fig. 5.17.

Petukhov's equation, developed in the 1960s, is superior to those developed earlier, such as the Sieder-Tate and McAdams correlations. However, Petukhov's equation is more cumbersome to use. Nonetheless, it is now the preferred correlation for convection in long tubes.

Forced Convection in Flow Normal to Single Tubes and Tube Banks This circumstance is encountered often, for example, in air flowing over a pipe carrying hot or cold fluid. Correlations of this phenomenon typically take the form $Nu_D = fn(Re_D, Pr)$.

Churchill and Bernstein [6] correlated data from many different sets of measurements. They suggested the following equation for the average Nusselt number:

$$\overline{Nu}_D = 0.3 + \frac{0.62\,Re_D^{1/2}\,Pr^{1/3}}{[1 + (0.4/Pr)^{2/3}]^{1/4}}\left[1 + \left(\frac{Re_D}{282,000}\right)^{5/8}\right]^{4/5} \quad (5.66)$$

as a general equation applicable in the range $10^2 < Re_D < 10^7$. For $Re_D < 4,000$, they suggested neglecting the last bracketed term in Eq. (5.66), giving

$$\overline{Nu}_D = 0.3 + \frac{0.62\,Re_D^{1/2}\,Pr^{1/3}}{[1 + (0.4/Pr)^{2/3}]^{1/4}} \quad (5.67)$$

In the range $20,000 < Re_D < 400,000$, Churchill and Bernstein suggested the direct use of Eq. (5.66). Properties are evaluated at $T_m = (T_w + T_b)/2$.

Tube Banks A tube bank or bundle is an array of parallel cylinders that exchange heat with a fluid flowing normal to them, parallel with them, or at some angle in between. The flow on the shell side of most shell-and-tube heat exchangers is generally normal to the tube bundles.

Figure 5.15 shows the two basic configurations of a tube bank in a cross-flow. In one the tubes are in line with the flow, and in the other the tubes are staggered in alternating rows. For either of these configurations, heat transfer data are correlated reasonably well with relations of the form

$$\overline{Nu}_D = C\,Re_D^n\,Pr^{1/3}$$

where the Reynolds number is based on the maximum average velocity, u_{max}, in the narrowest transverse area of the passage. Based on this idea, the Reynolds number is

$$Re_D = \frac{u_{max}D}{v}$$

Zukauskas [7] correlated a large number of data for tube-bank heat transfer. The resulting equation, Eq. (5.68), applies to a wide range of Pr and geometric factors S_T/D and S_L/D, which are illustrated in Fig. 5.15.

$$\overline{Nu}_D = Pr^{0.36}\,(Pr/Pr_w)^n fn(Re_D)$$
$$n = 0 \text{ for gases}$$
$$n = 1/4 \text{ for liquids} \quad (5.68)$$

The properties are evaluated at the fluid bulk temperature, except for Pr_w, which is evaluated at T_w.

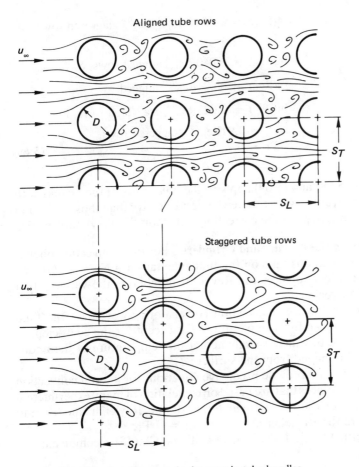

Figure 5.15 Aligned and staggered tube rows in tube bundles.

The function $\text{fn}(\text{Re}_D)$ takes the following forms for the various circumstances of flow and tube configurations

$10 \leqslant \text{Re}_D \leqslant 100$: $\quad \text{fn}(\text{Re}_D) = 0.8\text{Re}_D^{0.4} \quad$ aligned rows

$\qquad\qquad\qquad\qquad \text{fn}(\text{Re}_D) = 0.9\text{Re}_D^{0.4} \quad$ staggered rows

$100 < \text{Re}_D < 10^3$: $\quad$ treat tubes as though they were isolated

$10^3 \leqslant \text{Re}_D \leqslant 2 \times 10^5$: $\text{fn}(\text{Re}_D) = 0.27\text{Re}_D^{0.63} \quad$ aligned rows
$\qquad\qquad\qquad\qquad S_T/S_L < 0.7$

$\qquad\qquad\qquad\qquad$ For $S_T/S_L \geqslant 0.7$, heat exchange is much less effective. Therefore, tube banks are not designed in this range and no correlation is given.

$$\text{fn(Re}_D) = 0.35 \, (S_T/S_L)^{0.2} \text{Re}_D^{0.6} \text{ staggered rows,}$$
$$S_T/S_L \leq 2$$

$$\text{fn(Re}_D) = 0.040 \text{Re}_D^{0.4} \text{ aligned rows, } S_T/S_L \leq 2$$

$\text{Re}_D > 2 \times 10^5$:
$$\text{fn(Re}_D) = 0.021 \text{Re}_D^{0.84} \text{ aligned rows}$$

$$\underline{\text{fn(Re}_D) = 0.022 \text{Re}_D^{0.84}} \text{ staggered rows, } \text{Pr} > 1$$

$$\overline{\text{Nu}}_D = 0.91 \text{Re}_D^{0.84} \text{ staggered rows,}$$
$$\text{Pr} = 0.7 \tag{5.69}$$

Equations (5.68) and (5.69) apply to the *inner* rows of tube banks with many rows. The front rows are somewhat less effective than those deeper in the bank. Figure 5.16 gives the correction factor for "shallow" tube banks.

Free Convection around Plates and Cylinders For free-convection phenomena, the basic relations take on the fundamental form Nu = fn(Gr,Pr). The Grashof number replaces the Reynolds number as the variable that dictates the flow regime.

It is customary in free-convection correlations to evaluate the fluid properties at the mean film temperature, T_m, except for the coefficient of volume expansion, β, which is evaluated at the temperature of the undisturbed fluid far removed from the surface, T_f.

Churchill and Chu provide predictive equations for natural convection from isothermal surfaces that are based on correlation of many experimental data. The equations are written in terms of the Rayleigh and Prandtl numbers, where the Rayleigh number is Ra = GrPr.

For isothermal vertical plates, Churchill and Chu [8] recommend:

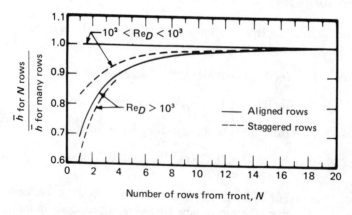

Figure 5.16 Correction for the heat transfer coefficients in the front rows of a tube bundle [7].

$$\overline{\text{Nu}}_L = 0.68 + 0.67\,\text{Ra}_L^{1/4}\left[1+\left(\frac{0.492}{\text{Pr}}\right)^{9/16}\right]^{-4/9} \qquad (5.70)$$

where $\overline{\text{Nu}}_L$ is based on the plate height, L. Equation (5.70) is accurate up to a Rayleigh number of about 10^9. For $\text{Ra} \geq 10^9$, Eq. (5.70) predicts a lower bound for experimental data.

For isothermal horizontal cylinders, Churchill and Chu [9] recommend:

$$\overline{\text{Nu}}_D = 0.36 + \frac{0.518\text{Ra}_D^{1/4}}{[1 + (0.559/\text{Pr})^{9/16}]^{4/9}} \qquad (5.71)$$

which is applicable in the range $10^{-4} \leq \text{Ra} \leq 10^9$. For Rayleigh numbers greater than 10^9, the equation

$$\overline{\text{Nu}}_D = \left(0.60 + 0.387\left\{\frac{0.518\text{Ra}_D^{1/4}}{[1 + (0.559/\text{Pr})^{9/26}]^{16/9}}\right\}^{1/6}\right)^2 \qquad (5.72)$$

is recommended. Actually, Eq. (5.72) is applicable for $\text{Ra} \geq 10^{-6}$, so that it could be used throughout the Rayleigh number range. It is more cumbersome than Eq. (5.71), however, so that Eq. (5.71) is preferred in the more limited Rayleigh number range.

Equation (5.70) can be used for vertical isothermal cylinders if the cylinder radius is large compared to the convection boundary-layer thickness. Lienhard [2] gives correction charts that allow Eq. (5.70) to be used to predict the heat transfer from a vertical cylinder for various Prandtl and Grashof numbers.

5.2.3 Radiation Heat Transfer

Radiation heat transfer is the most mathematically complicated of the three basic types of heat transfer because of the electromagnetic wave nature of thermal radiation. However, in certain applications, mainly at high temperatures, radiation is the dominant mode of heat transfer. Heat transfer in boiler and fired-heater enclosures is highly dependent on the radiative characteristics of the surface and the hot combustion gases.

It is known that for a body at temperature T radiating to its surroundings, the heat transfer is

$$\dot{Q} = \varepsilon\sigma A(T^4 - T_s^4) \qquad (5.73)$$

where ε is the emissivity of the surface, σ is the Stefan-Boltzmann constant, $\sigma = 0.1713 \times 10^{-8}$ Btu/hr ft^2 R^4, and T_S is the temperature of the surroundings. Temperature must be in absolute units, R and K. When $\varepsilon = 1$ for a surface, the surface is called a blackbody, i.e., a perfect emitter of thermal energy. Radiative properties of various surfaces are given in Appendix G.

The reader is directed to Siegel and Howell [10] for a comprehensive treatment of radiation in enclosures and radiation from luminous combustion zones. The development of radiative heat transfer relations required for a specific type of analysis will be introduced, when needed, in future chapters.

5.3 FLUID MECHANICS

The relations that govern the flow of fluids form the science that is called fluid mechanics. The behavior of the flowing fluid controls pressure drop (pumping power), mixing efficiency, and in some cases the efficiency of heat transfer. Hence, fluid mechanics is an integral part of most analyses that accompany an energy conservation program.

5.3.1 Fluid Dynamics

When a fluid flows, certain governing laws are applicable. For example, mass flowing in and out of control volumes must always be balanced against mass storage. In other words, conservation of mass must be satisfied.

In its most basic form, the so-called continuity equation (conservation of mass) is

$$\iint_{cs} \varrho(\bar{v} \cdot \bar{n}) \, dA + \frac{\partial}{\partial t} \iiint_{cv} \varrho \, dV = 0 \qquad (5.74)$$

where $\varrho(\bar{v} \cdot \bar{n})$ is the mass flow directed normal to the area A, V is volume, and ϱ is density. The $\varrho(\bar{v} \cdot \bar{n})$ terms are integrated over the control surface, while the $\varrho \, dV$ term depends on an integration over the control volume.

For a steady flow in a constant-area duct, the continuity equation simplifies to

$$\dot{m} = \varrho_f A_c V = \text{const.}$$

That is, the mass flow rate $\dot{m}$ is constant and is equal to the product of the fluid density ϱ_f, the duct cross section A_c, and the average fluid velocity V.

When the fluid is compressible and the flow is steady, one gets

$$\frac{\dot{m}}{\varrho_f} = \text{const.} = (VA_c)_1 = (VA_c)_2$$

where subscripts 1 and 2 refer to different points in a variable-area duct.

5.3.2 First Law — Fluid Dynamics

The first law of thermodynamics can be applied to fluid dynamical systems, such as duct flows. For cases where there is no heat transfer or chemical reaction, and where the internal energy of the fluid stream remains unchanged, the first law is

$$\frac{V_i^2 - V_e^2}{2g_c} + \frac{(z_i - z_e)}{g_c} g + \frac{p_i - p_e}{\varrho} + (w_p - w_f) = 0 \qquad (5.75)$$

where V is velocity, z is elevation, p is pressure, and g_c is the gravitational constant. The subscripts i and e refer to inlet and exit conditions, and w_p and w_f are the pump work and the work required to overcome friction in the duct, respectively.

The most difficult term to determine in Eq. (5.75) is the frictional work term, w_f, which depends on the fluid viscosity, the flow conditions, and the duct geometry. For simplicity, w_f is represented as

$$w_f = \frac{\Delta p_f}{\varrho} \qquad (5.76)$$

where Δp_f is the frictional pressure drop in the duct. Furthermore,

$$\frac{\Delta p_f}{\varrho} = \frac{f V^2 L}{2 g_c D} \qquad (5.77)$$

in a duct of length L and diameter D. The use of the friction factor[2] f is a convenient way to represent the differing influence of laminar and turbulent flows on the frictional pressure drop.

The character of the flow is determined by the Reynolds number, Re = $\varrho V D / \mu$, where x is the viscosity of the fluid. The Reynolds number represents the ratio of dynamic to viscous forces acting on the fluid.

Experiments have shown that if Re < 2,300, the flow is laminar. For larger Re, the flow is turbulent. Figure 5.17 shows how the friction factor depends on the Re value of the flow. For laminar flow, the f versus Re curve is single-valued and is equal to 64/Re. In the turbulent regime, the wall roughness, e, can affect the friction factor because of its effect on the velocity profile near the duct surface.

Equivalent Diameter   The equivalent diameter, D_e, can be used for noncircular ducts so that the relationships developed for round tubes can be used for both pressure drop and heat transfer. D_e is defined as

$$D_e = \frac{4 A_c}{P} \qquad (5.78)$$

where A_c is flow cross-sectional area and P is the "wetted" perimeter — i.e., that part of the flow cross section that touches the duct surfaces. For a circular system, $D_e = 4(\pi D^2/4\pi D) = D$, as it should. For an annular duct,

$$D_e = \frac{(\pi D_o^2/4 - \pi D_i^2/4)4}{\pi D_o + \pi D_i} = D_o - D_i$$

[2]This is called the Moody friction factor. The Fanning friction factor is four times the Moody factor.

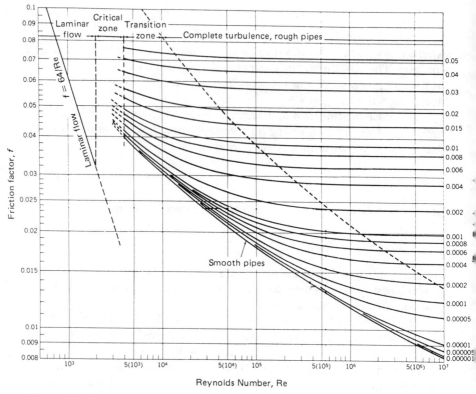

Figure 5.17 Moody friction factors for straight ducts, pipes, and tubes.

Pressure Drop in Ducts   Prediction of pressure drops in piping and duct networks is needed to determine the energy required for fluid circulation. The friction factor approach is adequate for straight sections of constant-area ducts. But valves, nozzles, elbows, and many other types of fittings are necessarily included in a network. These can be dealt with by defining an equivalent length L_e for the fitting. Table 5.2 shows L_e/D values for several common types of fittings. The pressure drop across these fittings is found by using this L_e/D in Eq. (5.77).

Pressure Drop across Tube Banks   Another commonly encountered phenomenon is the pressure drop caused by transverse flow across arrays of heat transfer tubes. One way to calculate this pressure drop is to find the "velocity head loss," N_v, through the tube banks, i.e.,

$$N_v = f N F_d \tag{5.79}$$

where f is the friction factor for the flow across the tubes, N is the number of tube rows crossed by the flow, and F_d is the depth factor. Figures 5.18 and 5.19 show the f factor and F_d relationships for pressure drop calculations.

Table 5.2 L_e/D for Screwed Fittings, Turbulent Flow Only[a]

Fitting	L_e/D
45° elbow	15
90° elbow, standard radius	31
90° elbow, medium radius	26
90° elbow, long sweep	20
90° square elbow	65
180° close return bend	75
Swing check valve, open	77
Tee (as el, entering run)	65
Tee (as el, entering branch)	90
Couplings, unions	Negligible
Gate valve, open	7
Gate valve, ¼ closed	40
Gate valve, ½ closed	190
Gate valve, ¾ closed	840
Globe valve, open	340
Angle valve, open	170

[a]Calculated from Crane Co. Technical Paper 409, May 1942.

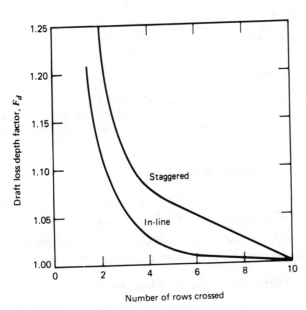

Figure 5.18 Depth factor for number of tube rows crossed in convection banks. Reprinted from *Steam: Its Generation and Use,* 38th ed., 1972, courtesy of Babcock and Wilcox.

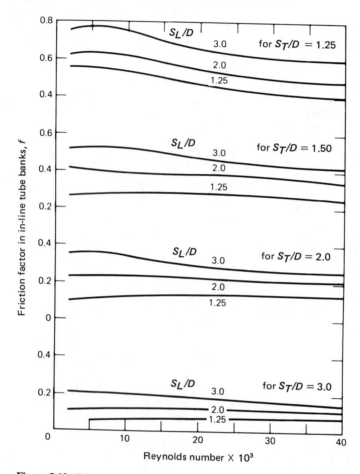

Figure 5.19 Friction factor f is affected by Reynolds number for various in-line tube patterns, cross-flow gas or air. D, tube diameter; S_L, gap distance perpendicular to the flow; S_T, gap distance parallel to the flow. Reprinted from *Steam: Its Generation and Use*, 38th ed., 1972, courtesy of Babcock and Wilcox.

For air, the pressure drop in inches of water is given by

$$\Delta p = N_v \frac{30}{p_{atm}} \frac{T}{1.75 \times 10^5} \left(\frac{G}{10^3} \right)^2 \tag{5.80}$$

where p_{atm} is the atmospheric pressure (inches of mercury), T is temperature (R), and G is the mass velocity (pounds mass per square foot per hour).

Bernoulli's Equation   There are some cases where the equation

$$\frac{p}{\varrho} + \frac{V^2}{2} + gz = \text{const.} \tag{5.81}$$

which is called Bernoulli's equation, is useful. Strictly speaking, this equation applies for inviscid, incompressible, steady flow along a streamline. However, even in viscous flow through a pipe the equation can be applied because of the confined nature of the flow. That is, the flow is forced to behave in a streamlined manner. Note that the first law of thermodynamics, Eq. (5.75), yields Bernoulli's equation if the friction work exactly equals the pump work.

5.3.3 Fluid Handling Equipment

Pumps, compressors, fans, and blowers are used to move gases and liquids through process networks and over heat exchanger surfaces. These devices are called fluid handlers. Equipment is selected by matching fluid handler characteristics to pumping power required to overcome the pressure drop in the circuit connected to the fluid handler.

Pumps used to transport liquids, and compressors, fans, and blowers used to circulate gases, have common performance features. For the purpose of illustration, a centrifugal pump will be used to discuss performance characteristics.

Centrifugal Machines Centrifugal machines operate on the principle of centrifugal acceleration of a fluid element in a rotating impeller/housing system to achieve a pressure gain and circulation. The characteristics that are important are flow rate, head, efficiency, and durability. The flow rate is the volumetric flow, $Q_f = VA_c$, usually given in cubic feet per minute (CFM). Head, h_p, is the pressure drop in a circuit that is equivalent to the height of a water column; the units are feet, indicating feet of water. The efficiency, n_p, is the ratio of the work imparted to the flow to the energy input of the pump. The flow rate (also called the capacity), the head, and the efficiency are related quantities and depend on the fluid behavior in the pump and the flow circuit. Durability is related to the wear, corrosion, and other factors that bear on a pump's reliability and lifetime.

Figure 5.20 shows the relation between flow rate and head selected for a centrifugal pump operating at constant speed with different impeller sizes and different efficiencies. Graphs of this type are called performance curves. The horsepower curves in Fig. 5.20 are for the pump motor and are called brake horsepower (BHP). Fluid horsepower (FHP) is the horsepower imparted to the fluid. The primary design constraint is a matching of flow rate to head. As the flow rate is increased, the allowable head must be reduced if other pump parameters are unchanged.

Analysis and experience have shown that there are scaling laws for centrifugal pump performance that predict the trends for changes in certain performance parameters. Basically, the scaling laws are:

Efficiency:

$$\eta_p = \text{fn}_1\left(\frac{Q_f}{D^3 n}\right)$$

Dimensionless head:

$$\frac{h_p g}{D^2 n^2} = \text{fn}_2\left(\frac{Q_f}{D^3 n}\right) \tag{5.82}$$

Dimensionless brake horsepower:

$$\frac{\text{BHP}g}{\gamma D^5 n^3} = \text{fn}_3\left(\frac{Q_f}{D^3 n}\right)$$

where D is the impeller diameter, n is the impeller rotary speed, g is gravitational acceleration, and γ is the specific weight of fluid.

The basic relationships (5.82) yield specific proportionalities such as:

$$Q_f \propto n \qquad h_p \propto n^2 \qquad \text{FHP} \propto n^3$$
$$Q_f \propto D^3 \qquad h_p \propto D^2 \qquad \text{FHP} \propto D^5$$

Density variations in pumps are negligible because liquids are incompressible. But for gas-handling equipment, density changes are very

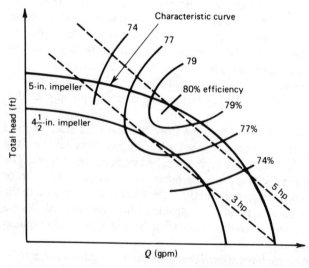

Figure 5.20 Performance curve for a centrifugal pump.

important. The scaling laws take the following forms for variable-density fluids:

$$\begin{aligned} h_p &\propto \varrho \\ \text{FHF} &\propto \varrho \end{aligned} \qquad \{Q_f, n \text{ constant}\}$$

$$\left.\begin{aligned} n \\ \text{FHP} \\ Q_f \end{aligned}\right\} \propto \varrho^{-1/2} \qquad \{h_p \text{ constant}\}$$

$$\left.\begin{aligned} n \\ Q_f \\ h_p \end{aligned}\right\} \propto \frac{1}{\varrho} \qquad \{\dot{m} \text{ constant}\}$$

$$\text{FHP} \propto \frac{1}{\varrho^2} \tag{5.83}$$

For centrifugal pumps, the following equations hold:

$$\text{FHP} = \frac{Q_f \varrho g h_p}{550 g_c} \tag{5.84}$$

$$\eta_p = \frac{Q_f \varrho g h_p/(550 g_c)}{\text{BHP}} = \frac{\text{FHP}}{\text{BHP}} \tag{5.85}$$

$$\eta_s = \eta_p \times \eta_m \tag{5.86}$$

where η_s is the system efficiency, η_m is the motor efficiency, Q_f is in cubic feet per second, ϱ is in pounds mass per cubic feet, and h_p is in pounds force per square foot. The motor-pump set should be selected so that at nominal operating conditions both the motor and the pump operate near their maximum efficiency.

For systems where two or more pumps are present, the following rules are helpful: To analyze pumps in parallel, add capacities at the same head. For pumps in series, add heads at the same capacity.

One notable difference exists between blower and pump performances; this is shown in Fig. 5.21. For blowers, the BHP continues to increase as the permissible head goes to zero, in contrast to the pump curve, where BHP approaches zero. The reason for this behavior is that the kinetic energy change of gases at high flow rates is quite significant for blowers but not for liquids in pumps.

Manufacturers of fluid-handling equipment provide excellent performance data for all types of equipment. Anyone considering replacement of

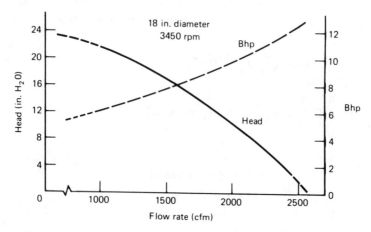

Figure 5.21   Variation of head and BHP with flow rate for a typical blower at constant speed.

equipment or a new installation should take full advantage of manufacturers' data.

Fluid-handling equipment operating on a principle other than centrifugal acceleration does not follow the centrifugal scaling laws. Evans [11] gives a thorough treatment of most types of fluid-handling equipment that are encountered in industrial application.

5.4   SUMMARY

This chapter summarizes the concepts and techniques in thermal-fluid science that are useful in energy conservation analyses. No example problems are presented in the chapter. Instead, the problems at the end of the chapter are typical of the kinds of calculations that might be required for energy conservation problems in later chapters and are presented so that those who have forgotten the material can use them for a review. It is recommended that most of these problems be solved by the student to ensure a sound understanding of thermal-fluid systems. The problems are grouped according to discipline: thermodynamics, heat transfer, and fluid mechanics.

NOMENCLATURE

A area
c specific heat
D diameter
D_e equivalent diameter

E	stored energy
e	wall roughness
F	force
F_d	tube bank depth factor
f	friction factor
G	heat generation rate per unit volume; mass velocity, $\dot{m}/A$
Gr	Grashof number
g	acceleration due to gravity
g_c	gravitational constant
H	enthalpy
h	specific enthalpy, H/m; heat transfer coefficient
h_p	pressure head, usually specified as water column height in feet
k	ratio of specific heats c_p/c_v; thermal conductivity
L	length
L_e	equivalent length
M	molecular weight
m	mass
$\dot{m}$	mass flow rate
N	number of tube rows
Nu	Nusselt number
N_v	velocity head loss
n	polytropic exponent; number of moles; rotary speed
P	wetted perimeter
P_r	reduced pressure, P/P_{cr}
p	pressure
Q	heat; volumetric flow rate
$\dot{Q}$	heat flux
q	heat flux density
R	gas constant
$\bar{R}$	universal gas constant, 1545.3 ft lb$_f$/lb-mole °R
Ra	Rayleigh number, GrPr
Re	Reynolds number
S	entropy
St	Stanton number
S_T, S_L	tube bank geometric factors
s	specific entropy, S/m
T	temperature
T_r	reduced temperature, T/T_{cr}
ΔT	temperature difference
U	internal energy
u	specific internal energy, U/m
V	velocity; volume
v	specific volume
W	work

w	work per unit mass, W/m
x	displacement; thermodynamic quality; mass fraction
y	mole fraction
Z	compressibility factor
z	elevation
α	thermal diffusivity
β	coefficient of thermal expansion
γ	specific weight
$\in$	emissivity
ϱ	density
η	efficiency
ϕ	relative humidity
ω	humidity ratio
τ	time
μ	viscosity
υ	kinematic viscosity, μ/ϱ
σ	Stefan-Boltzmann constant, 0.1713×10^{-8} Btu/(hr ft^2 °R^4)

Subscripts

b	bulk
cr	critical
cs	control surface
cv	control volume
f	liquid phase; fluid; friction
fg	difference between liquid and gas phase
g	gas phase
i	specie; inner
in	inlet
m	mean
out	outlet
p	constant pressure; pump
ref	reference value
rev	reversible
s	surroundings
th	thermal
v	constant volume
w	wall

Superscripts

—	molal quantities; mean value
*	low pressure

REFERENCES

1. W. C. Reynolds and H. C. Perkins, *Engineering Thermodynamics*, 2d ed., McGraw-Hill, New York, 1977.
2. J. H. Lienhard, *A Heat Transfer Textbook*, Prentice-Hall, Englewood Cliffs, N.J., 1980.
3. R. W. Fox and A.T. McDonald, *Introduction to Fluid Mechanics*, 2d ed., Wiley, New York, 1978.
4. L. S. Tong, *Boiling Heat Transfer and Two-Phase Flow*, Wiley, New York, 1965.
5. B. S. Petukhov, Heat Transfer and Friction in Turbulent Pipe Flow with Variable Physical Properties, in *Advances in Heat Transfer*, vol. 6, pp. 504–564, Academic Press, New York, 1970.
6. S. W. Churchill and M. Bernstein, A Correlating Equation for Forced Convection from Gases and Liquids to a Circular Cylinder in Crossflow, *J. Heat Transfer, Trans. ASME*, 99: 300–306, 1977.
7. A. Zukauskas, Heat Transfer from Tubes in Crossflow, in *Advances in Heat Transfer*, vol. 8, pp. 83–160, Academic Press, New York, 1971.
8. S. W. Churchill and H. H. S. Chu, Correlating Equations for Laminar and Turbulent Free Convection from a Vertical Plate, *Int. J. Heat Mass Transfer*, 18:1323–1329, 1975.
9. S. W. Churchill and H. H. S. Chu, Correlating Equations for Laminar and Turbulent Free Convection from a Vertical Plate, *Int. J. Heat Mass Transfer*, 18:1049–1053, 1975.
10. R. Siegel and J. R. Howell, *Thermal Radiation Heat Transfer*, 2d ed., Hemisphere, Washington, D.C., 1980.
11. F. L. Evans, Jr., *Equipment Design Handbook for Refineries and Chemical Plants*, vols. I and II, 2d ed., Gulf Publishing Co., Houston, Tex., 1980.

PROBLEMS

Thermodynamics

5.1 Find the volume in cubic feet required to contain 1 lb_m of water at the following conditions:

(a) $p = 14.7$ psia, $T = 212\,°FG$, x (quality) $= 0$

(b) $p = 14.7$ psia, $T = 212\,°F$, $x = 100\%$

(c) $p = 14.7$ psia, $T = 212\,°F$, $x = 50\%$

(d) $p = 100$ psia, $T = 350\,°F$

(e) $T = 500\,°F$, $h = 1286.0$ Btu/lb_m

(f) $p = 5,000$ psia, $T = 875\,°F$

5.2 What is the density (lb_m/ft^3) of air at 400 psig and 300°F? If the same amount of air existed in twice the volume and at the same temperature, what would its pressure be?

5.3 A tank having a volume of 10 ft^3 contains 3.0 lb_m of a mixture of liquid water and water vapor in equilibrium at 100 $lb_f/in.^2$. Calculate the volume and mass of the liquid and the volume and mass of vapor.

5.4 A refrigeration system with a volume of 0.90 ft^3 is evacuated and then charged with Freon-12. The temperature of the Fr-12 remains constant at 70°F. What is the mass of Fr-12 in the system when the pressure reaches 35 $lb_f/in.^2$ and the system is filled with saturated vapor? What fraction of the Fr-12 will exist as a liquid when 3 lb_m of Fr-12 has been placed in the system?

5.5 Saturated water vapor at 400°F is contained in a piston-cylinder system whose initial volume is 0.3 ft^3. The steam expands in a slow isothermal process to a pressure of 20 $lb_f/in.^2$. Find the work done on the piston during this process.

5.6 Freon-12 expands from a saturated liquid state at 200 psia across an expansion valve down to 10 psia. Assume that the expansion is a throttling process, i.e., h = constant. What are the temperature and quality of the Freon-12 after expansion?

5.7 Steam at 100 lb$_f$/in.^{2}, 400°F, enters an insulated nozzle at 200 ft/sec. It exhausts at 2,000 ft/sec and 20 lb$_f$/in.^{2}. Find the final temperature if the steam is superheated in the final state, and the quality if it is saturated.

5.8 Freon-12 vapor enters a compressor at 20 lb$_f$/in.^{2} and 40°F. The mass flow rate is 10 lb$_m$/min. Calculate the smallest diameter tubing that can be used if the Fr-12 velocity must not exceed 20 ft/sec.

5.9 Use the Mollier diagram (Appendix Fig. B.1) to find the end state (T, p) for steam expanding from p = 800 psia, T = 1,000°F, down to a moisture content of 6%. Assume an adiabatic, reversible expansion. What is the change in enthalpy between these two states?

5.10 A steam turbine accepts steam at p = 2,000 psia, T = 1050°F and expands it down to 1 psia. For a steam flow rate of 500,000 lb$_m$/hr and an efficiency of 85%, find the power output of the turbine in Btu's per hour, kilowatts, and horsepower.

5.11 A compressor operates with a compression ratio of 8:1. It produces 10 ft^3/min of compressed air with an efficiency of 80%. The intake is air at ambient pressure and temperature (14.7 psia, 70°F). What is the outlet air temperature and how much horsepower is required to drive the compressor?

5.12 A pressure vessel with a capacity of 200 ft^3 is evacuated; then it is connected by a valve to a line carrying superheated steam at 600 psia and 650°F. Neglecting heat transfer, what is the temperature of the steam in the vessel after the pressure is allowed to come to 600 psia? How much mass is in the vessel?

5.13 An engine that operates on the Carnot cycle accepts 1,000 Btu of heat from a reservoir at 600°F. The engine rejects heat to a reservoir at 80°F. Find the thermal efficiency of the cycle and the work done by the engine.

5.14 Find the thermal efficiency of a Carnot cycle operating between the temperature limits, 1,000 and 100°F. Compare this to the thermal efficiency of a Rankine cycle with no superheat that operates between the same temperature limits.

5.15 A Carnot cycle heat pump has ammonia as the working fluid. Heat is transferred from the ammonia during a process where ammonia changes from saturated vapor to saturated liquid at 100°F. Heat is picked up by the ammonia at 0°F. Show this cycle on a T-s diagram. What is the quality at the beginning and the end of the evaporation process at 0°F? What is the COP of this cycle as a refrigerator?

5.16 An open feedwater heater that heats water going to a boiler operates on the principle of mixing steam and water. For the conditions shown in the sketch, calculate the increase of entropy per hour, assuming a steady, adiabatic process.

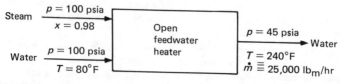

5.17 A heat pump is to be used for both home heating and cooling. When the outside air temperature is 100°F, 75,000 Btu/hr of cooling is needed to maintain the home interior at 70°F. When it is 40°F outside and 70°F inside, 75,000 Btu/hr of heating is needed. Estimate the COP for each of these cases for a system using Freon-12. Use a compressor efficiency of

75% for both cases. Neglect the temperature difference required between the Freon and air during phase change in the evaporator and condenser. Is the flow rate the same for the two modes of operation?

5.18 An air standard diesel cycle has a compression ratio of 14. The pressure at the beginning of the compression stroke is 14.7 psia. Find the difference in thermal efficiency for air temperatures of 100 and 0°F at the beginning of the compression stroke.

5.19 The compressor of an air-standard Brayton cycle operates on a compression ratio of 4:1. Air enters the compressor at 14.7 $lb_f/in.^2$, 70°F. The maximum cycle temperature is 1,600°F. The air flow rate is 20 lb_m/sec. Determine the compressor work, turbine work, and thermal efficiency of the cycle, assuming that specific heat is constant.

5.20 An insulated vessel of 2 ft^3 volume contains CO_2 at 200 $lb_f/in.^2$, 100°F. Additional CO_2 is fed into the vessel from a line at 1,200 $lb_f/in.^2$, 100°F. The valve is closed when $p = 1200$ $lb_f/in.^2$. Determine the amount of CO_2 added by using the compressibility charts.

5.21 Propane gas expands adiabatically in a turbine from 300°F, 400 $lb_f/in.^2$ down to 210°F, 50 $lb_f/in.^2$. Use the generalized charts to determine:

(a) The work done per pound mass of propane entering the turbine
(b) The increase of entropy per pound mass of propane flowing
(c) The isentropic turbine efficiency

5.22 Equal volumes of nitrogen and oxygen are mixed at a temperature of 150°F. What are the respective partial pressures of the nitrogen and oxygen, in pounds per square inch absolute?

5.23 Calculate the humidity ratio, dew point, mass of air, and mass of water vapor for 1,000 ft^3 of air at 14.7 $lb_f/in.^2$, 90°F, and 70% relative humidity.

5.24 Nitrogen and hydrogen are mixed adiabatically in a steady flow process, 3 lb_m of H_2 to 1 lb_m of N_2. Hydrogen enters at 20 psia, 100°F and the nitrogen at 20 psia, 500°F. The mixture exhausts at 14 psia. Find the final temperature of the mixture and the net entropy change per pound mass of mixture.

5.25 The following is the volumetric analysis of a gaseous mixture:

Component	Percent of volume
N_2	55
CO_2	23
CO	10
O_2	12

(a) Determine the mass fraction of each component.
(b) Calculate the mass of 500 ft^3 of the mixture when $p = 14.7$ psia and $T = 70°F$.
(c) Find the heat transfer if the mixture is heated in a steady process from 125 to 450°F.

5.26 Evaporative cooling is effective when temperature is high and humidity is low. This involves spraying water droplets into an air stream and allowing the mixture to cool as the water evaporates, as shown in the sketch. For atmospheric air at 100°F, 10% relative humidity with water at 60°F sprayed into it, what is the outlet temperature if the relative humidity rises to 40%? What is the relative humidity if the air is cooled to 75°F? (Note that an energy balance must be written because energy is consumed by latent and sensible means.)

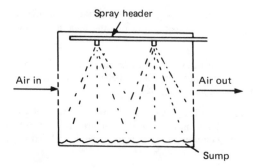

Heat Transfer

5.27 Show that the conduction equation applied to a plane slab with fixed surface temperatures gives a linear temperature profile. Also write an expression for heat transfer rate per unit area across the slab.

5.28 Using the result of Problem 5.27, show that for a wall composed of two layers of different materials, as shown in the sketch, the steady state heat transfer rate is

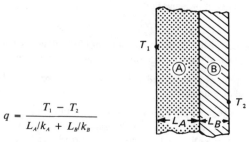

$$q = \frac{T_1 - T_2}{L_A/k_A + L_B/k_B}$$

5.29 Using the results of Problems 5.27 and 5.28, show that the steady heat transfer rate for a wall exposed to fluids at differing temperatures, such as that shown in the figure, is

$$q = \frac{T_{f1} - T_{f2}}{1/h_1 + L_A/k_A + L_B/k_B + L_C/k_C + 1/h_2}$$

The heat transfer coefficients are defined by $\dot{Q} = hA \, \Delta T$ at the inner and outer surfaces.

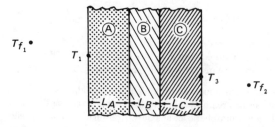

5.30 A plastic refrigerator wall is to be covered with a rigid polyurethane insulation with a thermal conductivity of 0.025 W/m °R. The refrigerator interior is maintained at −10 °C. The surface area of the wall is 100 m² and the cooling capacity of the refrigerator is 2 kW. Find the minimum thickness of insulation required to prevent condensation on the outer surface of the insulation for a dew point of the surrounding air of 15 °C.

5.31 A 1-kW electrical heater with a surface area of 0.1 m² is exposed to a 24°C fluid. Calculate the surface temperature of the heater for the following cases:

(a) The heater is in contact with air, $h = 30$ W/m² K.
(b) The heater is in contact with water, $h = 500$ W/m² K.
(c) The heater is boiling water, $h = 5,000$ W/m² K.

5.32 Calculate the heat flux through a plane brick wall ($k = 0.65$ W/m K) that is 7.5 cm thick and has interior and exterior temperatures of 24 and 40 °C, respectively. What is the centerline temperature of the brick?

5.33 A 1-cm-diameter aluminum wire carries a current of 1,000 amperes. The wire is covered by a 3-mm layer of insulation ($k = 0.10$ W/m K). If the outside of the insulation is 30°C, estimate its inside temperature. The resistance of the wire per unit length is 3.7×10^{-4} ohm/m.

5.34 Show that the Stanton number can be written in terms of the Nusselt and Prandtl numbers as St = Nu/(RePr). Rewrite Eq. (5.60) in terms of the Stanton number rather than the Nusselt number.

5.35 Water is flowing in a ¾-in., 14 gage copper tube at the rate of 25 ft/min. The tube's temperature is maintained by condensing steam at 1 psia on the outside. The tube is 10 ft long. Water enters the tube at 80°F. What is its outlet temperature?

5.36 Liquid ammonia is flowing in a 1-in., 12 gage copper tube at 3.5 ft/sec. The tube wall is maintained at 32°F. Calculate the average heat transfer coefficient for the case where the ammonia has been heated to 25°F after entering the tube at 0°F, using:

(a) the Sieder-Tate equation, Eq. (5.62)
(b) the Petukhov equation, Eq. (5.64)

How long would the tubes be for these two heat transfer coefficients?

5.37 Carbon dioxide flows through a 1.2-cm-square duct whose walls are isothermal at 20°C. The average temperature of the CO_2 is 120°C. Calculate the heat transfer rate per unit length, assuming fully developed laminar flow.

5.38 A square flat plate with a length of 20 in. is maintained at 250°F. Air at 70°F is forced over the plate at 70 ft/sec. Estimate the heat transfer rate from one side of the plate.

5.39 Air at 70°F flows normal to a 10-in. schedule 60 steel pipe carrying steam at 400°F. Air velocity is 15 mph. Calculate the annual heat loss from the pipe to the air for a 500-ft section of pipe.

5.40 Repeat Problem 5.39 for the case where air velocity is zero. What is the pipe surface temperature for this case?

5.41 A tube bank contains 10 rows of 3/4-in. stainless steel tubes. The tubes are staggered as shown in the sketch. Air at 70°F, 20 ft/sec is directed over the tubes, which are maintained at 250°F by hot oil flowing inside. Calculate the heat transfer coefficient for the tube bank.

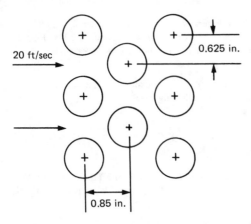

5.42 A 150-W light bulb emits as a blackbody into a vacuum. The filament is 0.15 mm in diameter by 7 cm long. Calculate the temperature of the filament, neglecting conduction from the filament.

5.43 Calculate the energy radiated from a roof made of a mixture of asphalt and crushed brick ($\varepsilon = 0.92$) when the roof is at 150°F and the air temperature is 40°F.

Fluid Mechanics

5.44 Show that the Reynolds number for pipe flow can be written as Re = GD/μ, where G is the mass velocity of flow in pounds mass per square foot per hour.

5.45 An incompressible fluid flows in a duct with an expansion section as shown in the sketch. In the first section, D_1, the Reynolds number is 4,000. Is the flow in section D_2 laminar or turbulent if $D_2 = 2D_1$?

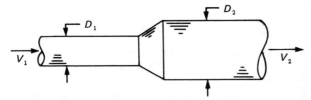

5.46 Air at 25 psia, 36°F, flows through an 8-in.-diameter duct. A pitot tube, located at the centerline, gives a manometer reading of 2 in. of water, which represents the difference between static and dynamic pressure. Calculate the volume flow rate in cubic feet per minute and the corresponding mass flow rate.

5.47 Chilled water at 40°F flows in a horizontal 10-in.-diameter cast iron pipe at 70 ft³/min. The pipe is 65 ft long with a standard 90° elbow and a half-open globe valve. Calculate the pressure drop in the pipe.

5.48 A particular flow network has 150 ft of 3-in. schedule 40 pipe, along with ten 90° standard radius elbows, one check valve, four tees (entering branch), and one gate valve, three-fourths open. Water at 65°F flows in the network at a rate of 250 gal/min. Calculate the frictional pressure drop in the piping network.

5.49 Through a manufacturing error, air-conditioning ducts of the cross-sectional shape shown in the sketch were delivered to a commercial building project. The project engineer decided to use them because the ductwork company provided a large discount from the estimated price. What is the equivalent diameter of the ducts?

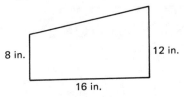

8 in. 12 in.

16 in.

5.50 A horizontal pipeline 6 miles long carries petroleum oil. The diameter of the pipe is 6 in. and the flow rate is 2.5 gal/sec. The viscosity of the oil is 0.045N sec/m² and the density of the oil is 890 kg/m³. Calculate the pressure drop in pounds per square inch for a cast iron pipe.

5.51 Air at $p = 1$ atm and 20 °C flows in a horizontal duct 100 m long. The volumetric flow rate is 10 m³/sec, and the pressure drop is not to exceed 300 Pa. Using sheet metal with $\varepsilon = 0.005$ m, would a square or a round duct use less metal?

5.52 A blower supplies air through an air-conditioning duct 4 × 8 in., 30 ft long. The blower imparts 0.30 kW to the flowing air. Find the volumetric flow rate of air through the duct, and the pressure just downstream of the blower.

SIX

ENERGY CONSERVATION IN COMBUSTION SYSTEMS

6.1 INTRODUCTION

Nearly half of the energy used in industry in the United States is consumed in combustion systems, either boilers or fired heaters, to provide process heat. In the late 1970s, about 9% of the nation's annual energy consumption went into the production of steam for industrial use and 10% went for direct process heat [1, 2]. Hence, improvements in the performance of fuel-burning systems are of prime importance in most industrial operations and can have a substantial impact on national energy consumption. From both the local and national perspectives, good energy conservation practice in industrial heating is especially important, since about 57% of this heat is produced by burning oil and gas [1]. The sharp rises in oil and gas prices in recent years and the likelihood of further increases in the future have provided an ample incentive to optimize combustion operations.

This chapter will introduce the basic principles of fuels and combustion and describe procedures for calculating boiler and heater performance and for carrying out tests of combustion equipment. Measures to improve combustion system performance will be discussed.

6.2 BOILER AND HEATER PERFORMANCE

6.2.1 Boiler Efficiency Defined

In evaluating the elements of combustion system performance, what causes this performance to degrade, and what can be done about it, it is helpful to think of a boiler or heater as a two-stage energy conversion system. Figure 6.1 illustrates the concept for a simple fire-tube boiler. Chemical energy in the fuel is released in the combustion process by rapid oxidation of the fuel, producing high-temperature products of combustion (flue gas) composed mostly of carbon dioxide, water vapor, nitrogen, and oxygen. The flue gas then passes into the heater section, transferring its energy by convection and radiation to the water to produce steam. The boiler efficiency, defined as the ratio of the useful heat in the product (in this case the heat content of the steam) to the energy content of the fuel, can be considered as the product of two components, the combustion efficiency and the heater efficiency, each reflecting one of the two stages of the energy conversion process:

$$\eta_B = \eta_C \times \eta_H$$

The combustion efficiency, η_c, reflects the extent to which the energy content of the fuel is converted to heat in the combustion gas. For a properly designed and operated burner, this efficiency should be very nearly 100%.

The heater efficiency represents the fraction of the heat content in the flue gas actually transferred to the steam, and this factor is typically well below 100% (70–80% is representative of an average industrial boiler).

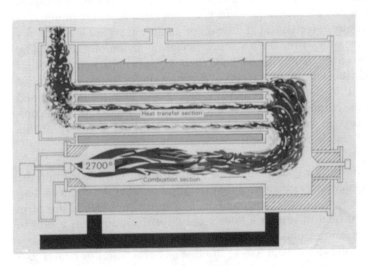

Figure 6.1 Concept of a boiler as an energy conservation system.

This factor is affected both by design, in terms of the amount of heat transfer surface provided, and by operating variables.

6.2.2 Factors Affecting Boiler Efficiency

Both operational and design parameters in combustion equipment affect its performance. In many cases, energy consumption can be reduced by simply recognizing those elements under the control of the operator and adjusting them to optimize efficiency; in other cases, equipment modifications must be carried out to realize improved performance. Some of the key variables affecting efficiency are as follows.

Fuel Condition To achieve maximum combustion efficiency, every molecule of combustible material in the fuel must react with the required number of oxygen molecules to effect complete chemical conversion. The energy content of unburned fuel is completely lost, regardless of how efficient the heater section is. Complete combustion requires proper treatment of the fuel before firing and good mixing with air in the combustion zone. Burning of gaseous fuels generally requires little attention to the condition of the fuel before firing, and good mixing is relatively easy to achieve. This is not necessarily the case with liquid and solid fuels. Fuel oil must be atomized to a fine mist to present a large surface area to the combustion air. Proper atomization is a function of the fuel viscosity, which in turn is strongly dependent on fuel temperature. Figure 6.2 illustrates this relationship. Simple rotary-type atomizers are particularly sensitive to fuel viscosity. While a medium-grade oil, such as No. 4, is low enough in viscosity to permit proper atomization at room temperature, a low-grade oil, such as No. 6, requires heating to properly atomize and, in fact, cannot even be easily pumped at room temperature. Mechanical and steam or air atomizers are less viscosity dependent, since they produce a stronger shearing action on the liquid by mechanical or aerodynamic means. Solid fuels, coals in particular, represent an especially critical problem in terms of preparation for firing, as they must first be reduced in size by mechanical grinding and then transported and injected into the combustion zone. The subject of coal preparation is a complicated one beyond the scope of this book; an excellent summary of this topic will be found in reference [3].

Excess Air Excess air is defined as the amount of air introduced into the combustion system over and above that required to meet the chemical balance oxygen requirements of the combustibles in the fuel. While this term will be covered quantitively in a later section, at this point it is useful to consider its meaning from a conceptual viewpoint.

Figure 6.3 shows the heat loss in the flue gas of a boiler or heater as a function of the amount of air used. If too little air is provided to oxidize all of

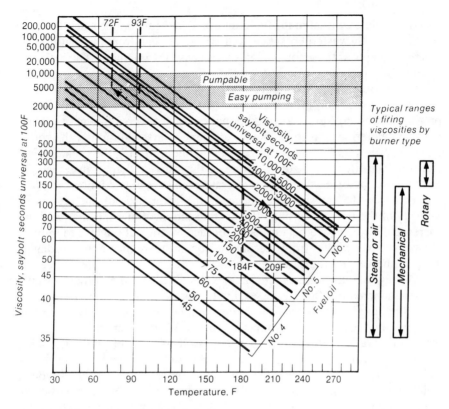

Figure 6.2 Atomization ranges for fuel oils.

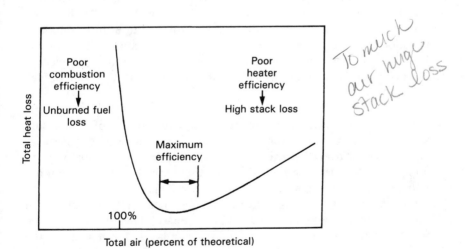

Figure 6.3 Energy loss as a function of excess air.

the combustible content of the fuel, unburned fuel goes up the stack. This is an exceptionally wasteful practice, and, as can be seen from the figure, the loss rises steeply as the amount of air drops below the minmum. This reflects a rapid drop in combustion efficiency.

On the other hand, once adequate oxygen has been provided to permit complete combustion, bringing in more air simply dilutes the combustion products without offering any additional energy conversion benefits. This has two deleterious effects on heater efficiency. First, dilution of the combustion gas lowers the temperature of the flame, reducing the radiant heat transfer. The effect can be quite strong, as the rate of radiant transfer varies as the fourth power of the flame temperature; thus, a little dilution can seriously reduce the transfer rate. Second, the additional air increases the velocity of the flue gas through the heater section. While this may slightly increase the convective heat transfer coefficient, the net effect is to reduce the convective transfer from the combustion gas by reducing the residence time of the gas in the heater section.

As Fig. 6.3 illustrates, increasing excess air above the optimum value causes an increase in flue gas losses, but the loss is not as severe as when too little air is used. For this reason, boilers and heaters are usually set to operate on the high side of the excess air curve. An important part of any energy conservation effort for combustion systems is determining the minimum air setting that will provide for complete combustion and maintaining system operation near that level at all times. Table 6.1 shows typical minimum excess air requirements for several common fuels. The particular value required for a given combustion system is a function not only of the fuel but also of the burner and combustion chamber design. Table 6.1 reflects the relative difficulty of adequately mixing combustion air with the various fuels; gas, which is easy to mix thoroughly, can be fired with a minimum of excess air, while more air must be provided to ensure complete combustion of liquid and solid fuels. Pulverized coal requires less air than stoker-fired coals, which are introduced into the furnace in larger pieces, since the former has a larger surface area for oxygen contact per unit of fuel fired.

Thermal Resistance of Heater Surfaces   The effectiveness of heater surfaces in transferring energy from the flue gas to the process fluid is a crucial component of the overall efficiency. The effectiveness is determined by the amount of heater surface provided and by the thermal resistance from the gas to the heated medium. This thermal resistance is a combination of several components. Starting with the flue gas, heat must be transferred through the "film resistance" from the gas to the solid heater surface, then through the metal wall to the process fluid side. A layer of soot or ash on the gas side can present an added source of resistance. The metal tube itself is generally highly conductive to heat in comparison with the other elements in the heat flow path, and for the most part the choice of tube wall thickness and material is determined by factors other than heat transfer limitations. In a steam boiler the boiling process is so effective that transfer from a clean tube surface to the water represents a very small resistance. This may not be the case, however, in a nonboiling process for low-conductivity fluids like hydrocarbons, and in this situation changing the flow conditions of the process fluid might significantly increase or decrease the process-side resistance. In either case, if a layer of mineral or organic scale should form on the process side, heat transfer can be significantly degraded. This ultimately shows up in the requirement for more fuel to obtain the same steam rate or process temperature.

Table 6.1 Usual Amount of Excess Air Supplied to Fuel-Burning Equipment

Fuel	Type of furnace or burners	Excess air (% by weight)
Pulverized coal	Completely water-cooled furnace for slag-tap or dry-ash removal	15–20
	Partially water-cooled furnace for dry-ash removal	15–40
Crushed coal	Cyclone furnace—pressure or suction	10–15
Coal	Spreader stoker	30–60
	Water-cooled vibrating-grate stoker	30–60
	Chain-grate and traveling-grate stokers	15–50
	Underfeed stoker	20–50
Fuel oil	Oil burners, register type	5–10
	Multifuel burners and flat-flame	10–20
Acid sludge	Cone and flat-flame type burners, steam-atomized	10–15
Natural, coke-oven, and refinery gas	Register-type burners	5–10
	Multifuel burners	7–12
Blast-furnace gas	Intertube nozzle-type burners	15–18
Wood	Dutch oven (10–23% through grates) and Hofft type	20–25
Bagasse	All furnaces	25–35
Black liquor	Recovery furnaces for kraft and soda-pulping processes	5–7

Source: [3].

✳ 6.3 PRINCIPLES OF COMBUSTION ANALYSIS

6.3.1 Basic Physical Laws Governing Combustion

From an energy utilization point of view, our main objective is to maximize for a given system the extent to which chemical energy in the fuel is converted into heat in the product. To accomplish this task, we must analyze the energy content of the combustion-side inputs (fuel and air) and outputs (especially the flue gas). Flue gas losses represent the primary source of inefficiency in a boiler or process heater, and by understanding the factors that contribute to these losses we can take action to reduce their magnitude.

Combustion processes are governed by several basic physical laws. The quantitative application of these laws will be taken up in a later section; here we will simply cite them and discuss their implications.

Conservation of Matter All matter that enters a system must leave or be stored inside; matter is never destroyed (at least not at the energy levels of interest to us here). In combustion analysis this principle comes into play in two ways. First, the total mass of fuel plus air (in pounds, for example) entering the system is the same as the total mass of flue gas exiting the stack. Second, individual atomic species are preserved, i.e., every atom of carbon entering the system leaves as an atom of carbon, every oxygen atom leaves as oxygen, and so on. Normally these atoms enter and leave in the form of chemical compounds such as methane (CH_4) or carbon dioxide (CO_2), or molecular species such as O_2 or N_2, and the molecular makeup changes from inlet to outlet. But conservation of matter requires that the same number of each type of atom that enters the process leave with the same identity. Thus, if we represent the conservation of matter principle in mathematical form, for N atomic species we will have N independent conservation relations.

Conservation of Energy Energy is neither lost nor destroyed, but it may be changed in form. Thus, we must be able to account for all energy flows in a system. Energy that flows into a combustion process in chemical form will be converted into thermal energy, which ultimately appears in the stack gas, in the process fluid being heated, in losses through the heater surface, and in blowdown or other discharges. Determination of boiler efficiency is largely a matter of careful energy accounting based on the principle of conservation of energy.

Ideal Gas Law While it may not be evident at first glance, all combustion processes occur in the gaseous phase. Liquid and solid fuels must be vaporized before chemical combination with oxygen takes place. Thus, all components taking part in a combustion reaction obey, to a good approximation, the ideal gas law, which states that the volume occupied by 1 mole of a gas (its molecular weight in mass units) is directly proportional to its temperature and inversely proportional to its pressure. In algebraic terms,

$$\bar{v} = \frac{\bar{R}T}{p}$$

where $\bar{v}$ is volume per mole (e.g., cubic feet per pound mole), p is pressure (e.g., pounds force per square foot), T is absolute temperature (e.g., degrees Rankine), and $\bar{R}$ is universal gas constant = 1545 ft lb_f/lb_m mole R.

For n moles of a gas, the ideal gas law can be expressed in terms of the total volume occupied by the gas,

$$PV = n\bar{R}T$$

Dalton's Law of Partial Pressures   Dalton's law states that in a mixture of ideal gases each component gas contributes a "partial pressure"; the sum of the partial pressures is equal to the total pressure of the mixture. Each partial pressure is the pressure that the component gas would exert if it alone occupied the entire volume of the mixture. To better understand Dalton's law and its implications, imagine a mixture of two gases, 1 and 2, with partial pressure p_1 and p_2, occupying a volume V at uniform temperature T. Let n_1 and n_2 be the numbers of moles of components 1 and 2, respectively, and n be the total number of moles in the mixture ($n = n_1 + n_2$). Applying the ideal gas law to each component and using the definition of partial pressure from Dalton's law,

$$\frac{p_1}{n_1} = \frac{\bar{R}T}{V}$$

$$\frac{p_2}{n_2} = \frac{\bar{R}T}{V}$$

Furthermore, we can write the ideal gas law for the mixture as a whole,

$$\frac{p}{n} = \frac{\bar{R}T}{V}$$

Since the right-hand side is the same for all three cases, or

$$\frac{p_1}{p} = \frac{n_1}{n} \qquad \frac{p_2}{p} = \frac{n_2}{n}$$

In other words, the partial pressure fraction of each component of a gas mixture is identical to the mole fraction of that component. As we will see later, this fact will be useful in combustion testing; by measuring partial pressures of gaseous components (e.g., in the flue gas of a boiler), we can infer from Dalton's law the makeup of the mixture in molar terms.

Amagat's Law of Partial Volumes Amagat's law is the volumetric equivalent of Dalton's law. It states that each component of a gas mixture may be thought of as contributing a partial volume; the sum of the partial volumes is the total volume of the mixture. Each partial volume is equal to the volume the particular component would occupy if it were compressed to the total pressure of the mixture while maintaining constant temperature. Following a development similar to the one above, we can easily show that for a gas mixture of i components

$$\frac{n_1}{n} = \frac{V_1}{V} \quad \frac{n_2}{n} = \frac{V_2}{V} \cdots \frac{n_i}{n} = \frac{V_i}{V}$$

that is, the mole fraction of each component is equal to its volume fraction in the mixture. One useful application of this result is in the analysis of gaseous fuels, in which the amount of each component is usually represented in terms of its percent by volume. From Amagat's law we can see that if every $100 \, ft^3$ of the mixture contains $20 \, ft^3$ of component i, then every 100 moles of the mixture will contain 20 moles of i.

$$\frac{n_i}{n} = \frac{V_i}{V} = \frac{20}{100}$$

Avogadro's Law Avogadro's law states that 1 mole of any ideal gas at a given temperature and pressure contains the same number of molecules as 1 mole of any other ideal gas under the same conditions. The importance of Avogadro's law is that it permits us to move from the microscopic representation of chemical reactions to the macroscopic level. For example, consider the simple combustion reaction

$$2H_2 + O_2 \rightarrow 2H_2O$$

This relation states that two molecules of hydrogen will react with one molecule of oxygen to form two molecules of water. It would be cumbersome indeed if we had to keep track of the number of molecules introduced into a combustion chamber to predict the resulting product mix. Avogadro's law tells us that if all constituents are at the same temperature and pressure, equal numbers of moles contain equal numbers of molecules. Hence we can interpret the coefficients in the chemical balance in macroscopic terms (moles) rather than molecular terms, i.e., 2 moles of hydrogen will react with 1 mole of oxygen to form 2 moles of water. We can easily translate this into mass terms. Since the molecular weight of molecular hydrogen (H_2) is 2, molecular oxygen (O_2) is 32, and water is 18, we can state that 4 lb_m of hydrogen (2 × 2 lb_m) and 32 lb_m of oxygen will produce 36 lb_m of water (2 × 18 lb_m).

Table 6.2 lists a number of chemical reactions commonly encountered in combustion. Most of these reactions involve the oxidation of hydrocarbons to carbon dioxide and water; where sulfur is present, sulfur dioxide is the most common product of combustion. Each reaction is shown in terms of its chemical equation, which implicitly states the conservation of matter for each atomic species, its molar equivalence (an implication of Avogadro's law), and its overall mass balance. We will make extensive use of all of the principles discussed above in performing combustion calculations in a subsequent section.

✳ 6.3.2 Air as a Source of Oxygen for Combustion

Since combustion is, by definition, oxidation of the components of a fuel, oxygen is a basic constituent in every combustion reaction. Pure oxygen, however, is used only in special applications, such as cutting and welding. In most industrial combustion, air provides the needed oxygen. Table 6.3 shows the composition of air generally accepted for combustion analysis. On a volumetric basis, air is approximately 21% oxygen, 78% nitrogen, and 1% inert components such as helium, neon, and argon. On a weight basis, approximately 23% is oxygen and 73% nitrogen.

Recall that from Amagat's law the volume fraction of each component of an ideal gas (such as air at atmospheric conditions) is also equal to its mole fraction. Therefore, 100 moles of air will contain approximately 21 moles of O_2 and 78 moles of N_2. Stating this another way, as shown in the lower part of Table 6.3, for every mole of oxygen required in a combustion process, 4.76 moles of air must be provided. Since volume ratios and molar ratios are equivalent, every cubic foot of oxygen will require 4.76 cubic feet of air; 3.76 moles (or cubic feet, as the case may be) of N_2 will "come along for the ride," even though it is not a major participant in the combustion process. (It should be noted that from an air pollution standpoint the formation of even trace amounts of oxides of nitrogen in combustion may be important. These, however, have a negligible influence on energy utilization efficiency and will not be addressed here.) On a mass basis, every 1 lb_m of oxygen requires 4.32 lb_m of dry air, 3.32 lb_m of which is nitrogen.

The equivalent molecular weight of a gas mixture such as air is simply the average of the molecular weights of the consitituents weighted by their mole fractions. Oxygen has a molecular weight of 32 and nitrogen, 28. Thus the equivalent molecular weight of air can be computed as

$$MW = (0.21)(32) + (0.78)(28) = 29$$

A lb_m-mole of air weighs about 29 lb_m.

The "dry air" described above is assumed to contain no water vapor.

Table 6.2 Common Chemical Reactions of Combustion

Combustible	Reaction	Moles	Pounds	Heat of combustion (high) (Btu/lb_m of fuel)
Carbon (to CO)	$2C + O_2 = 2CO$	$2 + 1 = 2$	$24\ lb_m + 32\ lb_m = 56\ lb_m$	4,000
Carbon (to CO_2)	$C + O_2 = CO_2$	$1 + 1 = 1$	$12 + 32 = 44$	14,100
Carbon monoxide	$2CO + O_2 = 2CO_2$	$2 + 1 = 2$	$56 + 32 = 88$	4,345
Hydrogen	$2H_2 + O_2 = 2H_2O$	$2 + 1 = 2$	$4 + 32 = 36$	61,100
Sulfur (to SO_2)	$S + O_2 = SO_2$	$1 + 1 = 1$	$32 + 32 = 64$	3,980
Methane	$CH_4 + 2O_2 = CO_2 + 2H_2O$	$1 + 2 = 1 + 2$	$16 + 64 = 80$	23,875
Acetylene	$2C_2H_2 + 5O_2 = 4CO_2 + 2H_2O$	$2 + 5 = 4 + 2$	$52 + 160 = 212$	21,500
Ethylene	$C_2H_4 + 3O_2 = 2CO_2 + 2H_2O$	$1 + 3 = 2 + 2$	$28 + 96 = 124$	21,635
Ethane	$2C_2H_6 + 7O_2 = 4CO_2 + 6H_2O$	$2 + 7 = 4 + 6$	$60 + 224 = 284$	22,325
Hydrogen sulfide	$2H_2S + 3O_2 = 2SO_2 + 2H_2O$	$2 + 3 = 2 + 2$	$68 + 96 = 164$	7,100

Table 6.3 Composition of Air

	Composition of dry air	
	% by volume	% by weight
Oxygen, O_2	20.99	23.15
Nitrogen, N_2	78.03	76.85
Inerts	0.98	—

Equivalent molecular weight of air = 29.0
Moisture — 1.3% by weight (standard for the boiler industry—ABMA)

$$\frac{\text{moles air}}{\text{mole oxygen}} = \frac{\text{ft}^3 \text{ air}}{\text{ft}^3 \text{ oxygen}} = \frac{100}{20.99} = 4.76$$

$$\frac{\text{moles } N_2}{\text{mole oxygen}} = \frac{79.01}{20.99} = 3.76$$

$$\frac{\text{lb}_m \text{ air (dry)}}{\text{lb } O_2} = \frac{100}{23.15} = 4.32$$

$$\frac{\text{lb}_m \text{ } N_2}{\text{lb}_m \text{ } O_2} = \frac{76.85}{23.15} = 3.32$$

Atmospheric air, in fact, always contains a small fraction of water vapor, the amount indicated by the humidity of the atmosphere at the particular location and time. Determination of this water vapor content from a psychrometric chart will be covered in a later section. In the absence of data, however, the American Boiler Manufacturers' Association has defined a "standard" moisture content of 1.3% by weight for use in combustion analysis.

6.4 COMBUSTION CONSTANTS

6.4.1 Tabulation of Combustion Parameters

Many of the concepts discussed in the preceding sections are summarized quantitatively in Table 6.4. This table contains constants for the combustion of a number of common fuel constituents with air and can be most useful in determining air requirements and flue gas composition. The first group of data includes the chemical formula for the substance, its molecular weight, its density (lb_m/ft^3), specific volume (ft^3/lb_m), and specific gravity (ratio of its density to the density of air).

The second data group designates the heats of combustion (more commonly termed heating values) of each substance. Heating value is the

Table 6.4 Combustion Constants

No.	Substance	Formula	Molecular weight	lb_m per ft³	ft³ per lb_m	Sp. gr. of air = 1.0000	Heat of combustion Btu per ft³ Gross (high)	Net (low)	Btu per lb_m Gross (high)	Net (low)	Required for combustion O_2	N_2	Air	Flue products CO_2	H_2O	N_2	Required for combustion O_2	N_2	Air	Flue products CO_2	H_2O	N_2
											For 100% total air moles per mole of combustible or ft³ per ft³ of combustible						For 100% total air lb_m per lb_m of combustible					
1	Carbon[a]	C	12.01	—	—	—	—	—	14,093	14,093	1.0	3.76	4.76	1.0	—	3.76	2.66	8.86	11.53	3.66	—	8.86
2	Hydrogen	H_2	2.016	0.0053	187.723	0.0696	325	275	61,095	51,623	0.5	1.88	2.38	—	1.0	1.88	7.94	26.41	34.34	—	8.94	26.41
3	Oxygen	O_2	32.00	0.0846	11.819	1.1053	—	—	—	—												
4	Nitrogen (atm)	N_2	28.01	0.0744	13.443	0.9718	—	—	—	—												
5	Carbon monoxide	CO	28.01	0.0740	13.506	0.9672	321	321	4,347	4,347	0.5	1.88	2.38	1.0	—	1.88	0.57	1.90	2.47	1.57	—	1.90
6	Carbon dioxide	CO_2	44.01	0.1170	8.548	1.5282	—	—	—	—												
Paraffin series																						
7	Methane	CH_4	16.04	0.0425	23.552	0.5543	1,012	911	23,875	21,495	2.0	7.53	9.53	1.0	2.0	7.53	3.99	13.28	17.27	2.74	2.25	13.28
8	Ethane	C_2H_6	30.07	0.0803	12.455	1.0488	1,773	1,622	22,323	20,418	3.5	13.18	16.68	2.0	3.0	13.18	3.73	12.39	16.12	2.93	1.80	12.39
9	Propane	C_3H_8	44.09	0.1196	8.365	1.5617	2,524	2,322	21,669	19,937	5.0	18.82	23.82	3.0	4.0	18.82	3.63	12.07	15.70	2.99	1.63	12.07
10	n-Butane	C_4H_{10}	58.12	0.1582	6.321	2.0665	3,271	3,018	21,321	19,678	6.5	24.47	30.97	4.0	5.0	24.47	3.58	11.91	15.49	3.03	1.55	11.91
11	Isobutane	C_4H_{10}	58.12	0.1582	6.321	2.0665	3,261	3,009	21,271	19,628	6.5	24.47	30.97	4.0	5.0	24.47	3.58	11.91	15.49	3.03	1.55	11.91
12	n-Pentane	C_5H_{12}	72.15	0.1904	5.252	2.4872	4,020	3,708	21,095	19,507	8.0	30.11	38.11	5.0	6.0	30.11	3.55	11.81	15.35	3.05	1.50	11.81
13	Isopentane	C_5H_{12}	72.15	0.1904	5.252	2.4872	4,011	3,708	21,047	19,459	8.0	30.11	38.11	5.0	6.0	30.11	3.55	11.81	15.35	3.05	1.50	11.81
14	Neopentane	C_5H_{12}	72.15	0.1904	5.252	2.4872	3,994	3,692	20,978	19,390	8.0	30.11	38.11	5.0	6.0	30.11	3.55	11.81	15.35	3.05	1.50	11.81
15	n-Hexane	C_6H_{14}	86.17	0.2274	4.398	2.9074	4,768	4,415	20,966	19,415	9.5	35.76	45.26	6.0	7.0	35.76	3.53	11.74	15.27	3.06	1.46	11.74
Olefin series																						
16	Ethylene	C_2H_4	28.05	0.0742	13.475	0.9740	1,604	1,503	21,636	20,275	3.0	11.29	14.29	2.0	2.0	11.29	3.42	11.39	14.81	3.14	1.29	11.39
17	Propylene	C_3H_6	42.08	0.1110	9.007	1.4504	2,340	2,188	21,048	19,687	4.5	16.94	21.44	3.0	3.0	16.94	3.42	11.39	14.81	3.14	1.29	11.39
18	n-Butene	C_4H_8	56.10	0.1480	6.756	1.9336	3,084	2,885	20,854	19,493	6.0	22.59	28.59	4.0	4.0	22.59	3.42	11.39	14.81	3.14	1.29	11.39
19	Isobutene	C_4H_8	56.10	0.1480	6.756	1.9336	3,069	2,868	20,737	19,376	6.0	22.59	28.59	4.0	4.0	22.59	3.42	11.39	14.81	3.14	1.29	11.39
20	n-Pentene	C_5H_{10}	70.13	0.1852	5.400	2.4190	3,837	3,585	20,720	19,359	7.5	28.23	35.73	5.0	5.0	28.23	3.42	11.39	14.81	3.14	1.29	11.39
Aromatic series																						
21	Benzene	C_6H_6	78.11	0.2060	4.852	2.6920	3,752	3,601	18,184	17,451	7.5	28.23	35.73	6.0	3.0	28.23	3.07	10.22	13.30	3.38	0.69	10.22
22	Toluene	C_7H_8	92.13	0.2431	4.113	3.1760	4,486	4,285	18,501	17,672	9.0	33.88	42.88	7.0	4.0	33.88	3.13	10.40	13.53	3.34	0.78	10.40
23	Xylene	C_8H_{10}	106.16	0.2803	3.567	3.6618	5,230	4,980	18,650	17,760	10.5	39.52	50.02	8.0	5.0	39.52	3.17	10.53	13.70	3.32	0.85	10.53
Miscellaneous gases																						
24	Acetylene	C_2H_2	26.04	0.0697	14.344	0.9107	1,477	1,426	21,502	20,769	2.5	9.41	11.91	2.0	1.0	9.41	3.07	10.22	13.30	3.38	0.69	10.22
25	Naphthalene	$C_{10}H_8$	128.16	0.3384	2.955	4.4208	5,854	5,854	17,303	16,708	12.0	45.17	57.17	10.0	4.0	45.17	3.00	9.97	12.96	3.43	0.56	9.97
26	Methyl alcohol	CH_3OH	32.04	0.0846	11.820	1.1052	868	767	10,258	9,066	1.5	5.65	7.15	1.0	2.0	5.65	1.50	4.98	6.48	1.37	1.13	4.98
27	Ethyl alcohol	C_2H_5OH	46.07	0.1216	8.221	1.5890	1,600	1,449	13,161	11,917	3.0	11.29	14.29	2.0	3.0	11.29	2.08	6.93	9.02	1.92	1.17	6.93
28	Ammonia	NH_3	17.03	0.0456	21.914	0.5961	441	364	9,667	7,985	0.75	2.82	3.57	—	1.5	3.32	1.41	4.69	6.10	—	1.59	5.51
29	Sulfur[a]	S	32.06	—	—	—	—	—	3,980	3,980	1.0	3.76	4.76	SO_2 1.0	—	3.76	1.00	3.29	4.29	SO_2 2.00	—	3.29
30	Hydrogen sulfide	H_2S	34.08	0.0911	10.979	1.1898	646	595	7,097	6,537	1.5	5.65	7.15	1.0	1.0	5.65	1.41	4.69	6.10	1.88	0.53	4.69
31	Sulfur dioxide	SO_2	64.06	0.1733	5.770	2.2640	—	—	—	—												
32	Water vapor	H_2O	18.02	0.0476	21.017	0.6215	—	—	—	—												
33	Air	—	18.02	0.0766	13.063	1.0000	—	—	—	—												

[a] Carbon and sulfur are considered as gases for molal calculations only.
[b] All gas volumes corrected to 60°F and 30 in. Hg dry.
Source: [3].

176

heat liberated when a substance completely reacts with oxygen to form CO_2 and water (and, in the case of sulfur-bearing compounds, SO_2). At combustion temperatures the water formed is, of course, in the vapor state. In certain types of fuel tests, however, the products of combustion are reduced to ambient temperature, leaving the water in liquid form. In these tests the latent heat of vaporization of water adds to the measured heat of the reaction. For clarity, therefore, two different heating values are designated, the higher (or gross) heating value, which includes the latent heat of the water vapor in the combustion products, and the lower (or net) heating value, which does not include the latent heat content. Heating values are always presented on a per-unit basis, usually either Btu's per pound mass of fuel or Btu's per cubic foot. Since water vapor is formed only when hydrogen is present, there is no difference between gross and net heating value for fuels such as carbon and carbon monoxide, which contain no hydrogen.

The third group of constants shows, on a molar or volumetric basis, the amounts of oxygen, nitrogen, and air used to burn the combustible substance, and the resulting products of combustion. To illustrate the origin of these constants and their meaning, consider the reaction for methane burning in pure oxygen:

$$CH_4 + 2O_2 \rightarrow CO_2 + 2H_2O$$

For each mole of methane burned, 2 moles of oxygen will be required, and 1 mole of carbon dioxide and 2 moles of water will be produced. Alternatively, by Amagat's law, for each cubic foot of methane, 2 ft^3 of oxygen will be required, and so on. If air is used to supply the oxygen, nitrogen will also be present in both the reactants and the products, in an amount of 3.76 moles (or cubic feet) for each mole (or cubic foot) of oxygen. Thus, the reaction in air can be written:

$$CH_4 + 2O_2 + 2(3.76)N_2 \rightarrow CO_2 + 2H_2O + 2(3.76)N_2$$

Even though the nitrogen plays no explicit role in the combustion reaction, it is important from the standpoint of combustion system design and analysis, since its mass is a major fraction of the overall mass of the reactants and products. The constants shown in this section of Table 6.4 simply give the values of the coefficients of each term in the combustion reaction for each substance of interest.

The last section of Table 6.4 shows similar constants, but converted into mass terms. For example, for the simple methane/oxygen reaction shown above, if each coefficient is multiplied by the molecular weight of the particular constituent, the equation can be rewritten:

$$(16.04 \text{ lb}_m)CH_4 + (64 \text{ lb}_m)O_2 \rightarrow (44.01 \text{ lb}_m)CO_2 + (36.04 \text{ lb}_m)H_2O$$

Dividing each term by 16.04 to normalize it to a "per pound of combustible" basis, we get the constants 3.99, 2.74, and 2.25 for oxygen, carbon dioxide, and water, as shown in the table. Similarly, we could include the nitrogen content of the air to get this constant as well. Although this section really does not provide any "new" data that are not directly derivable from other parts of the table, it is convenient to have the figures already available in mass terms for certain types of problems, as will be illustrated in the following section.

6.4.2 Application of Combustion Constants

Use of the combustion constants described above can best be illustrated through a couple of examples. The first of these is for a gaseous fuel in which the fuel analysis is given in volumetric terms for the compounds contained in the fuel (*proximate* analysis). In the second case a liquid fuel is used, and the analysis is presented in weight percentages on an elemental constituent basis (*ultimate* analysis).

Example 6.1 Natural Gas; Proximate Analysis

Suppose that a typical natural gas has the following analysis in percentages by volume:

Methane (CH_4)	84.1%
Ethane (C_2H_6)	6.7%
Carbon dioxide (CO_2)	0.8%
Nitrogen (N_2)	8.4%

Using the combustion constants in Table 6.4, we wish to determine the amount of theoretical air (i.e., not considering excess air) required for combustion of this fuel and the flue gas products produced. The calculations are summarized in Table 6.5.

We first list each fuel component and its percentage by volume (cubic feet of the component per 100 ft³ of fuel) and then determine the air requirement for each component. For methane, for example, we note from Table 6.4 that 9.53 ft³ of air is required to burn 1 ft³ of methane. For the methane in 100 ft³ of our fuel, therefore, we will require 84.1 × 9.53, or 801.5 ft³ of air. Similarly, for the ethane, 16.68 ft³ of air is required per cubic foot of ethane for a total of 111.8 ft³. CO_2 and N_2 do not burn, but are simply passed through into the flue gas; therefore, no air is required for these constituents. (However, when we consider energy balances later, the energy required to heat these constituents up to the flue gas temperature will be important, so we must be careful to account for them in the mass balance here.) Summing the air requirements determined above, the total theoretical air for combustion is 913.3 ft³ per 100 ft³ of fuel.

The flue gas composition can now be determined. From Table 6.4, each cubic foot of methane will produce 1 ft³ of CO_2, 2 ft³ of H_2O, and 7.53 ft³ of N_2 (carried through from the combustion air). Multiplying each constant by 84.1, we determine the flue gas products produced by the methane: 84.1, 168.2, and 633.3 ft³ of CO_2, H_2O, and N_2, respectively per 100 ft³ of fuel. The products of combustion for ethane are determined in a similar way. Note that the flue gas constants for CO_2 and N_2 in the fuel are just 1.0, since these components pass through directly. Summing up the components, it is interesting to note that the total volume of flue gas is about 10 times that of the fuel itself and that over 70% is nitrogen, most of which is "extra baggage" brought in with the combustion air.

Example 6.2 Fuel Oil; Ultimate Analysis

In this example the analysis of the fuel is presented in weight percent rather than volume

Table 6.5   Combustion Products Determination (Material Balances)

Example 6.1: Natural Gas, Proximate Analysis
Fuel analysis (percent by volume):

Methane (CH_4)	84.1
Ethane (C_2H_6)	6.7
Carbon dioxide (CO_2)	0.8
Nitrogen (N_2)	8.4

Constituent	$ft^3/100\ ft^3$ fuel	ft^3 air/ 100 ft^3 fuel	Flue gas, $ft^3/100\ ft^3$ fuel CO_2	H_2O	N_2
CH_4	84.1	84.1 × 9.53 = 801.5	84.1 × 1 = 84.1	84.1 × 2 = 168.2	84.1 × 7.5 = 633.3
C_2H_6	6.7	6.7 × 16.68 = 111.8	6.7 × 2 = 13.4	6.7 × 3 = 20.1	6.7 × 13.1 = 88.3
CO_2	0.8	—	0.8 × 1 = 0.8	—	—
N_2	8.4	—	—	—	8.4 × 1 = 8.4
Total	100 ft^3 fuel	913.3 ft^3 air	98.3 ft^3 CO_2	188.3 ft^3 H_2O	730 ft^3 N_2

percent. Also, the ultimate analysis of the fuel is used, i.e., the fuel constituents are presented in terms of elemental components, such as carbon, hydrogen, and oxygen, rather than compounds such as methane and ethane. This type of analysis is more typical of liquid and solid fuels, whereas proximate analysis on a volumetric basis is more common for gaseous fuels. For the particular fuel oil to be analyzed here, the analysis, in percent by weight, is as follows:

Sulfur (S)	2.3%
Hydrogen (H_2)	9.7%
Carbon (C)	85.6%
Oxygen (O_2)	2.0%
Ash	0.4%

Table 6.6 lists the constituents and their percentages (this time in pounds mass per 100 lb_m of fuel). From Table 6.4, the air required for each constituent is determined and multiplied by the weight of that constituent to determine its total contribution to the overall theoretical air requirement. When the fuel contains oxygen, however, it must be treated a little differently. Every pound of oxygen contained in the fuel means one less pound required externally from combustion air. Hence, for every pound of oxygen, 4.32 lb less air is required.

Determination of the products in the flue gas is similar to Example 6.1, except that the section of Table 6.4 headed "lb_m per lb_m of combustible" is used, rather than the constants used above. Note that the oxygen in the fuel contributes a negative value of −3.32 lb_m per 100 lb_m of fuel to the nitrogen in the flue gas, since this oxygen reduces the combustion air requirement. Also, the sulfur in the fuel produces SO_2; this flue gas component is listed in the CO_2 column of Table 6.4 to save space, since it is only produced in the case of sulfur-bearing fuels.

Table 6.6 Combustion Products Determination (Material Balances)

Example 6.2: Fuel Oil, Ultimate Analysis
Fuel analysis (percent by weight):
Sulfur (S)   2.3
Hydrogen (H_2)   9.7
Carbon (C)   85.6
Oxygen (O_2)   2.0
Ash   0.4

Constituent	lb_m/100 lb fuel	Air (lb_m)/100 lb_m fuel	Flue gas lb_m/100 lb_m fuel			
			CO_2	H_2O	N_2	SO_2
S	2.3	$2.3 \times 4.29 = 9.88$	—	—	$2.3 \times 3.29 = 7.6$	$2.3 \times 2 = 4.6$
H_2	9.7	$9.7 \times 34.34 = 333$	—	$9.7 \times 8.94 = 87$	$9.7 \times 26.41 = 256$	—
C	85.6	$85.6 \times 11.53 = 987$	$85.6 \times 3.66 = 313$	—	$85.6 \times 8.86 = 758$	—
O_2	2.0	$-2.0 \times 4.32 = -8.64$	—	—	$-2 \times 3.32 = -6.64$	—
Ash	0.4	—	—	—	—	—
Total	100	1321 lb_m air	313 lb_m CO_2	87 lb_m H_2O	1015 lb_m N_2	4.6 lb_m SO_2

6.5 FUELS AND THEIR PROPERTIES (skip)

The physical and chemical properties of fuels have a profound influence on the efficiency of combustion. Most industrial fuels are complex mixtures of gases, liquids, or solids. To determine the air requirements for proper combustion, the products of combustion, and the heat liberated, it is necessary to know in detail the composition of the fuel and its heating values; proper preparation for firing requires knowledge of density and viscosity. By convention, properties of different types of fuels are presented in different ways. In this section we examine the properties of many of the common fuels used in industrial boilers and process heaters.

6.5.1 Gaseous Fuels

Table 6.7 lists the properties of a number of representative natural gases in the United States. The analysis (i.e., composition) of each gas is shown in two ways: the proximate (or constituent) analysis and the ultimate (or elemental) analysis. The former is the most common for gaseous fuels and is always presented on a volumetric basis. All of the gases shown are so-called dry gas. When applied to natural gas, the terms wet and dry refer not to water content but to the presence of hydrocarbon liquids entrained in the gas. Natural gas produced in conjunction with petroleum reservoirs is typically wet at the wellhead, while gas from non-oil-producing reservoirs is dry. In either case, liquids are usually separated out at the wellhead before gas is delivered to the pipeline.

Referring to the constituent analysis, it is seen that the predominant components of all natural gas are methane and ethane, with small amounts of other combustible and inert components. The ultimate analysis is simply the elemental breakdown by weight of the compounds listed above, with carbon and hydrogen as the principal components. From the table of combustion constants (Table 6.4), it will be noted that the higher heating value of hydrogen is more than four times that of carbon; thus, it is not surprising that the higher heating value for the fuel as a whole is correlated closely with its hydrogen content. On a volumetric basis, higher heat values for natural gas are typically on the order of 1,000 to 1,100 Btu/ft^3; as a rule, therefore, it takes about a thousand cubic feet to produce a million Btu's.

Table 6.7 also shows the specific gravity of each gas, i.e., its density (lb$_m$/ft^3) divided by the density of air. Note that the density of natural gas is a little over half that of air.

Table 6.8 shows characteristics of fuels produced in the Lurgi process, the oldest and most widely used commercial process for coal gasification. While the gas composition varies slightly with the type of fuel used, carbon monoxide and hydrogen are the principal combustible components in the steam-oxygen process. With air as an oxidizer in the gasifier, these compo-

Table 6.7 Selected Samples of Natural Gas from U.S. Fields

	Sample no. and source of gas				
	1 Pennsylvania	2 Southern California	3 Ohio	4 Louisiana	5 Oklahoma
Analyses					
Constituents, percent by volume					
H_2 Hydrogen	—	—	1.82	—	—
CH_4 Methane	83.40	84.00	93.33	90.00	84.10
C_2H_4 Ethylene	—	—	0.25	—	—
C_2H_6 Ethane	15.80	14.80	—	5.00	6.70
CO Carbon monoxide	—	—	0.45	—	—
CO_2 Carbon dioxide	—	0.70	0.22	—	0.80
N_2 Nitrogen	0.80	0.50	3.40	5.00	8.40
O_2 Oxygen	—	—	0.35	—	—
H_2S Hydrogen sulfide	—	—	0.18	—	—
Ultimate, percent by weight					
S Sulfur	—	—	0.34	—	—
H_2 Hydrogen	23.53	23.30	23.20	22.68	20.85
C Carbon	75.25	74.72	69.12	69.26	64.84
N_2 Nitrogen	1.22	0.76	5.76	8.06	12.90
O_2 Oxygen	—	1.22	1.58	—	1.41
Specific gravity (relative to air)	0.636	0.636	0.567	0.600	0.630
Higher heat value					
Btu/ft^3 at 60°F and 30 in. Hg	1,129	1,116	964	1,002	974
Btu/lb$_m$ of fuel	23,170	22,904	22,077	21,824	20,160

Source: [3].

Table 6.8 Typical Lurgi Gas Composition

Gas component	Anthracite	Steam-oxygen process, gas produced (vol. %) Bituminous coal	Lignite	Steam-air process, gas produced (vol. %)
CO_2	28.5	27.5	32.2	13.3
CO	21.0	21.0	18.1	13.3
H_2	42.9	41.0	37.0	19.6
CH_4	7.0	8.8	12.0	5.5
N_4	0.4	0.4	0.3	37.5
H_2O				10.1
COS				0.1
H_2S				0.6
Heating value, Btu	295			180

nents are diluted by the presence of a large fraction of nitrogen, and a significant amount of water is also produced. Fuel heating values are low, typically ranging from 150 to 300 Btu/ft^3. All coal gasification processes involve partial combustion of the combustibles in the coal, in some cases followed by a "shift reaction" with steam to obtain a higher level of hydrogen and hydrocarbons in the product. Heating values are generally lower than those for natural gas, although the magnitude may vary widely depending on the degree of shift used in the process.

6.5.2 Liquid Fuels

Most liquid fuels are complex mixtures of hydrocarbons. This complexity makes it desirable to classify fuel oils in general categories, or grades, whose properties fall within certain specified limits. Table 6.9 shows a breakdown of fuel oil properties by oil grade.

Numbers 1 and 2 fuel oils are called light distillates because they primarily contain the lighter hydrocarbons separated out in the distillation of

Table 6.9 Range of Analyses of Fuel Oils

	Grade of fuel oil				
	No. 1	No. 2	No. 4	No. 5	No. 6
Weight, percent					
Sulfur	0.01–0.5	0.05–1.0	0.2–2.0	0.5–3.0	0.7–3.5
Hydrogen	13.3–14.1	11.8–13.9	(10.6–13.0)[a]	(10.5–12.0)[a]	(9.5–12.0)[a]
Carbon	85.9–86.7	86.1–88.2	(86.5–89.2)[a]	(86.5–90.2)[a]	(86.5–90.2)[a]
Nitrogen	Nil–0.1	Nil–0.1	—	—	—
Oxygen	—	—	—	—	—
Ash	—	—	0–0.1	0–0.1	0.01–0.5
Gravity					
Deg. API	40–44	28–40	15–30	14–22	7–22
Specific	0.825–0.806	0.887–0.825	0.966–0.876	0.972–0.922	1.022–0.922
lb$_m$ per gallon	6.87–6.71	7.39–6.87	8.04–7.30	8.10–7.86	8.51–7.68
Pour point, °F	0 to −50	0 to −40	−10 to +50	−10 to +80	+15 to +85
Viscosity					
Centistokes at 100 °F	1.4–2.2	1.9–3.0	10.5–65	65–200	260–750
SUS at 100 °F	—	32–38	60–300	—	—
SSF at 122 °F	—	—	—	20–40	45–300
Water and sediment, vol. %	—	0–0.1	Trace to 1.0	0.05–1.0	0.05–2.0
Heating value Btu per lb$_m$/gross (calculated)	19,670–19,860	19,170–19,750	18,280–19,400	18,100–19,020	17,410–18,990

[a]Estimated.
Source: [3].

crude oil. Grades 4 through 6 are termed residual oils, since they are mainly composed of the heavier components left in the bottom of the distillation column. Light distillate oils tend to be higher in hydrogen content and lower in sulfur than residual oils, and contain no ash (inorganic mineral residue). Because they are hydrogen-rich, they are also slightly higher in heating value. As a rule, the higher heating values of fuel oils are 18,000–19,000 Btu/lb_m.

The density of fuel oil is a good indicator of its grade; Fig. 6.4 shows the correlation between higher heating value and density, alternatively expressed in pounds per gallon, specific gravity (i.e., density relative to pure water), and API degrees. The latter measure, established by the American Petroleum Institute, is commonly used for petroleum products. Its magnitude varies inversely with the density, i.e., a higher density fluid will have a lower gravity in degrees API.

The viscosity of fuel oil is an extremely important property, as it affects both the ability to pump the fuel to the burner and the fuel's atomizing characteristics. As shown in Table 6.9, viscosity varies widely with oil grade; while distillate oils have viscosities not unlike that of water, heavy residuals are extremely viscous and usually must be heated to permit proper handling. Figure 6.5 shows the variation of viscosity with temperature. Taking the range of 2,000–5,000 SUS as a good estimate for easy pumpability, it is evident that the lighter oils can be satisfactorily delivered to the burner even at low temperatures, while heavy oils, such as No. 6, would typically need to be heated to 100°F or more for pumping.

6.5.3 Properties of Coal

Coal is one of the most abundant fossil fuels, particularly in the United States. Although at one time it was the most common industrial fuel, the artificially low price of oil and gas over the past several decades has significantly reduced its use in many industrial sectors. As liquid and gaseous fuels have become more scarce, however, the importance of coal has again begun to increase.

Coal is one of the most complicated of the commonly used fuels, both because handling solids is considerably more difficult than handling liquids and gases and because its chemical makeup is so variable. Efficient combustion of coal requires detailed knowledge of its properties and an understanding of how these properties affect its burning and heat transfer characteristics.

Table 6.10 shows the classification system for coals set up by the American Society for Testing of Materials (ASTM). While other classification systems are in use, this one is most commonly used in the United States. Coals are divided in a general way into classes and into groups within each class. The physical and chemical properties of a coal determine its

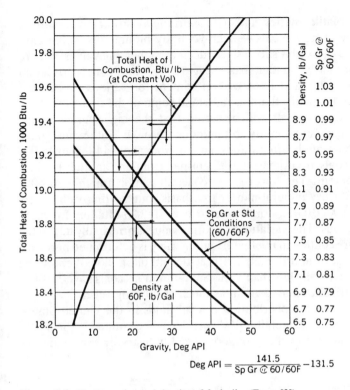

$$\text{Deg API} = \frac{141.5}{\text{Sp Gr @ 60/60F}} - 131.5$$

Figure 6.4 Heating value and density of fuel oils. (From [3])

classification. Properties of primary interest from a combustion standpoint are the fixed carbon, volatile matter, and heating value. Volatile matter consists of hydrocarbons and other gases (not including water vapor) that are driven off when the coal is heated in a standard test. The particular constituents of volatile matter vary considerably from coal to coal but generally consist of hydrogen, oxygen, carbon monoxide, methane, and other hydrocarbons. Fixed carbon is the solid combustible residue that remains when volatile components have been removed. These properties may be presented on either a dry or moist mineral-matter-free basis; in Table 6.10 they are shown on a dry basis. This means that inherent moisture contained in the coal is driven off prior to determination of the combustible components; further, the weight of noncombustible mineral matter, which forms a residue of ash after combustion of the coal, is subtracted in determining the percentage of combustible components (hence, the term mineral-matter-free).

Anthracitic coals tend to be high in fixed carbon and low in volatile matter. Because of their high carbon content, these coals are used almost exclusively in the production of coke for steelmaking, and they have a heating value roughly that of pure carbon. Bituminous coals have a higher proportion of volatile matter and a proportionally lower fixed carbon content. The lower-grade coals — high-volatility bituminous, subbituminous, and lignitic — exhibit successively decreasing heating values and are distinguished primarily on that basis. Figure 6.6 shows the correlation between heating value and fixed carbon content for various classes and grades of coal and explains the reason behind some of the figures that are missing from Table 6.10. It will be noted that the higher-grade coals, ranging from medium-volatility bituminous through anthracite, have approximately constant heating values ranging from about 14,000 to 15,000 Btu/lb$_m$. These coals are primarily distinguished from one another by their fixed carbon content. The lower-grade coals all exhibit roughly 50–60% fixed carbon on a dry mineral-matter-free basis, but heating values vary from around 7,000 Btu/lb$_m$ for lignite to over 14,000 Btu/lb$_m$ for high-volatility bituminous. This variation in heating value is principally due to the inherent moisture content in the fuel. Heating value is determined here on a moist basis. Classes I and II coals (anthracitic and bituminous) tend to be low in moisture, typically less than 10%. Subbituminous coals have 15–30% moisture, while lignites may contain as much as 40% moisture.

Figure 6.7 shows the distribution of coal fields in the United States. Traditionally, high-grade metallurgical coals have been mined in the northeastern states and sold to the contiguous steel industry or exported. The

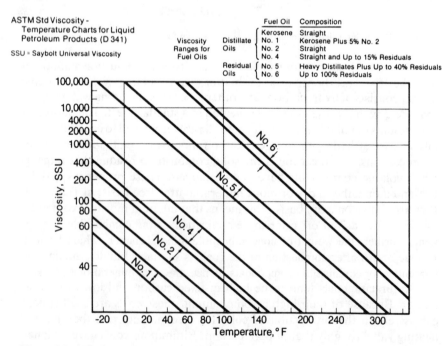

Figure 6.5 Viscosity of fuel oils. (From [5]).

ble 6.10   Classification of Coals by Rank[a]

ss and group	Fixed carbon limits, % (dry, mineral-matter-free basis)		Volatile matter limits, % (dry, mineral-matter-free basis)		Calorific value limits, Btu/lb_m (moist,[b] mineral-matter-free basis)		Agglomerating character
	Equal to or greater than	Less than	Equal to or greater than	Less than	Equal to or greater than	Less than	
Anthracitic							
. Meta-anthracite	98	—	—	2	—	—	Nonagglomerating
. Anthracite	92	98	2	8	—	—	
. Semianthracite[c]	86	92	8	14	—	—	
Bituminous							
. Low-volatile bituminous coal	78	86	14	22	—	—	
. Medium volatile bituminous coal	69	78	22	31	—	—	
. High-volatile A bituminous coal	—	69	31	—	14,000[d]	—	Commonly agglomerating[e]
. High-volatile B bituminous coal	—	—	—	—	13,000[d]	14,000	
. High-volatile C bituminous coal	—	—	—	—	11,500	13,000	
					10,500	11,500	Agglomerating
Subbituminous							
. Subbituminous A coal	—	—	—	—	10,500	11,500	
. Subbituminous B coal	—	—	—	—	9,500	10,500	Nonagglomerating
. Subbituminous C coal	—	—	—	—	8,300	9,500	
Lignitic							
. Lignite A	—	—	—	—	6,300	8,300	
. Lignite B	—	—	—	—	—	6,300	

This classification does not include a few coals, principally nonbanded varieties, which have unusual sical and chemical properties and which come within the limits of fixed carbon or calorific value of the h-volatile bituminous and subbituminous ranks. All of these coals either contain less than 48% dry, eral-matter-free fixed carbon or have more than 15,500 moist, mineral-matter-free Btu per pound.

Moist refers to coal containing its natural inherent moisture but not including visible water on the surface of coal.

If agglomerating, classify in low-volatile group of the bituminous class.

Coals having 69% or more fixed carbon on the dry, mineral-matter-free basis shall be classified by fixed bon, regardless of calorific value.

It is recognized that there may be nonagglomerating varieties in these groups of the bituminous class, and there notable exceptions in high-volatile C bituminous group.

ource: Reprinted from *ASTM Standards* D 388, Classification of Coals by Rank.

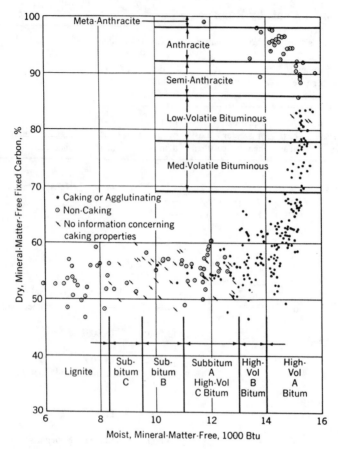

Figure 6.6 Correlation of fixed carbon with heating value for coals. (From [3])

primary sources of coal for boiler fuel have been the Appalachian region and Illinois. With the resurgence of coal as an important industrial energy source, major deposits of subbituminous coal and lignite in the western and southern states are becoming increasingly important.

Table 6.11 shows a breakdown of properties for typical U.S. coals as mined. Note that the terms proximate and ultimate analysis are used somewhat differently with respect to coal than with liquid and gaseous fuels. Proximate analysis in this case refers to the major constituent categories rather than specific chemical compounds, since the latter are difficult to determine accurately for solid fuels. Ultimate analysis still refers to an elemental breakdown, with the exception of moisture and ash, both of which are removed prior to chemical analysis of the combustible part of the fuel. Coal ash is generally a complex mixture of inorganic compounds. While the detailed composition is important in the design of coal combustion systems,

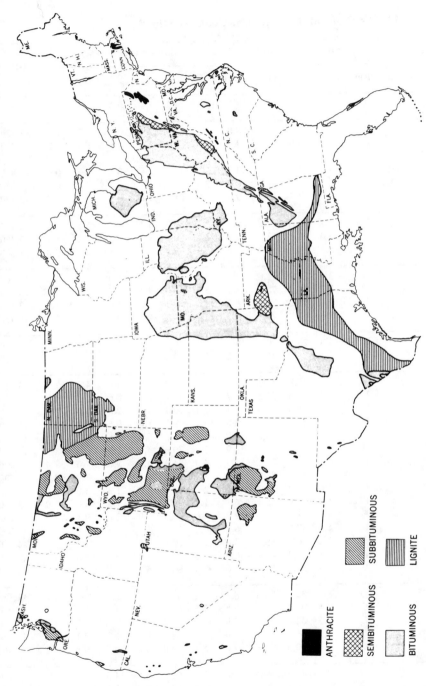

Figure 6.7 Location of coal fields in the United States. (From [5])

ANTHRACITE

SEMIBITUMINOUS

BITUMINOUS

SUBBITUMINOUS

LIGNITE

Table 6.11 Analyses of Typical U.S. Coals as Mined

State	Rank	Proximate analysis (%)				Ultimate analysis (%)						HHV (Btu/lb$_m$)
		H_2O	VM	FC	ASH	H_2O	C	H_2	S	O_2	N_2	
RI	A	13.3	2.5	65.3	18.9	13.3	64.2	0.4	0.3	2.7	0.2	9,313
CO	B	2.5	5.7	83.8	8.0	2.5	83.9	2.9	0.7	0.7	1.3	13,720
NM	B	2.9	5.5	82.7	8.9	2.9	82.3	2.6	0.8	1.3	1.2	13,340
PA												
Orchard Bed	B	5.4	3.8	77.1	13.7	5.4	76.1	1.8	0.6	1.8	0.6	11,950
Mammoth Bed	B	2.3	3.1	87.7	6.9	2.3	86.7	1.9	0.5	0.9	0.8	13,540
Holmes Bed	B	4.9	3.7	82.2	9.2	4.9	81.6	1.8	0.5	1.3	0.7	12,820
AR	C	2.1	9.8	78.8	9.3	2.1	80.3	3.4	1.7	1.7	1.5	13,700
PA	C	3.0	8.4	78.9	9.7	3.0	80.2	3.3	0.7	2.0	1.1	13,450
VA	C	3.1	10.6	66.7	19.6	3.1	70.5	3.2	0.6	2.2	0.8	11,850
AR	D	3.4	16.2	71.8	8.6	3.4	79.6	3.9	1.0	1.8	1.7	13,700
MD	D	3.2	18.2	70.4	8.2	3.2	79.0	4.1	1.0	2.9	1.6	13,870
OK	D	2.6	16.5	72.2	8.7	2.6	80.1	4.0	1.0	1.9	1.7	13,800
WV	D	2.7	17.2	76.1	4.0	2.7	84.7	4.3	0.6	2.2	1.5	14,730
PA	E	3.3	20.5	70.0	6.2	3.3	80.7	4.5	1.8	2.4	1.1	14,310
VA	E	3.1	21.8	67.9	7.2	3.1	80.1	4.7	1.0	2.4	1.5	14,030
AL	F	5.5	30.8	60.9	2.8	5.5	80.3	4.9	0.6	4.2	1.7	14,210
CO	F	1.4	32.6	54.3	11.7	1.4	73.4	5.1	0.6	6.5	1.3	13,210
KS	F	7.4	31.8	52.4	8.4	7.4	70.7	4.6	2.6	5.0	1.3	12,670
KY	F	3.1	35.0	58.9	3.0	3.1	79.2	5.4	0.6	7.2	1.5	14,290
MO	F	5.4	32.1	53.5	9.0	5.4	71.6	4.8	3.6	4.2	1.4	12,990
NM	F	2.0	33.5	50.6	13.9	2.0	70.6	4.8	1.3	6.2	1.2	12,650
OH	F	4.9	36.6	51.2	7.3	4.9	71.9	4.9	2.6	7.0	1.4	12,990
OK	F	2.1	35.0	57.0	5.9	2.1	76.7	4.9	0.5	7.9	2.0	13,630
PA	F	2.6	30.0	58.3	9.1	2.6	76.6	4.9	1.3	3.9	1.6	13,610
TN	F	1.8	35.9	56.1	6.2	1.8	7.7	5.2	1.2	6.0	1.9	13,890
TX	F	4.0	48.9	34.9	12.2	4.0	65.5	5.9	2.0	9.1	1.3	12,230
UT	F	4.3	37.2	51.8	6.7	4.3	72.2	5.1	1.1	9.0	1.6	12,990
VA	F	2.2	36.0	58.0	3.8	2.2	80.6	5.5	0.7	5.9	1.3	14,510
WA	F	4.3	37.7	47.1	10.9	4.3	68.9	5.4	0.5	8.5	1.5	12,610
WV	F	2.4	33.0	60.0	4.6	2.4	80.8	5.1	0.7	4.8	1.6	14,350
IL	G	8.0	33.0	50.6	8.4	8.0	68.7	4.5	1.2	7.6	1.6	12,130
KY	G	7.5	37.7	45.3	9.5	7.5	66.9	4.8	3.5	6.4	1.4	12,080
MO	G	10.5	32.0	44.6	12.9	10.5	63.4	4.2	2.5	5.2	1.3	11,300
OH	G	8.2	36.1	48.7	7.0	8.2	68.4	4.7	1.2	9.1	1.4	12,160
WY	G	5.1	40.5	49.8	4.6	5.1	73.0	5.0	0.5	0.6	1.2	12,960
IL	H	12.1	40.2	39.1	8.6	12.1	62.8	4.6	4.3	6.6	1.0	11,480
IN	H	12.4	36.6	42.3	8.7	12.4	63.4	4.3	2.3	7.6	1.3	11,420
IA	H	14.1	35.6	39.3	11.0	14.1	58.5	4.0	4.3	7.2	0.9	10,720
MI	H	12.4	35.0	47.0	5.6	12.4	65.8	4.5	2.9	7.4	1.4	11,860

Table 6.11   Analyses of Typical U.S. Coals as Mined (*Continued*)

CO	I	19.6	30.5	45.9	4.0	19.6	58.8	3.8	0.3	2.2	1.3	10,130
WY	I	23.2	33.3	39.7	3.8	23.2	54.6	3.8	0.4	3.2	1.0	9,420
ND	J	34.8	28.2	30.8	6.2	34.8	42.4	2.8	0.7	2.4	0.7	7,210
TX	J	33.7	29.3	29.7	7.3	33.7	42.5	3.1	0.5	2.1	0.8	7,350

[a]A is the theoretical air required for combustion under stoichiometric conditions (no excess air).
[b]Rank key: A, meta-anthracite; B, anthracite; C, semianthracite; D, low-vol. bituminous; E, med.-vol. bituminous; F, high-vol. bituminous A; G, high-vol. bituminous B; H, high-vol. bituminous C; I, subbituminous; J, lignite.
Source: [5].

from the standpoint of energy utilization, only the total ash content by weight is needed. For a more detailed discussion of coal ash considerations, references [3] and [4] are recommended.

6.5.4   Properties of Industrial Waste and By-product Fuels

In many industrial processes, by-products are produced with sufficiently high fuel value to make them economic as supplementary energy sources. In most cases these by-products are burned within the plant; with rising prices of conventional fuels, municipal and industrial wastes are now also attracting interest as purchased fuels. This section will deal with the most common waste and by-product fuels.

Wood wastes have commonly been used as fuel sources in the lumber and pulp and paper industries for some time. Table 6.12 shows typical analyses of dry wood products. The ultimate analysis of most woods shows them to be approximately 50% carbon, 6% hydrogen, and 40–45% oxygen by weight, with higher heating values typically in the range of 8,500–9,000 Btu/lb$_m$.

In lumber production, roughly 50% of the original timber is removed prior to cutting into usable lumber. The by-products, in the form of sawdust, bark, and chips, are termed *hog fuel*. Typical properties of hog fuel are shown in Table 6.13. This fuel is normally about 50% moisture as received and has an ultimate analysis and heating value roughly the same as that shown for dry wood.

By-product gases are used extensively as a fuel source in iron- and steelmaking. An important raw material input to the blast furnace is coke, produced by lean combustion of high-grade coal in a coke oven. Large quantities of off-gas are produced; Table 6.14 shows characteristics of typical coke-oven gases. Hydrogen and methane account for approximately

90% of the combustible content of coke-oven gas, with carbon monoxide, heavier hydrocarbons, and inert components comprising the balance. Heating values typically are in the range of 500–600 Btu/ft^3, or about 20,000 Btu/lb$_m$.

In the blast furnace, iron oxide is reduced to molten iron with carbon monoxide, produced by combustion in the coke bed. As shown in Table 6.15, the combustible component of blast furnace gas consists primarily of CO with a small fraction of hydrogen, the balance being mostly nitrogen and carbon dioxide. A typical higher heating value for blast furnace off-gas is on the order of 100 Btu/ft^3, about one-tenth that of natural gas.

Oil refining also produces substantial quantities of combustible by-product gases. Table 6.16 shows typical analyses for a variety of refinery gases. As might be expected because of the great diversity of processes in a

Table 6.12   Typical Analyses of Dry Wood

	C	H$_2$	S	O$_2$	N$_2$	Ash	HHV (Btu/lb$_m$)
Softwoods[a]							
Cedar, white	48.40	6.37	—	44.46	—	0.37	8,400[b]
Cypress	54.98	6.54	—	38.08	—	0.40	9,870[b]
Fir, Douglas	52.3	6.3	—	40.5	0.1	0.8	9,050
Hemlock, western	50.4	5.8	0.1	41.4	0.1	2.2	8,620
Pine pitch	59.00	7.19	—	32.68	—	1.13	11,320[b]
white	52.55	6.08	—	41.25	—	0.12	8,900[b]
yellow	52.60	7.02	—	40.07	—	0.31	9,610[b]
Redwood	53.5	5.9	—	40.3	0.1	0.2	9,040
Hardwoods[a]							
Ash, white	49.73	6.93	—	43.04	—	0.30	8,920[b]
Beech	51.64	6.26	—	41.45	—	0.65	8,760[b]
Birch, white	49.77	6.49	—	43.45	—	0.29	8,650[b]
Elm	50.35	6.57	—	42.34	—	0.74	8,810[b]
Hickory	49.67	6.49	—	43.11	—	0.73	8,670[b]
Maple	50.64	6.02	—	41.74	0.25	1.35	8.580
Oak black	48.78	6.09	—	44.98	—	0.15	8,180[b]
red	49.49	6.62	—	43.74	—	0.15	8,690[b]
white	50.44	6.59	—	42.73	—	0.24	8,810[b]
Poplar	51.64	6.26	—	41.45	—	0.65	8,920[b]

[a]The terms "hard" and "soft" wood, contrary to popular conception, have no reference to the actual hardness of the wood. According to the *Wood Handbook,* prepared by the Forest Products Laboratory of the U.S. Department of Agriculture, hardwoods belong to the botanical group of trees that are broad-leaved, whereas softwoods belong to the group that have needle or scalelike leaves, such as evergreens: cypress, larch and tamarack are exceptions.

[b]Calculated from reported high heating value of kiln-dried wood assumed to contain 8% moisture.

Source: [5].

Table 6.13 Analysis of Hog Fuels

Kind of fuel	Western hemlock	Douglas fir	Pine sawdust
H_2O as received	57.9	35.9	—
H_2O air-dried	7.3	6.5	6.3
Proximate analysis, dry fuel			
VM	74.2	82.0	79.4
FC	23.6	17.2	20.1
Ash	2.2	0.8	0.5
Ultimate analysis, dry fuel			
Hydrogen	5.8	6.3	6.3
Carbon	50.4	52.3	51.8
Nitrogen	0.1	0.1	0.1
Oxygen	41.4	40.5	41.3
Sulfur	0.1	0	0
Ash	2.2	0.8	0.5
HHV, Btu/lb_m dry	8620	9050	9130

Source: [5].

refinery, depending on the product mix desired, there is substantial variation in the composition and heating value of the product gases. The predominant constituents range from methane (CH_4) to *n*-butane (C_4H_{10}) and
hydrogen. Depending on the particular mix of hydrocarbons, higher heating
values range from around 500 to nearly 3,000 Btu/ft^3.

Municipal solid waste has received increasing attention in recent years
as a potential fuel source and is now being used commercially for utility
steam generation. It has proved economical to recover metallic components
by magnetic or air classification prior to burning; in some cases, glass may
also be removed. Table 6.17 shows properties of typical solid waste. Municipal refuse is moderately high in moisture content (20–30%) and high in
ash-producing inorganic matter (e.g., glass). These factors produce a relatively low heating value of 3,000–6,500 Btu/lb_m. While this low heating
value means that it is not economical to transport solid waste for very long
distances, its natural availability in urban areas can, in many cases, make it
quite competitive with coal.

6.6 COMBUSTION SYSTEM EFFICIENCY CALCULATIONS

In this section we will discuss a systematic approach to the calculation of
energy losses in boilers and process heaters. As discussed earlier, the overall
efficiency of a combustion system is defined as the heat content of the output
product (e.g., steam or process fluid) divided by the chemical energy
content of the fuel being fired. Ideally, one could determine efficiency by

Table 6.14   Characteristics of Typical Coke-Oven Gases at 60°F and 30 in. Hg, Dry

	Percent by volume								Density	HHV	
CO_2	O_2	N_2	CO	H_2	CH_4	C_2H_4	C_6H_6	(lb_m/ft^3)	(Btu/ft^3)[a]	(Btu/lb_m)
1.8	0.2	3.4	6.3	53.0	31.6	2.7	1.0	0.0298	596	20,010
1.4	0.5	4.2	5.1	57.4	28.5	2.9	—	0.0263	539	20,490
2.6	0.6	3.7	6.1	47.9	33.9	5.2	—	0.0316	603	19,070
3.13	—	—	11.93	42.16	37.14	4.76	0.88	0.0359	663	18,500
0.1	—	2.4	6.8	27.7	50.0	13.0	—	0.0393	829	21,100
0.75	—	12.1	6.0	53.0	28.15	—	—	0.0291	477	16,390

[a]If gas is saturated with moisture at 60°F and 30.0 in. Hg, reduce by 1.74%.
Source: [5].

Table 6.15 Characteristics of Typical Blast-Furnace Gases at 60°F and 30 in. Hg, Dry

	Percent by volume					Density	HHV	
CO_2	N_2	CO	H_2	CH_4		(lb_m/ft^3)	$(Btu/ft^3)^a$	(Btu/lb_m)
14.5	57.5	25.0	3.0	—		0.0779	90.1	1150
13.0	57.6	26.2	3.2	—		0.0771	94.6	1219
15.59	59.28	23.35	1.7	0.08		0.0792	81.4	1021
8.7	56.5	32.8	1.8	0.2		0.0762	113.3	1478
5.7	59.0	34.0	1.3	—		0.0753	113.6	1498
6.0	60.0	27.0	2.0	5.0		0.0734	144.0	1950

^{a}If gas is saturated with moisture at 60°F and 30.0 in. Hg, reduce by 1.74%.
Source: [5].

Table 6.16 Characteristics of Typical Refinery and Oil Gases at 60°F and 30 in. Hg, Dry

Percent by volume	A	B	C	D	E	F	G	H
O_2	—	—	—	—	—	2.3	0.9	0.1
N_2	—	—	—	—	—	8.7	8.4	2.7
CO_2	—	—	—	—	3.3	—	2.2	1.0
CO	—	—	—	—	1.5	—	14.3	6.8
H_2	—	—	—	—	5.6	—	50.9	59.2
H_2S	2.18	—	—	—	—	—	—	—
CH_4	41.62	4.30	92.10	5.0	30.9	30.3	15.9	25.4
C_2H_6	20.91	82.70	1.90	12.0	19.8	13.4	5.0	—
C_3H_8	19.72	13.00	4.50	30.0	38.1	19.1	—	—
C_4H_{10}	9.05	—	1.30	34.0	0.0	14.7	—	—
C_5H_{12}	6.52	—	0.02	19.0	—	1.8	—	—
$C_3H_3H_6$	—	—	—	—	0.2	9.7	2.4	4.8
Density, lb_m/ft^3	0.08676	0.08377	0.04845	0.13760	0.08102	0.09232	0.03631	0.02756
HHV								
Btu/ft^3	1,898	1,858	1,136	2,988	1,696	1,844	519	586
Btu/lb_m	21,880	22,170	23,460	21,720	20,930	19,970	14,300	21,270

Source: [5].

Table 6.17 Analyses of Typical Residential Solid Waste with Magnetic Metals Removed

	Percent as received
Proximate analysis	
H_2O	19.7–31.3
Ash	9.4–26.8
VM	36.8–56.2
FC	0.6–14.6
HHV, Btu/lb_m	3,100–6,500
Ultimate analysis	
H_2O	19.7–31.3
Ash	9.4–26.8
Carbon	23.4–42.8
Hydrogen	3.4–6.3
Nitrogen	0.2–0.4
Chlorine	0.1–0.9
Sulfur	0.1–0.4
Oxygen	15.4–31.9

Source: [5].

measuring these two quantities directly, i.e., by metering the flow and thermodynamic properties of the incoming fuel and outgoing product. In practice, however, it is usually difficult to obtain these measurements without installing expensive permanent metering devices in the system. Fortunately, it is usually relatively simple to determine efficiency indirectly by determination of the characteristics of the flue gas and, in the case of a boiler, the blowdown water released to keep dissolved solids concentrations to an acceptable level. Given this basic information, physical principles discussed earlier can be applied to estimate energy losses with reasonable accuracy.

Boiler and heater losses include several components. The largest of these, typically amounting to 15–20% loss in industrial combustion systems, is the heat and combustible material content in the flue gas and solid refuse discharged from the heater. Surface heat losses (often termed radiation losses) are typically on the order of 1–2%, and blowdown losses typically less than 1%. Because they are so dominant, stack and combustible losses will be treated here in some detail.

6.6.1 Molal Basis Calculation Method

In this section we concentrate on the *molal basis* calculation method for determining stack and combustible losses. This approach, which is extensively developed in reference [3], is closely related to the standard ASME Performance Test Code for efficiency of steam generating units and is especially well suited for in-service tests of combustion equipment. Other calculation methods are also available, some of which are more convenient than the molal method for rough predictive estimates. Detailed descriptions of these alternative approaches are given in references [3] and [5].

A standard calculation outline for the molal method is shown in Tables 6.18A–C. Before considering the procedure in detail, it will be useful to review some pertinent definitions. *Theoretical air* describes the quantity of air needed to exactly meet the oxygen requirements of all the fuel components burned. *Excess air,* as discussed earlier, is the additional air provided, usually expressed in percentage terms, to ensure complete combustion under less than perfect mixing conditions. *Total air* is the sum of theoretical and excess air.

Orsat analysis is a volumetric analysis of the components of the flue gas, specifically CO_2, O_2, CO, and, by difference, N_2, all expressed on a percentage basis. Of the several available test methods, an Orsat analysis provides the most complete picture of flue gas composition. However, as will be shown in a subsequent section, other less complete flue gas analyses may be used without seriously compromising the accuracy of the test. For most common low-nitrogen fuels, excess air can be determined from an Orsat analysis using the relation

Table 6.18A Input Data for Combustion Calculations—Molal Basis

Fuel:	Fuel unit: 100 lb_m for solid or liquid fuels
	100 moles for gaseous fuels

(a) Fuel analysis as fired, % by weight (S and L) or by volume (G). For gaseous fuels, convert analysis by compounds to elemental analysis.

(b) Fuel heating value (Btu/lb_m for S and L, Btu/ft³ for G):

(c) Combustible in refuse, % C:

(d) Carbon unburned, lb_m/100 lb_m of fuel:

$C\,(lb_m/100\,lb_m) = \%$ ash in fuel $\times \dfrac{\%\ C}{100 - \%\ C} =$

(e) Flue gas Orsat analysis (percent by volume). Skip if percent excess air is known.

CO_2: O_2:   CO: N_2:

(f) % Excess air (assigned or determined from Eq. (6.1) or Fig. 6.8):

(g) Distribution of carbon in fuel. Skip if no CO in flue gas and no carbon in refuse:

Moles C in fuel = (lb_mC/100 lb_m fuel)/12 =	(g1)
Moles C in refuse = Line (d)/12 =	(g2)
Moles C burned = Line (g1) – Line (g2) =	(g3)
Fraction of C burned to $CO_2 = \dfrac{\%\ CO_2\ \text{in Orsat}}{\%\ CO_2 + \%\ CO} =$	(g4)
Moles C to CO_2 = Line (g4) × Line (g3) =	(g5)
Moles C to CO = Line (g3) – Line (g5) =	(g6)

(h) Combustion air ambient conditions (two out of three required):

Dry-bulb temperature, t_1:   Wet-bulb temperature:

Relative humidity (%):

(i) Humidity ratio (from psychrometric chart, Fig. 6.9):

=   (lb_mH$_2$O/lb_m dry air)

(j) Flue gas exit temperature, t_2:

Table 6.18B Mass Balance for Molal Basis Combustion Calculations

	Fuel, O_2, and air per unit of fuel				Fuel gas composition, moles per fuel unit					
Fuel constituent	Per fuel unit, lb_m	Mol. wt. divisor	Moles fuel constituent	O_2 multiplier	(a) O_2 moles theoretically required	(b) CO_2 + SO_2	(c) O_2	(d) N_2	(e) H_2O	(f) CO
1 C to CO_2	—	12	—	1	—	—				
2 C to CO	—	12	—	0.5	—					—
3 CO to CO_2	—	28	—	0.5	—	—				
4 C unburned	—	12	—							
5 H_2	—	2	—	0.5	—				—	
6 S	—	32	—	1	—	—				
7 O_2 (deduct)	—	32	—	-1	—					
8 N_2	—	28	—		0			—		
9 CO_2	—	44	—		0	—				
10 H_2O	—	18	—		0				—	
11 Ash	—									
12 Sum	—				—	—	—	—	—	—

13	O_2 (theoretical) reqd. = O_2 line 12a	—
14	O_2 (excess) = $\dfrac{\% \text{ E.A.}}{100} \times O_2$, line 12a	—
15	O_2 (total) supplied = lines 13 + 14	—
16	N_2 supplied = 3.76 × O_2, line 15	—
17	Air (dry) supplied = $O_2 + N_2$	—
18	H_2O in air = moles dry air × line i × 29/18	—
19	Air (wet) supplied = lines 17 + 18	—
20	Flue gas constituents = lines 12 to 18, total	— — — — —
21	Total flue gas (sum of lines 20b–20f):	Dry flue gas:

$$\text{\% excess air} = 100 \times \frac{O_2 - CO/2}{0.264\, N_2 - (O_2 - CO/2)} \tag{6.1}$$

Alternatively, Fig. 6.8 can be used to relate excess air to the flue gas analysis.

To determine the moisture content of incoming air, the psychrometric chart (Fig. 6.9) is useful. This is a plot of the specific humidity, or moisture content per pound of dry air, as a function of the dry-bulb temperature, the temperature that would be determined by a standard thermometer in the intake air stream. Relative humidity, a quantity related to the partial pressure of water vapor in the air and readily available from meteorological data for a given locale, is a parameter on the psychrometric chart. An alternative parameter is the wet-bulb temperature, a quantity directly measurable with a sling psychrometer.

Table 6.18C Computation of Stack and Combustible Losses; Combustion Calculations for Molal Basis

	$CO_2 + SO_2$	O_2	N_2	H_2O	CO	Total
22 Flue gas constituents						
23 Mc_p, mean, t_2 to t_1' (for $t_1' = 80°F$; see Fig. 6.9)	—	—	—	—	—	—
24 In dry flue gas = moles each, line 20 $\times Mc_p \times (t_2 - t_1')$	—	—	—		—	—
25 In H_2O in air = moles H_2O, line 18 $\times Mc_p \times (t_2 - t_1')$				—		—
26 In sensible heat, H_2O in fuel = moles, lines (5e + 10e) $\times Mc_p \times (t_2 - t_1')$						
27 In latent heat, H_2O in fuel = moles, lines (5e + 10e) $\times 1,040 \times 18$						
28 Total in wet flue gas						
29 Due to carbon in refuse = line d $\times 14,100$						
30 Due to unburned CO in flue gas = moles C to CO $\times 12 \times 9,755$						
31 Total flue gas losses + unburned combustible = lines 28 + 29 + 30						
32 Heat value of fuel unit = $\begin{cases} 100 \times \text{line b for solid and liquid fuels} \\ 394 \times \text{line b} \times 100 \text{ for gaseous fuels} \end{cases}$						
33 Stack and combustible loss, percent of heat input = $100 \times$ line 31 $\div$ line 32						

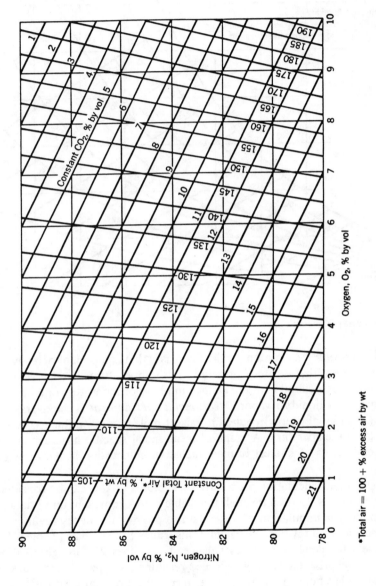

Figure 6.8 Determination of total air from Orsat analysis. (From [3])

201

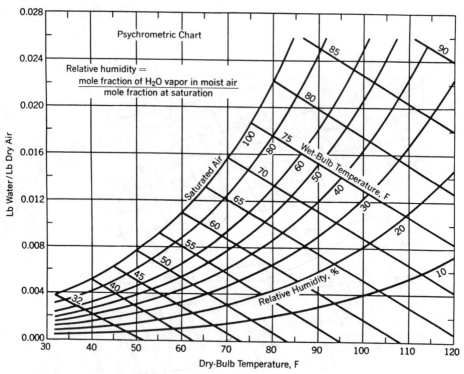

Figure 6.9 Psychrometric chart.

The *sensible heat* content of the flue gas is that component of its overall energy content that is directly related to its temperature. *Latent heat* is the component associated with the change in phase of the water content of the flue gas from liquid to vapor. *Mean molal specific heat* is the average, or mean, quantity of energy required to raise 1 mole of a given flue gas component 1 °F in temperature. A graph of mean molal specific heats for the flue gas components of interest in most industrial combustion is shown in Fig. 6.10.

The molal calculation procedure of Tables 6.18A–C can be considered in several parts. Lines a through j in Table 6.18A represent input data and preliminary manipulation of these data to provide the necessary inputs to the mass and energy balance calculations. Lines 1 through 21 (Table 6.18B) represent the combustion mass balance, while lines 22 through 33 (Table 6.18C) comprise the energy balance, resulting in a percentage stack and combustible energy loss. In general terms the procedure consists of the following steps.

1. A unit of fuel is chosen as the basis for calculation. For solid and liquid fuels, the standard basis is 100 lb. For gaseous fuels, 100 moles is used as

the standard unit. It will be recalled from Amagat's law that an analysis of a gaseous mixture on a volumetric percentage basis can be directly interpreted as an equivalent analysis on a molal basis. Hence, a fuel containing 85% methane by volume will contain 85 moles of methane per 100 moles of fuel. Since analyses of gaseous fuels are typically presented on a volume basis, the justification for using the molal fuel unit is clear.

2. The fuel analysis is broken down into its elemental components, including both combustible and noncombustible constituents. If the given fuel analysis is an ultimate analysis, this step is already taken care of. If it is approximate analysis, as in the case of natural gas, this must first be converted to an elemental basis. This is illustrated in Example 6.4.

3. If combustible matter is present in the refuse (usually as carbon) or in the flue gas (generally CO), these constituents are subtracted from the combustible content of the fuel. The remaining combustible content actually takes part in the combustion reaction.

4. If the fuel analysis is on a weight basis, each constituent is converted to an appropriate molal quantity by dividing by its molecular weight.

5. The required number of moles of oxygen for each combustible component is determined from the chemical equation for oxidation of that

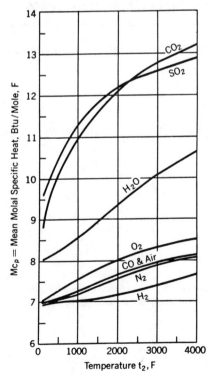

Figure 6.10 Mean molal specific heat of gases between final temperature (t_2) and 80 °F at standard atmospheric pressure. (From [3])

particular component, and the corresponding number of moles of combustion product in the flue gas is determined.

6. The oxygen requirements for all the combustibles are totaled to give the required theoretical oxygen for combustion. Based on the quantity of excess air used, as determined from the flue gas analysis, the total oxygen for the test is determined. The amount of nitrogen brought in with this oxygen is added to yield the total dry air requirement.

7. From a psychrometric chart, the amount of water vapor present in the combustion air is determined.

8. All components of the flue gas are added to give the total quantity of flue gas produced. This completes the mass balance portion of the calculations.

9. For the given stack gas temperature, the mean molal specific heat of each flue gas component is determined. By multiplying the number of moles of each component by its mean molal specific heat and the temperature rise from ambient to stack conditions, the sensible heat content of the flue gas is calculated.

10. The latent heat content of H_2O in the flue gas, due to both H_2O in the fuel and H_2O produced by combustion of hydrogen, is determined.

11. The heat value of unburned combustibles in the refuse and flue gas is determined.

12. All components of stack and combustible loss are totaled and divided by the heat content of one fuel unit to normalize to a percentage loss basis.

The calculation procedure described above is illustrated in the following examples.

Example 6.3

Lignite is burned in an industrial boiler; the ultimate analysis of the fuel on a weight basis is as follows:

Carbon	60%
Hydrogen	2
Sulfur	1
Oxygen	1
Nitrogen	1
Water	30
Ash	5
Total	100%

The higher heating value of the fuel is 8200 Btu/lb$_m$ (as fired). An Orsat analysis taken during a boiler test yields the following dry flue gas analysis (percentages by volume):

CO_2	15.2%
O_2	3.6
CO	0.5
N_2	80.7

Table E6.3A Input Data for Combustion Calculations—Molal Basis

Fuel: Lignite	Fuel unit: 100 lb_m for solid or liquid fuels
	100 moles for gaseous fuels

(a) Fuel analysis as fired, percent by weight (S and L) or by volume (G). For gaseous fuels, convert analysis by compounds to elemental analysis.

C	60%	N_2	1
H_2	2	H_2O	30
S	1	Ash	5
O_2	1		

(b) Fuel heating value (Btu/lb_m for S and L, Btu/ft^3 for G): 8,200

(c) Combustible in refuse, % C: 10

(d) Carbon unburned, lb_m/100 lb_m of fuel:

$$C\ (lb_m/100\ lb_m) = \%\ \text{ash in fuel} \times \frac{\%\ C}{100 - \%\ C} = 0.55$$

(e) Flue gas Orsat analysis (percent by volume). Skip if percent excess air is known.

CO_2: 15.2 $\qquad$ O_2: 3.6 $\qquad$ CO: 0.5 $\qquad$ N_2: 80.7

(f) % Excess air (assigned or determined from Eq. (6.1) or Fig. 6.8): 18.6%

(g) Distribution of carbon in fuel. Skip if no CO in flue gas and no carbon in refuse:

$$\text{Moles C in fuel} = (lb_m C/100\ lb_m\ \text{fuel})/12 = \frac{60}{12} = 5.0 \tag{g1}$$

$$\text{Moles C in refuse} = \text{Line (d)}/12 = \frac{0.55}{12} = 0.046 \tag{g2}$$

$$\text{Moles C burned} = \text{Line (g1)} - \text{Line (g2)} = 4.95 \tag{g3}$$

$$\text{Fraction of C burned to } CO_2 = \frac{\%\ CO_2\ \text{in Orsat}}{\%\ CO_2 + \%\ CO} = 0.97 \tag{g4}$$

$$\text{Moles C to } CO_2 = \text{Line (g4)} \times \text{Line (g3)} = \cancel{4.97}\ 4.74 \tag{g5}$$

$$\text{Moles C to CO} = \text{Line (g3)} - \text{Line (g5)} = 0.16 \tag{g6}$$

(h) Combustion air ambient conditions (two out of three required):

Dry-bulb temperature, t_1': 80 °F $\qquad$ Wet-bulb temperature:

Relative humidity (%): 70%

(i) Humidity ratio (from psychrometric chart, Fig. 6.9):

$$= 0.015\ (lb_m H_2O/lb_m\ \text{dry air})$$

(j) Flue gas exit temperature, t_2: 425°F

The solid refuse from the bottom of the boiler is analyzed and found to contain 10% carbon by weight. During the test the ambient air is at 80 °F (dry bulb) and 70% relative humidity, and the fuel gas temperature is 425 °F.

Determine the stack and combustible losses for the boiler and the resulting boiler efficiency.

The input data and preliminary calculations for this case are shown in Table E6.3A. For lignite, 100 lb_m is taken as the basis for calculation, and thus the fuel analysis in percent by weight can be interpreted as lb_m of each component per fuel unit. In line d, the amount of carbon left unburned in the refuse per fuel unit is calculated; this value is used in line g to determine the amount of carbon actually burned. Since the Orsat analysis indicates that some CO is present in the fuel gas, some of the carbon in the fuel is only partially oxidized, while the rest is completely burned to CO_2. Lines g4–g6 proportion the carbon accordingly, based on the relative amounts of CO and CO_2 in the Orsat. From the psychrometric chart, the humidity ratio (lb_m water/lb_m of dry air) is determined in line i from the given ambient conditions.

Table E6.3B shows the mass balance calculations. In lines 1 and 2, the molal quantities of carbon burned to CO_2 and CO are carried over from lines g5 and g6, respectively. If desired, the mass equivalents of these quantities can be back-calculated by multiplying moles by molecular weight. For all of the other fuel components, the masses per 100 lb_m of fuel are carried over directly from the fuel analysis and converted to molal quantities. The O_2 required to burn each component and the resulting flue gas contribution can then be computed. Note that CO_2 and SO_2 are lumped together in the combustion products. Generally, the amount of SO_2 produced is very small relative to CO_2, which is the principal flue gas constituent in this case. Furthermore, it will be noted from Fig. 6.10 that the molal specific heats of CO_2 and SO_2 are nearly the same; thus, for energy balance purposes, little error in the flue gas loss is incurred by lumping these two components together.

In lines 13–19 the contribution to the flue gas from the air required for combustion is computed. It is interesting to observe that nitrogen brought in with atmospheric oxygen accounts for nearly three-fourths of the total mass of the flue gas.

The combustion losses due to each of the components in the flue gas and in the refuse are computed in Table E6.3C. The large amount of nitrogen that must be heated produces about three-fourths of the dry flue gas loss. The next most important losses are due to latent heat of the moisture in the fuel (usually quite high for lignite) and incomplete combustion (as indicated by CO in the flue gas). Overall, about 20% of the heating value of the fuel is lost up the stack and in the refuse.

Example 6.4

Natural gas composed of 83% methane (CH_4), 16% ethane (C_2H_6), and 1% nitrogen (N_2) is burned with 15% excess air, using atmospheric air at 90°F dry-bulb and 80°F wet-bulb temperatures. The flue gas exhaust temperature is 390°F. Determine the stack losses under these conditions.

The input data and preliminary calculations for this example are shown in Table E6.4A. For natural gas, there is no refuse, and since the percentage of excess air is given in lieu of an Orsat analysis, we must assume no combustible in the flue gas. Thus, lines c, d, e, and g may be ignored. The fuel analysis, however, is given in proximate terms (i.e., hydrocarbon compounds), and for the mass balance these compounds must be broken down into their respective carbon and hydrogen contributions, as shown in line (a). For gases, the calculations are based on a fuel unit of 100 moles, which makes conversion of the molal composition of the fuel to moles of carbon and hydrogen quite straightforward. One mole of methane, for example, contains 1 mole of carbon and 2 moles of hydrogen, so for every

e E6.3B   Mass Balance for Molal Basis Combustion Calculations

	Fuel, O_2, and air per unit of fuel					Flue gas composition, moles per fuel unit				
tuent	Per fuel unit lb_m	Mol. wt. divisor	Moles fuel constituent	O_2 multiplier	(a) O_2 Moles theoretically required	(b) CO_2 + SO_2	(c) O_2	(d) N_2	(e) H_2O	(f) CO
to CO_2	57.5	12	4.79	1	4.79	4.79				
to CO	1.85	12	0.16	0.5	0.08					0.16
O to CO_2	—	28	—	0.5	—	—				
unburned	0.55	12	0.046							
2	2	2	1.0	0.5	.5				(1.0) →	
	1	32	0.031	1	.631	0.031	~~0.031~~			
2 (deduct)	1	32	0.031	−1	−0.031					
2	1	28	0.036	0	0			0.036		
O_2	—	44	—	0	0	—				
2O	30	18	1.67	0	0				1.67	
sh	5									
um	100				5.37	4.82	—	0.036	2.67	0.16

		(a)	(b)	(c)	(d)	(e)	(f)
2 (theoretical) reqd. = O_2 line 12a		5.37					
2 (excess) = $\dfrac{\% \text{ E.A.}}{100} \times O_2$, line 12a		1.02		1.02			
2 (total) supplied = lines 13 + 14		6.39					
2 supplied = 3.76 × O_2, line 15		24.03			24.03		
ir (dry) supplied = $O_2 + N_2$		30.42					
2O in air = moles dry air × line i × $\dfrac{29}{18}$		0.74				0.74	
ir (wet) supplied = lines 17 + 18		31.16					
ue gas constituents = lines 12 to 18, total			4.82	1.02	24.07	3.41	0.16

tal flue gas (sum of lines 20b to 20f): 33.48   Dry flue gas: 30.07

$$33.48 - H_2O = 30.07$$

Table E6.3C Computation of Stack and Combustible Losses; Combustion Calculations for Molal Basis

		CO$_2$ + SO$_2$	O$_2$	N$_2$	H$_2$O	CO	Total
22	Flue gas constituents						
23	Mc_p, mean, t_2 to t'_1 (for t'_1 = 80°F; see Fig. 6.9)	10.0	7.2	7.0	8.2	7.0	
24	In dry flue gas = moles each, line 20 × Mc_p × ($t_2 - t'_1$)	16,629	2,534	58,129	—	386	77,678
25	In H$_2$O in air = moles H$_2$O, line 18 × Mc_p × ($t_2 - t'_1$)						2,093
26	In sensible heat, H$_2$O in fuel = moles, lines (5e + 10e) × Mc_p × ($t_2 - t'_1$)						7,553
27	In latent heat, H$_2$O in fuel = moles, lines (5e + 10e) × 1,040 × 18						49,982
28	Total in wet flue gas						137,306
29	Due to carbon in refuse = line d × 14,100						7,755
30	Due to unburned CO in flue gas = moles C to CO × 12 × 9,755						18,730
31	Total flue gas losses + unburned combustible = lines 28 + 29 + 30						163,791
32	Heat value of fuel unit = $\begin{cases}100 \times \text{line b for solid and liquid fuels} \\ 394 \times \text{line b} \times 100, \text{for gaseous fuels}\end{cases}$						820,000
33	Stack and combustible loss, percent of heat input = 100 × line 31 ÷ line 32						20%

Table E6.4A   Input Data for Combustion Calculations — Molal Basis

Fuel: Natural gas   Fuel unit: 100 lb_m for solid or liquid fuels
                         100 moles for gaseous fuels

(a) Fuel analysis as fired, % by weight (S and L) or by volume (G). For gaseous fuels, convert analysis by compounds to elemental analysis.

Moles/100 moles fuel	$\rightarrow$	Moles C	Moles H_2	Moles N_2
CH_4	83	83	166	
C_2H_6	16	32	48	
N_2	1			1
Total	100	115	214	1

(b) Fuel heating value (Btu/lb_m for S and L, $\boxed{\text{Btu/ft}^3 \text{ for G)}}$: 1123

(c) Combustible in refuse, % C: N.A.

(d) Carbon unburned, lb_m/100 lb_m of fuel:

$$C \, (lb_m/100 \, lb_m) = \% \text{ ash in fuel} \times \frac{\% \, C}{100 - \% \, C} = \text{N.A.}$$

(e) Flue gas Orsat analysis (percent by volume). Skip if percent excess air is known.

CO_2: O_2:   CO: N_2:

(f) % Excess air (assigned or determined from Eq. (6.1) or Fig. 6.8): 15% assigned

(g) Distribution or carbon in fuel. Skip if no CO in flue gas and no carbon in refuse: N.A.

Moles C in fuel = (lb_mC/100 lb_m fuel)/12 =   (g1)

Moles C in refuse = Line (d)/12 =   (g2)

Moles C burned = Line (g1) – Line (g2) =   (g3)

Fraction of C burned to CO_2 = $\dfrac{\% \, CO_2 \text{ in Orsat}}{\% \, CO_2 + \% \, CO}$ =   (g4)

Moles C to CO_2 = Line (g4) × Line (g3) =   (g5)

Moles C to CO = Line (g3) – Line (g5) =   (g6)

(h) Combustion air ambient conditions (two out of three required):

Dry-bulb temperature, t_1': 90 °F   Wet-bulb temperature: 80°F ·

Relative humidity (%):

(i) Humidity ratio (from psychrometric chart, Fig. 6.9):

= 0.020 (lb_mH$_2$O/lb_m dry air)

(j) Flue gas exit temperature, t_2: 390°F

unit, the methane contributes 83 moles of carbon and 166 moles of hydrogen. Other hydrocarbons in the fuel can be similarly broken down. Compounds such as CO_2 and H_2O, which appear explicitly in the mass balance, need not be reduced to elemental form.

6.6.2 Radiation Losses (omit)

Heat loss to the environment from the boiler or heater surface, usually termed radiation loss even though it occurs through the combined mechanisms of radiation and convection, is extremely difficult to measure directly. This loss is relatively small compared with the flue gas loss, as pointed out above, and hence it is usually sufficient to estimate it. This can be done by direct heat transfer calculations based on the surface area of the

Table E6.4B Mass Balance for Molal Basis Combustion Calculations

	Fuel, O_2, and air per unit of fuel					Flue gas composition, moles per fuel unit				
Fuel constituent	Per fuel unit lb_m	Mol. wt. divisor	Moles fuel constituent	O_2 multiplier	(a) O_2 Moles theoretically required	(b) CO_2 + SO_2	(c) O_2	(d) N_2	(e) H_2O	(f) CO
1 C to CO_2	—	12	115	1	115	115				
2 C to CO	—	12	—	0.5	—					—
3 CO to CO_2	—	28	—	0.5	—	—				
4 C unburned	—	12	—							
5 H_2	—	2	214	0.5	107				214	
6 S	—	32	—	1	—	—				
7 O_2 (deduct)	—	32	—	−1	—	—				
8 N_2	—	28	1		0			1		
9 CO_2	—	44	—		0	—				
10 H_2O	—	18	—		0				—	
11 Ash	—									
12 Sum	—				222	115	—	1	214	—

13 O_2 (theoretical) reqd. = O_2 line 12a	222				
14 O_2 (excess) $= \dfrac{\% \ E.A.}{100} \times O_2$, line 12a	33.3	33.3			
15 O_2 (total) supplied = lines 13 + 14	255.3				
16 N_2 supplied = 3.76 × O_2, line 15	959.9			959.9	
17 Air (dry) supplied = O_2 + N_2	1,215.2				
18 H_2O in air = moles dry air × line i × $\dfrac{29}{18}$	39.2				39.2
19 Air (wet) supplied = lines 17 + 18	1,254.4				
20 Flue gas constituents = lines 12 to 18, total		115	33.3	960.9	253.2	—

21 Total flue gas (sum of lines 20b to 20f): 1,362.4   Dry flue gas: 1,109.2

Table E6.4C Computation of Stack and Combustible Losses; Combustion Calculations for Molal Basis

		$CO_2 + SO_2$	O_2	N_2	H_2O	CO	Total
22	Flue gas constituents						
23	Mc_p, mean, t_2 to t_1' (for $t_1' = 80\,°F$; see Fig. 6.9)	9.5	7.3	7.0	8.2	—	
24	In dry flue gas = moles each, line 20 × Mc_p × $(t_2 - t_1)$	327,750	72,927	2,017,890	—	—	2,418,567
25	In H_2O in air = moles H_2O, line 18 × Mc_p × $(t_2 - t_1)$						96,432
26	In sensible heat, H_2O in fuel = moles, lines (5e + 10e) × Mc_p × $(t_2 - t_1)$						526,440
27	In latent heat, H_2O in fuel = moles, lines (5e + 10e) × 1,040 × 18						4,006,080
28	Total in wet flue gas						7,047,519
29	Due to carbon in refuse = line d × 14,100						—
30	Due to unburned CO on flue gas = moles C to CO × 12 × 9,755						
31	Total flue gas losses + unburned combustible = lines 28 + 29 + 30						7,047,519
32	Heat value of fuel unit = $\begin{cases} 100 \times \text{line b for solid and liquid fuels} \\ 394 \times \text{line b} \times 100, \text{for gaseous fuels} \end{cases}$						44,246,200
33	Stack and combustible loss, % of heat input = 100 × line 31 ÷ line 32						15.9%

unit and the surface and ambient temperatures or by using published correlations relating radiation loss to overall boiler operating conditions. Figure 6.11 shows radiation loss as a percentage of gross heat input for a boiler as a function of its output capacity and operating firing rate. These curves were developed by the American Boiler Manufacturers' Association from data on a number of relatively small boilers and are widely used in standard performance testing. While the figures are considered to be somewhat conservative, they are adequate for energy management purposes. Consider, for example, a boiler rated at 4,000 million Btu/hr. When operating at full load, the radiation loss for this boiler would be approximately 0.2% of the gross heat input. When operating at 50% load, or 2,000 million Btu/hr, the radiation loss is roughly twice as high on a percent-of-heat-input basis. This is not surprising when one considers that as long as the steam temperature conditions in the boiler waterwalls are constant, the external shell will remain at essentially constant temperature regardless of the firing rate. Thus, the absolute value of the radiation loss is roughly constant and represents an increasing percentage of the heat input as the boiler load is reduced.

6.6.3 Blowdown Losses

As steam is generated, dissolved impurities in the makeup water tend to accumulate in the boiler drum. To keep these impurities within acceptable limits, water is either periodically or continuously withdrawn from the drum and replaced with fresh makeup. This represents an energy loss, since the water in the drum is at hot saturation conditions corresponding to the steam pressure, while the makeup water is at a low temperature. Blowdown loss, which typically amounts to less than 1% of fuel input, can be measured directly by the same technique used for determining condensate flow, as described in Chapter 7. Alternatively, blowdown rate can be estimated from the mass balance on boiler water impurities. This procedure is described in reference [6].

6.7 TESTING COMBUSTION EQUIPMENT EFFICIENCY

6.7.1 Elements of a Boiler or Heater Test Program

Careful testing and evaluation of performance of combustion equipment is an important prerequisite to deciding on a course of action for efficiency improvement. The equipment should be thoroughly inspected and its operation observed for signs of obvious functional problems. When these have been corrected, a test can be conducted to quantify efficiency and determine the magnitude of possible improvement. Finally, various energy

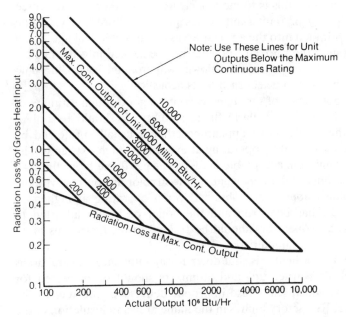

Figure 6.11 Radiation loss for waterwall boilers (From American Boiler Manufacturers' Association [5])

conservation options can be evaluated from the standpoint of energy savings and economic feasibility. The first two steps will be discussed in this section, and the last in the final section of this chapter.

6.7.2 Inspecting and Observing Boiler and Heater Operation

A number of areas of boiler and heater operation can and should be checked and corrected before running a quantitative efficiency test. All mechanical linkages should be checked for tightness and smooth operation. This is particularly important in the fuel and air control systems. Simple combustion controls use a cam arrangement to adjust air and fuel flow rates to approximately correct proportions for various firing rates. Substantial error in the fuel-to-air ratio can be introduced by slack in the linkages connecting the cam system to the air damper and fuel valve. The result is either too much excess air or not enough; in either case, efficiency is adversely affected.

Similarly, fouling of the air intake dampers with dirt can reduce the airflow for a given damper setting. Air intake screens and dampers should be regularly inspected and cleaned.

Several observations should be made while the equipment is in normal operation. Signs of air leakage into or out of the shell should be investigated.

A convenient way to do this is to move a cigarette or other smoke source around the periphery of joints and access panels. Alternatively, a fine powder can be sprinkled into the air near areas of suspected leakage. In an induced-draft system the smoke or powder will be sucked into the heater, while in a forced-draft system it will be blown away from the leak. Ultrasonic devices can also be used to detect such leaks acoustically.

The visual appearance of the flame can give a good rough estimate of combustion conditions, particularly flame stability. It should be noted, however, that the stable flame appearance that is usually preferred by operators is often achieved by operating at excessively high oxygen levels. For example, at minimum acceptable excess air, a gas flame typically has a lazy rolling appearance and a somewhat yellowish color, in contrast with the short, bushy, bluish appearance obtained when operating at higher air levels. Once a test has been run and the boiler or heater adjusted to minimum excess air, operators should be trained to recognize this more desirable visual flame characteristic.

When the boiler or heater is shut down for maintenance, several areas should be checked. Burner tips and atomizers should be inspected for fouling and wear, which can cause improper delivery of the fuel to the combustion zone. Refractory linings in the flame area and insulation of the shell should be checked for deterioration and replaced if necessary.

Perhaps the most serious point of efficiency loss detectable during a maintenance inspection is fouling of the boiler or heater heat transfer surfaces. On the gas side, soot accumulation can occur, even in a gas-fired unit that has been temporarily converted to oil-firing during a gas curtailment. In fact, this situation is quite common, since often the control system, which is set up for proper excess air with gas, provides inadequate air for fuel oil, resulting in smoky operation and accumulation of a carbon residue on the gas side. On the water side, deposition of a crystalline scale deposit can occur due to improper boiler water treatment. In process heaters a similar deposit often forms due to polymerization or coking of hydrocarbon process fluids. In any case, these deposits on both the gas and process fluid sides present an added heat transfer resistance, and additional fuel is needed to provide the necessary process heat input.

6.7.3 Boiler and Heater Testing

Data Requirements   In Section 6.6 a procedure was outlined for calculating the energy losses in combustion equipment. Several inputs are required for the calculation; in this section we will discuss how these inputs are obtained.

Fuel characteristics, in particular the heating value and the proximate or ultimate analysis, are important factors. Heating value is usually easy to obtain from fuel billing records, since most industrial fuels are priced on a heat equivalent basis. On gas bills, this may be represented as a "Btu factor," which is multiplied by the price per thousand cubic feet. For example, the gas bill may indicate a base price of \$4.50 per 1,000 ft^3 with a Btu factor of 1.047. The base price is generally predicated on a heating value of 1,000 Btu/ft^3; thus, a good estimate for the actual heating value is 1,047 Btu/ft^3. While gas utility rate provisions vary from place to place, gas supplies to industrial customers are usually sampled and measured with a calorimeter about once a month to confirm that the heating value is within certain contractually guaranteed limits. A gas analysis may also be obtained from the gas supplier, and with proper planning it may be possible to schedule a test to coincide with the supplier's sampling.

Fuel oils are specified by grade, and the specifications on different grades are generally sufficient to provide the needed information for industrial boiler tests. Table 6.9 can be used for this purpose. A more accurate estimate of heating value can be obtained by measuring the specific gravity and using the correlation given in Fig. 6.4.

Coal properties are generally more difficult to obtain unless all the coal comes from a single mine. This information should be requested from the coal supplier. If more than one source is used, it may be necessary to submit a fuel sample to a testing laboratory for direct determination of the heating value and an analysis of constituents.

Combustion air conditions can be measured directly with a thermometer and a sling psychrometer in the vicinity of the air intake. The moisture in the air, however, has a relatively small effect on boiler or heater losses, and for most industrial tests it should be sufficient to use local weather data, corrected to conditions at the boiler. The temperature should actually be measured at the intake; if it is significantly higher than the outside air temperature, the specific humidity of the outside air should be used rather than taking the ambient relative humidity (typically supplied by the weather service) and applying it at the intake temperature. Suppose, for example, the outside air is at 40°F and 70% relative humidity, while the temperature near the boiler intake is at 90°F. From the psychrometric chart, Fig. 6.9, the specific humidity of the outside air is 0.0035 lb$_m$ moisture/lb$_m$ dry air, and since this value does not change as the air is heated to 90°F, the intake air has the same moisture content. However, if the same relative humidity of 70% is assumed at 90°F, a moisture content of 0.022 lb$_m$ moisture/lb$_m$ dry air is obtained from the chart. Even though the effect of this large error in moisture on the overall boiler efficiency calculation is relatively small, it is easy enough to correct as outlined above.

Perhaps the single most important variable in the boiler test is the flue gas temperature. Fortunately, it is also one of the easiest to determine, but precautions must be taken to ensure the accuracy of the measurement. A simple thermometer with a bimetallic element can be used when the installation is visually accessible. When remote sensing is required, a thermocouple element with a dial-type readout or permanent recording device is usually used. In either case, some care must be exercised in installing the sensor. If the stack is not insulated, or the installation point is too close to the last heat transfer surface, the flue gas may not be well mixed and a significant temperature gradient may exist across the stack. To check this, the sensor should extend well into the stack and temperature measurements should be checked at several points across the diameter to see whether substantial variation exists. If it does another site should be selected or an averaging procedure used to correct the indicated temperature. Another reason for using a long sensor is to minimize conduction error in the measurement. If a short sensor length is used, heat conduction through the protecting sheath to the outside air can reduce the temperature at the sensor, causing it to read lower than the actual flue gas temperature. A rough rule of thumb is to use a minimum stem length of 10–20 times the sheath diameter to minimize conduction error. While it is desirable to use a sensor that extends as close to the stack center as possible, this is not always achievable in practice, particularly for large-diameter stacks.

An additional consideration in selecting a location for measuring flue gas temperature is avoiding radiation error. If the sensor is within the line of sight of the furnace flame or radiant walls, thermal radiation directly to the sensing element will cause it to read higher than the actual gas temperature at that point. Thus, the measuring location should be sufficiently far downstream to prevent the sensor from "seeing" the radiant zone.

Next to stack gas temperature, the flue gas analysis is the most critical piece of data in a combustion test. Specifically, the percentage of excess air and quantity of combustibles in the flue gas must be determined. Traditionally, the Orsat analyzer (see Fig. 2.9) has been used to determine the flue gas analysis. An Orsat analysis has the advantage that all components are determined from a single gas sample; its main disadvantage is the incovenience of handling the various chemical solutions; also, the accuracy of the resulting data is to some extent dependent on the skill of the technician carrying out the test.

A more convenient method is the use of a single-component chemical analyzer like the one shown in Fig. 2.10. This device works on the same principle as an Orsat analyzer, but measures only one component of the flue gas, either oxygen or carbon dioxide. Excess air can be correlated directly to either of these constituents for a given fuel. Figure 6.12 shows the relation between excess air and flue gas carbon dioxide or oxygen for various fuels. Note that the excess air/oxygen curves show much less variation from fuel to fuel than do the excess air/CO_2 curves. Thus, if a single measurement is to be made to determine excess air, oxygen is the preferred measurement. When using a single-component O_2 analyzer, CO in the flue gas must be measured separately, using a device such as the one shown in Fig. 2.11 or an electronic analyzer (Fig. 2.12). Because the electronic analyzer continuously draws a sample through a built-in pump and has a relatively short response time of

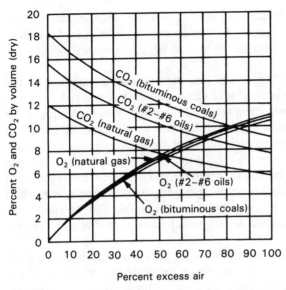

Figure 6.12 Oxygen and carbon dioxide in flue gas versus excess air for various fuels.

several seconds, it is particularly useful for making on-line boiler adjustments; in contrast, with the wet analyzers described above, adjustments must be made by "cut-and-try" due to the lengthy procedure (at best several minutes) involved in determining the composition of the sample.

Boiler blowdown rate can be measured directly by using a weigh-bucket technique, as described in Chapter 7 for measuring condensate from steam-heated equipment. Alternatively, blowdown rate can be estimated from the mass balance of dissolved solids in the boiler makeup water. This requires sampling of the makeup and blowdown waters and determination of the dissolved solids in each.

Procedural Considerations While it is sometimes possible to test combustion equipment performance while the unit is in production operation, it is usually necessary to interfere with normal operation at least to a minor degree in order to maintain a constant test condition long enough for the required measurements. Thus, it is important that the test procedure be organized and systematic to minimize disruption of production.

Boilers and heaters should always be tested at several firing rates ranging from full load to 50% load or below. Figure 6.13 illustrates a typical variation in boiler losses with load. Maximum efficiency is generally achieved at or near full load; at reduced load, efficiency tends to drop off, particularly below about 50% of rated capacity. The extent of this efficiency drop is dependent on a number of factors; the type of control system is perhaps the most significant. Simple controls of the mechanical linkage type generally provide too much excess air at low loads to compensate for inherent inaccuracy in the control settings. By testing over a range of loads, a clear picture of the actual efficiency of the system during its normal operational variations can be obtained.

In the preceding section it was pointed out that installation of the temperature sensor is important and that significant errors can be introduced if the sensor is improperly placed. The same is true for sampling of the flue gas. A stainless steel sampling tube should always be used; carbon steel, copper, and brass are unsatisfactory, since these metals can react with oxygen in the sample, producing a reading lower than the actual flue gas value. The entry of the sampling tube into the stack should be well sealed to avoid infiltration of air into the sampling region; this is especially important in induced-draft systems. Samples should never be taken at the stack exit; recirculation of atmospheric air at this point makes oxygen readings meaningless.

In systems using liquid and gaseous fuels, the fuel pressure just ahead of the burner can be used as a rough measure of the firing rate, since fuel flow is approximately proportional to the square root of the pressure at the burner. (To be more precise, the flow is proportional to the square root of the pressure *difference* between the burner orifice and the combustion chamber; as an approximation, it is assumed that the combustion chamber pressure is small compared with the burner pressure.) Thus, for example, if the burner

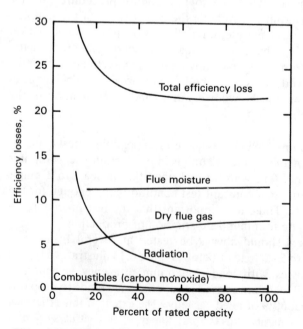

Figure 6.13   Effect of firing rate on boiler losses.

pressure on a gas-fired boiler is measured as 20 in. of water column with a simple manometer when the system is at 100% load, the load at 15 in. of water pressure is

$$\sqrt{\frac{15}{20}} = 0.866 \approx 87\% \text{ load}$$

The data and procedures described in this section provide the necessary inputs for a detailed determination of boiler or heater losses, using the molal calculation method covered earlier. In the following section, techniques for improving combustion system efficiency will be discussed.

6.8 EFFICIENCY IMPROVEMENT OF COMBUSTION SYSTEMS

Once the performance of a boiler or process heater has been determined, a number of measures are possible to improve it. These range from simple operating adjustments that may be implemented at little or no cost to major capital improvements. In this section, these options for performance improvement will be covered in detail.

6.8.1 Operating and Maintenance Improvements

Operating Strategy for Multiple Units   When several boilers or heaters are in use in a plant, reductions in overall fuel consumption can often be achieved by good operating strategy. As load increases, the most efficient units should be brought on line first and should be loaded up to their maximum efficiency points. As load varies, the variation of efficiency with load discussed in Section 6.7.3 should be kept in mind. It will be recalled from Fig. 6.13 that overall efficiency for a boiler tends to be relatively flat down to about 50% load, but drops off rapidly below that level. If, for example, two boilers are available, each with a rated capacity of 100,000 lb_m/hr, and a total steam demand of 140,000 lb_m/hr is required, it may be advantageous to operate both boilers at 70% load rather than operate one at 100% and the other at 40%.

In a very complex operation, such as a paper mill where some units are fired on waste fuels and others on purchased fuel, and where cogeneration of steam and power is practiced extensively, operating strategy can be extremely important. In such cases, detailed mathematical models can be developed and implemented on computers to provide continuous real-time optimization of system operations. The integration of such computer-based controls with measured real-time performance data will be discussed further in Section 6.8.2.

Minimizing Excess Air by Control Adjustment   As discussed earlier, excess air has a significant effect on boiler and heater efficiency. Figure 6.14 shows the percent improvement in efficiency that can typically be achieved by reducing excess air for given stack temperature. A boiler operating at a stack temperature of 500°F, for example, loses about 0.075% efficiency for each 1% of excess air above that required for complete combustion. A reduction in excess air from 35 to 10%, therefore, would produce an overall efficiency improvement of nearly 2%, assuming no reduction in stack temperature. In fact, the improvement will usually be higher, since in most cases reducing excess air will also cause a drop in stack temperature.

One approach to reduction of excess air is to install a better control system; this will be discussed in detail in a later section. However, significant improvements can often be achieved by simply "tuning" the existing control system. Figure 6.15 shows a typical curve of CO content versus excess oxygen for a boiler or heater with a simple parallel-positioning control system. As the O_2 level is reduced at a given firing setting, CO levels remain acceptably low until some minimum excess air is reached, beyond which combustion is incomplete and CO rises sharply. The sharpness of this "knee" in the O_2/CO curve will depend on the type of control system and burner being used. A CO level of about 400 ppm (0.04% concentration) is typically considered an acceptable threshold level.

This minimum percentage of oxygen can be determined for a given boiler or heater using the flue gas analysis techniques discussed in Section 6.7.3. While an Orsat or single-component wet test device can be used, an electronic analyzer that simultaneously reads out O_2 and combustibles (CO) almost instantly is much more effective for this type of procedure. The boiler or heater is held at a fixed firing rate (i.e., constant fuel flow) while the stack damper or fan inflow damper mechanism is adjusted to reduce airflow.

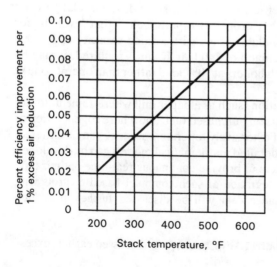

Figure 6.14 Curve showing percent efficiency improvement for each 1% reduction in excess air.

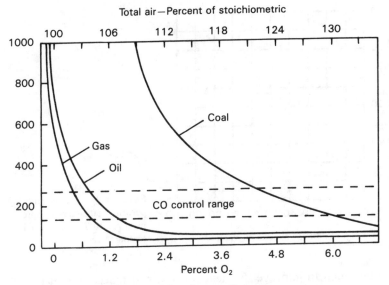

Figure 6.15 Relation between CO and excess O_2.

When a rapid increase in CO concentration in the flue gas is noted, the adjustment is backed off slightly to allow for variation in operating conditions and fixed at that level. Care must be taken to ensure that the setting is satisfactory at all firing rates, since typically boilers and heaters require a higher percentage of excess air at low-fire conditions than at high-fire. Simple parallel-positioning controls use a cam adjustment to accommodate this variation; adjustment of these controls generally requires some skill and should be carried out by a trained professional.

Reducing Heat Transfer Resistance High stack temperature in a boiler or heater indicates that the heat exchanger section of the system is not transferring enough of the energy contained in the combustion gas to the product being produced. Figure 6.16 shows the strong effect of stack temperature on overall efficiency of a combustion system. A boiler operating at 20% excess air, for example, will realize about 0.25% improvement in efficiency for every 10°F drop in stack temperature. Thus, an action that reduces stack temperature from 500 to 400°F will increase efficiency by 2.5%.

The rate of heat transfer is affected by both the heat transfer surface area and the heat transfer resistance from the gas to the product. Figure 6.17 shows a schematic cross section of a boiler tube. Deposits of soot on the fire side of the tube and scale on the water side act as thermal resistors in series with the metal tube wall. When scaling and soot deposition are extensive, the tube wall resistance is, in fact, small by comparison. While a shortage of heat exchange area can only be remedied by adding a heat recovery unit, several

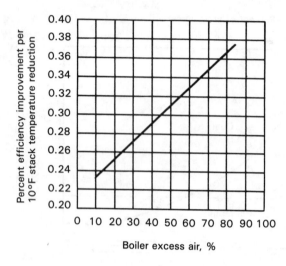

Figure 6.16 Curve showing percent efficiency improvement per 10 °F drop in stack temperature.

steps can be taken to improve the heat transfer resistance without major capital expenditure.

Reducing Fire-side Resistance Deposition of soot on boiler tubes is common in boilers and heaters fired with heavy liquid or solid fuels and uncommon in gas-fired systems. In recent years, even gas-fired units have become subject to fire-side fouling when operated temporarily on oil during periods of gas curtailment. The presence of soot deposits is readily determined during a regular maintenance inspection. These deposits, if due to infrequent oil firing, can be simply removed with a mechanical or pneumatic cleaning device.

If the boiler or heater is to be regularly fired on heavy liquid or solid fuels, soot blowers should be considered. Two basic types of soot blowers are in general use. A rotary soot blower periodically directs a jet of compressed air or steam against the tube surfaces while rotating the nozzle row to cover a wide area. The retracting-type soot blower works in a similar fashion, but traverses the tube rows rather than rotating the nozzles.

Another approach to reducing soot deposition effects is the use of fuel additives. This measure has been found effective in some cases where heavy

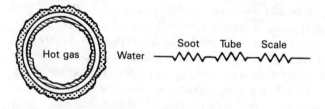

Figure 6.17 Cross section of boiler tube with thermal resistance analogy.

fuels (e.g., low-grade oils) are used regularly. It should be noted, however, that heavy smoking of oil combustion systems usually indicates that all is not right with the fuel atomization or combustion air controls. Additives are generally not the most economic permanent cure for these problems. They can be useful, however, in reducing corrosive effects of certain low-concentration combustion products and in minimizing emission problems in boilers and heaters.

One measure that has been found to be cost effective in reducing fire-side heat transfer resistance in older fire-tube boilers is the addition of turbulence promoters. These are twisted metal strips that are installed in selected portions of the fire-tube to cause swirling of the hot combustion gases in the tubes. The swirling motion enhances heat transfer from the gas to the tube wall. Reductions of 100°F or more in stack temperature are not unusual in fire-tube boilers manufactured before about 1970. In newer boilers these devices are usually not needed, as additional heat exchanger area has already been provided, which largely negates their usefulness. The installation of turbulence promoters should be undertaken by a specialist, since they affect the distribution of flue gas flow through the tubes. Proper length and placement of the strips are important, and some adjustment of the combustion air control system may also be required, since the flow resistance of the tubes is increased.

Reducing Water-side Resistance Crystalline deposits on the water side of boiler tubes can seriously degrade performance. These deposits form due to the presence in the boiler water of "inverse solubility salts," compounds whose solubility in water decreases with increasing temperature. Inverse solubility salts of importance in boiler operation include silica and carbonates of magnesium, calcium, and sodium. When they are present in solution in the boiler water, precipitation of these salts can occur on the hot boiler tubes, forming scale. The degradation in boiler efficiency depends on the type of scale as well as its thickness, since different substances have different thermal conductivities. In addition to reducing boiler efficiency, impurities in boiler water can create other problems as well. Table 6.19 lists various water contaminants and their effects on boiler operation. The most notable of these effects are corrosion of boiler tubes and priming and foaming, which produce carryover of liquid into the steam lines.

Figure 6.18 shows the reduction in overall boiler efficiency that can occur due to scaling of boiler tubes. For example, a 1/32-in. layer of hard iron-silica scale can degrade boiler efficiency by about 7%. Moreover, such a scale layer can deposit very rapidly at the elevated temperature of boiler tubes, and even when one starts with a clean surface a high thermal resistance may build up within hours. Scale growth rate tends to diminish with time; scale builds rapidly, then levels off to a more or less steady condition. Thus, it is not valid to assume that the scale layer on a boiler surface builds

Table 6.19 Water Impurities and Their Effects on Boilers

Contaminant	Composition	Source	Effect on boilers
Calcium bicarbonate	$Ca(HCO_3)_2$	Mineral deposits	Scale
Calcium carbonate	$CaCO_3$	Mineral deposits	Scale
Calcium chloride	$CaCl_2$	Mineral deposits	Scale
Calcium sulfate	$CaSO_4$	Mineral deposits	Scale and corrosion
Carbonic acid	H_2CO_3	Adsorption from the atmosphere, mineral deposits, decomposition of organic matter	Corrosion
Free acids	HCl, H_2SO_4	Mine drainage and industrial wastes	Corrosion
Magnesium bicarbonate	$Mg(HCO_3)_2$	Mineral deposits	Scale
Magnesium carbonate	$MgCO_3$	Mineral deposits	Scale
Magnesium chloride	$MgCl_2$	Industrial wastes	Scale and corrosion
Oil and grease	—	Domestic and industrial wastes	Corrosion, deposits, priming, and foaming
Organic matter and sewage	—		Corrosion, deposits, priming, and foaming
Oxygen	O_2	From atmosphere	Corrosion
Silica	SiO_2	Mineral deposits	Scale
Sodium bicarbonate	$NaHCO_3$	Mineral deposits	Priming, foaming, and embrittlement
Sodium carbonate	Na_2CO_3	Mineral deposits	Priming, foaming, and embrittlement
Sodium chloride	$NaCl$	Sewage, industrial wastes, and mineral deposits	Inert, but may be corrosive under some conditions
Suspended solids	—	Surface drainage and industrial wastes	Priming, foaming, sludge, or scale

Source: [6].

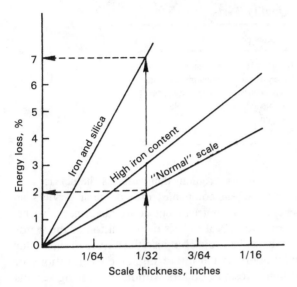

Figure 6.18 Boiler losses due to scale deposits.

up at a uniform rate over a long period of time between cleanings. Unfortunately, the scaling process becomes energy-wasteful in a short time.

Inspection of the water side of a boiler is a crucial step in an energy conservation program. If scale is present, even in small quantities, samples should be carefully removed and preserved for chemical analysis. Samples should also be taken from the mud and steam drums of water-tube boilers to permit a scale-removal specialist to determine the composition of the scale and assess how it is being formed. A badly scaled boiler can be successfully cleaned, either chemically or mechanically, if the scale composition is known. A word of caution is in order, however. Most really effective boiler cleaning procedures utilize acids, which can corrode tube surfaces and cause expensive damage if not properly handled. This is a job best handled by a specialist.

Once the boiler has been cleaned, steps should be taken to prevent the recurrence of scaling. Samples should be taken of the boiler water (by sampling from the blowdown line), raw feedwater entering the existing water treatment system, treated feedwater, and condensate being returned to the boiler feedwater tank. These samples may be sent to a commercial water treatment laboratory for analysis and recommendations on appropriate measures for water quality maintenance. Inexpensive water test kits can be purchased to monitor important water quality parameters such as pH and hardness on a continuing basis. Table 6.20 lists some of the basic tests and the species they indicate.

Water Treatment Systems Selection of the best type of water treatment sys-

Table 6.20 Basic Water Quality Tests

Test	Indicates presence of
Alkalinity	Bicarbonates, carbonates, hydroxides
Total hardness	Calcium and magnesium carbonates
Turbidity	Suspended solids, organic matter
Dissolved solids	Soluble salts of Mg, Ca, Na
Oxygen	Oxygen
Silica	Silica

tem for each application is a complex problem that requires the services of an expert. Most large water treatment companies provide such services at little or no cost, and independent water treatment consultants may also be employed to help with the problem. Not only is the condition of the water itself an important factor in these decisions, but also the nature of the boiler application. Low-pressure boilers can tolerate higher concentrations of dissolved impurities than high-pressure boilers without producing serious scaling problems. Water treatment can be considered in two phases: pretreatment of feedwater before delivery to the boiler and treatment of water in the boiler drum.

Boiler pressure is an important determinant in setting guidelines for acceptable levels of contaminants in boiler feedwater. Table 6.21 shows recommended feedwater impurity guidelines as a function of pressure established by the American Society of Mechanical Engineers (ASME).

Table 6.22 summarizes the types of pretreatment systems commercially available and their effect on feedwater quality. Aeration of makeup water can reduce concentrations of iron, manganese, and hydrogen sulfide by oxidation, and can reduce carbon dioxide content by mechanical agitation. However, aeration also saturates the feedwater with oxygen, and steps must

Table 6.21 Allowable Impurities in Boiler Feedwater

Drum pressure (psig)	Iron (ppm Fe)	Copper (ppm Cu)	Total hardness (ppm $CaCO_3$)
0–300	0.100	0.050	0.300
301–450	0.050	0.025	0.300
451–600	0.030	0.020	0.200
601–750	0.025	0.020	0.200
751–900	0.020	0.015	0.100
901–1000	0.020	0.015	0.050
1001–1500	0.010	0.010	0.0
1501–2000	0.010	0.010	0.0

Table 6.22   Feedwater Pretreatment Systems

1. Aeration	Reduce iron, manganese, hydrogen sulfide, carbon dioxide
2. Prechlorination	Reduce organic matter and microorganisms
3. Clarification	Reduce color and suspended matter
4. Cold process softening	Reduce suspended matter, iron, calcium, magnesium, bicarbonate, and possibly some silica
5. Cation exchange, sodium cycle	Reduce calcium and magnesium
6. "Split stream" cation exchangers, sodium and hydrogen cycles with degasifier	Reduce calcium, magnesium, sodium, and bicarbonate
7. Cation exchanger, sodium cycle followed by anion exchanger, chloride cycle for dealkalizing or hydroxide cycle for dealkalizing and desilicizing	Reduce calcium, magnesium, bicarbonate, and, if needed, silica
8. Hot lime followed by filters and cation exchangers, sodium cycle	Reduce suspended matter, iron, calcium, magnesium, bicarbonate, and silica
9. Demineralization	Reduce all cations, anions, and silica to very low concentrations

Source: [6].

be taken to deaerate the water before its introduction into the boiler. Chlorination is effective in reducing organic matter and microorganisms, and water containing a high content of suspended matter may be clarified by use of chemical clarifiers and by settling. Cold process softeners and ion exchange columns are effective in reducing hardness components and, to some extent, silica. Silica is an especially "bad actor" since it is relatively resistant to chemical removal once it has formed a hard scale on boiler tubes and it has a high thermal resistance. For removal of silica to low concentration levels, hot lime softeners and demineralization columns are recommended.

Boiler Drum Water Treatment   Once feedwater has been delivered to the boiler, special consideration must be given to water chemistry, since the boiling process increases concentrations of impurities in the boiler water by evaporation of the pure component. Further, the mechanism of scale formation on high-temperature surfaces dictates higher standards for impurity levels than in the makeup water. Table 6.23 indicates the guidelines established by the ASME for boiler water. Note again that the allowable concentrations are strongly dependent on boiler pressure.

Concentration of contaminants in boiler water is controlled by blowdown of a certain amount of water from the boiler drum, either periodically or continuously. Recovery of heat from boiler blowdown will be discussed in a subsequent section; the relative merits of continuous and periodic blowdown will be covered here.

In small systems and older boilers, conventional practice has been to blow down boiler water at periodic intervals by manually or automatically opening a blowdown valve. This drains off a quantity of water with high contaminant levels, allowing it to be replaced with lower-concentration makeup water and thereby diluting the contaminant concentrations in the boiler drum. Because the concentration of impurities rises and falls periodically, as shown in Fig. 6.19, impurity levels after blowdown must be considerably below allowable limits in order to keep the maximum concentration (just before blowdown) within tolerances. As a result, the average concentration level over a period of time is well below allowable levels for scale-free operation. This is wasteful of both expensive treated water and energy, since it is generally impractical to recover heat in a periodic blowdown system.

While blowdown is the primary means of maintaining acceptable impurity levels, the effectiveness of blowdown in minimizing scaling and corrosion can be enhanced by the use of chemical additives for boiler water treatment. The two most common additives are phosphates and chelates, used with other compounds such as organics and iron dispersants. Phosphates act to precipitate calcium and magnesium ions into an easily removable sludge before they form scale on the boiler tubes. Depending on the chemistry of the water, the sludge itself may have a tendency to bake onto the tube surfaces; this can be inhibited by the addition of organics and polymers that

Table 6.23 Allowable Impurities in Boiler Water

Drum pressure (psig)	Silica (ppm SiO$_2$)	Total alkalinity[a] (ppm CaCO$_3$)	Specific conductance (micromhos/cm)
0–300	150	700	7,000
301–450	90	600	6,000
451–600	40	500	5,000
601–750	30	400	4,000
751–900	20	300	3,000
901–1000	8	200	2,000
1001–1500	2	0[b]	150
1501–2000	1	0[b]	100

[a]Alkalinity not to exceed 10% of specific conductance.
[b]Minimum level of OH alkalinity in boilers below 1000 psi must be individually specified with regard to silica solubility and other components of internal treatment.

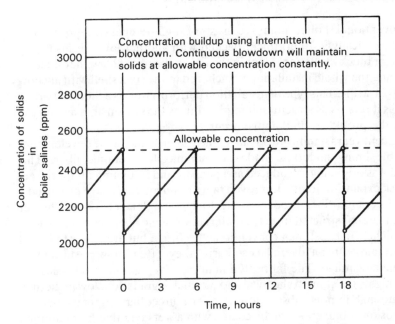

Figure 6.19 Dissolved solids concentration in boiler water for periodic versus continuous blowdown.

modify the sludge characteristics to make it less adherent. Chelates are organic salts that, when introduced into the boiler water, convert scale-forming salts into soluble heat-stable complexes. As with phosphate treatment, auxiliary compounds are usually introduced to prevent such undesirable secondary effects as corrosion and priming.

Maintenance and Replacement of Refractory Linings In Section 6.6 it was noted that for a boiler or process heater in good condition, surface radiation losses typically account for 1–2% of the fuel consumption. For an older system, these losses may be considerably higher. Conventional firebrick, under the effects of repeated heating and cooling, can become cracked, and separation can occur at the seams of the shell, allowing heat to leak through. The outside surface temperature is a good indicator of this problem, and techniques such as infrared imaging or scanning with a surface thermometer may be useful in inspecting the surface for evidence of lining deterioration. In units that undergo frequent cycling, such as heat-treating furnaces and brick kilns, a substantial amount of energy may be stored in the firebrick lining itself; this heat is lost every time the furnace is shut down. In induced-draft systems where the firebox is at negative pressure, leakage of air through openings in the shell dilutes the flue gas, reducing its heating effectiveness and increasing stack losses. In such instances, relining the boiler or heater may be worthwhile.

Conventional boiler linings consist of high-temperature refractory firebrick, typically 9–10 in. thick, backed with another layer of insulating firebrick or block insulation. Gaps in the firebrick are sealed with refractory cement and the whole installation is anchored to the steel shell with anchoring studs or strips. This type of installation performs well as a thermal barrier as long as it retains its structural integrity, but as the system ages and cracks appear, the problems cited above occur.

In recent years significant advances have been made in the development of flexible lining materials made of alumina-silica ceramic fiber. This material was originally introduced as a packing and insulating fill for refractory brick expansion joints. Improved fabrication methods have produced a range of ceramic fiber products, and today complete boiler and heater linings can be constructed from the material. Ceramic fiber materials tend to have a higher first cost than conventional firebrick, but are easier to install and maintain. On an overall cost basis, they often show a substantial economic advantage. The fabricated materials are light in weight, easy to handle, highly resistant to chemical and thermal degradation, low in thermal mass, and high in noise absorption. Areas of direct flame impingement or other erosion-prone areas can be coated with a wet layer that hardens upon drying, providing a wear-resistant coating.

Ceramic fiber linings have also been very effective in stacks and hot-air ducts, such as those used in air preheaters. Excellent resistance to chemical attack is a primary asset in this application, since condensation of sulfuric acid vapors can cause serious corrosion problems in uninsulated stacks and ducts.

6.8.2 Capital Improvements for Improved Energy Efficiency

Improved Combustion Controls Combustion control — that is, the proper proportioning of fuel and air over the entire operating range of the boiler or heater — has already been shown to have a major effect on combustion efficiency. In the preceding section, adjustment of existing controls was discussed; in this section, we will consider replacement or upgrading of existing controls.

The simplest type of boiler control system, and that most commonly in use on small industrial boilers and heaters, is the parallel-positioning or "jackshaft" system; see Fig. 6.20. In the jackshaft controller, steam pressure or process fluid temperature is the control signal. As pressure or temperature drops, indicating a demand for more heat, a master regulator adjusts the position of the jackshaft linkage connected in parallel to the fuel valve and the damper on the combustion air fans. Because of the fixed relationship between the position of the fuel valve and the air damper, this type of system has very limited flexibility to handle variations in air and fuel conditions. The varying air requirements with changes in firing rate can be accommodated to some extent by using a cam mechanism on the fuel valve,

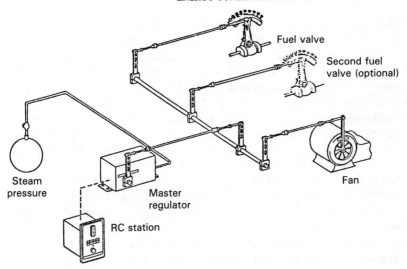

Figure 6.20 Common parallel-positioning (jackshaft) boiler controls.

such that it has a nonlinear characteristic with jackshaft position. A further refinement is the use of a second fuel valve with a different flow characteristic in parallel with the primary valve. However, despite mechanical refinements, the performance of this type of system is limited by the lack of any type of direct sensing of combustion conditions in setting the air-fuel ratio.

A number of factors affect the accuracy within which a given excess air setting can be maintained by the control system. These include variations in fuel and air properties, changes in the pressure-flow characteristics of the fans, valves, and ducts, and basic inaccuracy of the control system itself.

With simple nonfeedback controls, these inaccuracies can total on the order of ±30%, and to ensure that the minimum acceptable excess air is available under all operating conditions, it is necessary to set the controls to provide approximately 40% or more excess air. Compared with operation at minimum excess air, this represents an efficiency loss of about 5%.

A major improvement can be made by simply providing the operators with some means of directly measuring oxygen in the flue gas, permitting them to periodically adjust the regulator manually to maintain a given oxygen level. This essentially "closes the control loop," using the human operator as the controller. For continuous monitoring of oxygen, the wet analyzers discussed earlier are not satisfactory, and even the electronic sampling analyzer is not well suited for use on a continuous basis. Solid-state zirconium oxide sensors permanently installed in the stack are preferred for continuous oxygen monitoring. With oxygen trim, most of the random sources of error inherent in the jackshaft system are eliminated, permitting excess air to be reduced to within a few percent of minimum.

Operator control of combustion systems works satisfactorily so long as the boiler or heater runs at a more or less constant rate most of the time and can be constantly attended. When loads fluctuate significantly, as in plants where batch processing is used, or when the system is normally unattended, some form of automatic feedback control is required. One form of analog control actually meters fuel and air flow directly and maintains a predetermined relationship between the flows for each operating setting. This system is considerably more accurate than a jackshaft control, since it directly senses and controls the most relevant quantities. When coupled to an oxygen sensor, the direct fuel-metered system is capable of holding a minimum oxygen setting with a high degree of accuracy.

All of the oxygen-based systems discussed above depend on the basic premise that a predetermined level of flue-gas oxygen is optimum for all operating situations. This is true enough so long as fuel composition and physical properties (such as viscosity) are relatively constant. In some instances, however, multiple fuels may be fired in the same boiler or heater at different times. This is common, for example, where waste fuels are fired to supplement purchased fuels in pulp and paper mills, chemical plants, and oil refineries. In recent years, a new type of control system based on sensing of CO in the flue gas has been developed that is essentially insensitive to the type of fuel being fired. Figure 6.21 shows the relationship between boiler efficiency, excess oxygen (i.e., O_2 in the flue gas), and CO for a typical boiler or heater. As noted earlier, as excess air drops too low, CO rises sharply and boiler efficiency falls off rapidly. The peak of the efficiency curve typically occurs at about 150–300 ppm CO, a very low concentration. This concentration occurs at different excess air levels for different fuels.

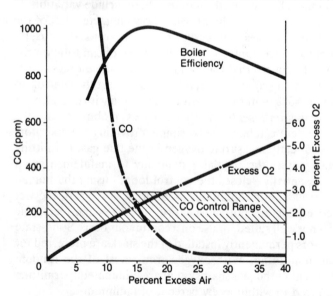

Figure 6.21   Relation between CO, excess O_2, and boiler efficiency.

A CO-based control system continuously measures the low CO levels in the flue gas (Fig. 6.22) using infrared absorption, and controls excess air automatically to keep the CO concentration in the 150–300-ppm range. This type of system thus automatically adjusts to the fuel being fired and can even handle multiple fuels fired simultaneously. Infrared sensing is expensive compared with oxygen sensing, and so these systems are generally economic only in large boilers or heaters where the firing conditions are unusual, as described above. In spite of their high first costs, they have typically shown very competitive payback periods in the types of installations for which they are best suited.

Low Excess Air Burners   The greatly increased interest in high-efficiency combustion systems since the mid-1970s has produced significant improvements in engineering of burners to permit operation at lower excess air levels. Burners are now available that are designed for firing of heavy oils with excess air on the order of 3%. This very high performance is achieved by a combination of carefully controlled aerodynamics in the combustion zone and improved atomization of the fuel. Cost savings accrue due to both excess air reduction and reduction in the steam or compressed air required for atomization. The diffuser section of the burner is designed to produce a high rate of recirculation in the exit region (Fig. 6.23). This enhances mixing of fuel and air and rapidly vaporizes fine fuel droplets by contact with the hot combustion gas. These burners are generally tailored to the specific combustion system to which they are fitted and are adjusted at the site to produce near-optimum flame conditions. At present they are rather expensive for use in small boilers and heaters but have proved effective in large units. As they come into more common use, the cost will undoubtedly drop and the principle of their operation will be incorporated in less expensive combustion systems as well.

Heat Recovery in Combustion Systems   Most combustion equipment built before the mid-1970s was designed for minimum lifetime cost in a cheap fuel market. When heat is absorbed from flue gas, the incremental cost of each additional Btu becomes greater as successively larger heat transfer areas are required. For a given fuel price, there is a point at which the additional cost of recovering heat cannot be justified in terms of fuel savings. With the dramatic rise in fuel prices since 1974, the economically optimum heat transfer area for a boiler or heater has become much greater. In new combustion equipment, additional area is now being incorporated within the unit; in older equipment, the additional area must be added externally in a retrofitted heat recovery unit.

Existing boilers and process heaters can be upgraded in heat transfer efficiency with reasonable economic return if flue gas temperatures exceed

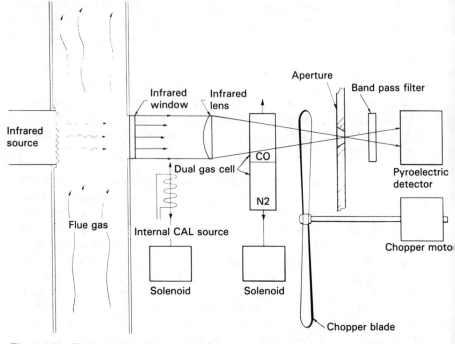

Figure 6.22   Typical stack installation for flue gas analyzer.

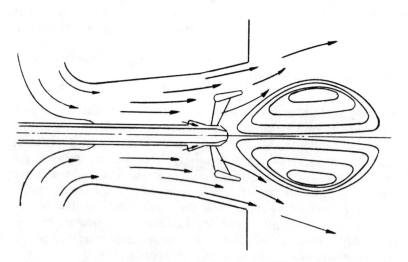

Figure 6.23 Recirculating combustion zone in low-excess-air burner.

about 450°F, or even lower in the case of large systems. Several factors affect the profitability of a heat recovery system, including the cost of fuel and its escalation rate, the age of the unit and its anticipated remaining life, and the operating conditions it serves. Three general types of systems are commonly used for waste heat recovery in boilers.

If stack temperatures exceed about 800–900°F, it is feasible to add additional steam generating capacity with a waste heat boiler. This is one of the more expensive types of heat recovery system; generally waste heat boilers are economic only in larger systems or when a demand for steam exists that would otherwise necessitate purchase of an additional boiler.

For stack temperatures below 500–600°F, an air preheater or economizer (which preheats boiler feedwater or process fluid) is generally a more attractive investment. Economizers, such as the unit shown in Fig. 6.24, are the less expensive of the two, because the low heat transfer resistance offered by the liquid on the feedwater or process side of the unit means less required heat transfer area. They are also relatively compact and in many cases can be installed directly in the stack without extensive modification. Installation costs are generally the lowest for all the types of heat recovery systems; this is quite important in overall economic terms, since installation can easily double or even triple the cost of a system. Efficiency improvements of 3–7% are common for economizer units installed on package boilers, often producing payback periods of 1–2 years.

Air preheaters require more heating surface to make up for the greater resistance to heat transfer in a gas-to-air unit and hence are more expensive than economizers for the same Btu recovery rate. The two most common types of air preheaters are the tubular or recuperative types and the rotary regenerator or "heat wheel."

Tubular preheaters contain a fixed bundle of tubes through which the flue gas passes (Fig. 6.25). Incoming combustion air passes over the outsides of the tubes in a single- or multipass arrangement. Small units of this type can be installed directly in the stack; larger units are installed externally and connected through a system of ducts. This type of unit is mechanically simple and easy to maintain.

Rotary air preheaters, as shown in Fig. 6.26, are more complex but can offer a significant advantage in terms of compactness. A large wheel containing a packing of metal or ceramic rotates slowly between the ducts containing hot and cold gas. As the wheel passes through the hot duct, the packing is heated and stores thermal energy. This energy is transferred to the incoming air as the wheel rotates to the cold side. A typical rate of rotation is 1–3 revolutions per minute, and thus very little mechanical power is required to drive the unit. Heat wheels are not suitable for stack mounting, but because of their compact design can often be used when space is limited. They are primarily suited for large systems and are extensively used in the electric power industry.

Figure 6.24 Economizer. (Courtesy Voss Finned Tube Products)

A third type of preheater that has attracted attention in recent years is the heat pipe exchanger, shown in Fig. 6.27. The heat pipe carries heat down its length with very little drop in temperature, since the primary mechanism for heat transport is through the latent heat of the vapor contained inside, and thus it has a very high effective thermal conductivity. The primary resistance to heat transfer in a preheater, however, is from the gases on each side of the exchanger to the heat pipe itself. Hence the main advantage of the

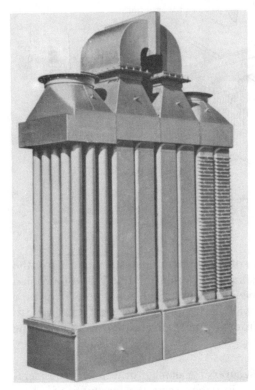

Figure 6.25 Tubular air preheater. (Courtesy Thermal Transfer Corp., Division of Kleinewefers, Monroeville, Pa.)

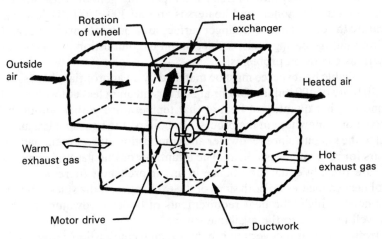

Figure 6.26 Rotary air heater.

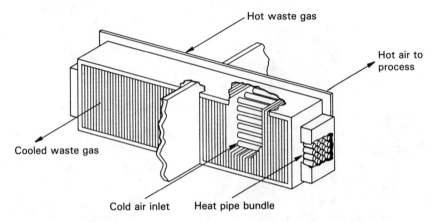

Figure 6.27 Heat pipe air preheater.

heat pipe exchanger is uniformity of the temperature distribution across the duct. This eliminates cold spots and can permit the flue gas to be cooled somewhat more than in a conventional tubular recuperator. Heat pipe exchangers can also be used to advantage in situations where it is desirable to physically separate the flue gas from the fluid being heated, as when heating a flammable process liquid.

Design Considerations in Heat Recovery A number of technical considerations must be taken into account in determining the feasibility of a heat recovery unit. The lower limit of flue gas temperature generally considered practical in industrial boilers and heaters is around 300–350°F. Below this temperature the cost of heat exchanger surface area becomes excessive, and when firing sulfur-bearing fuels, condensation of combustion products can cause serious corrosion problems.

The available heat source must be matched by an appropriate heat sink. The availability of 100 million Btu/hr at 200°F is meaningless unless there is a demand for that quantity of heat at that temperature, either within the boiler itself or somewhere in the process. In boilers, the heat sink is usually internal to the system, since feedwater and combustion air represent natural reservoirs for receiving waste heat. The quantities of flue gas and combustion air are roughly equal, and hence in an air preheater there is a good match of heat capacities on both sides of the exchanger. If the stack temperature is not too high, the heat requirements of boiler feedwater are also usually well matched to the available source.

Retrofit of a heat recovery unit in an existing combustion system is an economic decision. While recovery is among the most expensive investments for energy conservation, it often yields the greatest savings. Heat recovery systems are relatively complex, requiring special controls and careful attention to structural considerations. In addition to fuel savings and equipment costs, engineering, maintenance, tax consequences, and peripheral system modifications must be taken into account. For example, installation of a stack-mounted economizer will impose an extra pressure drop on the flue gas which must be compensated for by the fans. An air preheater may necessitate changes in the existing burners to accommodate heated air. Other considerations include avoidance of corrosion in the stack due to cooling of the combustion products below their dew point and planning of space and access for maintenance of the heat recovery system. A careful engineering study is usually justified.

Blowdown Heat Recovery The energy contained in boiler blowdown water is much smaller in magnitude than that in the flue gas, but can nevertheless be significant and worth recovering. The simplest arrangement for this purpose is a hairpin-type shell-and-tube heat exchanger of the type shown in Fig. 6.28, which transfers heat directly to the incoming makeup water. Blowdown rate is controlled by a thermostatic valve in the blow-off line leading to the discharge. If several boilers are used, a separate exchanger is usually installed for each boiler.

If the quantity of blowdown water is large and if there is an application for low-pressure steam, the exchanger may be preceded by a flash tank in which steam is generated by flashing as the high-pressure saturated water is throttled to lower pressure. With the flash tank arrangement, it is possible to feed a single heat recovery system from more than one boiler. In high-pressure boilers it may even be feasible to incorporate more than one flash tank, each operating at a different pressure to feed into the various steam headers.

6.9   CONCLUSION

This chapter has emphasized energy conservation considerations in fuel-

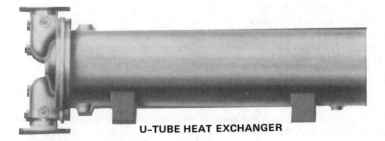

U-TUBE HEAT EXCHANGER

Figure 6.28   Blowdown heat recovery unit. (Courtesy Crane Co., Cochrane Division, King of Prussia, Pa.)

burning systems, often the most energy-intensive part of an industrial plant. To maintain a sound program of energy management for combustion equipment, it is necessary to understand the many factors affecting efficiency and to know how to evaluate actual performance of the equipment in operation. Although it is not possible to treat this complex subject exhaustively in a single chapter, the methods and concepts presented here should provide a good starting point for most purposes.

REFERENCES

1. Roger W. Sant, *The Least-Cost Energy Strategy,* Carnegie-Mellon University Press, Pittsburgh, Pa., 1979
2. Oak Ridge Associated Universities, *Industrial Energy Use Data Book,* ORAU-160, Oak Ridge, Tenn., 1980.
3. Babcock and Wilcox Co., *Steam: Its Generation and Use,* 38th ed., New York, 1972.
4. R. B. Knust, T. B. Hamilton, and R. P. Hensel, Large Steam Generator Design as Influenced by Low Rank Coal, presented at the American Power Conference, Chicago, April 1968.
5. J. G. Singer, *Combustion: Fossil Power Systems,* 3d ed., Combustion Engineering, Inc., Windsor, Conn., 1981.
6. Betz Laboratories, Inc., *Handbook of Industrial Water Conditioning,* 8th ed., Trevose, Pa., 1980.

PROBLEMS

6.1 Starting with the combustion equations in Table 6.2, verify the volumes and masses for reactants and products shown in Table 6.4 for methane and acetylene.

6.2 A natural gas sample of volume 2 ft^3 is drawn at 50 psia and 80°F. The sample is found to contain 85% methane, 12% ethane, 2% nitrogen, and 1% carbon dioxide by volume. How many moles of each constituent are in the sample, and what is the partial pressure of each?

6.3 Using the combustion data in Table 6.4, determine the theoretical air requirements (in both cubic feet and pounds mass) and the quantities (by both volume and mass) of each of the combustion products for the following fuels burned with 0% excess air:

(a) 300 ft^3 of Ohio natural gas (Table 6.7).
(b) 150 lb$_m$ of No. 2 fuel oil (Table 6.9); assume values of the ultimate analysis within the specified ranges which total to 100%.
(c) 1000 lb$_m$ of Texas lignite (Table 6.11).
(d) 100 lb$_m$ of refinery gas A (Table 6.16).
(e) 1 ton of municipal garbage (Table 6.17); assume values of the ultimate analysis within the specified ranges which total to 100%.

6.4 Octane (C_8H_{18}) is burned with 0% excess air at a constant pressure of 1 atm. The products of combustion are cooled to 90°F. How many pounds mass of water are produced per pound mass of fuel? If ambient air with a relative humidity of 90% and temperature of 80°F is used for combustion, how many pounds mass of water will be condensed per pound mass of fuel burned when the combustion products are cooled to 80°F?

6.5 A sample of by-product gas from a chemical plant has the following volumetric analysis: 13% CO_2, 17% CO, 20% H_2, 5% CH_4, 10% H_2O, 25% N_2, 7% C_2H_6, and 3% H_2S. Determine the ultimate analysis for this fuel in percent by weight.

6.6 A high-volatility Montana bituminous coal, whose properties are given in Table 6.11, is to be burned in a boiler with 40% excess air. The flue gas temperature is predicted to be 550°F when the ambient air is at 90°F and 70% relative humidity. Assuming no combustible material in the refuse, determine the stack and combustible losses for the boiler.

6.7 Louisiana natural gas is burned in a process heater with ambient air at 70°F and 65% relative humidity. The flue gas temperature is 395°F, and an Orsat analysis gives the following flue gas analysis: 10% CO_2, 4% O_2, 1% CO, and 85% N_2. Determine the stack and combustible losses for the heater.

6.8 Hog fuel derived from pine sawdust is used in a furniture-manufacturing operation for the power boiler. The ultimate analysis of the dry fuel as given by Table 6.13 is: H_2, 6.3%; C, 51.8%; N_2, 0.1%; O_2, 41.3%; and ash, 0.5%.

   The higher heating value of the fuel is 9,130 Btu/lb$_m$. The boiler stack is monitored for CO_2 and O_2, and an O_2 content of 5% and a CO_2 content of 12% (both by volume) are measured on a day when the ambient air is 80°F and 60% relative humidity. The flue gas temperature is measured at 350°F. What is the boiler efficiency?

6.9 A marine boiler on a banana boat burns No. 6 fuel oil with an ultimate analysis of 1.5% sulfur, 10.0% hydrogen, 88.0% carbon, and 0.5% ash. The higher heating value is 18,500 Btu/lb$_m$. The boiler is operated at 15% excess air on a day when the temperature at the air intake is 80°F and the relative humidity is 80%; the stack temperature is 450°F. Determine the efficiency.

6.10 Natural gas with a proximate analysis of 84.1% methane, 6.7% ethane, 0.8% carbon dioxide, and 8.4% nitrogen is burned with 120% theoretical air. Calculate the percent CO_2 and O_2 by volume in the flue gas and compare your results with Fig. 6.12.

6.11 No. 1 fuel oil with an ultimate analysis as follows is burned with atmospheric air at 80°F and 50% relative humidity: sulfur, 0.5%; hydrogen, 13.3%; carbon, 86.1%; oxygen, 0.1%.

(a) Find the Orsat analysis assuming 100% theoretical air.
(b) For 115% theoretical air, find the amount of water vapor in the products of combustion (lb_m H_2O/lb_m fuel).
(c) If the products of part (b) are cooled to 80°F, how much water would be condensed per pound mass of fuel?

6.12 A liquid fuel has an ultimate analysis of 84% carbon, 13% hydrogen, and 3% oxygen. When this fuel is burned, the Orsat analysis of the gaseous combustion products is 11.5% CO_2, 4.1% O_2, and 0.9% CO, taken at a total pressure of 14.5 psia.

(a) Determine the amount of air supplied per pound of fuel.
(b) Find the dew point of the water in the flue gas.

SEVEN

ENERGY CONSERVATION IN STEAM AND CONDENSATE SYSTEMS

7.1 INTRODUCTION

Nearly half of the energy used by industry goes into the production of process steam, approximately the same total energy usage as that required to heat all the homes and commercial buildings in the United States. Why is so much of our energy resource expended for the generation of industrial steam?

Steam is one of the most abundant, least expensive, and most effective heat transfer media obtainable. Water is found everywhere and requires little modification from its raw state to make it directly usable in process equipment. During boiling and condensation, if the pressure is held constant and both water and steam are present, the temperature also remains constant. Further, the temperature is uniquely fixed by the pressure; hence, by maintaining constant pressure, which is an easy parameter to control, excellent control of process temperature also can be maintained. The conversion of a liquid to a vapor absorbs large quantities of heat in each pound of water. The resulting steam is easy to transport, and, because it is so energetic, small quantities of it can move large amounts of heat. This means that relatively inexpensive pumping and piping can be used compared to that needed for other heating media.

This chapter is reproduced in part from the chapter titled Steam and Condensate Systems by P. S. Schmidt in *Energy Management Handbook,* Wayne Turner, ed., John Wiley & Sons, New York, 1982, by permission.

Lastly, the process of heat transfer by condensation, as in the jacket of a steam-heated vessel, is extremely efficient. High rates of heat transfer can be obtained with small equipment, saving both space and capital. For these reasons, steam is widely used as the heating medium in thousands of industries.

Before the mid-1970s, many steam systems in industry were energy-wasteful. This was not necessarily bad design for the times, because energy was so cheap that it was logical to save first cost, even at the expense of a considerable increase in energy requirements. But things have changed, and today it makes good sense to explore every possibility for improving the energy efficiency of steam systems.

This chapter will introduce some of the language commonly used in dealing with steam systems and will define the basic design constraints and applicability of some of the vast array of manufacturers' data and literature that appear in the marketplace. The various factors that produce inefficiency in steam system operations and some of the measures that can be taken to improve this situation also will be discussed. Simple calculation methods will be introduced to estimate the quantities of energy that may be lost and that may be partially recoverable by the implementation of energy conservation measures. Some important energy conservation areas pertinent to steam systems are covered in other chapters in this book and will not be repeated here. In particular, the reader is referred to Chapters 6 (Boilers), 10 (Industrial Insulation), and 12 (Cogeneration).

7.1.1   Components of Steam and Condensate Systems

Figure 7.1 shows a schematic of a typical steam system in an industrial plant. The boiler, or steam generator, produces steam at the highest pressure (and therefore the highest temperature) required by the various processes. This steam is carried from the boiler-house through large steam mains to the process equipment area. Here, transfer lines distribute the steam to each piece of equipment. If some processes require lower temperatures, the steam may be throttled to lower pressure through a pressure-regulating valve (designated PRV on the diagram) or through the back-pressure turbine. Steam traps located on the equipment allow condensate to drain back into the condensate return line, where it flows back to the condensate receiver. Steam traps also perform other functions, such as venting air from the system on start-up, which will be discussed in detail later.

The system shown in Fig. 7.1 is, of course, highly idealized. In addition to the components shown, other elements, such as strainers, check valves, and pumping traps, may be utilized. In some plants, condensate may simply be released to a drain and not returned; the potential for energy conservation through recovery of condensate will be discussed later. Also, in the system shown, the flash steam produced in the process of throttling across

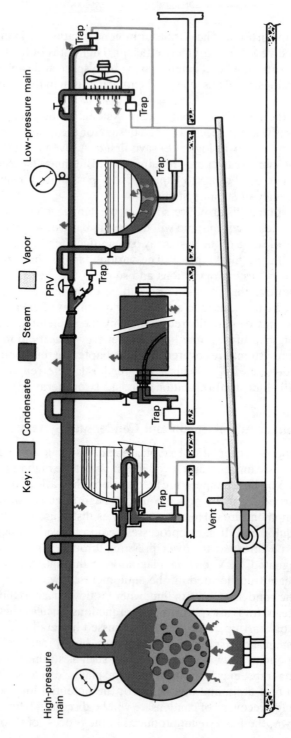

Key:

■ Condensate
■ Steam
□ Vapor

High-pressure main

PRV

Low-pressure main

Trap

Vent

Figure 7.1 Typical steam system components. (Courtesy, Armstrong International)

the steam traps is vented to the atmosphere. Prevention of this loss represents an excellent opportunity to save energy and money.

7.1.2 Energy Conservation Opportunities in Steam Systems

Many opportunities for energy savings exist in steam system operations, ranging from simple operating procedure modifications to major retrofits requiring significant capital expenditures. Table 7.1 shows a checklist of energy conservation opportunities applicable to most steam systems. It is helpful for energy conservation engineers to maintain such a running list applicable to their own situations. Ideas are frequently presented in the technical and trade literature, and plant operators often make valuable contributions, since they are the people closest to the problem.

To "sell" such improvements to plant management and operating personnel, it is necessary to demonstrate the economic value of a project or operating change. The following sections discuss the thermal properties of steam, how to determine the steam requirements of plant equipment, how to estimate the amount of steam required to make up for system losses, how to assign a dollar value to this steam, and various approaches to alleviating the losses.

7.2 THERMAL PROPERTIES OF STEAM

7.2.1 Definitions and Terminology

Before discussing numerical calculations of steam properties for various applications, it will be useful to review certain terms commonly used in steam system operations.

British thermal unit (Btu): One Btu is the amount of heat required to raise 1 pound of water 1°F in temperature. To get a perspective on this quantity, 1 cubic foot of natural gas at atmospheric pressure will release about 1,000 Btu's when burned in a boiler with no losses. This same 1,000 Btu's will produce a little less than 1 pound of steam at atmospheric conditions, starting from water at room temperature.

Boiling point: The boiling point is the temperature at which water begins to boil at a given pressure. The boiling point of water at sea-level atmospheric pressure is about 212°F. At high altitude, where the atmospheric pressure is lower, the boiling point is lower. Conversely, the boiling point of water rises with increasing pressure. In steam systems we usually refer to the boiling point as the saturation temperature.

Absolute and gage pressure: In steam system literature, two different pressures are frequently used. The *absolute pressure*, designated in pounds per square inch absolute (psia), is the true force per unit of area (e.g., pounds per square inch) exerted by the steam on the wall of the pipe or

Table 7.1 Checklist of Energy Conservation Opportunities in Steam and Condensate Systems

General operations

1. Review operation of long steam lines to remote single-service applications. Consider relocation or conversion of remote equipment, such as steam-heated storage tanks.
2. Review operation of steam systems used only for occasional service, such as winter-only steam tracing lines. Consider use of automatic controls, such as temperature-controlled valves, to ensure that the systems are used only when needed.
3. Implement a regular steam leak survey and repair program.
4. Publicize to operators and plant maintenance personnel the annual cost of steam leaks and unnecessary equipment operations.
5. Establish a regular steam-use monitoring program, normalized to production rate, to track progress in reduction of steam consumption. Publicize on a monthly basis the results of this monitoring effort.
6. Consider revision of the plantwide steam balance in multipressure systems to eliminate venting of low-pressure steam. For example, provide electrical backup for currently steam-driven pumps or compressors to permit shutoff of turbines when excess low-pressure steam exists.
7. Check actual steam usage in various operations against theoretical or design requirement. Where significant disparities exist, determine the cause and correct it.
8. Review pressure level requirements of steam-driven mechanical equipment to evaluate the feasibility of using lower pressure levels.
9. Review temperature requirements of heated storage vessels and reduce to minimum acceptable temperatures.
10. Evaluate production scheduling of batch operations and revise if possible to minimize to start-ups and shutdowns.

Steam trapping

1. Check sizing of all steam traps to ensure that they are adequately rated to provide proper condensate drainage. Also review types of traps in various services to ensure that the most efficient trap is being used for each application.
2. Implement a regular steam trap survey and maintenance program. Train maintenance personnel in techniques for diagnosing trap failure.

Condensate recovery

1. Survey condensate sources presently being discharged to waste drains for feasibility of condensate recovery.
2. Consider opportunities for flash steam utilization in low-temperature processes presently using first-generation steam.
3. Consider pressurizing atmospheric condensate return systems to minimize flash losses.

Mechanical drive turbines

1. Review mechanical drive standby turbines presently left in the idling mode and consider the feasibility of shutting down standby turbines.
2. Implement a steam turbine performance testing program and clean turbines on regular basis to maximize efficiency.
3. Evaluate the potential for cogeneration in multipressure steam systems presently using large pressure-reducing valves.

Insulation

1. Survey surface temperatures using infrared thermometry or thermography on insulated equipment and piping to locate areas of insulation deterioration. Maintain insulation on a regular basis.
2. Evaluate insulation of all uninsulated lines and fittings previously thought to be uneconomic. Rising energy costs have made insulation of valves, flanges, and small lines desirable in many cases where it was previously unattractive.
3. Survey the economics of retrofitting additional insulation on presently insulated lines and upgrade insulation if economical.

vessel containing it. However, we usually measure pressures with sensing devices that are exposed to the atmosphere outside and, that therefore, register an indication not of the true force inside the vessel but of the difference between that force and the force exerted by the outside atmosphere. This difference is the *gage pressure*, designated in pounds per square inch gage (psig). Since atmospheric pressure at sea level is usually around 14.7 psi, we can obtain the absolute pressure by simply adding 14.7 to the gage pressure reading. In tables of steam properties it is more common to see pressures listed in psia, and hence it is necessary to make the appropriate correction to the pressure indicated on a gauge.

Saturated and superheated steam: If cold water is put into a boiler and heated, its temperature will begin to rise until it reaches the boiling point. If we continue to heat the water, rather than continuing to rise in temperature, it begins to boil and produce steam. As long as the pressure remains constant, the temperature will remain at the saturation temperature for the given pressure, and the more heat added, the more liquid will be converted to steam. We call this boiling liquid a *saturated liquid*, and refer to the steam so generated as *saturated vapor*. We can continue to add more heat and will simply generate more saturated vapor (or simply *saturated steam*) until the water is completely boiled off. At this point, if we continue to add heat, the steam temperature will begin to rise once more. We call this *superheated steam*. This chapter will concentrate primarily on the behavior of saturated steam, which is the steam condition most commonly encountered in industrial process heating applications. Superheated steam is common in power generation and is often produced in industrial systems when cogeneration of power and process heat is used (Chapter 12).

Sensible and latent heat: Heat input that is directly registered as a change in temperature of a substance is called *sensible heat*, because we can "sense" it with our sense of touch or with a thermometer. For example, the heating of water mentioned above before it reaches the boiling point would be sensible heating. When the heat goes into the conversion of a liquid to a vapor in boiling, or vice versa in the process of condensation, it is termed *latent heat*. Thus, when a pound of steam condenses on a heater surface to produce a pound of saturated liquid at the same temperature, we say that it has released its latent heat. If the condensate cools further, it is releasing sensible heat.

Enthalpy: The total energy content of a flowing medium, usually expressed in Btu's per pound, is termed its *enthalpy*. The enthalpy of steam at any given condition takes into account both latent and sensible heat as well as the "mechanical" energy content reflected in its pressure. Hence, steam at 500 psia and 600°F will have a higher enthalpy than steam at the same temperature but at 300 psia. Also, saturated steam at any temperature and pressure has a higher enthalpy than condensate at the same conditions due to the latent heat content of the steam.

Specific volume: The specific volume of a substance is the amount of space (e.g., cubic feet) occupied by 1 pound of the substance. This term will become important in some of our later discussions, because steam normally occupies a much greater volume for a given mass than water (i.e., it has a much greater specific volume), and this must be taken into account when considering the design of condensate return systems.

Entropy: Entropy is a thermodynamic property, defined by the second law of thermodynamics, which is useful in accounting for irreversibilities in processes. In application, isentropic processes (i.e., processes in which entropy stays constant) in turbines, pumps, and compressors define the ideal to which real machines are compared. Turbine efficiency, for example, is defined as the ratio of actual power output to the power that could be produced if the steam were expanded isentropically from the given inlet state to the given exhaust pressure.

Condensate: Condensate is the liquid produced when steam condenses on a heater surface. As will be shown later, condensate still contains a significant fraction of its energy and can be returned to the boiler to conserve fuel.

Flash steam: When hot condensate at its saturation temperature corresponding to the elevated pressure in a heating vessel rapidly drops in pressure when passing through a steam trap or a valve, it is suddenly at a temperature above the saturation temperature for the new pressure. Steam is thus generated which absorbs sufficient energy to drop the temperature of the condensate to the appropriate saturation level. This is called *flash steam*, and the pressure reduction process is called *flashing*. In many condensate return systems, flash steam is simply released to the atmosphere, but it may have practical applications in energy conservation.

Boiler efficiency: Boiler efficiency is the percentage of the energy released in the burning of fuel in a boiler which actually goes into the production of steam. The remaining percentage is lost through radiation from the boiler surfaces, blowdown of the boiler water to maintain satisfactory impurity levels, and loss of the hot flue gas up the stack. While this chapter will not discuss in detail the subject of boiler efficiency, which is covered in Chapter 6, it is important to recognize that this parameter relates the energy savings obtainable by conserving steam to the fuel savings obtainable at the boiler, a relation of obvious economic importance. Thus, if we save 100 Btu of steam energy and have a boiler with an efficiency of 80%, the actual *fuel* energy saved would be 100/0.80, or 125 Btu. Because boilers always have an efficiency less than 100% (usually around 75–80%) there is a built-in "amplifier" of any energy savings effected in the steam system.

7.2.2 Properties of Saturated Steam

In calculating the energy savings obtainable through various measures, it is important to understand the quantitative thermal properties of steam and condensate. Table 7.2 is an abbreviated compilation of the properties of saturated steam. More complete tables are contained in Appendix B.

Columns 1 and 2 of Table 7.2 list various pressures, either in gage (psig) or absolute (psia). Note that these two pressures always differ by about 15 psi (14.7 to be more precise). Remember that the former represents the pressure indicated on a normal pressure gauge, while the latter represents the true pressure inside the line. Column 3 shows the saturation temperature corresponding to each of these pressures. Note, for example, that at an absolute pressure of 14.696 (the normal pressure of the atmosphere at sea level) the saturation temperature is 212°F — the figure we are all familiar with.

Suppose that a pressure of 150 psi is indicated on the pressure gauge on a steam line. What is the temperature inside the line? Coming down column 1, we find a pressure of 150, and moving over to column 3, we note that the corresponding steam temperature at this pressure is about 366°F.

Column 4 lists the sensible heat of the saturated liquid in Btu's per pound of water. We can see at the head of the column that this sensible heat is designated as 0 at a temperature (column 3) of 32°F. This is an arbitrary reference point, and therefore the heat indicated at any other temperature tells us the amount of heat added to raise the water from an initial value of 32°F to that temperature. For example, referring back to our 150-psig steam, the water contains about 338.5 Btu per pound; starting from 32°F, 10 lb of water would contain 10 times this number, or about 3385 Btu. We can also substract one number from another in this column to find the amount of heat necessary to raise the water from one temperature to another. If the water started at 101.74°F, it would contain a heat of 69.7 Btu/lb, and to raise it from this temperature to 366°F would require 338.5 – 69.7, or about 268.8 Btu for each pound of water. Column 5 shows the latent heat content of a pound of steam for each pressure. For our example of 150 psig, we can see that it takes about 857 Btu to convert each pound of saturated water into saturated steam. Note that this is a much larger quantity than the heat content of the water alone, confirming the earlier observation that steam is a very effective carrier of heat; in this case, each pound can give up 857 Btu when condensed on a surface back to saturated liquid. Column 6, the enthalpy of the saturated steam, is simply the sum of columns 4 and 5, since each pound of steam contains both the latent heat required to vaporize the water and the sensible heat required to raise the water to the boiling point in the first place.

Column 7 shows the specific volume of the saturated steam at each pressure. Note that as the pressure increases, the steam is compressed; that is, it occupies less space per pound. Steam at 150 psig occupies only 2.75 ft^3/lb; if released to atmospheric pressure (0 psig) it would expand to nearly 10 times this volume. By comparison, saturated liquid at atmospheric pressure has a specific volume of only 0.017 ft^3/lb (not shown in Table 7.2), and it changes only a few percent over the entire pressure range of interest

Table 7.2 Abbreviated Thermodynamic Properties of Saturated Steam

(1) Gage pressure	(2) Absolute pressure (psia)	(3) Steam temp. (°F)	(4) Enthalpy of saturated liquid (Btu/lb)	(5) Latent heat (Btu/lb)	(6) Enthalpy of steam (Btu/lb)	(7) Specific volume (ft³/lb)
Inches of vacuum						
29.743	0.08854	32.00	0.00	1075.8	1075.8	3306.00
29.515	0.2	53.14	21.21	1063.8	1085.0	1526.00
27.886	1.0	101.74	69.70	1036.3	1106.0	333.60
19.742	5.0	162.24	130.13	1001.0	1131.1	73.52
9.562	10.0	193.21	161.17	982.1	1143.3	38.42
7.536	11.0	197.75	165.73	979.3	1145.0	35.14
5.490	12.0	201.96	169.96	976.6	1146.6	32.40
3.454	13.0	205.88	173.91	974.2	1148.1	30.06
1.418	14.0	209.56	177.61	971.9	1149.5	28.04
psig						
0.0	14.696	212.00	180.07	970.3	1150.4	26.80
1.3	16.0	216.32	184.42	967.6	1152.0	24.75
2.3	17.0	219.44	187.56	965.5	1153.1	23.39
5.3	20.0	227.96	196.16	960.1	1156.3	20.09
10.3	25.0	240.07	208.42	952.1	1160.6	16.30
15.3	30.0	250.33	218.82	945.3	1164.1	13.75
20.3	35.0	259.28	227.91	939.2	1167.1	11.90
25.3	40.0	267.25	263.03	933.7	1169.7	10.50
30.3	45.0	274.44	243.36	928.6	1172.0	9.40
40.3	55.0	287.07	256.30	919.6	1175.9	7.79
50.3	65.0	297.97	267.50	911.6	1179.1	6.66
60.3	75.0	307.60	277.43	904.5	1181.9	5.82
70.3	85.0	316.25	286.39	897.8	1184.2	5.17
80.3	95.0	324.12	294.56	891.7	1186.2	4.65
90.3	105.0	331.36	302.10	886.0	1188.1	4.23
100.0	114.7	337.90	308.80	880.0	1188.8	3.88
110.3	125.0	344.33	315.68	875.4	1191.1	3.59
120.3	135.0	350.21	321.85	870.6	1192.4	3.33
125.3	140.0	353.02	324.82	868.2	1193.0	3.22
130.3	145.0	355.76	327.70	865.8	1193.5	3.11
140.3	155.0	360.50	333.24	861.3	1194.6	2.92
150.3	165.0	365.99	338.53	857.1	1195.6	2.75
160.3	175.0	370.75	343.57	852.8	1196.5	2.60
180.3	195.0	379.67	353.10	844.9	1198.0	2.34
200.3	215.0	387.89	361.91	837.4	1199.3	2.13
225.3	240.0	397.37	372.12	828.5	1200.6	1.92
250.3	265.0	406.11	381.60	820.1	1201.7	1.74
	300.0	417.33	393.84	809.0	1020.8	1.54
	400.0	444.59	424.00	780.5	1204.5	1.16
	450.0	456.28	437.20	767.4	1204.6	1.03
	500.0	467.01	449.40	755.0	1204.4	0.93
	600.0	486.21	471.60	731.6	1203.2	0.77
	900.0	531.98	526.60	668.8	1195.4	0.50
	1200.0	567.22	571.70	611.7	1183.4	0.36
	1500.0	596.23	611.60	556.3	1167.9	0.28
	1700.0	613.15	636.30	519.6	1155.9	0.24
	2000.0	635.82	671.70	463.4	1135.1	0.19
	2500.0	668.13	730.60	360.5	1091.1	0.13
	2700.0	679.55	756.20	312.1	1068.3	0.11
	3206.2	705.40	902.70	0.0	902.7	0.05

here. Thus, a pound of saturated liquid condensate at 212°F will expand more than 1600 times in volume in being converted to a vapor. This shows that piping systems for the return of condensate from steam-heated equipment must be sized primarily to accommodate the large volume of flashed vapor and that the volume occupied by the condensate itself is relatively small.

The steam tables can be a valuable tool in estimating energy savings, as illustrated in the following example.

Example 7.1

A 100-ft run of 6-in. steam piping carries saturated steam at 95 psig. Tables obtained from an insulation manufacturer indicate that the heat loss from this piping run is presently 110,000 Btu/hr. With proper insulation, the manufacturer's tables indicated that this loss could be reduced to 500 Btu/hr. How many pounds per hour of steam saving does this installation represent? If the boiler is 80% efficient, what would be the resulting fuel savings?

From the insulation manufacturer's data, we can find the reduction in heat loss:

$$\text{Heat loss reduction} = 110,000 - 500 = 109,500 \text{ Btu/hr}$$

From Table 7.2 at 95 psig (halfway between 90 and 100), the total enthalpy of the steam is about 1,188.4 Btu/lb. The steam saving is therefore:

$$\text{Steam saving} = \frac{109,500 \text{ Btu/hr}}{1,188.4 \text{ Btu/lb}}$$

$$\cong 92 \text{ lb/hr}$$

Assume that condensate is returned to the boiler at about 212°F; thus, the condensate has a heat content of about 180 Btu/lb. The heat required to generate 95-psig steam from this condensate is 1,188.4 − 180, or 1,008.4 Btu/lb. If the boiler is 80% efficient, therefore;

$$\text{Fuel saving} = \frac{1,008.4 \text{ Btu/lb} \times 92 \text{ lb/hr}}{0.80}$$

$$\cong 1.16 \text{ million Btu/hr}$$

7.2.3 Properties of Superheated Steam and Subcooled Liquid

If additional heat is added to saturated steam with no liquid remaining, it begins to superheat and the temperature will rise. Table 7.3 shows some thermodynamic properties of superheated steam; more extensive tables are included in Appendix B. Unlike the saturated steam of Table 7.2, where each pressure had only a single temperature (the saturation temperature) associated with it, superheated steam may exist, for a given pressure, at any temperature above the saturation temperature. Thus properties must be tabulated as functions of both temperature and pressure, rather than pressure alone. With this exception, the values in the superheated steam table may be used exactly like those in Table 7.2

Note the vertical line in each column of Table 7.3. Properties listed to the left of the line are for subcooled liquid, i.e., liquid cooled below satura-

Table 7.3 Thermodynamic Properties of Superheated Steam and Subcooled Liquid

Absolute pressure (psi)		Temperature (°F)														
		100	200	300	400	500	600	700	800	900	1000	1100	1200	1300	1400	1500
1	v	0.0161	392.5	452.3	511.9	571.5	631.1	690.7								
	h	68.00	1150.2	1195.7	1241.8	1288.6	1336.1	1384.5								
5	v	0.0161	78.14	90.24	102.24	114.21	126.15	138.08	150.01	161.94	173.86	185.78	197.70	209.62	221.53	233.45
	h	68.01	1148.6	1194.8	1241.3	1288.2	1335.9	1384.3	1433.6	1483.7	1534.7	1586.7	1639.6	1693.3	1748.0	1803.5
10	v	0.0161	38.84	44.98	51.03	57.04	63.03	69.00	74.98	80.94	86.91	92.87	98.84	104.80	110.76	116.72
	h	68.02	1146.6	1193.7	1240.6	1287.8	1335.5	1384.0	1433.4	1483.5	1534.6	1586.6	1639.5	1693.3	1747.9	1803.4
15	v	0.0166	0.0166	29.899	33.963	37.985	41.986	45.978	49.964	53.946	57.926	61.905	65.882	69.858	73.833	77.807
	h	68.04	168.09	1192.5	1239.9	1287.3	1335.2	1383.8	1433.2	1483.4	1534.5	1586.5	1639.4	1693.2	1747.8	1803.4
20	v	0.0166	0.0166	22.356	25.428	28.457	31.466	34.465	37.458	40.447	43.435	46.420	49.405	52.388	55.370	58.352
	h	68.05	168.11	1191.4	1239.2	1286.9	1334.9	1383.5	1432.9	1483.2	1534.3	1586.3	1639.3	1693.1	1747.8	1803.3
40	v	0.0166	0.0166	11.036	12.624	14.165	15.685	17.195	18.699	20.199	21.697	23.194	24.689	26.183	27.676	29.168
	h	68.10	168.15	1186.6	1236.4	1285.0	1333.6	1382.5	1432.1	1482.5	1533.7	1585.8	1638.8	1692.7	1747.5	1803.0
60	v	0.0161	0.0166	7.257	8.354	9.400	10.425	11.438	12.446	13.450	14.452	15.452	16.450	17.448	18.445	19.441
	h	68.15	168.20	1181.6	1233.5	1283.5	1332.3	1381.5	1431.3	1481.8	1533.2	1585.3	1638.4	1692.4	1747.1	1802.8
80	v	0.0161	0.0166	0.0175	6.218	7.018	7.794	8.560	9.319	10.075	10.829	11.581	12.331	13.081	13.829	14.577
	h	68.21	168.24	269.74	1230.5	1281.3	1330.9	1380.5	1430.5	1481.1	1532.6	1584.9	1638.0	1692.0	1746.8	1802.5
100	v	0.0161	0.0166	0.0175	4.935	5.588	6.216	6.833	7.443	8.050	8.655	9.258	9.860	10.460	11.060	11.659
	h	68.26	168.29	269.77	1227.4	1279.3	1329.6	1379.5	1429.7	1480.4	1532.0	1584.4	1637.6	1691.6	1746.5	1802.2
120	v	0.0161	0.0166	0.0175	4.0786	4.6341	5.1637	5.6831	6.1928	6.7006	7.2060	7.7096	8.2119	8.7130	9.2134	9.7130
	h	68.31	168.33	269.81	1224.1	1277.4	1328.1	1378.4	1428.8	1479.8	1531.4	1583.9	1637.1	1691.3	1746.2	1802.0
140	v	0.0161	0.0166	0.0175	3.4661	3.9526	4.4119	4.8585	5.2995	5.7364	6.1709	6.6036	7.0349	7.4652	7.8946	8.3233
	h	68.37	168.38	269.85	1220.8	1275.3	1326.8	1377.4	1428.0	1479.1	1530.8	1583.4	1636.7	1690.9	1745.9	1801.7

Subcooled Liquid

Superheated Vapor

		Subcooled Liquid			Superheated Vapor												
160	v	0.0161	0.0166	0.0175	3.0060	3.4413	3.8480	4.2420	4.6295	5.0132	5.3945	5.7741	6.1522	6.5293	6.9055	7.2811
	h	68.42	168.42	269.89	1217.4	1273.3	1325.4	1376.4	1427.2	1478.4	1530.3	1582.9	1636.3	1690.5	1745.6	1801.4
180	v	0.0161	0.0166	0.0174	2.6474	3.0433	3.4093	3.7621	4.1084	4.4505	4.7907	5.1289	5.4657	5.8014	6.1363	6.4704
	h	68.47	168.47	269.92	1213.8	1271.2	1324.0	1375.3	1426.3	1477.7	1529.7	1582.4	1635.9	1690.2	1745.3	1801.2
200	v	0.0161	0.0166	0.0174	2.3598	2.7247	3.0583	3.3783	3.6915	4.0008	4.3077	4.6128	4.9165	5.2191	5.5209	5.8219
	h	68.52	168.51	269.96	1210.1	1269.0	1322.6	1374.3	1425.5	1477.0	1529.1	1581.9	1635.4	1689.8	1745.0	1800.9
250	v	0.0161	0.0166	0.0174	0.0186	2.1504	2.4662	2.6872	2.9410	3.1909	3.4382	3.6837	3.9278	4.1709	4.4131	4.6546
	h	68.66	168.63	270.05	375.10	1263.5	1319.0	1371.6	1423.4	1475.3	1527.6	1580.6	1634.4	1688.9	1744.2	1800.2
300	v	0.0161	0.0166	0.0174	0.0186	1.7665	2.0044	2.2263	2.4407	2.6509	2.8585	3.0643	3.2688	3.4721	3.6746	3.8764
	h	68.79	168.74	270.14	375.15	1257.7	1315.2	1368.9	1421.3	1473.6	1526.2	1579.4	1633.3	1688.0	1743.4	1799.6
350	v	0.0161	0.0166	0.0174	0.0186	1.4913	1.7028	1.8970	2.0832	2.2652	2.4445	2.6219	2.7980	2.9730	3.1471	3.3205
	h	68.92	168.85	270.24	375.21	1251.5	1311.4	1366.2	1419.2	1471.8	1524.7	1578.2	1632.3	1687.1	1742.6	1798.9

Source: [2].

tion temperature for the given pressure. Values listed to the right are in the superheat region.

Example 7.2

Suppose that in the preceding example, the steam line is carrying superheated steam at 250 psia (235 psig) and 400°F (note that both temperature and pressure must be specified for superheated steam). For the same reduction in heat loss (108,500 Btu/hr), how many pounds per hour of steam are saved?

From Table 7.3, the enthalpy of steam at 250 psia and 400°F is 1,263.5 Btu/lb. Thus:

$$\text{Steam saving} = \frac{109,500 \text{ Btu/hr}}{1,263.5 \text{ Btu/lb}}$$

$$= 86.6 \text{ lb/hr}$$

7.2.4 Heat Transfer Characteristics of Steam

As mentioned in Section 7.1, steam is one of the most effective heat transfer media available. The rate of heat transfer from a fluid medium to a solid surface (such as the surface of a heat exchanger tube or a jacketed heating vessel) can be expressed by Newton's law of cooling:

$$q = h(T_f - T_s)$$

where q is the rate of heat transfer per unit of surface area (e.g., Btu/hr ft^2), h a proportionality factor called the heat transfer coefficient, T_f the temperature of the fluid medium, and T_s the temperature of the surface.

Table 7.4 shows the orders of magnitude of h for several heat transfer media. Condensation of steam can be several times as effective as the flow of water over a surface for the transfer of heat and may be a thousand times more effective than a gaseous heating medium such as air.

In a heat exchanger the overall effectiveness must take into account the fluid resistances on both sides of the exchanger and the conduction of heat through the tube wall. These effects are generally lumped into a single overall heat transfer coefficient, U, defined by

$$q = U(T_{f1} - T_{f2})$$

where q is defined as before, U is the conductance, and T_{f1} and T_{f2} are the temperatures of the two fluids. In addition, there is a tendency for fluids to deposit fouling layers of crystalline, particulate, or organic matter on transfer surfaces, which further impedes the flow of heat. This impediment is characterized by a fouling resistance, which for design purposes, is usually incorporated as an additional factor in determining the overall coefficient (see Section 8.3).

Table 7.5 illustrates typical values of U (not including fouling) and the fouling resistance for exchangers employing steam on the shell side versus

Table 7.4   Orders of Magnitude of Heat Transfer Coefficients

Heating process	Order of magnitude of h (Btu/hr ft^2 °F)
Free convection, air	1
Forced convection, air	5–10
Forced convection, water	250–1,000
Condensation, steam	5,000–10,000

exchangers using a light organic liquid (such as a typical heat transfer oil). The 30–50% higher U values for steam translate directly into a proportionate reduction in required heat exchanger area for the same fluid temperatures. Furthermore, fouling resistances for the steam-heated exchangers are 50–100% lower than for similar service with an organic heating medium, since pure steam contains no contaminants to deposit on the exchanger surface. From the design standpoint, this means that the additional heating surface incorporated to allow for fouling need not be as great. From the operating viewpoint, it translates into energy conservation, since more heat can be transferred per hour in the exchanger for the same fluid conditions, or the same heating duty can be met with a lower fluid temperature difference, if the fouling resistance is lower.

7.3   ESTIMATING STEAM USAGE AND ITS VALUE

To properly assess the worth of energy conservation improvements in steam systems, it is first necessary to determine how much steam is actually required to carry out a desired process, how much energy is being wasted through various system losses, and the dollar value of these losses. Such

Table 7.5   Comparison of Steam and Light Organics as Heat Exchange Media

Shell-side fluid	Tube-side fluid	Typical U (Btu/hr ft^2 °F)	Typical fouling resistance (hr ft^2 °F/Btu)
Steam	Light organic liquid	135–190	0.001
Steam	Heavy organic liquid	45–80	0.002
Light organic liquid	Light organic liquid	100–130	0.002
Light organic liquid	Heavy organic liquid	35–70	0.003

information will be needed to determine the potential gains achievable with insulation, repair or improvement of steam traps, and condensate recovery systems.

7.3.1 Determining Steam Requirements

Several approaches can be used to determine process steam requirements. In order of increasing reliability, they include the use of steam consumption tables for typical equipment, detailed system energy balances, and direct measurement of steam and/or condensate flows. The choice of which method to use depends on how critical the steam-using process is to the plant's overall energy consumption and how the data are to be applied.

For applications not requiring a high degree of accuracy, such as developing rough estimates of the distribution of energy within a plant, steam consumption tables have been developed for various kinds of process equipment. Table 7.6 shows steam consumption tables for a number of typical industrial and commercial applications. To illustrate, suppose we want to estimate the steam used in a soft drink bottling plant to wash 2,000 bottles per minute. From Table 7.6 we see that, typically, a bottle washer uses about 310 lb of 5-psig steam per hour for each 100 bottles per minute of capacity. For a washer with a 2,000 bottle/min capacity we would use about 20 times this value, or 6200 lb of steam per hour. Referring to Table 7.2, we see that 5-psig steam has a total enthalpy of about 1,156 Btu/lb. The hourly heat usage of this machine, therefore, would be approximately $1,156 \times 6,200$, or a little over 7 million Btu/hr. Remember that this is the heat content of the steam itself and not the fuel heat content required at the boiler, since the boiler efficiency has not yet been taken into account.

Note that most of the entries in Table 7.6 show the steam consumption "in use," and not the peak steam consumption during all phases of operation. These figures are fairly reliable for equipment that operates on a more or less steady basis; however, they may be quite low for batch processing operations, or operations where the load on the equipment fluctuates significantly during its operation. For this reason, steam equipment manufacturers recommend that estimated steam consumption values be multiplied by a factor of safety, typically between 2 and 5, to ensure that the equipment will operate properly under peak load conditions. This can be quite important from the standpoint of energy efficiency. For example, if a steam trap is sized for average load conditions only, during start-up or heavy load operations condensate will tend to back up into the heating vessel, reducing the effective area for condensation and hence reducing its heating capacity. For steam traps and condensate return lines, the incremental cost of slight oversizing is small, and factors of safety are used to ensure that the design is conservative enough to guarantee rapid removal of the condensate. Gross oversizing of steam traps, however, can also cause excessive steam loss. This

Table 7.6 Typical Steam Consumption Rates for Industrial and Commercial Equipment

Type of installation	Description	Typical pressure (psig)	Steam consumption (lb/hr) in use
Bakeries	Dough room trough, 8 ft long	10	4
	Oven, white bread, 120-ft^2 surface	10	29
Bottle washing	Soft drinks: per 100 bottles/min	5	310
Dairies	Pasteurizer, per 100 gal heated 20 min	15–75	232 (max)
Dishwashers	Dishwashing machine	15–20	60–70
Hospitals	Sterilizers, instrument, per 100 in.^{3} approximately	40–50	3
	Sterilizers, water, per 10 gal, approximately	40–50	6
	Disinfecting ovens, double door, 50–100 ft^3, per 10 ft^3 approximately	40–50	21
Laundries	Steam irons, each	100	4
	Starch cooker, per 10-gal cap	100	7
	Laundry presses, per 10-in. length, approximately	100	7
	Tumblers, 40-in. per 10-in. length, approximately	100	38
Plastic molding	Each 12- to 15-ft^2 platen surface	125	29
Paper	Corrugators per 1000 ft^2	175	29
manufacture	Wood pulp paper, per 100 lb paper	50	372
Restaurants	Standard steam tables, per foot length	5–20	36
	Steam-jacketed kettles, 25-gal stock	5–20	29
	Steam-jacketed kettles, 60-gal stock	5–20	58
	Warming ovens, per 20 ft^3	5–20	29
Silver mirroring	Average steam tables	5	102
Tire shops	Truck molds, large	100	87
	Passenger molds	100	29

point will be discussed more in the section on steam traps, where appropriate factors of safety for specific applications are given.

The second and generally more accurate approach to estimating steam requirements is by direct energy balance calculations for the process. A comprehensive discussion of energy balances for complex equipment is beyond the scope of this section; we will, however, illustrate a simple energy balance for equipment involving the heating of a single product.

The energy balance concept states simply that any energy put into a system with steam must be either absorbed by the product and/or the equipment itself, dissipated to the environment, carried out with the product, or carried out in the condensate. In Section 7.2 the concepts of sensible and latent heat were discussed in the context of heat absorption by water in the production of steam. We can extend these concepts to consider heat absorption by any material, such as the heating of air in a dryer or the evaporation of water in the process of condensing milk.

The *sensible* heat requirement of any process is defined in terms of the specific heat of the material being heated. Table 7.7 gives specific heats for a number of common substances. The specific heat specifies the number of Btu's required to raise 1 lb of a substance through a temperature rise of 1 °F. Remember that for water, 1 Btu is by definition the amount of heat required to raise 1 lb by 1°F. The specific heat of water, therefore, is exactly 1 (at least near normal ambient conditions). To see how the specific heat can be used to calculate steam requirements for the sensible heating of products, consider the following example.

Example 7.3

A paint dryer requires about 3,000 CFM of 200°F air, which is heated in a steam-coil unit. How many pounds of 50-psig steam does this unit require per hour?

The density of air at temperatures of several hundred degrees or below is about 0.075 lb/ft³. The number of pounds of air passing through the dryer is, then:

$$3{,}000 \text{ ft}^3/\text{min} \times 60 \text{ min/hr} \times 0.075 \text{ lb/ft}^3 = 13{,}500 \text{ lb/hr}$$

Suppose the air enters the steam-coil unit at 70°F. Its temperature will then be raised by 200 − 70 = 130°F. From Table 7.7 the specific heat of air is 0.24 Btu/lb °F. The energy required to provide this temperature rise is, therefore,

Table 7.7   Specific Heats of Common Materials

Material	Btu/lb °F	Material	Btu/lb °F
	Solids		
Aluminum	0.22	Iron, cast	0.49
Asbestos	0.20	Lead	0.03
Cement, dry	0.37	Magnesium	0.25
Clay	0.22	Porcelain	0.26
Concrete, stone	0.19	Rubber	0.48
Concrete, cinder	0.18	Silver	0.06
Copper	0.09	Steel	0.12
Glass, common	0.20	Tin	0.05
Ice, 32°F	0.49	Wood	0.32–0.48
	Liquids		
Acetone	0.51	Milk	0.90
Alcohol, methyl, 60–70°F	0.60	Naphthalene	0.41
Ammonia, 104°F	1.16	Petroleum	0.51
Ethlyene glycol	0.53	Soybean oil	0.47
Fuel oil, sp. gr. 86	0.45	Tomato juice	0.95
Glycerine	0.58	Water	1.00
	Gases (Constant-pressure specific heats)		
Acetone	0.35	Carbon dioxide	0.20
Air, dry, 32–392°F	0.24	Methane	0.59
Alcohol	0.45	Nitrogen	0.24
Ammonia	0.54	Oxygen	0.22

$$13,500 \text{ lb/hr} \times 0.24 \text{ Btu/lb °F} \times 130°F = 421,200 \text{ Btu/hr}$$

Employing the energy balance principle, whatever energy is absorbed by the product (air) must be provided by an equal quantity of steam energy, less the energy contained in the condensate.

From Table 7.2 the total enthalpy per pound of 50-psig steam is about 1,179.1 Btu, and the condensate (saturated liquid) has an enthalpy of 267.5 Btu/lb. Each pound of steam, therefore, gives up $1,179.1 - 267.5$, or 911.6 Btu (i.e., its latent heat) to the air. The steam required is, therefore,

$$\frac{\text{Heat required by air per hour}}{\text{Heat released per pound of steam}} = \frac{421,200}{911.6} = 462 \text{ lb/hr}$$

To illustrate how latent heat comes into play in the steam requirements of a process, consider another example, this time involving a steam-heat evaporator.

Example 7.4

A milk evaporator uses a steam-jacketed kettle in which milk is batch-processed at atmospheric pressure. The kettle has a capacity of 1,500 lb per batch. Milk is heated from 80 to 212°F, where 25% of its mass is then driven off as vapor. Determine the amount of 15-psig steam required per batch, not including the heating of the kettle itself.

We must first heat the milk from 70 to 212°F (sensible heating) and then evaporate off 0.25 × 1,500 = 375 pounds of water.

From Table 7.7, the specific heat of milk is 0.90 Btu/lb °F. The sensible heat requirement is, therefore,

$$0.90 \times 1,500 \text{ lb} \times (212 - 80)°F = 178,200 \text{ Btu/batch}$$

In addition, we must provide the latent heat to vaporize 375 lb or water at 212°F. From Table 7.2, 970.3 Btu/lb is required. The total latent heat is, therefore,

$$375 \times 970.3 = 363,863 \text{ Btu/batch}$$

and the total heat input is

$$363,863 + 178,200 = 542,063 \text{ Btu/batch}$$

This heat must be supplied as the latent heat of 15-psig steam, which, from Table 7.2, is about 945 Btu/lb. The total steam requirement is then

$$\frac{542,063 \text{ Btu/batch}}{945 \text{ Btu/lb}} = 574 \text{ lb of 15-psig steam/batch}$$

We could have also determined the start-up requirement to heat the steel kettle, if we could estimate its weight, by using the specific heat of 0.12 Btu/lb °F for steel, as shown in Table 7.7.

7.3.2 Estimating Surface and Leakage Losses

In addition to the steam required to carry out a process, heat is lost through the surfaces of pipes, storage tanks, and jacketed heater surfaces, and steam is lost through malfunctioning steam traps and leaks in flanges, valves, and other fittings. Estimation of these losses is important, because fixing them can often be the most cost-effective energy conservation measure available.

Figure 7.2 illustrates the annual heat loss, based on operation for 24 hr per day, 365 days per year, for bare steam lines at various pressures. For example, a 100-ft run of 6-in. line operating at 100 psig will lose about 1400 million Btu per year. The economic return on an insulation retrofit can easily be determined with price data obtained from an insulation contractor.

Figure 7.3 can be used to estimate heat losses from flat surfaces at elevated temperatures or from already insulated piping runs for which the outside jacket surface temperature is known. It shows the heat flow per hour per square foot of exposed surface area as a function of the difference in temperature between the surface and the surrounding air. Note that the nature of the surface significantly affects the magnitude of the heat loss. This is because thermal radiation, which is strongly dependent on the character of the radiating surface, plays an important role in heat loss at elevated temperatures, as does convective heat loss to the air.

Another important source of energy loss in steam systems is leakage from components such as loose flanges or malfunctioning steam traps. Figure 7.4 permits estimation of this loss for steam at various pressures leaking through holes. The heat losses are represented in million Btu's per year, based on full-time operation. Using Fig. 7.4, we see that a stuck-open steam trap with a 1/8-in. orifice would waste about 600 million Btu/yr of

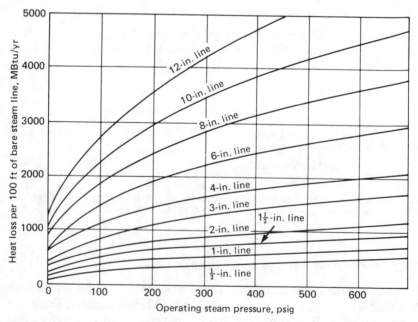

Figure 7.2 Heat loss from bare steam lines. (From [3].)

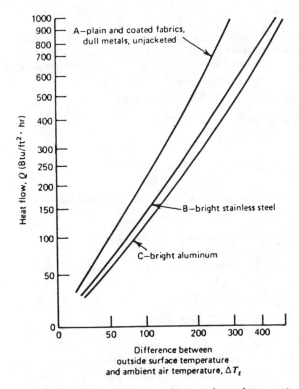

Figure 7.3 Heat losses from surfaces at elevated temperatures.

steam energy when leaking from a 100-psig line. Fig. 7.4 can also be used to estimate magnitudes of leakage from other sources of more complicated geometry. It is necessary first to determine an approximate area of leakage (in square inches) and then to calculate the equivalent hole diameter represented by that area. The example below illustrates this calculation.

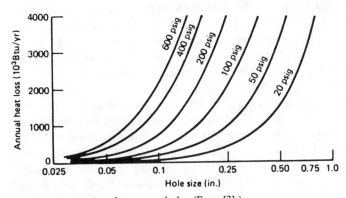

Figure 7.4 Heat loss from steam leaks. (From [3].)

Example 7.5

A flange on a 200-psig steam line has a leaking gasket. The maintenance crew, looking at the gasket, estimates that it is about 0.020 in. thick and that it is leaking from about 1/8-in. of the periphery of the flange. Estimate the annual heat loss in the steam if the line is operational 8,000 hr per year.

The area of the leak is a rectangle 0.020 in. wide and 1/8 in. long:

$$\text{Leak area} = 0.020 \times 1/8 = 0.0025 \text{ in.}^2$$

An equivalent circle will have an area of $\pi D^2/4$, so if

$$\frac{\pi D^2}{4} = 0.0025 \qquad D = 0.056 \text{ in.}$$

From Fig. 7.4, this leak, if occurring year-round (8,760 hr), would waste about 200 million Btu/yr of steam energy. For actual operation:

$$\text{Heat lost} = \frac{8,000}{8,760} \times 200 \text{ million} = 182.6 \text{ million Btu/yr}$$

Since the boiler efficiency has not been considered, the actual fuel wastage would be about 25% greater.

(SKP) ## 7.3.3 Measuring Steam and Condensate Rates

In maintaining steam systems at peak efficiency, it is often desirable to monitor continuously the rate of steam flow through the system, particularly at points of major usage such as steam mains. Figure 7.5 shows one of the most common types of flow-metering devices, the calibrated orifice. This is a sharp-edged restriction that causes the steam flow to neck down and then reexpand after passing through the orifice. As the steam accelerates to pass through the restriction, its pressure drops, and this pressure drop, if measured, can be easily related to the flow rate. The calibrated orifice is one of a class of devices known as obstruction flow meters, all of which work on the same principle of restricting the flow and producing a measurable pressure drop. Other types of obstruction meters are the ASME standard nozzle and the venturi. Orifices, while simple to manufacture and relatively easy to install (e.g., between flanges), are also subject to wear, which ultimately causes them to give unreliable readings. Nozzles and venturis, while more expensive initially, tend to be more resistant to erosion and wear, and also produce less permanent pressure drop once the steam reexpands to fill the pipe. Care must be exercised in the installation of all these devices, since turbulence and flow irregularities produced by valves, elbows, and fittings immediately upstream of the obstruction will produce erroneous readings.

Figure 7.6 shows another type of flow-metering device used for steam, called an annular averaging element. The annular element principle is somewhat different from that of the devices discussed above; the element averages the pressure produced when steam strikes the holes facing into the flow direction and subtracts from this average impact pressure a static

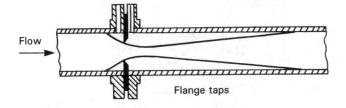

Flow →

Flange taps

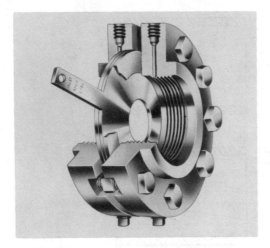

Figure 7.5 Orifice flow meter. (Courtesy, Bristol Division, ACCO.)

pressure sensed by a tube facing downstream. As with obstruction-type flow meters, the flow is related to this pressure difference.

A device that does not utilize pressure drop for steam metering applications is the vortex shedding flow meter, illustrated in Fig. 7.7. A solid bar extends through the flow, and as steam flows around the bar, vortices are shed alternately from one side to the other in its wake. As the vortices shift from side to side, the frequency of shedding can be detected with a thermal or

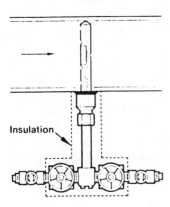

Insulation

Figure 7.6 Annular averaging element installed in steam line. (Courtesy, Dietrich Standard Corp.)

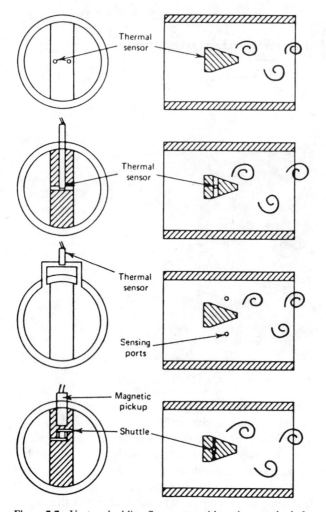

Figure 7.7 Vortex shedding flow meter with various methods for sensing fluctuations.

magnetic detector. This frequency varies directly with the rate of flow. The vortex shedding meter is quite rugged, since the only function of the object extending into the flow stream is to provide an obstruction to generate vortices; hence it can be made of heavy-duty stainless steel. Also, vortex shedding meters tend to be relatively insensitive to variations in the steam properties, since they produce a pulsed output rather than an analog signal.

The target flow meter (not shown) is also suitable for some steam applications. This type of meter uses a target, such as a small cylinder, mounted on the end of a metal strut that extends into the flow line. The strut is gaged to measure the force on the target, and if the properties of the fluid are accurately known, this force can be related to flow velocity. The target meter is especially useful when only intermittent measurements are needed, as the unit can be hot-tapped in a pipe through a ball valve and withdrawn when not in use. The requirement for accurate property data limits the usefulness of this type of meter in situations where steam conditions vary considerably, especially where high moisture is present.

The devices discussed above are useful in permanent installations for continuous or periodic measurements of steam flow; there is no simple way to directly measure steam flow on a spot-check basis without cutting into the system. There is, however, a relatively simple indirect method, illustrated in Fig. 7.8, for determining the rate of steam usage in systems with unpressurized condensate return lines or open systems in which condensate is dumped to a drain. If a drain line is installed after the trap, condensate may be caught in a barrel and the weight of a sample measured over a known period of time. Precautions must be taken to ensure that the flash steam, generated when the condensate drops in pressure as it passes through the trap, does not bubble out of the barrel. This would be both a safety hazard and a source of error in the measurements due to the loss of mass in vapor form. The barrel should be partially filled with cold water prior to the test so that flash steam will condense as it bubbles through the water. An energy balance can also be made on the water at the beginning and end of the test by measuring its temperature, and with proper application of the steam tables, a check can be made to ensure that the trap is not blowing through.

Figure 7.9 illustrates another instrument that can be used to monitor condensate flow on a regular basis, the rotameter. A rotameter indicates the flow rate of the liquid by the level of a specially shaped float that rises in a calibrated glass tube such that its weight exactly balances the drag force of the flowing condensate.

Measurements of this type can be very useful in monitoring system

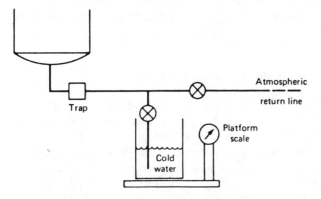

Figure 7.8 Weigh-bucket technique for condensate measurement.

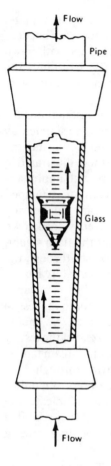

Figure 7.9 Liquid rotameter.

performance, since any unusual change in steam or condensate rate not associated with a corresponding change in production rate, would tend to indicate an equipment malfunction producing poor efficiency.

7.3.4 Computing the Dollar Value of Steam

In analyzing energy conservation measures for steam systems, it is important to establish a value for steam in dollars per pound; this value will depend on the steam pressure, the boiler efficiency, and the price of fuel.

Steam may be valued from two points of view. The more common approach, termed the *enthalpy method*, takes into account only the heating capability of the steam and is most appropriate when steam is used primarily for process heating. The enthalpy method is illustrated by the following example.

Example 7.6

An oil refinery produces 200-psig saturated steam in a large boiler. Some of the steam is used directly in high-temperature processes, and some of it is let down to 30 psig through regulating valves for use at lower temperatures. The feedwater is added to the boiler at about 160°F. The boiler efficiency has been determined to be 82%, and boiler fuel is priced at $2.20/million Btu. Establish the values of 200-psig and 30-psig steam (in dollars per pound).

Using the enthalpy method, we determine the increased heat content for each steam pressure required from the boiler if feedwater enters at 160°F. From Table 7.2:

Total enthalpy of steam at 200 psig = 1,199.3 Btu/lb
Total enthalpy of steam at 30 psig = 1,172 Btu/lb
Enthalpy of feedwater at 160°F ≅ 130 Btu/lb (about the same as saturated liquid at 160°F)
Heat added per pound of 200-psig steam = 1,199.3 − 130 = 1,069.3 Btu/lb
Heat added per pound of 30-psig steam = 1,042 Btu/lb
Fuel Btu's required per pound at 200 psig = 1,069.3/0.82 = 1,304 Btu
Fuel Btu's required per pound at 30 psig = 1,042/0.82 = 1,271 Btu

With boiler fuel priced at $2.20/million Btu, or 0.22¢ per thousand Btu's:

Value of 200-psig steam = 0.22 × 1.304 = 0.29¢/lb or $2.90/thousand lb
Value of 30-psig steam = 0.22 × 1.271 = 0.28¢/lb or $2.80/thousand lb

Another approach to the valuation of steam, the *availability* or *entropy* method, takes into account not only the heat content of the steam but also its power-producing potential if it were expanded through a steam turbine. This method is most applicable in plants where cogeneration (simultaneous generation of power and use of steam for process heat) is practiced. The availability method will not be explicitly covered here; the equivalent energy and cost accounting concepts are illustrated in Chapter 12.

7.4 STEAM TRAPS AND THEIR APPLICATION Skip

7.4.1 Functions of Steam Traps

Steam traps are important elements of steam and condensate systems and may represent a major energy conservation opportunity (or problem, as the case may be). The basic function of a steam trap is to allow condensate formed in the heating process to be drained from the equipment. This must be done speedily to prevent backup of condensate in the system.

Inefficient removal of condensate produces two adverse effects. First, if condensate is allowed to back up in the steam chamber, it cools below the steam temperature as it gives up sensible heat to the process and reduces the effective potential for heat transfer. Since condensing steam is a much more effective heat transfer medium than stagnant liquid, the area for condensation is reduced, and the efficiency of the heat transfer process deteriorates. This results in longer cycle times for batch processes or lower throughput rates in continuous heating processes. In either case, inefficient condensate removal almost always increases the amount of energy required.

A second reason for efficient removal of condensate is the avoidance of "water hammer" in steam systems. This phenomenon occurs when slugs of liquid become trapped between steam pockets in a line. The steam, which has a much larger specific volume, can accelerate these slugs to high velocity, and when they strike an obstruction, such as a valve or an elbow, they produce an impact force not unlike that of hitting the element with a hammer (hence the name). Water hammer can be extremely damaging to equipment, and proper design of trapping systems is necessary to avoid it.

The second crucial function of a steam trap is to facilitate the removal of air from the steam space. Air can leak into the steam system when it is shut down, and some gas is always liberated from the water in the boiling process and carried through the steam lines. Air mixed with steam occupies some of the volume that would otherwise be filled by the steam itself. Each of these components, air and steam, contributes its share to the total pressure exerted in the system; it is a fundamental thermodynamic principle that, in a mixture of gases, each component contributes to the pressure in the same proportion as its share of the volume of the space. For example, consider a steam system at 100 psia (note that in this case it is necessary to use *absolute* pressures), in which 10% of the volume is air instead of steam. From thermodynamics, 10% of the pressure, or 10 psia, is contributed by the air, and 90%, or 90 psia, by the steam. Referring to Table 7.2, the corresponding steam temperature is between 316 and 324°F, or approximately 320°F. If the air were not present, the steam pressure would be 100 psia, corresponding to a temperature of about 328°F, so the presence of air in the system reduces the temperature for heat transfer. This means that more steam must be generated to do a given heating job. Table 7.8 indicates the temperature reduction caused by the presence of air in various quantities at given pressures (psig) and shows that the effective temperature may be seriously degraded.

In operation, the situation is usually even worse than indicated in Table

Table 7.8   Temperature Reduction Caused by Air in Steam Systems

Pressure (psig)	Temperature of steam, no air present	Temperature of steam mixed with various percentages of air (by volume)		
		10%	20%	30%
10	240.1	234.3	228.0	220.9
25	267.3	261.0	254.1	246.4
50	298.0	291.0	283.5	275.1
75	320.3	312.9	204.8	295.9
100	338.1	330.3	321.8	312.4

7.8. We have considered the temperature reduction assuming that the air and steam are uniformly mixed. In fact, on a real heating surface, as air and steam move adjacent to the surface, the steam is condensed out into a liquid while the air stays behind in the form of vapor. In the region very near the surface, therefore, the air occupies an even larger fraction of the volume than in the steam space as a whole, and acts effectively as an insulating blanket on the surface. Suffice it to say that air is an undesirable parasite in steam systems, and its removal is important for proper operation.

Oxygen and carbon dioxide, in particular, have another adverse effect — corrosion in condensate and steam lines. Oxygen in condensate produces pitting or rusting of the surface, which can contaminate the water and make it undesirable as boiler feed, and carbon dioxide in solution with water forms carbonic acid, which is highly corrosive to metallic surfaces. These components must be removed from the system, partially by good steam trapping and partially by proper deaeration of condensate, as discussed in a later section.

7.4.2 Types of Steam Traps and Their Selection

Various types of steam traps are on the market, and the selection of the best trap for a given application is an important one. Many manufacturers produce several types of traps for specific applications, and manufacturers' representatives should be consulted in arriving at a choice. This section will give a brief introduction to the subject and comment on its relevance to improved energy utilization in steam systems.

Steam traps may be classified into three general groups: mechanical traps, based on the density difference between condensate and steam or air; thermostatic traps, based on the difference in temperature between steam, which stays close to its saturation temperature, and condensate, which cools rapidly; and thermodynamic traps, based on the difference in flow properties between liquids and vapors.

Figures 7.10 and 7.11 show two types of mechanical traps in common use for industrial applications. Figure 7.10 illustrates the principle of the bucket trap. In the trap illustrated, an inverted bucket is placed over the inlet line, inside an external chamber. The bucket is attached to a lever arm, which opens and closes a valve as the bucket rises and falls in the chamber. As long as condensate flows through the system, the bucket has a negative buoyancy, since liquid is present both inside and outside the bucket. The valve is open and condensate is allowed to drain continuously to the return line. As steam enters the trap it fills the bucket, displacing condensate, and the bucket rises, closing off the valve. Noncondensable gases, such as air and carbon dioxide, bubble through a small vent hole and collect at the top of the trap, to be swept out with flash steam the next time the valve opens. Steam may also leak through the vent, but it is condensed on contact with the cool

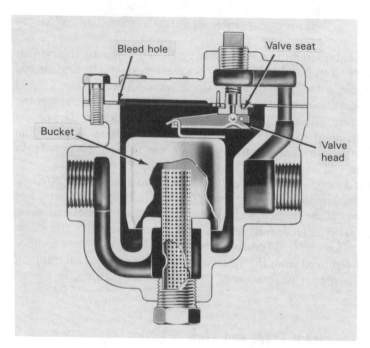

Figure 7.10 Inverted-bucket steam trap. (Courtesy, Sarco Company, Inc.)

Figure 7.11 Float-and-thermostatic steam trap. (Courtesy, Sarco Company, Inc.)

chamber walls and collects as condensate in the chamber. The vent hole is quite small, so the rate of steam loss through this leakage action is not excessive. As condensate again begins to enter the bucket, it loses buoyancy and begins to drop until the valve opens and again discharges condensate and trapped air.

The float-and-thermostatic trap, illustrated in Fig. 7.11, works on a similar principle. In this case, instead of a bucket, a buoyant float rises and falls in the chamber as condensate enters or is discharged. The float is attached to a valve, similar to the one in a bucket trap, which opens and closes as the ball rises and falls. Since there is no natural vent in this trap and the ball cannot distinguish between air and steam, which have similar densities, special provision must be made to remove air and other gases from the system. This is usually done by incorporating a small thermostatically actuated valve in the top of the trap. At low temperatures, the valve bellows contracts, opening the vent and allowing air to be discharged to the return line. When steam enters the chamber, the bellows expands, sealing the vent. Some float traps are also available without this thermostatic air vent feature; external provision must then be made to permit proper air removal from the system. The float-and-thermostatic trap permits continuous discharge of condensate, unlike the bucket trap, which is intermittent. This can be an advantage in certain applications.

Figure 7.12 illustrates a thermostatic steam trap. In this trap a temperature-sensitive bellows expands and contracts in response to the temperature of the fluid in the chamber surrounding the bellows. When condensate

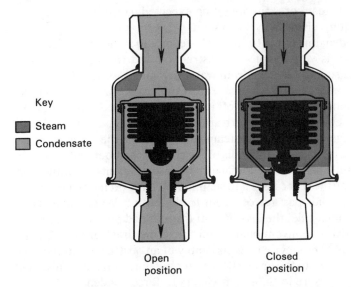

Key

■ Steam

▨ Condensate

Open position

Closed position

Figure 7.12 Thermostatic steam trap. (Courtesy, Armstrong International)

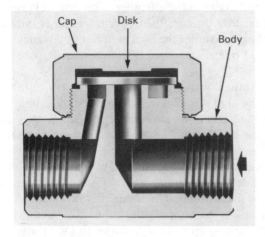

Cap   Disk   Body

Figure 7.13 Disk or thermodynamic steam trap. (Courtesy, Sarco Company, Inc.)

surrounds the bellows, it contracts, opening the drain port. As steam enters the chamber, the elevated temperature causes the bellows to expand and seal the drain. Since air also enters the chamber at a temperature lower than that of steam, the thermostatic trap is naturally self-venting, and it also is a continuous drain trap. The bellows in the trap can be partially filled with a fluid and sealed, so that an internal pressure is produced that counterbalances the external pressure imposed by the steam. This feature makes the bellows-type thermostatic trap somewhat self-compensating for variations in steam pressure. Another type of thermostatic trap uses a bimetallic element. This type of trap is not well suited for applications in which significant variations in steam pressure might be expected, since it is responsive only to temperature changes in the system.

The thermodynamic or controlled disk steam trap is shown in Fig. 7.13. This type of trap is very simple in construction and can be made quite compact and resistant to damage from water hammer. A small disk covers the inlet orifice in a thermodynamic trap. Condensate or air, moving at relatively low velocity, lifts the disk off its seat and is passed through to the outlet drain. When steam enters the trap, it passes through at high velocity because of its large volume. As the steam passes through the space between the disk and its seat, it strikes the walls of the control chamber and produces a rise in pressure. This pressure imbalance between the outside of the disk and the side facing the seat causes it to snap shut, sealing off the chamber and preventing the further passage of steam to the outlet. When condensate again enters the inlet side, the disk lifts off the seat and permits its release.

An alternative to conventional steam traps, the drain orifice, is illustrated in Fig. 7.14. This device consists simply of an obstruction to the flow of condensate, similar to the orifice flow meter but much smaller. This small hole allows the pressure in the steam system to force condensate to drain continuously into the lower pressure return system. Obviously, if steam,

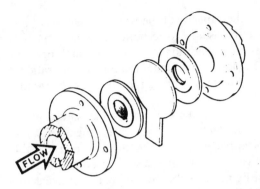

Figure 7.14 Drain orifice. (Courtesy, Flexitallic Gasket Company)

rather than condensate, enters it will also pass through the orifice and be lost. The strategy of using drain orifices is to select an orifice size that permits condensate to drain at such a rate that live steam seldom enters the system. Even if steam does occasionally pass through, the small size of the orifice limits the steam leakage rate to a value much less than would occur in a "stuck-open" malfunction of one of the types of traps discussed above. Drain orifices can be successfully applied in systems that have a well-defined and relatively constant condensate load. They are not suited for use where the condensate load may vary widely with operating conditions.

As mentioned above, a number of operating requirements must be taken into consideration in selecting the appropriate trap for a given application. Table 7.9 lists these considerations and presents one manufacturer's performance ratings for the traps discussed above. In selecting a trap for a particular application, assistance should be obtained from manufacturers' representatives, since a large body of experience in actual service has been accumulated over the years.

7.4.3 Considerations in Steam Trap Sizing

As mentioned earlier in this section, good energy conservation practice demands the efficient removal of condensate from process equipment. It is thus necessary to ensure that traps are properly sized for the given condensate load. Grossly oversized traps waste steam through excessive surface heat loss and internal venting, while undersized traps permit accumulation of condensate with a resultant loss in heat transfer effectiveness, of the equipment.

Steam traps are sized on the basis of two specifications: the condensate load (e.g., in pounds per hour or gallons per minute) and the pressure differential across the trap in pounds per square inch gage. Section 7.3 covered various methods for estimating condensate loads expected under normal operating conditions.

It is good practice to choose the capacity of the trap on the basis of this expected load multiplied by a factor of safety to account for peaks at start-up and fluctuations in normal operation conditions. It is not unusual for start-up condensate loads to be three to four times higher than steady operational loads, and in some applications they may range up to 10 times the steady-state load.

Table 7.10 presents typical factors of safety for condensate capacity recommended by steam trap manufacturers. These are ranges of safety factors to consider in various applications. While there is considerable variation in the recommended values, in both energy and economic terms the cost of oversizing is ordinarily not prohibitive, and conservative safety

Table 7.9 Comparison of Steam Trap Characteristics

Characteristic	Inverted bucket	Float-and-thermostatic	Disk	Bellows thermostatic
A Method of operation	Intermittent	Continuous	Intermittent	Continuous[a]
B Energy conservation (time in service)	Excellent	Good[b]	Poor	Fair
C Resistance to wear	Excellent	Good	Poor	Fair
D Corrosion resistance	Excellent	Good	Excellent	Good
E Resistance to hydraulic shock	Excellent	Poor	Excellent	Poor
F Vents air and CO_2 at steam temperature	Yes	No	No	No
G Ability to vent air at very low pressure (¼ psig)	Poor	Excellent	NR[c]	Good
H Ability to handle start-up air loads	Fair	Excellent	Poor	Excellent
I Operation against back pressure	Excellent	Excellent	Poor	Excellent
J Resistance to damage from freezing[d]	Good	Poor	Good	Good
K Ability to purge system	Excellent	Fair	Excellent	Good
L Performance on very light loads	Excellent	Excellent	Poor	Excellent
M Reponsiveness to slugs of condensate	Immediate	Immediate	Delayed	Delayed
N Ability to handle dirt	Excellent	Poor	Poor	Fair
O Comparative physical size	Large[e]	Large	Small	Small
P Ability to handle "flash steam"	Fair	Poor	Poor	Poor
Q Mechanical failure (open-closed)	Open	Closed	Open[f]	Closed[g]

[a]Can be intermittent on low load.
[b]Excellent when "secondary steam" is utilized.
[c]Not recommended for low-pressure operations.
[d]Cast iron traps not recommended.
[e]In welded stainless steel construction—medium.
[f]Can fail closed due to dirt.
[g]Can fail open due to wear.
Source: [1].

Table 7.10   Typical Factors of Safety for Steam Traps (Condensate Flow Basis)

Application	Factor of safety
Autoclaves	3–4
Blast coils	3–4
Dry cans	2–3
Dryers	3–4
Dry kilns	3–4
Fan system heating service	3–4
Greenhouse coils	3–4
Hospital equipment	2–3
Hot-water heaters	4–6
Kitchen equipment	2–3
Paper machines	3–4
Pipe coils (in still air)	3–4
Platen presses	2–3
Purifiers	3–4
Separators	3–4
Steam-jacketed kettles	4–5
Steam mains	3–4
Submerged surfaces	5–6
Tracer lines	2–3
Unit heaters	3–4

factors are usually used. The exception to this rule is in the sizing of disk-type traps, which may not function properly if loaded considerably below design. Drain orifices must also be sized close to normal operating loads. Again, the advice of the manufacturer should be solicited for the specific application in mind.

The other important design specification is the pressure differential over which the trap will operate. Since pressure is the driving force that moves condensate through the trap and on to the receiver, trap capacity will increase, for a given trap size, as the pressure increases. The trap operating pressure differential is not simply the boiler pressure. on the upstream side of the trap, steam pressure may drop through valves and fittings and through heat transfer passages in the process equipment. Thus, the appropriate upstream pressure is the pressure at the trap inlet, which, to a reasonable approximation, can usually be taken as the process steam pressure at the equipment. Back-pressure on the outlet side of the trap must also be considered. This includes the receiver pressure (if the condensate return system is pressurized), the pressure drop associated with flash steam and condensate flow through the return lines, and the head of water associated with risers if the trap is located at a point below the condensate receiver. Condensate return lines are usually sized for a given capacity to maintain no greater than 5,000 ft/min velocity of the flash steam. Table 7.11 shows the expected

Table 7.11 Condensate Capacities and Pressure Drops for Return Lines

Return line capacity in lb/hr with pressure drop in psi for 100 ft of pipe at a velocity of 5000 ft/min

Supply press, psig →	5	15		30			60				100					250					
Return press, psig →	0	0	5	0	5	10	0	5	10	20	0	5	10	20	30	0	5	10	20	30	50
Pipe size (in.)[a]																					
1/2	1425	590	1335	360	640	1055	235	370	535	1010	180	270	370	615	955	115	165	215	325	450	760
	4.0	4.0	5.3	4.0	5.3	6.5	4.0	5.3	6.5	8.9	4.0	5.3	6.5	8.9	11.3	4.0	5.3	6.5	8.9	11.3	15.9
3/4	2495	1035	2340	635	1125	1855	415	650	940	1779	310	470	645	1085	1675	200	285	375	570	795	1330
	2.35	2.35	3.14	2.35	3.14	3.88	2.35	3.14	3.88	5.32	2.35	3.14	3.88	5.32	6.72	2.35	3.14	3.88	5.32	6.72	9.49
1	4045	1680	3790	1030	1820	3005	760	1055	1520	2865	505	765	1045	1755	2715	325	465	605	925	1285	2155
	1.53	1.53	2.04	1.53	2.04	2.51	1.53	2.04	2.51	3.44	1.53	2.04	2.51	3.44	4.36	1.53	2.04	2.51	3.44	4.36	6.15
1¼	7000	2905	6565	1780	3150	5200	1155	1830	2635	4960	875	1320	1810	3035	4695	560	800	1050	1600	2225	3735
	0.95	0.95	1.26	0.95	1.26	1.55	0.95	1.26	1.55	2.13	0.95	1.26	1.55	2.13	2.69	0.95	1.26	1.55	2.13	2.69	3.80
1½	9530	3955	8935	2425	4290	7080	1575	2490	3585	6750	1190	1795	2465	4135	6395	760	1090	1430	2175	3025	5080
	0.73	0.73	0.97	0.73	0.97	1.20	0.73	0.97	1.20	1.64	0.73	0.97	1.20	1.64	2.07	0.73	0.97	1.20	1.64	2.07	2.93
2	15710	6525	14725	3995	7070	11670	2595	4105	5910	11125	1965	2960	4060	6810	10540	1255	1800	2355	3585	4990	8375
	0.48	0.48	0.64	0.48	0.64	0.79	0.48	0.64	0.79	1.08	0.48	0.64	0.79	1.08	1.37	0.48	0.64	0.79	1.08	1.37	1.93
2½	22415	9305	21005	5700	10085	16650	3705	5855	8430	15875	2800	4225	5795	9720	15035	1790	2565	3360	5115	7120	11950
	0.36	0.36	0.48	0.36	0.48	0.59	0.36	0.48	0.59	0.81	0.36	0.48	0.59	0.81	1.03	0.36	0.48	0.59	0.81	1.03	1.45
3	34610	14370	32435	8800	15570	25710	5720	9045	13020	24515	4325	6625	8950	15005	23220	2765	3965	5185	7900	10990	18450
	0.26	0.26	0.34	0.26	0.34	0.42	0.26	0.34	0.42	0.58	0.26	0.34	0.42	0.58	0.73	0.26	0.34	0.42	0.58	0.73	1.03
3½	46285	19220	43380	11765	20825	34385	7650	12095	17410	32785	5785	8725	11970	20070	31050	3695	5300	6940	10565	14700	24675
	0.21	0.21	0.27	0.21	0.27	0.34	0.21	0.27	0.34	0.46	0.21	0.27	0.34	0.46	0.59	0.21	0.27	0.34	0.46	0.59	0.83
4	59595	24745	55855	15150	26815	44275	9850	15575	22415	42210	7450	11235	15410	25840	39980	4760	6825	8935	13600	18925	31770
	0.17	0.17	0.23	0.17	0.23	0.28	0.17	0.23	0.28	0.38	0.17	0.23	0.28	0.38	0.49	0.17	0.23	0.28	0.38	0.49	0.25
5	93655	38890	87780	23810	42140	69580	15480	24475	35230	66335	11705	17660	24220	40610	62830	7475	10725	14040	21375	29745	49930
	0.12	0.12	0.16	0.12	0.16	0.20	0.12	0.16	0.20	0.20	0.12	0.16	0.20	0.20	0.17	0.12	0.16	0.20	0.23	0.17	0.11
6	135245	56160	126760	34385	60855	100480	22350	35345	50875	95795	16905	25500	34975	58645	90735	10800	15490	20270	30865	42950	72105
	0.10	0.10	0.13	0.10	0.13	0.04	0.10	0.13	0.04	0.05	0.10	0.13	0.13	0.05	0.01	0.10	0.13	0.04	0.05	0.13	0.01
8	234195	97245	219505	59540	105380	173995	38705	61205	88095	165880	29270	44160	60565	101550	157115	18700	26820	35105	53450	74375	124855
	0.02	0.02	0.02	0.02	0.02	0.01	0.02	0.02	0.01	0.01	0.02	0.02	0.01	0.01	0.01	0.02	0.02	0.01	0.01	0.01	0.01

[a]Schedule 40 pipe.

Table 7.12   Typical Pressure-Capacity Specifications for Steam Traps

Differential pressure (psi)	Capacity (lb/hr) for model			
	A	B	C	D
5	450	830	1600	2900
10	560	950	1900	3500
15	640	1060	2100	3900
20	680	880	1800	3500
25	460	950	1900	3800
30	500	1000	2050	4000
40	550	770	1700	3800
50	580	840	1900	4100
60	635	900	2000	4400
70	660	950	2200	3800
80	690	800	1650	4000
100	640	860	1800	3600
125	680	950	2000	3900
150	570	810	1500	3500
200	—	860	1600	3200
250	—	760	1300	3500
300	—	510	1400	2700
400	—	590	1120	3100
450	—	—	1200	3200

Source: Armstrong International.

pressure drop per 100 ft of return line that can be expected under design conditions. A 60-psig system, for example, returning condensate to an unpressurized receiver (0 psig) through a 2-in. line, would have a return line pressure drop of just under 1/2 psi per 100-ft run, and the condensate capacity of the line would be about 2,600 lb/hr. The pressure head produced by a vertical column of water is about 1 psi per 2 ft of rise. These components can be summed to estimate the back-pressure on the system, and the appropriate pressure for sizing the trap is then the difference between the upstream pressure and the back-pressure.

Table 7.12 shows a typical pressure-capacity table from a manufacturer's catalog. The use of such a table is illustrated by the following example.

Example 7.7

A steam-jacketed platen press in a plastic lamination operation uses about 500 lb/hr of 30-psig steam in normal operation. A 100-ft run of 1-in. pipe returns condensate to a receiver pressurized to 5 psig; the receiver is located 15 ft above the level of the trap. From the capacity-differential pressure specifications in Table 7.12, select a suitable trap for this application.

Using a factor of safety of 3 from Table 7.10, a trap capable of handling $3 \times 500 = 1,500$ lb/hr of condensate will be selected.

To determine the system back-pressure, add the receiver pressure, the piping pressure drop, and the hydraulic head due to the elevation of the receiver.

Entering Table 7.11 at 30-psig supply pressure and 5-psig return pressure, the pipe pressure drop for a 1-in. pipe is just slightly over 2 psi at a condensate rate somewhat higher than our 1,500 lb/hr; a 2-psi pressure drop is a reasonable estimate.

The hydraulic head due to the 15-ft riser is 1/2 psi/ft $\times$ 15 = 7.5 psi. The total back pressure is therefore

$$5 \text{ psi (receiver)} + 7.5 \text{ psi (riser)} + 2 \text{ psi (pipe)} = 14.5 \text{ psi}$$

or the differential pressure driving the condensate flow through the trap is $30 - 14.5 = 15.5$ psi.

From Table 7.12, we see that a model C trap will handle 2,100 lb/hr at 15-psi differential; this would be the correct choice.

7.4.4   Maintaining Steam Traps for Efficient Operation

Steam traps can and do malfunction in two ways. They can stick in the closed position, causing condensate to back up into the steam system, or they can stick open, allowing live steam to discharge into the condensate system. The former type of malfunction is usually quickly detectable, since flooding of a process heater with condensate will usually so degrade its performance that the failure is soon evidenced by a significant change in operating conditions. This type of failure can have disastrous effects on equipment by producing damaging water hammer and causing process streams to back up into other equipment. Because of these potential problems, steam traps are often designed to fail in the open position; for this reason, they are among the biggest energy wasters in an industrial plant. Broad experience in large process plants using thousands of steam traps has shown that, typically, from 15 to 60% of the traps in a plant may be blowing through, wasting enormous amounts of energy. Table 7.13 shows the cost of wasted 100-psig steam (typical of many process plant conditions) for leak diameters characteristic of steam trap orifices. At higher steam pressures, the leakage would be even greater; the loss rate does not go down in direct proportion at lower steam pressures, but declines at a rate proportional to the *square root* of the pressure. For example, a 1/8-in. leak in a system at 60 psig, instead of the 100 psig shown in the table, would still waste over 75% of the steam rate shown (the square root of 60/100). The cost of wasted steam far outweighs the cost of proper maintenance to repair the malfunctions, and comprehensive steam trap maintenance programs have proven to be among the most attractive energy conservation investments available in large process plants. Most types of steam traps can be repaired, and some have inexpensive replaceable elements for rapid turnaround.

A major problem facing the energy conservation engineer is diagnosis of open traps. The fact that a trap is blowing through can often be detected by a rise in temperature at the condensate receiver, and it is quite easy to

Table 7.13   Annual Cost of Steam Leaks

Leak diameter (in.)	Steam wasted per month (lb)[a]	Cost per month[b]	Cost per year[b]
1/16	13,300	$ 40.00	$ 480.00
1/8	52,200	156.00	1,890.00
1/4	209,000	626.00	7,800.00
1/2	833,000	2,500.00	30,000.00

[a]Based on 100-psig differential across orifice.
[b]Based on steam value of $3.00 per thousand pounds. Cost will scale in direct proportion for other steam values.

monitor this simple parameter. There are also several direct methods for checking trap operation. Figure 7.15 shows the simplest approach for open condensate systems where traps drain directly to atmospheric pressure. In proper, normal operation, a stream of condensate drains from the line along with a lazy cloud of flash steam, produced as the condensate throttles across the trap. When the trap is blowing through, a well-defined jet of live steam will issue from the line with either no condensate or perhaps a condensate mist associated with steam condensate at the periphery of the jet.

Visual observation is less convenient in a closed condensate system but can be utilized if a test valve is placed in the return line just downstream of

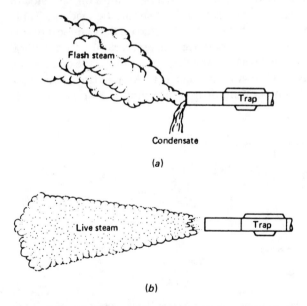

Figure 7.15   Visual observation of steam trap operation in open systems.

the trap, as shown in Fig. 7.16. This system has the added advantage that the test line may be used to measure condensate discharge rate as a check on equipment efficiency, as discussed earlier. An alternative in closed condensate return systems is to install a sight-glass just downstream of the trap. These are relatively inexpensive and permit quick visual observation of trap operation without interfering with normal production.

Another approach to steam trap testing is to listen to the sound of the trap during operation. Table 7.14 describes the sounds made by various types of traps during normal and abnormal operation. This method is most effective with disk-type traps, although it can be used to some extent with the other types as well. An industrial stethoscope can be used to listen to the trap, although under many conditions the characteristic sound will be masked by noises transmitted from other parts of the system. Ultrasonic detectors may be used effectively in such cases; these devices are, in effect, electronic stethoscopes with acoustic filtering to make them sensitive to sound and vibration only in the very high frequency range. Steam blowing through a trap emits a very high pitched sound produced by intense turbulence at the trap orifice, in contrast to the lower-pitched and lower-intensity sound of liquid flowing through. Ultrasonic methods can, therefore, give a more reliable measure of steam trap performance than conventional "listening" devices.

A third approach to steam trap testing makes use of the drop in saturation temperature associated with pressure drop across the trap. Condensate tends to cool rapidly in contact with uninsulated portions of the return line, accentuating the temperature difference. If the temperature on each side of the trap is measured, a sharp temperature drop should be evident. Table 7.15 shows typical temperatures that can be expected on the condensate side for various condensate pressures. In practice, the temperature drop method can be rather uncertain because of the range of temperatures the condensate may exhibit and because, in blowing through a stuck-open trap, live steam will itself undergo some temperature drop. For example, 85-psig saturated

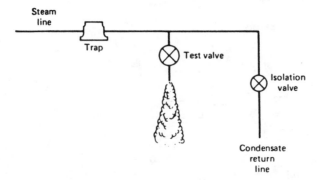

Figure 7.16 Visual observation of steam trap operation in closed systems.

Table 7.14 Operating Sounds of Various Types of Steam Traps

Trap	Proper operation	Malfunctioning
Disk-type (impulse or thermodynamic)	Opening and snap-closing of disk several times per minute	Rapid chattering of disk as steam blows through
Mechanical-type (bucket)	Cycling sound of the bucket as it opens and closes	Fails open—sound of steam blowing through
Thermostatic-type	Sound of periodic discharge if medium to high load; possibly no sound if light load, throttled discharge	Fails closed—no sound

steam blowing through an orifice to 15 psig will drop from 328 to about 300°F, and may then cool further by radiation and convection from uninsulated surfaces. From Table 7.15, the expected condensate-side temperature is about 215–238°F for this pressure. Thus, while the difference is still substantial, misinterpretation is possible, particularly if accurate measurements of the steam and condensate pressures on each side of the trap are not available.

The most successful programs of steam trap diagnosis utilize a combination of these methods, coupled with a regular maintenance program to ensure that the traps are kept in proper operating condition.

This section has discussed the importance of good steam trap performance to successful energy conservation in steam systems. Traps must be properly selected and installed for the given service and appropriately sized to ensure efficient removal of condensate and gases. Once the traps are in service, expenditures for regular monitoring and maintenance are easily repaid in fuel savings.

Table 7.15 Typical Pipe Surface Temperatures for Various Operating Pressures

Operating pressure (psig)	Typical line temperatures (°F)
0	190–210
15	215–238
45	248–278
115	295–330
135	304–340
450	395–437

7.5 CONDENSATE RECOVERY

Condensate from steam systems is wasted, or at least used inefficiently, in many industrial operations. Yet improvements in the condensate system can offer the greatest savings of any of the measures discussed in this chapter. This section will present methods for estimating the potential energy and mass savings achievable through good condensate recovery and the considerations involved in condensate recovery system design.

7.5.1 Estimating of Heat and Mass Losses in Condensate Systems

The saturated liquid condensate produced when steam condenses on a heating surface still retains a significant fraction of the energy contained in the steam itself. Referring to Table 7.2, for example, it is seen that at a pressure of about 80 psig, each pound of saturated liquid contains about 295 Btu, or nearly 25% of the original energy contained in the steam at the same pressure. In some plants this condensate is simply discharged to a wastewater system, which is wasteful not only of energy but also of water and the expense of boiler feedwater treatment. Even if condensate is returned to a receiver at atmospheric pressure, a considerable fraction is lost in the form of flash steam. Table 7.16 shows the percentage condensate loss due to flashing from systems at the given steam pressure to a flash tank at a lower pressure. For example, in the 80-psig system discussed above, if condensate is returned to a vented receiver rather than discharged to a drain, nearly 12% is vented to the atmosphere as flash steam. Thus, about 3% of the original steam energy (0.12×0.25) goes up the vent pipe. Table 7.16 shows

Table 7.16 Percentage of Mass Converted to Flash Steam in a Flash Tank

Steam pressure (psig)	Percent flash for flash tank pressure (psig)										
	0	2	5	10	15	20	30	40	60	80	100
5	1.7	1.0	0								
10	2.9	2.2	1.4	0							
15	4.0	3.2	2.4	1.1	0						
20	4.9	4.2	3.4	2.1	1.1	0					
30	6.5	5.8	5.0	3.8	2.6	1.7	0				
40	7.8	7.1	6.4	5.1	4.0	3.1	1.3	0			
60	10.0	9.3	8.6	7.3	6.3	5.4	3.6	2.2	0		
80	11.7	11.1	10.3	9.0	8.1	7.1	5.5	4.0	1.9	0	
100	13.3	12.6	11.8	10.6	9.7	8.8	7.0	5.7	3.5	1.7	0
125	14.8	14.2	13.4	12.2	11.3	10.3	8.6	7.4	5.2	3.4	1.8
160	16.8	16.2	15.4	14.1	13.2	12.4	10.6	9.5	7.4	5.6	4.0
200	18.6	18.0	17.3	16.1	15.2	14.3	12.8	11.5	9.3	7.5	5.9
250	20.6	20.0	19.3	18.1	17.2	16.3	14.7	13.6	11.2	9.8	8.2
300	22.7	21.8	21.1	19.9	19.0	18.2	16.7	15.4	13.4	11.8	10.1
350	24.0	23.3	22.6	21.6	20.5	19.8	18.3	17.2	15.1	13.5	11.9
400	25.3	24.7	24.0	22.9	22.0	21.1	19.7	18.5	16.5	15.0	13.4

that this loss could be halved by operating the flash tank at a pressure of 30 psig, providing a low-temperature steam source for potential use in other parts of the process. Flash steam recovery will be discussed in more detail later.

Even when condensate is fully recovered by one of the methods to be described below, heat losses can still occur from uninsulated or poorly insulated return lines. These losses can be recovered cost-effectively by the proper application of thermal insulation.

7.5.2 Methods of Condensate Heat Recovery

Several options are available for recovery of condensate, ranging in cost and complexity from simple and inexpensive to elaborate and costly. Which option is best depends on the amount of condensate to be recovered, other uses for its energy, and the potential cost savings relative to other possible investments.

The simplest option, which can be utilized if condensate is presently being discharged, is the installation of a vented flash tank to collect condensate from various points of formation and cool it sufficiently to allow it to be delivered back to the boiler feed tank. Figure 7.17 schematically illustrates such a system. It consists of a series of collection lines tying the points of condensate generation to the flash tank, which allows the liquid to separate from the flash steam; the flash steam is vented to atmosphere through an open pipe. Condensate may be gravity-drained through a strainer and a trap. To avoid further generation of flash steam, a cooling leg may be incorporated to cool the liquid below its saturation temperature.

Flash tanks must be sized to produce proper separation of the flash steam from the liquid. As condensate is flashed, steam will be generated rather violently, and as vapor bubbles burst at the surface, liquid may be entrained and carried out through the vent. This is a nuisance, and it can be a safety hazard if the vent is located near personnel or equipment. Table 7.17 permits the estimation of flash tank size required for a given application. Strictly speaking, flash tanks must be sized on the basis of volume; however, if a typical length-to-diameter ratio of about 3:1 is assumed, flash tank dimensions can be represented as the product of diameter and height, which has the units of square feet (area), even though this particular product has no direct physical significance. Consider, for example, the sizing of a vented flash tank for the collection of 80-psig condensate at a rate of about 3,000 lb/hr. For a flash tank pressure of 0 psig (atmospheric pressure), the product of diameter and length is about 2.5 per 1,000 lb. Therefore, a diameter times length of 7.5 ft^2 would be needed for this application. A tank 1.5 ft in diameter and 5 ft long would be satisfactory. Of course, for flash tanks, as for other condensate equipment, conservative design would suggest the use of an appropriate safety factor.

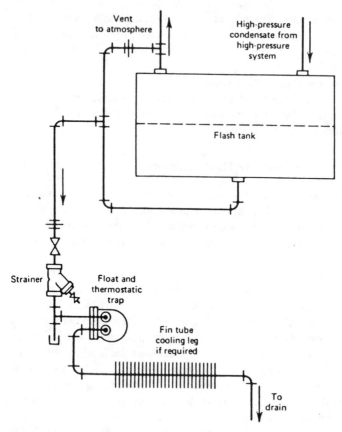

Figure 7.17 Flash tank vented to atmosphere. (Courtesy, Sarco Company, Inc.)

As noted above, venting of flash steam to the atmosphere is a wasteful process, and if significant amounts of condensate are to be recovered it may be desirable to attempt to utilize this flash steam. Figure 7.18 shows a modification of the simple flash tank system to accomplish this. Rather than venting to the atmosphere, the flash tank is pressurized and flash steam is piped to a low-pressure steam main, where it can be utilized for process purposes. From Table 7.17 it will be noted that the flash tank can be smaller in physical size at elevated pressure, although, of course, it must be properly designed for pressure containment. If the 80-psig condensate in the example above were flashed in a 15-psig tank, only about 2.7 ft^2 of diameter × length would be required. A tank 1 ft in diameter and 3 ft long could be utilized. Atmospheric vents are usually provided for automatic pressure relief and to allow manual venting if desired.

If pressurized flash steam is to be used, the cost of piping to set up a low-pressure steam system may be significant, particularly if the flash tank is remote from the potential low-pressure steam applications. Thus, it is desirable to plan such a system to minimize these piping costs by generating the flash steam near its point of use. Figure 7.19 illustrates such an application. Here, an air heater having four sections formerly used 100-psi steam in all sections. Because the temperature difference between the steam and the cold incoming air is larger than the difference at the exit end, the condensate load would be unevenly balanced between the four sections, with the heaviest load in the first section; lower temperature steam could be utilized here. In the revised arrangement, 100-psi steam is used in the last three sections; condensate is drained to a 5-psi flash tank, where low-pressure steam is generated and piped to the first section, substantially reducing the overall steam load to the heater. Note that for backup purposes a pressure-controlled reducing valve has been incorporated to supplement the low-pressure flash steam at light load conditions. This example shows how flash steam can be used directly without an expensive piping system to distribute it. A similar approach could apply to adjacent pieces of equipment in a multiple-batch operation.

While flash steam in a low-pressure system may appear to be an almost "free" energy source, its practical application involves a number of problems that must be carefully considered. These are all essentially economic in nature.

Table 7.17   Flash Tank Sizing

Steam pressure (psig)	Flash tank area[a] for flash tank pressure (psig)										
	0	2	5	10	15	20	30	40	60	80	100
400	5.41	4.70	3.89	3.01	2.44	2.03	1.49	1.15	0.77	0.56	0.42
350	5.14	4.45	3.66	2.84	2.28	1.91	1.38	1.07	0.70	0.51	0.37
300	4.86	4.15	3.42	2.62	2.11	1.75	1.26	0.96	0.62	0.44	0.31
250	4.41	3.82	3.12	2.39	1.91	1.56	1.11	0.85	0.52	0.37	0.25
200	3.98	3.40	2.80	2.12	1.68	1.37	0.97	0.72	0.43	0.28	0.18
175	3.75	3.20	2.61	1.95	1.57	1.26	0.87	0.64	0.38	0.23	0.15
160	3.60	3.08	2.50	1.86	1.46	1.19	0.80	0.59	0.34	0.21	0.12
150	3.48	2.98	2.41	1.80	1.40	1.14	0.77	0.56	0.31	0.19	0.10
140	3.36	2.86	2.31	1.72	1.35	1.08	0.72	0.52	0.29	0.16	0.08
130	3.24	2.76	2.23	1.65	1.29	1.02	0.67	0.49	0.26	0.14	0.07
120	3.12	2.65	2.15	1.57	1.22	0.97	0.64	0.44	0.23	0.12	0.04
110	2.99	2.52	2.05	1.50	1.15	0.91	0.58	0.40	0.20	0.09	0.02
100	2.85	2.41	1.92	1.40	1.07	0.85	0.53	0.36	0.16	0.06	
90	2.68	2.26	1.81	1.30	0.99	0.77	0.48	0.31	0.13	0.05	
80	2.52	2.12	1.67	1.18	0.90	0.68	0.42	0.25	0.09		
70	2.34	1.95	1.55	1.08	0.81	0.61	0.35	0.20	0.04		
60	2.14	1.77	1.39	0.96	0.70	0.52	0.27	0.14			
50	1.94	1.59	1.22	0.81	0.58	0.41	0.20	0.08			
40	1.68	1.36	1.02	0.67	0.44	0.30	0.11				
30	1.40	1.10	0.81	0.50	0.29	0.16					
20	1.06	0.81	0.55	0.28	0.12						
12	0.75	0.48	0.28								
10	0.62	0.42	0.23								

[a]Flash tank area (ft^2) = diameter × length of horizontal tank for 1000 lb condensate per hour being discharged.

Source: [14].

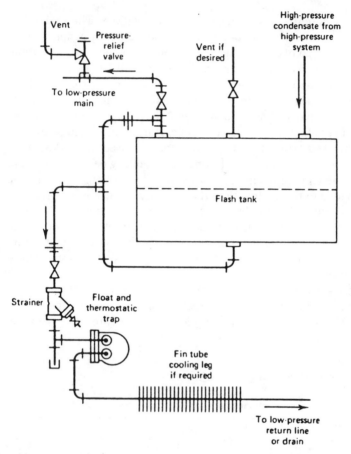

Figure 7.18 Pressurized flash tank discharging to low-pressure steam system. (Courtesy, Sarco Company, Inc.)

The quantity of condensate and its pressure (yielding a given quantity of flash steam) must be sufficiently large to provide a significant amount of available energy at the desired pressure. System costs do not go up in simple proportion to capacity. Rather, there is a large initial cost for piping, flash tank installation, and other system components, and therefore the overall cost per unit of heat recovered becomes significantly less as the system becomes larger. The nature of the condensate-producing system itself is also important. For example, if condensate is produced at only two or three points from large steam users, the cost of the condensate collection system will be considerably less than that of a system in which there are many small users.

Another important consideration is the potential for application of the flash steam. The availability of 5,000 lb per hour of 15-psig steam is mean-

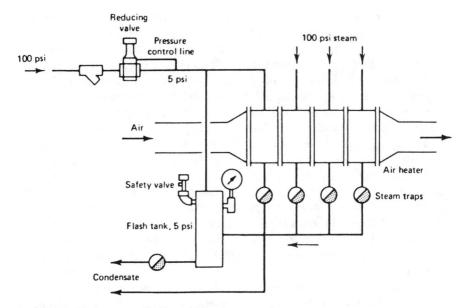

Figure 7.19 Flash steam utilization within a process unit.

ingless unless there is a need for a heat source of this magnitude near 250°F. Thus, potential uses must be properly matched to the available supply. Flash steam is most effectively utilized when it can supplement an existing low-pressure steam supply rather than provide the sole source of heat to equipment. Not only must the total average quantity of flash steam match the needs of the process, but the time variations of source and user must be taken into account, since steam cannot be economically stored for use at a later time. Thus, flash steam might not be a suitable heat source for sequential batch processes in which the number of operating units is small and significant fluctuations in steam demand exist.

When considering the possible conversion to low-pressure steam of an existing piece of equipment presently operating on high-pressure steam, it is important to recognize that steam pressure can have a significant effect on equipment operation. Since a reduction in steam pressure also means a reduction in temperature, a unit may not have adequate heating surface area to provide the necessary heat capacity to the process at reduced pressure. Existing steam distribution piping may not be adequate, since steam is lower in density at low pressure than at high pressure. Typically, larger piping is required to transport the low-pressure vapor at acceptable velocities. While one might expect the heat losses from the pipe surface to be less with low-pressure steam because of its lower temperature, this may not be the case if a larger pipe (and hence larger surface area) is needed to handle the low-pressure vapor. This requirement will also make insulation more expensive.

When flash steam is used in a piece of equipment, the resulting low-pressure condensate must still be returned to a receiver for delivery back to the boiler. Flash steam will again be produced if the receiver is vented, although somewhat less than in the flashing of high-pressure condensate. This flash steam and that produced from the flash tank condensate draining into the receiver will be lost unless some additional provision is made for its recovery, as shown in Fig. 7.20. In this system, rather than venting to the atmosphere, the steam rises through a cold water spray, which condenses it. This spray might be boiler makeup water, in which case the energy of the flash steam is used for makeup preheat. Not only the heat content but also the mass of the flash steam is saved, reducing makeup water requirements and saving the incremental costs of makeup water treatment. This system has the added advantage that it has a deaerating effect on the condensate and feedwater. If the cold spray is metered in order to produce a temperature in the tank above about 190°F, dissolved gases in the condensate and feedwater, particularly oxygen and CO_2, will come out of solution, and since they are not condensed by the cold water spray, they will be released through the atmospheric vent. As with flash steam systems, this system (usually termed a *barometric condenser* or *spray deaerator*) requires careful consideration to ensure its proper application. The system must be compatible with the boiler feedwater system, and controls must be provided to coordinate boiler makeup demands with the condensate load.

An alternative approach to the barometric condenser is shown in Fig. 7.21. In this system, condensate is cooled by passing it through a submerged coil in the flash tank before it is flashed. This reduces the amount of flash steam generated. Cold water makeup (possibly boiler feedwater) is regulated by a temperature-controlled valve.

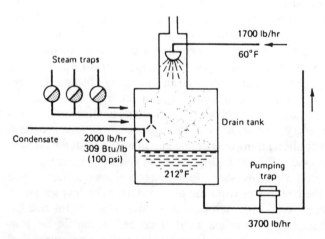

Figure 7.20 Flash steam recovery in spray tank.

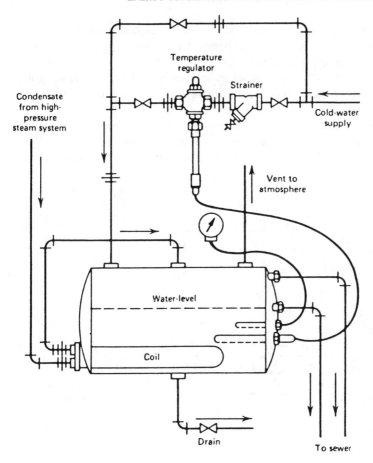

Figure 7.21 Flash tank with condensate precooling. (Courtesy, Sarco Company, Inc.)

The systems described above have one feature in common. In all cases, the final condensate state is atmospheric pressure, which may be required to permit return of the condensate to the existing boiler feedwater makeup tank. If condensate can be returned at an elevated pressure, a number of advantages may be realized.

Figure 7.22 shows schematics of two pressurized condensate return systems. Condensate is returned, in some cases without the need for a steam trap, to a high-pressure receiver, which routes the condensate directly back to the boiler. The boiler makeup unit and/or deaerator feeds the boiler in parallel with the condensate return unit, and appropriate controls must be incorporated to coordinate the operation of the two units. Systems such as the one shown in Fig. 7.22a are available for condensate pressures up to about 15 psig. For higher pressures, the unit can be used in conjunction with a flash tank, as shown in Fig. 7.22b. This system would be suitable where an application for 15-psig steam is available. These elevated-pressure systems represent an attractive option in relatively low pressure applications, such as steam-driven absorption chillers. When considering them, care must be taken to ensure that dissolved gases in the boiler makeup are at suitable levels to avoid corrosion, since the natural deaeration effect of atmospheric venting is lost.

One of the key engineering considerations in the design of all the above systems is the problem of pumping high-temperature condensate. To understand the nature of the problem, it is necessary to introduce the

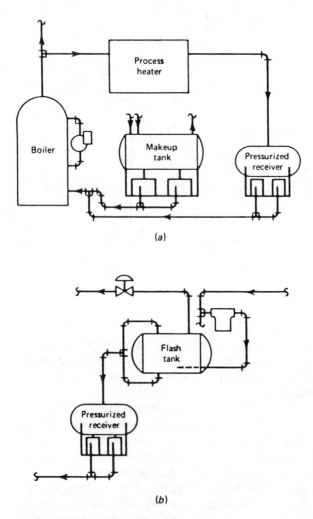

(a)

(b)

Figure 7.22 Pressurized condensate receiver systems. (a) Low-pressure process requirements; (b) flash tank for use with high-pressure systems.

concept of net positive suction head (NPSH) for a pump. This means the amount of static fluid pressure that must be provided at the inlet side of the pump to ensure that no vapor will be formed as the liquid passes through the pump mechanism, a phenomenon known as cavitation. As liquid moves into the pump inlet from an initially static condition, it accelerates and its pressure drops rapidly. If the liquid is at or near its saturation temperature in the stationary condition, this sudden drop in pressure will produce boiling and the generation of vapor bubbles. Vapor can also be generated by air coming out of solution at reduced pressure. These bubbles travel through the pump impeller, where the fluid pressure rises, causing the bubbles to collapse. The inrush of liquid into the vapor space produces an impact on the impeller surface that can have an effect comparable to sandblasting. Clearly, this is deleterious to the impeller and can cause rapid wear. Most equipment operators are familiar with the characteristic grinding sound of cavitation in pumps when air is advertently allowed to enter the system, and the same effect can occur due to steam generation in high-temperature condensate pumping.

To avoid cavitation, manufacturers specify a minimum pressure above saturation that must be maintained on the inlet side of the pump, so that even when the pressure drops through the inlet port, saturation or deaeration conditions will not occur. This minimum pressure requirement is termed the net positive suction head.

For condensate applications, special low-NPSH pumps have been designed. Figure 7.23 illustrates the difference between a conventional pump and a low-NPSH pump. In the conventional pump (Fig. 7.23*a*), fluid on the suction side is drawn directly into the impeller, where the rapid pressure drop occurs in the entry passage. In the low-NPSH pump (Fig. 7.23*b*), a small "preimpeller" provides an initial pressure boost to the incoming fluid, with relatively little drop in pressure at the entrance. This extra stage of pumping essentially provides a greater head to the entry passage of the main impeller, so that the system pressure at the suction side of the pump can be much closer to saturation conditions than that required for a conventional pump. Low-NPSH pumps are higher in price than conventional centrifugal pumps, but they can greatly simplify the problem of design for high-temperature condensate return and in some cases can actually reduce overall system costs.

An alternative device for the pumping of condensate, called a pumping trap, utilizes the pressure of the steam itself as the driving medium. Figure 7.24 illustrates the mechanism of a pumping trap. Condensate enters the inlet side and rises in the body until it activates a float-operated valve, which admits steam or compressed air into the chamber. A check valve prevents condensate from being pushed back through the inlet port, and another check valve allows the steam or air pressure to drive it out through the exit side. When the condensate level drops to a predetermined position, the

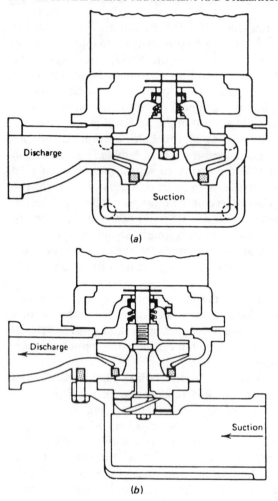

Figure 7.23 Conventional and low-NPSH pumps. (Courtesy, ITT Corp., Fluid Handling Division)

steam or air valve is closed, allowing the pumping cycle to start again. Pumping traps have certain inherent advantages over electrically driven pumps for condensate return applications. They have no NPSH requirement and hence can handle condensate at virtually any temperature without regard to pressure conditions. They are essentially self-regulating, since the condensate level itself determines when the trap pumps; thus, no auxiliary electrical controls are required for the system. This has another advantage in environments where explosion-proofing is required, as in refineries and chemical plants. Electrical lines need not be run to the system, since it utilizes steam as a driving force, and the steam line is usually close at hand.

Pumping traps operate more efficiently with compressed air, if available, because when steam is introduced into the chamber, some of it condenses before its pressure can drive the condensate out. Thus, for the same pressure, more steam is required to give the same pumping capacity as compressed air. The disadvantages of pumping traps are their mechanical complexity, which makes them susceptible to maintenance problems, and the fact that they are available only in limited capacities.

The engineering of a complete condensate recovery system from scratch can be an involved process, requiring the design of tanks, plumbing, controls, and pumping devices. For large systems, there is little alternative to engineering and fabricating the system to the specific plant requirements. For small-to-moderate-capacity applications, however, packaged systems incorporating all of the above components are commercially available. Figures 7.25a and b show examples of two such systems, the former using electrically driven low-NPSH pumps and the latter utilizing a pumping trap (the lower unit in the figure).

7.5.3 Overall Planning Considerations in Condensate Recovery Systems

As mentioned earlier, condensate recovery systems require careful engineering to ensure that they are compatible with overall plant operations, that are safe and reliable, and that they can actually achieve their energy efficiency potential. In this section, a few of the overall planning factors that should be considered are enumerated and discussed.

1. *Availability of adequate condensate sources:* An energy audit should

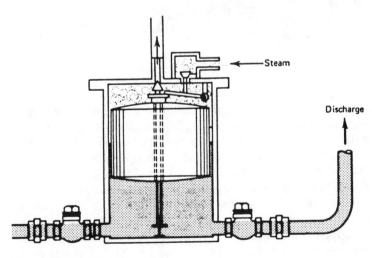

Figure 7.24 Pumping trap.

Figure 7.25   Pressurized condensate return systems. (*a*) Unit with low-NPSH pumps.

be performed to collect (a) detailed information on the quantity of condensate available from all of the various steam-using sources in the plant and (b) the relevant data associated with these sources. Such data include condensate pressure, quantity, and source location relative to other steam-using equipment and to the boiler room. Certain other information may also be pertinent. For example, if the condensate is contaminated by contact with other process streams, it may be unsuitable for recovery. It is not valid to assume that steam used in the process automatically results in recoverable condensate. Stripping steam used in refining of petroleum and in other separation processes is a good example.

   2. *Survey of possible flash steam applications:* The process should be

surveyed in detail to assess what applications presently using first-generation steam might be adaptable to the use of flash steam or to heat recovered from condensate. Temperatures and typical heat loads are necessary but not sufficient. Heat transfer characteristics of the equipment itself may be important. A skilled heat transfer engineer, given the present operating characteristics of a process heater, can make reasonable estimates of the heater's capability to operate at a lower pressure.

Figure 7.25 (*Continued*) (*b*) Unit with pumping trap. (Courtesy, ITT Corp., Fluid Handling Division)

3. *Analysis of condensate and boiler feedwater chemistry:* To ensure that conditions in the boiler are kept in a satisfactory state to avoid scaling and corrosion, a change in feedwater treatment may be required when condensate is recovered and recycled. Water samples should be obtained from the present condensate drain, from the boiler blowdown, from the incoming-water source, and from the outlet of the present feedwater treatment system. With this information, a water treatment specialist can analyze the overall water chemical balance and ensure that the treatment system is properly configured to maintain good boiler water conditions.

4. *Piping systems:* A layout of the present steam and condensate piping system is needed to assess the need for new piping and the adequacy of existing runs. This permits the designer to select the best locations for flash tanks and minimize the need for extensive new piping.

5. *Economic data:* The bottom line in any energy recovery project, including condensate recovery, is its profitability. To analyze the profitability of the system it is necessary to estimate the quantity of heat recovered, converted into its equivalent fuel usage, the costs and cost savings associated with the water treatment system, and the savings in water cost. In addition to the basic capital and installation costs of the condensate recovery equipment, there may be additional costs associated with modification of existing equipment to make it suitable for flash steam utilization, and there will almost certainly be a cost of lost production during installation and checkout of the new system.

7.6 SUMMARY

This chapter has discussed a number of considerations in effecting energy conservation in industrial and commercial steam systems. Good energy management begins by improving the operation of existing systems and then progresses to evaluation of system modifications to maximize energy efficiency. Methods and data have been presented to assist the energy conservation engineer in estimating the potential for savings by improving steam system operations and by implementing system design changes. As with all industrial and commercial projects, expenditure of capital must be justified by reductions in operating expense. With energy costs rising substantially more rapidly than the cost of most equipment, it is clear that energy conservation will continue to provide ever more attractive investment opportunities.

REFERENCES

1. Armstrong Machine Works, Steam Conservation Guidelines for Condensate Drainage, Bulletin M101, Three Rivers, Mich., 1976.

2. Babcock and Wilcox Corp., *Steam: Its Generation and Use*, 38th ed., New York, 1972.
3. Department of Commerce, National Bureau of Standards, *Energy Conservation Program Guide for Industry and Commerce*, NBS Handbook 115, Washington D.C., 1974.
4. Department of Energy, United Kingdom, *Utilization of Steam for Process and Heating*, Fuel Efficiency Booklet 3, London, 1977.
5. Department of Energy, United Kingdom, *Flash Steam and Vapour Recovery*, Fuel Efficiency Booklet 6, London, 1977.
6. Department of Energy, United Kingdom, *How to Make the Best Use of Condensate*, Fuel Efficiency Booklet 9, London, 1978.
7. Masahiro Ida, Utilization of Low Pressure Steam and Condensate Recovery, TLV Technical Information Bulletin 5123, TLV Company, Ltd., Tokyo, 1975.
8. David L. Isles, Maintaining Steam Traps for Best Efficiency, *Hydrocarbon Process*. 56 (1): Jan. 1977.
9. J. M. Jesionowski, W. E. Danekind, and W. R. Rager, Steam Leak Evaluation Technique: Quick and Easy to Use, *Oil Gas J*. 74(2), Jan. 12, 1976.
10. Donald W. Kern, *Process Heat Transfer*, McGraw-Hill, New York, 1950.
11. Donald May, First Steps in Cutting Steam Costs, *Chem. Eng*. 80(26), Nov. 12, 1973.
12. E. S. Monroe, Jr., Select the Right Steam Trap, *Chem. Eng*. 83(1), Jan. 5, 1976.
13. Joseph Mower, Conserving Energy in Steam Systems, *Plant Eng*. 32(8), Apr. 13, 1978.
14. Sarco Company, Inc., *Hook-up Designs for Steam and Fluid Systems*, 5th ed., Allentown, Pa., 1975.

PROBLEMS

7.1 Two thousand lb_m/hr of fuel oil (S.G. = 0.86) must be heated from 40°F storage temperature to 120°F to permit proper atomization. How many lb_m/hr of 300-psig saturated steam will be required for the oil heater?

7.2 One thousand ft^3/min of nitrogen for purging a distillation column enters a steam heater at 70°F and leaves at 400°F. The nitrogen pressure is approximately 300 psia in the heater. Superheated steam at 350 psia and 500°F is used to heat the nitrogen, and condensate leaves the heater subcooled at 100°F. Calculate the required rate of steam flow.

7.3 A desuperheater for a styrene plant brings 100,000 lb_m/hr of superheated steam at 300 psia and 500°F to the saturated vapor state at the same pressure by injecting cold water into the steam line. Purified water is available at 100°F and atmospheric pressure. Determine the enthalpy change per pound to pump the water up to the desuperheater pressure, assuming a 65% pump efficiency, the mass flow of water required to desuperheat the steam, and the total pump power required.

7.4 A truck tire retreading operation utilizes 50 molds, each processing 5 tires per hour. Each tire contains 90 lb of rubber and 10 lb of steel cord. A steel mold weighs 300 lb. In each mold cycle the mold and tire are heated from 90°F to 300°F using 100-psig saturated steam. Neglecting surface heat losses from the molds, estimate the hourly steam capacity required for the plant, and compare your result with values determined from Table 7.6.

7.5 In a dairy, milk is pasteurized in a 1,000-gal stainless steel batch pasteurizer by raising its temperature from 40°F to 200°F, where it is held at temperature for 20 min. The pasteurizer surface, which is covered with a fabric insulation coating, has an area of about 100 ft^2; the surface temperature when pressurized with steam is 110°F and the room temperature is 70°F. The heated mass of the pasteurizer is about 500 lb of steel. Determine the amount of 20-psig saturated steam required to run two batches per hour, and compare your results with an estimate based on Table 7.6. Use Fig. 7.3 to estimate the surface loss during the holding period.

7.6 The weigh-tank technique illustrated in Fig. 7.8 is to be used to measure the rate of 50-psig steam consumption for a wood pulping unit of 1,500 lb_m capacity in a paper mill. The tank is initially filled about halfway with 250 lb_m of water at 75°F. After closing the condensate return line valve, the condensate is allowed to flow into the weigh-tank for 1 min., at which time the total water mass is found to be 350 lb_m and the water temperature is 165°F. Compare the measured steam consumption rate with the typical value shown in Table 7.6, and based on an energy balance for the tank, determine whether the steam trap is functioning properly (i.e., passing only 50-psig condensate and no live steam).

7.7 From ideal gas mixture relations (see Chapter 6), it has been shown that for a two-component gas mixture the ratio of the partial volumes of the components is equal to the ratio of the partial pressures. Determine the temperature of a mixture of 80% steam and 20% air by volume at a pressure of 75 psig, assuming that the steam is saturated at its partial pressure. Compare your results with Table 7.8.

7.8 Fifteen thousand lb/hr of condensate at 90 psig from a cottonseed oil extraction unit is being delivered to a vented (i.e., atmospheric pressure) flash tank. How many lb/hr of treated water and how many Btu/hr of energy are being lost to the atmosphere in this installation? Calculate your results from basic thermodynamic principles, assuming constant enthalpy throttling through the steam trap that precedes the flash tank and negligible pressure drop in the condensate return line. Compare your answer with an estimate based on Table 7.16.

7.9 Five thousand lb_m/hr of 60-psig condensate from a small distillation column is to be delivered to a 10-psig flash tank.

(a) Using basic energy and mass balances, assuming adiabatic throttling, determine the amount of 10-psig steam produced per hour and compare your results with Table 7.16.
(b) Calculate directly the diameter of the condensate return line required such that the vapor velocity in the line does not exceed 5,000 ft/min. For this calculation, you may neglect the volume of the liquid condensate in the line; justify this assumption. Compare your result with Table 7.11; as a rough estimate, the nominal size for Schedule 40 pipe can be considered about equal to its inside diameter for pipe sizes above 1 inch.
(c) From Table 7.11, estimate the pressure drop for an 80-ft-long condensate return line.

7.10 Using Table 7.17, size a horizontal flash tank for the system described in Problem 7.9. Assume a length-to-diameter ratio of 3:1 for the tank.

7.11 In Problem 7.8, how much mass and energy could be saved if a 30-psig flash tank were used and the condensate returned to the boiler at this pressure, even if the flash steam were vented to atmosphere through a pressure regulating valve? How much additional energy would be saved if the flash steam were utilized in another process?

7.12 Estimate the dimensions of a horizontal flash tank required to meet the conditions specified in Problem 7.8, maintaining a length-to-diameter ratio of 3:1. Resize the tank for the elevated pressure conditions of Problem 7.11. For these pressures, tank cost is roughly proportional to total surface area. What is the approximate percentage cost differential between the two tanks?

7.13 The atmosphere spray tank shown in Fig. 7.19 is used to recover the flash steam mass from a 120-psig return line delivering 3,500 lb_m/hr of condensate to the tank. How much 70°F, 50-psig water is required to condense all the flash steam, such that the water leaving the tank is subcooled by 10°F at 10 psig? (*Practical point of interest:* the requirement of 10°F subcooling would typically be set by the specifications on the condensate return pump to prevent cavitation.)

7.14 Consider the system shown in Fig. 7.18 for utilizing flash steam within a heater unit. Suppose the unit, with steam conditions as indicated in the figure, is used to heat 2,000 ft^3/min of air at 1 atm from 70°F to 175°F in a textile dryer. Estimate the amount of 100-psig steam required before the conversion to flash steam recovery and the amount required after modification to use 5-psig steam in the first stage. Assume that all condensate leaves the heater sections as saturated liquid.

7.15 Suppose the system shown in Fig. 7.18 is used to heat air from 100°F to 300°F for a paint-curing oven. The heat exchanger sections are in crossflow, with both fluids unmixed. Originally, when 100-psig steam was used directly in the first section, the air temperature rise in that section was 75°F. What percent increase in heat-exchanger area would be required to produce the same heating duty in section 1 if that section were converted over to the use of 5-psig flash steam? Refer to Chapter 8 for data on heat-exchanger design parameters.

7.16 Develop an audit plan for flash steam utilization in a petroleum refinery using the process description in Section 15.6. The plan should define potential sources and sinks for flash steam and possible match-ups between process units. A checklist of required data to evaluate these match-ups thermodynamically should be prepared, with possible sources of data listed (see Chapter 2). A checklist of other considerations pertinent to technical and economic feasibility evaluation (e.g., distance between units, steam contamination, etc.) should also be developed.

7.17 Develop an audit plan like the one described above for the low-density polyethylene plant shown in Section 15.5.

7.18 Develop an audit plan like the one described above for the integrated paper mill shown in Section 15.4.

EIGHT

HEAT EXCHANGERS

Heat exchangers are devices for exchanging energy between two or more fluids. They are important in an extremely wide range of industrial applications. For example, heat exchangers are used in the power, process, refining, cryogenic, metals, and manufacturing industries. They are also key components in many commercial products, such as air-conditioning and heating systems, refrigeration devices, automobiles, and aircraft.

Heat exchangers have been in use for a long time. A major impetus for their development was the development of the steam engine, which led to a need to transfer heat effectively between water and gases. Automotive and aircraft power plants, which require cooling, spurred activity in the early twentieth century. Heat exchanger use in industry today is so widespread that it defies documentation.

The theory of heat exchangers is well developed and is based primarily on the first law of thermodynamics. The science of heat transfer is needed to describe local and overall heat transfer behavior within an exchanger. The methods of fluid mechanics are used in calculating the mechanical behavior of an exchanger, including pressure drop and, in some cases, vibration caused by flow instabilities. In this chapter we present only a cursory view of heat exchanger theory since it is well documented in many textbooks and handbooks (see references [1] and [2]).

When energy was cheap, little attention was given to designing heat exchangers that made optimal use of available energy resources. The primary constraints were the *duty* of an exchanger — the amount of heat exchange required — and the surface area required to accommodate the duty. The duty is a thermal constraint, while the size is an economic constraint since the costs of materials and fabrication are directly related to the heat transfer area.

Potential energy savings, along with economic incentives, have led to increased efforts to produce heat exchanger configurations that can be used to:

- Reduce the size of an exchanger for a specified heat duty
- Upgrade the capacity of an existing exchanger
- Reduce the approach temperature difference for process streams
- Reduce pressure drop and pumping power

Some aspects of these efforts will be presented in this chapter.

8.1 CLASSIFICATION OF HEAT EXCHANGERS

The simplest heat exchanger is an open reservoir in which mixing of fluids occurs at different thermal conditions. Heat exchangers of this type are called direct-contact exchangers. Open feedwater heaters and cooling towers are examples of these exchangers. An indirect type is one in which heat is transferred between hot and cold fluids across a solid surface. The heat transfer surfaces in indirect-type heat exchangers can be modified in order to achieve more effective heat transport; hence, these exchangers will be emphasized in this chapter.

Shah [3] points out that heat exchangers can be classified in several ways, e.g., according to (1) transfer process, (2) flow arrangement, (3) construction, (4) surface compactness, (5) heat transfer mechanisms, and (6) number of fluids. Figure 8.1 illustrates the various classifications and subtypes delineated by Shah. Several of these classifications are self-explanatory.

8.1.1 Flow Arrangement

Single-pass exchangers are the simplest type and provide the basis for heat exchanger theory. Figure 8.2 shows a double-pipe system through which the hot and cold fluids pass only once in counterflow. Figure 8.3 shows the temperature distributions for counterflow exchangers with different thermal characteristics for the two streams. By reversing the flow direction on the cold side, we create the parallel-flow case. The temperature behavior for that case is illustrated in Fig. 8.4, which shows the influence of the thermal capacities of the streams, $C = \dot{m}c_p$.[1]

---

[1] Symbols are defined in the Nomenclature section at the end of the chapter.

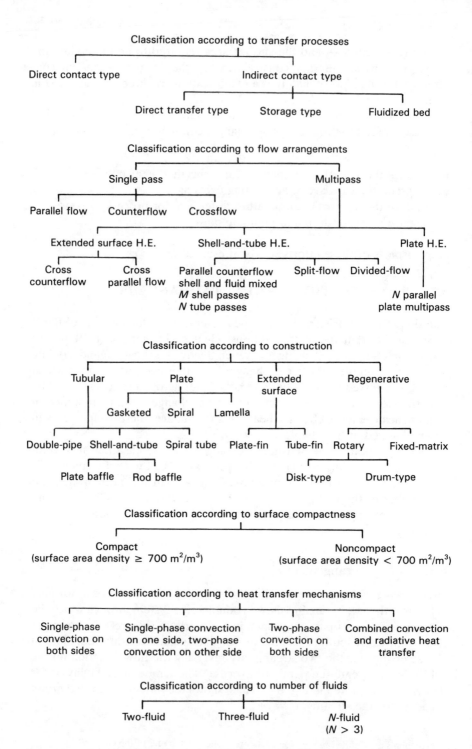

Figure 8.1 Classification of heat exchangers. (From Shah [3]. Reprinted with permission.)

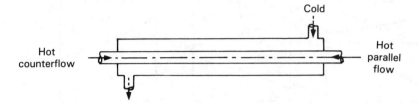

Figure 8.2 Double-pipe heat exchanger.

There are distinct differences in heat transfer performance between counterflow and parallel-flow exchangers. The counterflow arrangement allows the cold inlet to achieve temperatures greater than the hot outlet. This is thermodynamically impossible in the parallel-flow exchanger. Large temperature differences at the inlet of parallel exchangers can induce high thermal stresses in structural parts. However, even in light of these disadvantages, parallel flow is sometimes used for the following reasons:

1. The required heat transfer area may be less than for a counterflow exchanger.
2. Piping may be suited only to parallel flow.
3. Tube wall temperatures are more uniform than in counterflow.
4. The actual heat transfer compared to the theoretical maximum heat transfer is low but can be kept nearly constant over wide ranges of flow rate.
5. It can provide early initiation of nucleate boiling for phase change applications.

From Figs. 8.3 and 8.4 we see that the choice of the appropriate temperature difference is not obvious for the equation,

$$Q = UA \, \Delta T \qquad (8.1)$$

which describes the heat transfer within a heat exchanger. The overall heat

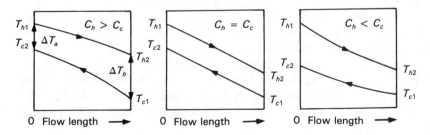

Figure 8.3 Temperature distributions in a counterflow heat exchanger.

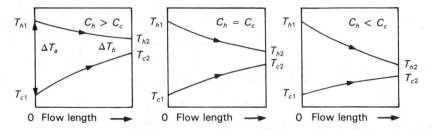

Figure 8.4 Temperature distributions in a parallel-flow heat exchanger.

transfer coefficients for plane and tubular configurations are shown in Fig. 8.5. For tubular configurations the coefficient U must be based on a specific area. Usually the outside area, A_o, is used; however, the relation $U_oA_o = U_iA_i$ makes it possible to base analyses on either the inside or outside tube area. The coefficient U is usually taken to be uniform throughout the exchanger and is calculated with the help of convective heat transfer correlations, given in Chapter 5.

Typical overall coefficients for shell-and-tube exchangers are shown in Table 8.1. The fouling resistances[2] are also included in this table. Typical coefficients for clean (as fabricated) surfaces are shown for comparison.

Log-Mean Temperature Difference The counterflow exchanger allows a simple thermodynamic analysis showing that the correct temperature difference is the so-called log-mean temperature difference

$$\Delta T_{lm} = \frac{\Delta T_a - \Delta T_b}{\ln |\Delta T_a/\Delta T_b|}$$ (8.2)

where a and b refer to respective ends of the exchanger as shown in Figs. 8.3 and 8.4. The derivation of this equation is omitted here; details may be found in any elementary heat transfer text, e.g., Lienhard [5]. It can easily be shown that the ΔT_{lm} for the parallel-flow exchanger is exactly the same as Eq. (8.2). For the case where $\Delta T_a = \Delta T_b$ (counterflow), L'Hospital's rule applied to ΔT_{lm} shows that the appropriate ΔT is ΔT_a (or ΔT_b).

A crossflow exchanger yields temperature distributions that are more complicated. Figure 8.6 shows one such configuration along with its temperature distributions at inlet and exit ports.

For compactness, many exchangers have multiple passes of hot and/or cold streams. Figure 8.7 shows a typical system with the hot fluid flowing through an outer case known as a shell, while the cold fluid flows through a U-tube contained within the shell. The cold fluid makes two "passes,"

[2] The effect of fouling resistance is discussed in Section 8.3.

Table 8.1 Typical Design Coefficients for Shell-and-Tube Exchangers

Fluid 1	Fluid 2	Fouling resistance (hr ft² °F/Btu)	$U_o]^{a,b}$ (Btu/hr ft² °F)	$U_o]^c$ (Btu/hr ft² °F)
Water[c,d]	Water	0.0015	250–300	400–545
Water	Gas, about 10 psig	0.001	15–20	15–20
Water	Gas, about 100 psig	0.001	30–40	32–42
Water	Gas, about 1000 psig	0.001	60–100	64–111
Water	Light organic liquids[e]	0.0015	125–175	154–238
Water	Medium organic liquids[f]	0.002	75–125	88–167
Water	Heavy organic liquids[g]	0.0025	40–75	44–92
Water	Very heavy organic liquids[h]	0.004		
	Heating		10–40	10–48
	Cooling		5–15	5–15
Steam	Gas, about 10 psig	0.0005	15–20	15–20
Steam	Gas, about 100 psig	0.0005	35–45	36–46
Steam	Gas, about 1000 psig	0.0005	70–110	73–116
Steam	Light organic liquids	0.001	135–190	156–234
Steam	Medium organic liquids	0.0015	80–135	91–170
Steam	Heavy organic liquids	0.002	45–80	46–95
Steam	Very heavy organic liquids	0.0035	15–45	16–54
Steam (no noncondensables)	Water	0.001	300–400	425–660
Light organic liquids	Light organic liquids	0.002	100–130	125–175
Light organic liquids	Medium organic liquids	0.0025	70–100	85–133
Light organic liquids	Heavy organic liquids	0.003		
	Heating		40–75	46–97
	Cooling		45–50	47–59
Light organic liquids	Very heavy organic liquids	0.004		
	Heating		20–50	21–63
	Cooling		5–25	5–25
Medium organic liquids	Medium organic liquids	0.003	50–80	60–105
Medium organic liquids	Heavy organic liquids	0.0035		
	Heating		30–50	36–60
	Cooling		15–35	16–40

Table 8.1 Typical Design Coefficients for Shell-and-Tube Exchangers (*Continued*)

Fluid 1	Fluid 2	Fouling resistance (hr ft² °F/Btu)	$U_o]^{a,b}$ (Btu/hr ft² °F)	$U_o]^c$ (Btu/hr ft² °F)
Medium organic liquids	Very heavy organic liquids	0.0045		
	Heating		15–30	16–35
	Cooling		5–25	5–28
Heavy organic liquids	Heavy organic liquids	0.005	10–30	10–35
Heavy organic liquids	Very heavy organic liquids	0.006	5–15	5–16
Gas, about 10 psig	Gas, about 10 psig	0	10–15	—
Gas, about 10 psig	Gas, about 100 psig	0	15–20	—
Gas, about 10 psig	Gas, about 1000 psig	0	15–25	—
Gas, about 100 psig	Gas, about 100 psig	0	20–30	—
Gas, about 100 psig	Gas, about 1000 psig	0	25–35	—
Gas, about 1000 psig	Gas, about 1000 psig	0	35–60	—
Water	Condensing light organic vapors, pure components[i]	0.001	150–200	175–250
Water	Condensing medium organic vapors, pure components[i]	0.001	100–150	110–175
Water	Condensing heavy organic vapors, pure component[i]	0.002	75–100	88–125

[a] The total fouling resistance and the overall heat transfer coefficient are based on the total outside tube area.

[b] Allowable pressure drops on each side are assumed to be about 10 psi except for (1) low-pressure gas and condensing vapor, where the pressure drop is assumed to be about 5% of the absolute pressure, and (2) heavy organics, where the allowable pressure drop is assumed to be about 20–30 psi.

[c] Aqueous solutions give approximately the same coefficients as water.

[d] Liquid ammonia gives about the same results as water.

[e] Light organic liquids include liquids with viscosities less than about 0.5 cp.

[f] Medium organic liquids include liquids with viscosities between about 0.5 and 1.5 cp.

[g] Heavy organic liquids include liquids with viscosities greater than 1.5 cp but not over 50 cp.

[h] Very heavy organic liquids include liquids having viscosities greater than about 50 cp. Estimation of coefficients for these materials is very uncertain.

[i] These values may be used for vapor mixtures when the condensing range of the vapor is less than half of the temperature difference between the outlet coolant and the vapor. If the condensing range is greater than this, or if significant amounts of noncondensable gas are present, the coefficient should be reduced toward the values shown for gas cooling; in these cases the accuracy of the estimation is very uncertain.

Source: Adapted from [4]. Reprinted with permission.

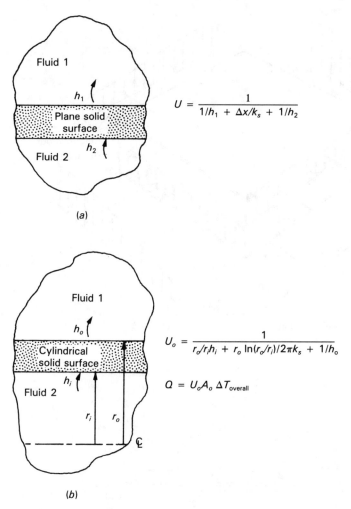

Figure 8.5 Overall heat transfer coefficients for heat exchange surfaces: (*a*) plane boundary; (*b*) tubular boundary.

requiring two curves to represent the total temperature behavior of the cold fluid, as shown in Fig. 8.7*b*. This type of exchanger is also known as a 1–2 heat exchanger. The baffles serve to increase the heat transfer between the flows by channeling the shell flow over the tubes at right angles.

Multiple-Pass Correction Factors The ΔT_{lm} must be corrected for multi-pass systems to account for the different flow paths. The heat transfer equation becomes

$$Q = UA\, \Delta T_{lm} F \qquad (8.3)$$

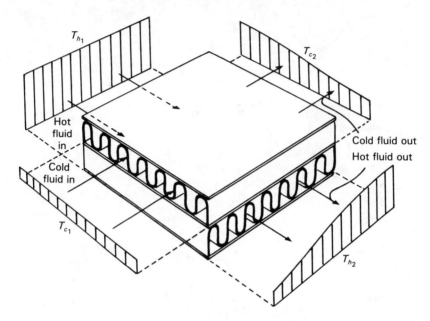

Figure 8.6 Temperature distribution in crossflow heat exchanger.

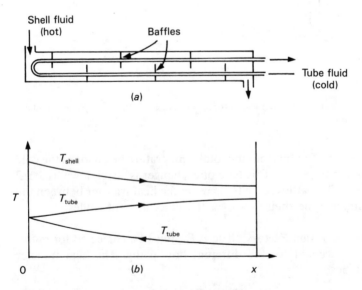

Figure 8.7 (a) A 1–2 heat exchanger (one shell pass and two tube passes); (b) corresponding temperature distribution.

When Eq. (8.3) is used, ΔT_{lm} is evaluated for the equivalent counterflow single-pass exchanger with the same hot and cold temperatures. The correction factor F was shown by Bowman et al. [6] to be a function of heat exchanger effectiveness, ε, and capacity ratio, C_c/C_h, for all cases. The effectiveness is defined as

$$\varepsilon = \frac{Q_{actual}}{Q_{max}} = \frac{(\dot{m}c_p)_c(T_{c2} - T_{c1})}{(\dot{m}c_p)_c(T_{h1} - T_{c1})} = \frac{T_{c2} - T_{c1}}{T_{h1} - T_{c1}} \tag{8.4}$$

In other words, if the cold outlet temperature, T_{c2}, were brought to the hot inlet temperature, T_{h1}, the maximum Q would be realized. In accordance with the first law of thermodynamics, the capacity ratio is defined as

$$R = \frac{C_c}{C_h} = \frac{T_{h1} - T_{h2}}{T_{c2} - T_{c1}} \tag{8.5}$$

Figure 8.8 shows the correction factor F based on values recommended by the Tubular Equipment Manufacturers Association (TEMA) for shell-and-tube exchangers and the curves of Bowman et al. for crossflow exchangers. On these plots, only the R parameter lines up to $R = 1.0$ are included, although R can be greater than unity according to Eq. (8.5). Shamsundar [7] has shown that

$$F(\varepsilon, R) = F(\varepsilon R, 1/R) \tag{8.6}$$

for exchangers that require a correction factor for the log-mean temperature difference. Both the TEMA and Bowman charts contain sets of curves for $R > 1$ in addition to those shown in Fig. 8.8. However, Eq. (8.6) makes the $R > 1$ curves redundant, because it represents a form of reciprocity between the parameters for ε and εR, and for R and $1/R$. Use of only the $R \leqslant 1$ curves can actually result in slightly more accurate values of F, because when ε is small (for $R > 1$), the R curves are nearly vertical. Examination of the $R \leqslant 1$ curves in Fig. 8.8 reveals that some curvature exists over the appropriate ε change.

To illustrate the use of the reciprocity rule of Eq. (8.6), let us consider a one-pass crossflow exchanger with both passes unmixed, with $\varepsilon = 0.23$ and $R = 4$. The equivalent effectiveness is $\varepsilon R, = 0.92$ and the equivalent capacity ratio is $1/R = 0.25$. Figure 8.8c gives $F = F(0.92, 0.25) = 0.75$. The reader is invited to verify that the reciprocity rule does indeed work by using a chart with the full range of R values plotted on it; e.g., see reference [5].

For exchangers in which one of the fluids undergoes either evaporation or condensation at constant temperature, no correction is needed and $F = 1$.

The technique for calculating Q requires that we know all the inlet and outlet temperatures for an exchanger, its surface area, and the overall heat transfer coefficient. In some cases the temperatures are not all known. This requires a trial-and-error solution or the use of an alternative method, which will be discussed later.

8.1.2 Construction

We have already discussed the shell-and-tube construction. The crossflow exchange of Fig. 8.6 is a plate type, with the streams separated by planar sections. "Extended surface" refers to exchangers in which fins are used to augment heat transfer. This can be done on all surfaces. Figure 8.9 shows two configurations for plate-fin extended-surface exchangers. A tube-fin exchange surface is illustrated in Fig. 8.10. Individual tubes can also be

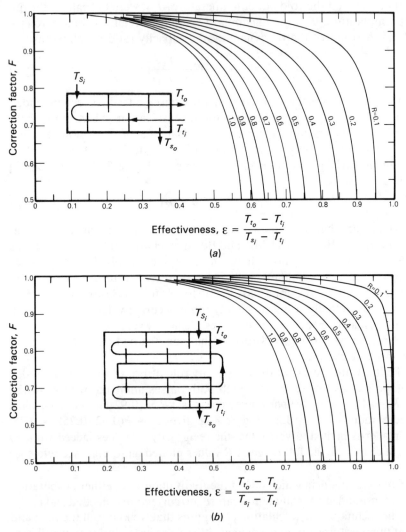

Figure 8.8 LMTD correction factor, F, for (a) an exchanger with 1 shell pass and 2, 4, 6, . . . tube passes (the reciprocity rule for $R > 1.0$); (b) an exchanger with 2 shell passes and 4 or more tube passes.

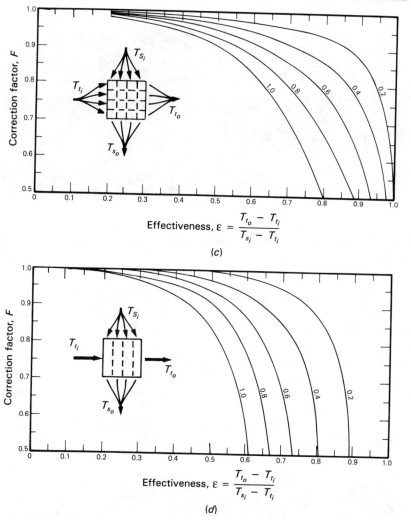

Figure 8.8 (*Continued*)   (*c*) A 1-pass crossflow exchanger with both passes unmixed; (*d*) a 1-pass crossflow exchanger with one pass mixed.

finned as shown in Fig. 8.11. Systems having surfaces like those shown in Fig. 8.11 are classed generally as recuperators; they are steady-state heat exchangers with no mixing of the streams.

Regenerative heat exchangers involve storage of heat in an element as the hot stream passes over it and subsequent loss of the heat as the cold stream passes over it. This is done by rotating the heat storage element through the two streams, or by holding the element stationary and directing the hot and cold streams alternatively over it. These systems will be discussed further in Chapter 9.

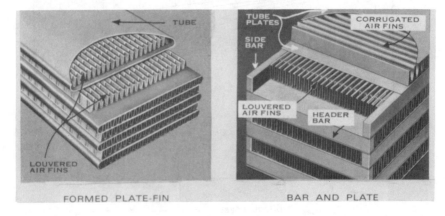

Figure 8.9 Plate-fin heat exchangers. (Courtesy, Harrison Radiator Division, GMC, Lockport, New York)

Shell-and-tube exchangers are very rugged and can be custom designed for practically any operating condition, such as high pressure and temperature, and highly corrosive atmospheres and fluids. They are used extensively in the refining and petrochemical industries and as steam generators, water heaters, and oil coolers in power plants. Most shell-and-tube exchangers use internal baffles to direct the shell flow at nearly right angles to the tube bank because this results in less required heat transfer area. Some baffle arrangements are shown in Fig. 8.12.

A recent development is the so-called rod-baffled shell-and-tube exchanger. In this case transverse rods rather than solid baffle plates are

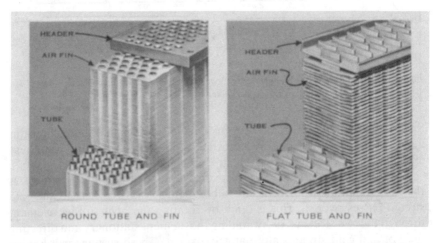

Figure 8.10 Tube-fin heat exchangers. (Courtesy, Harrison Radiator Division, GMC, Lockport, New York)

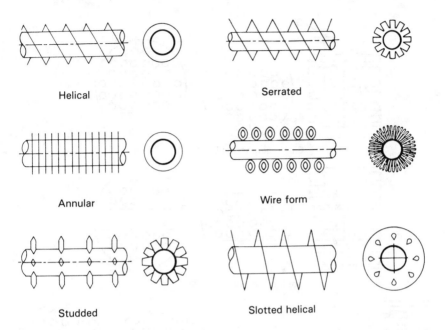

Figure 8.11 Examples of externally finned tubes.

used to support the tubes. Figure 8.13 shows the arrangement. The shell flow in this configuration is basically parallel to the tubes, resulting in lower pressure drop. Flow over the rods gives excellent mixing because of vortex shedding and the heat transfer is relatively efficient. Flow-induced vibrations are virtually eliminated by the rod support of the tubes, which is another advantage of this construction.

Plate exchangers are used in paper mills, in the dairy, beverage, food, and pharmaceutical industries, and in the synthetic rubber industry. They compete with shell-and-tube exchangers in some refining and petrochemical applications. Sealing gasket materials limit the maximum temperatures to about 530°F (275°C), and plate exchangers are usually used below 300°F (150°C) to avoid expensive, exotic gasket materials. Pressure is limited to about 300 psig. Pinhole leaks can develop in the plates and are hard to detect. Plate exchangers are somewhat limited in size, although some of the larger units have about 16,000 ft^2 of surface.

On the positive side, the plate exchanger is easily disassembled for inspection and cleaning. Leakage from one fluid to the other cannot occur unless a plate develops a hole. Gasket leaks go to the outside of the exchanger. Plate exchangers achieve uniformity of heat transfer, which is advantageous in applications such as cooking, sterilization, and pasteurization. Hot and cold spots are virtually eliminated. The high turbulence caused by the plates also reduces fouling.

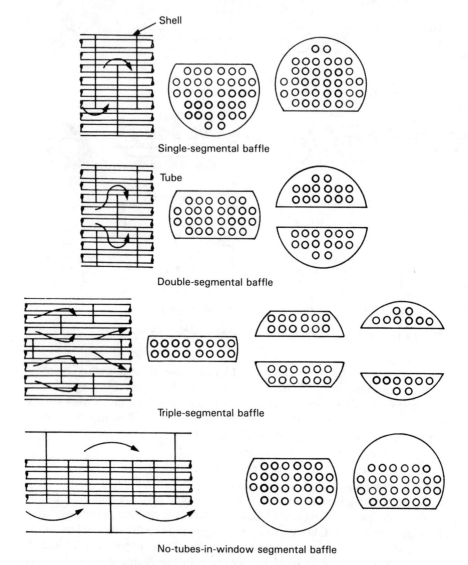

Figure 8.12 Plate baffle types for shell-and-tube heat exchangers.

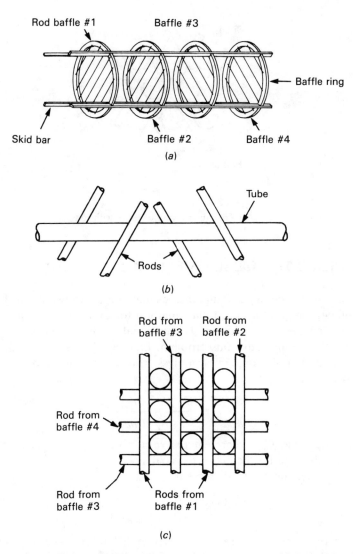

Figure 8.13 Details of the rod-baffled shell-and-tube heat exchanger. (*a*) Four rod baffles held by skid bars (no tubes shown); (*b*) tube supported by four rods; (*c*) square layout of tubes with rods.

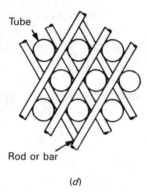

Tube

Rod or bar

(d)

Figure 8.13 (*Continued*)   (*d*) Triangular layout of tubes with rods.

8.2   THE NTU-ε DESIGN METHOD

In some cases the outlet temperatures may not be known a priori, and the log-mean temperature difference method for exchanger prediction would require an iterative solution. Dimensional analysis suggests a different approach that alleviates this problem. Lienhard [5] uses the π-theorem to show that the following dimensionless groups can be formed. The capacity ratio is

$$C_R \equiv \frac{C_{\min}}{C_{\max}} = \frac{(\dot{m}c_p)_{\min}}{(\dot{m}c_p)_{\max}} \tag{8.7}$$

where $C_{\min}$ and $C_{\max}$ represent the respective $\dot{m}c_p$ products regardless of whether they are for the hot or the cold side.

The exchanger effectiveness is

$$\varepsilon = \frac{C_h (T_{h1} - T_{h2})}{C_{\min} (T_{h1} - T_{c1})} \equiv \frac{C_c (T_{c2} - T_{c1})}{C_{\min} (T_{h1} - T_{c1})} \tag{8.8}$$

where the subscript 1 refers to inlet and 2 refers to outlet quantities. Equation (8.8) represents the ratio of actual heat transfer to the thermodynamically limited maximum heat transfer that could be achieved only in a counterflow exchanger with infinite area. The first definition is for $C_h = C_{\min}$ and the second is for $C_c = C_{\min}$.

The number of heat transfer units, NTU, is

$$\text{NTU} = \frac{AU}{C_{\min}} \tag{8.9}$$

An energy balance written for a specific exchanger configuration will yield a relationship between ε and NTU with C_R as a parameter. Table 8.2 shows these relationships for several configurations. The capacity ratio C_R is limited to the extremes zero and unity. When there is phase change in one side of the exchanger, as in condensers and boilers, C_R is zero. Table 8.2 shows the appropriate expressions for these limits.

It is clear that ε can be calculated if NTU and C_R are known, allowing the design process to proceed without prior knowledge of the outlet temperatures. Figure 8.14 gives plots of the ε-NTU behavior of several exchanger configurations. The asymptotic nature of the plots shows that little increase in effectiveness is achieved above NTU ≈ 3. The most economical configuration for an exchanger usually occurs for $1 < $ NTU $ < 3$. Increasing NTU primarily involves increasing the area, since C_{min} and U are approximately constant for a particular configuration.

Table 8.2 Heat Exchanger Effectiveness Relations

$$N = \text{NTU} = \frac{UA}{C_{min}} \qquad C_R = \frac{C_{min}}{C_{max}}$$

Flow geometry	Relation
Double pipe	
Parallel flow	$\varepsilon = \dfrac{1 - \exp\left[-N(1 + C_R)\right]}{1 + C_R}$
Counterflow	$\varepsilon = \dfrac{1 - \exp\left[-N(1 - C_R)\right]}{1 - C_R \exp\left[-N(1-C_R)\right]}$
Counterflow, $C_R = 1$	$\varepsilon = \dfrac{N}{N + 1}$
Crossflow	
Both fluids unmixed	$\varepsilon = 1 - \exp\left[\dfrac{\exp\left(-NC_R n\right) - 1}{C_R n}\right]$ where $n^a = N^{-0.22}$
Both fluids mixed	$\varepsilon = \left[\dfrac{1}{1 - \exp\left(-N\right)} + \dfrac{C_R}{1 - \exp\left(-NC_R\right)} - \dfrac{1}{N}\right]^{-1}$
C_{max} mixed, C_{min} unmixed	$\varepsilon = (1/C_R)\{1 - \exp\left[-C_R(1 - e^{-N})\right]\}$
C_{max} unmixed, C_{min} mixed	$\varepsilon = 1 - \exp\{-(1/C_R)[1 - \exp(-NC_R)]\}$
Shell and tube	
One shell pass, 2, 4, 6 tube passes	$\varepsilon = 2\left\{1 + C_R + (1 + C_R^2)^{1/2}\dfrac{1 + \exp\left[-N(1 + C_R^2)^{1/2}\right]}{1 - \exp\left[-N(1 + C_R^2)^{1/2}\right]}\right\}^{-1}$
All exchangers with $C_R = 0$	$\varepsilon = 1 - e^{-N}$

^{a}Approximate value.

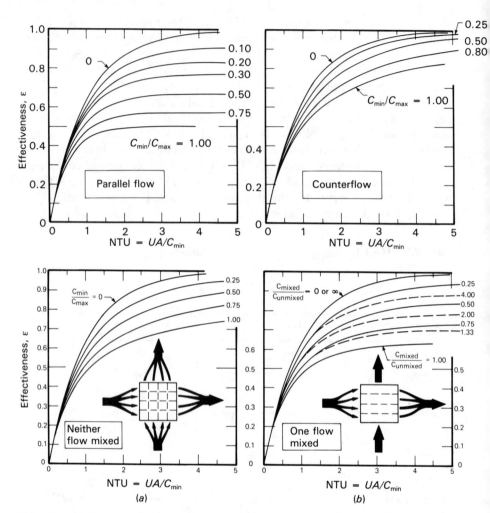

Figure 8.14 Effectiveness of parallel and counterflow heat exchangers and some other heat exchanger configurations. (*a*) Crossflow exchanger, neither fluid mixed; (*b*) crossflow exchanger, one fluid mixed.

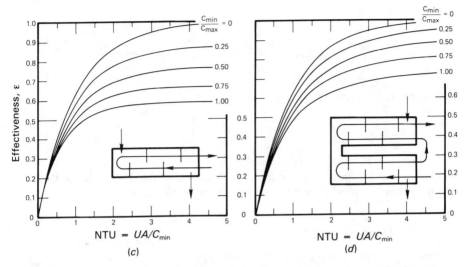

NTU $= UA/C_{min}$

(c)

NTU $= UA/C_{min}$

(d)

Figure 8.14 Effectiveness of parallel and counterflow heat exchangers and some other heat exchanger configurations (*Continued*). (*c*) One-shell-pass, two-tube-pass exchanger (can also be used for 4, 6, 8, 10, 12 tube passes with maximum error in ε of 0.040 at $C_{min}/C_{max} = 1$ and large NTU); (*d*) two-shell-pass, four tube-pass exchanger (can also be used for 4, 6, 8, ... tube passes with accuracy if there are equal numbers of tube passes in each shell pass).

8.3 FOULING

The term *fouling* is used in the literature to mean any undesirable deposit on a heat transfer surface that increases the resistance to heat transmission. In practice, industrial heat exchangers rarely operate with nonfouling fluids. Cryogenic heat exchangers remain relatively clean—they may be the only exception to the above statement.

Fouling is one of two major unsolved problems in heat exchangers, the other being flow-induced vibrations. However, in the past decade the mechanisms involved in fouling have come under increased scrutiny. The role played by operating parameters that are likely to affect fouling, such as pressure, temperature, and flow rate, has also been examined.

The total resistance to heat transfer for a cylindrical tube with fouling at both the inner and outer surfaces is given by

$$\frac{R_T}{A_o} = \frac{1}{h_i A_i} + \frac{R_i}{A_i} + \frac{\Delta r}{k_t A_{ml}} + \frac{R_o}{A_o} + \frac{1}{h_o A_o} \qquad (8.10)$$

The temperature behavior for such a system is depicted in Fig. 8.15. The factors R_i and R_o are known as *fouling factors*.

The overall U for fouled conditions can be expressed in terms of a clean exchanger coefficient and a term dependent upon the overall fouling resistance. Based on the outside area, A_o, it is given by

$$\left[\frac{1}{U_o}\right]_f = \underbrace{\frac{1}{h_i}\frac{A_o}{A_i} + \frac{\Delta r}{k_t}\frac{A_o}{A_{ml}} + \frac{1}{h_o}}_{(1/U_o)_c} + \underbrace{R_i\frac{A_o}{A_i} + R_o}_{R_f} \qquad (8.11)$$

The overall fouling resistance is then

$$R_f = \left[\frac{1}{U_o}\right]_f - \left[\frac{1}{U_o}\right]_c \qquad (8.12)$$

Fouling is a function of time, fluid/surface combination, temperature, and concentration, among other factors. With our current knowledge of the phenomenon, predicting the rate of fouling is far from simple.

8.3.1 Types of Fouling

Bott [8] has classified fouling into five types:

- Particulate fouling: The deposition of particles suspended in fluid streams (e.g., sand, muds, corrosion debris, dust in gas streams).
- Reaction fouling: The fouling caused by a chemical reaction at the heat transfer surface, namely cracking and polymerization of hydrocarbons.
- Solubility fouling: The precipitation and deposition of dissolved material caused by solubility changes with temperature (e.g., wax from kerosene, calcium carbonate and calcium sulfate from cooling waters).
- Corrosion fouling: The special case where in situ deposits are formed with the assistance of the heat exchanger surface material.
- Biological fouling: The growth of microorganisms on heat transfer surfaces (e.g., slime formation in cooling towers).

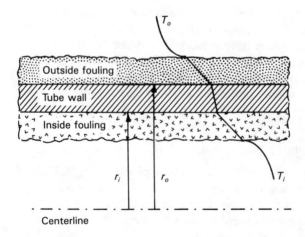

Figure 8.15 Schematic of a fouled tube.

Regardless of the type of fouling, its occurrence leads to great costs on a worldwide basis. The major effects of fouling are increases in capital expenditure caused by oversizing of exchangers, higher energy costs associated with poorer performance of equipment, maintenance costs involved in cleaning, and loss of production during plant shutdown for cleaning operations.

Fouling is usually observed to develop in one of three ways. Figure 8.16 shows *linear, falling-rate,* and *asymptotic* fouling curves. The asymptotic curve implies that after a certain operating period fouling is fully developed, and if the equipment is designed with that in mind, there would be no further concern with fouling. The effect of asymptotic fouling on the overall coefficient U is shown in Fig. 8.17. Linear fouling implies that the fouling thickness increases indefinitely. However, both the linear and falling-rate curves may represent the early states of asymptotic fouling. Thus care should be exercised when interpreting the results of fouling tests.

8.3.2 The Nature of Fouling

A widely accepted description of the nature of fouling was given by Kern and Seaton [9]. They expressed the net rate of addition of fouling resistance to a surface as the difference between a uniform deposition rate X_d and a function X_r. The removal function is usually taken to be proportional to the existing fouling resistance, as $X_r = BR_f$, where B is a rate constant. under these assumptions, Kern and Seaton found that the fouling resistance varies as

$$R_f = R_f^* \left[1 - \exp(-\beta\tau) \right] \tag{8.13}$$

where R_f^* is ϕ_d/β, representing the asymptotic value of fouling resistance.

Much of the recent work on fouling has focused on predicting the removal rate of the deposit fouling layer. It has been shown that the higher the wall shear stress, the lower R_f^* will be. This implies that the fluid flow

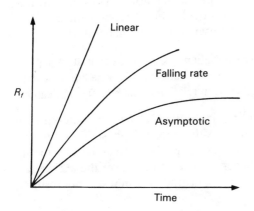

Figure 8.16 Linear, falling-rate, and asymptotic fouling curves.

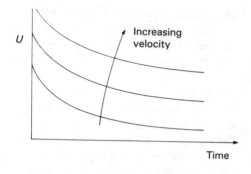

Figure 8.17 Effect of asymptotic fouling on the overall heat transfer coefficient.

"tears" away deposited materials. The characteristic time required to approach an asymptotic value decreases with mechanical strength of the deposited layer. This is further evidence that the shear stress exerted on the layer is responsible for the removal of materials. Apparently, if the shear stress exceeds the yield strength of the deposit, the deposit will be broken off and carried away.

Collier [10], O'Callaghan [11], and Pinheiro [12] give excellent reviews of recent research on the nature of fouling. Although the fouling problem is unresolved, there are practices that will minimize the effects of fouling. Brief descriptions of some of these follow.

Maintaining High Velocities Higher flow velocities have two conflicting effects that tend to partially offset each other. The rate of deposition may increase with velocity, while the rate of removal also increases with velocity. Figure 8.18 shows the effects of velocity on calcium carbonate scaling from cooling water. It is clear that competing effects are at work but that, at higher velocities, the removal overcomes the deposition effects.

Design for Asymptotic Fouling Oversizing the required heat transfer area is an acceptable practice that offsets the future loss in heat transfer efficiency. Values of R_f^* must be obtained from experiments rather than from calculation. At start-up, the equipment will provide above-design heat transfer. To avoid any potential problems, the user should be made aware of this.

Periodic Cleaning In situations where fouling is known to occur at an approximately fixed rate, equipment can be cleaned on a specific schedule. Cleaning is costly and this cost must be considered in relation to the benefits of cleaning.

Temperature Redistribution Fouling deposits usually form preferentially where the surface-to-fluid temperature difference is highest. This tends to lower the heat flux in those parts of the exchanger because of the additioning fouling, causing the heat flux to become more uniform over the exchanger. The result is that the overall performance degradation may be underestimated by an average value of R_f if the heat transfer in the clean exchanger is dominated by local heat flux behavior. A sectional treatment of the fouled exchanger may be called for because of the surface temperature redistribution.

8.4 HEAT TRANSFER AUGMENTATION

Improving heat transfer performance is known as enhancement, intensification, or augmentation. This generally means an increase in heat transfer coefficient. An operating exchanger is of fixed size and the range of ΔT is also limited. The basic equation of heat exchanger performance, $Q = UA\,\Delta T$, shows that any increases in heat transfer must be achieved by controlling the heat transfer coefficient U.

8.4.1 Augmentation Techniques

Techniques for augmenting heat transfer can be classified according to whether or not they require external power. Those that do are called *active*, and those that do not are called *passive*. Brief descriptions of some of these techniques are given by Bergles [15] and are summarized below.

Passive Techniques   *Treated surfaces* are produced by fine-scale alteration

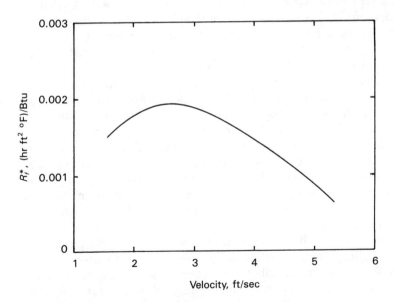

Figure 8.18 Asymptotic fouling resistance versus velocity for plain 0.5-in. outer diameter tubes with $CaCO_3$ scaling. (From Watkinson et al. [13])

of the surface finish or coating (continuous or discontinuous). They are used mostly for boiling and condensing, because the roughness height of the surface treatment is below that which effects single-phase heat transfer.

Rough surfaces are produced in many ways with the surface relief ranging from random sand grain type to discrete protuberances. The configuration is generally chosen to promote turbulence rather than increase the heat transfer surface area. This technique is normally used with single-phase flow.

Extended surfaces are routinely used in many exchangers. Of particular interest are new types of extended surfaces, such as integral inner-fin tubing, and improvement of heat transfer coefficients on extended surfaces by shaping or interrupting the surfaces.

Swirl flow devices have inserts for forced flow that create rotating and/or secondary flow. The types of inserts used include inlet vortex generators, twisted-tape inserts, and axial-core inserts with a screw-type winding.

Surface tension devices consist of wicking or grooved surfaces to direct the flow of liquid during boiling or condensing.

Additives for liquids include solid particles and gas bubbles in single-phase flows and liquid trace additives for boiling systems.

Additives for gases are liquid droplets or solid particles, either dilute phase (gas-solid suspensions) or dense phase (fluidized beds).

Active Techniques *Mechanical aids* involve stirring by mechanical means or by rotating the surface. Surface "scraping," widely used for viscous liquids in the chemical process industry, can also be applied to duct flow gases. Equipment with rotating heat exchanger ducts is available commercially.

Surface vibration at either low or high frequency has been used primarily to improve single-phase heat transfer.

Fluid vibration is the more practical type of vibration enhancement due to the mass of most heat exchangers. The vibrations range from pulsations of about 1 Hz to ultrasound. Single-phase fluids are of primary concern.

Electrostatic fields (DC or AC) are applied in many different ways to dielectric fluids. Generally speaking, electrostatic fields can be directed to cause greater bulk mixing of fluid in the vicinity of a heat transfer surface.

Injection involves supplying gas to a flowing liquid through a porous heat transfer surface or by injecting a similar fluid upstream of the heat transfer surface. Surface degassing of liquids can produce augmentation similar to gas injection.

Suction involves vapor removal in nucleate or film boiling, or fluid withdrawal in single-phase flow, through a porous heated surface.

Some of the effects described above are inherent in heat exchange equipment and provide "natural" augmentation of heat transfer. Roughness of fabricated surfaces, degassing of liquids, vibration from rotating machinery, flow oscillations, and electrical fields may be present in commercial exchangers.

8.4.2 Augmentation Decision-Making

Assuming that augmentation techniques are available, what are the conditions that would lead an energy manager to implement them, either for new equipment or in retrofitting old equipment? Basically, the decision to augment depends on positive answers to the following questions:

1. Does the heat exchange application represent significant potential savings of energy and/or dollars?
2. Are techniques available for the desired application?
3. Are techniques available in commercial equipment and have they been proved in the field?
4. Does an economic study of the proposed augmentation show it to be competitive with other proposed projects?

For two-fluid exchangers, the thermal resistance ratio between the streams is the primary factor in deciding whether augmentation would be desirable. Clearly, augmentation for the stream with the higher resistance should be considered. This is especially true if there is one stream that dominates the total resistance.

Methods of assessing the thermal-hydraulic consequences of augmentation for the tube side of a single-pass shell-and-tube heat exchanger have been suggested. Table 8.3 lists sets of criteria for judging the effects of in-tube augmentation. Take, as an example, criterion 4, where the number of tubes, their length, and the heat duty Q are fixed and the goal is to reduce the pumping power, P, through augmentation. The important parameter then is P_a/P_b. To reduce pumping power the flow rate $\dot{m}$ and/or Δp would have to be reduced. If techniques for doing this are available, the evaluation process could then proceed.

In-Tube Augmentation Determination of the feasibility of augmentation may be illustrated by selecting a case in which one of the heat transfer coefficients is low and, in fact, is the limiting factor in the overall heat transfer coefficient. Consider a single-pass shell-and-tube exchanger in which ammonia is to be condensed in the shell and water is the coolant flowing in the tubes. The limiting resistance to heat transfer is on the tube side, so that in-tube augmentation is the logical choice. We shall assume constant heat duty and pumping power. How much can the area be decreased by in-tube augmentation? This situation corresponds to criterion 5 in Table 8.3.

The heat transfer equation for the baseline case, Q_b, and the augmented case, Q_a, becomes

$$Q_a = Q_b = U_b A_b \, \Delta T_{lm} = U_a A_a \, \Delta T_{lm}$$

The in-tube coefficient is limiting, so that $U_b \propto h_{b,i}$ and $U_a \propto h_{a,i}$. Using this

Table 8.3 Performance Evaluation Criteria for Single-Pass, Shell-and-Tube Exchangers with In-Tube Augmentation

Criterion	Fixed parameters					Thermal-hydraulic goal				Consequences					Note
	Geom.	$\dot{m}$	Δp	P	Q	Geom.↓	$\Delta p\downarrow$ $P\downarrow$	$Q\uparrow$	Parameter of interest	Geom.	$\dot{m}$	Δp	P	Q	
1	n,L	X						X	Q_a/Q_b			↑	↑		f
2	n,L		X					X	Q_a/Q_b		→		→		a,f
3	n,L			X				X	Q_a/Q_b		→	←			f
4	n,L				X		X		P_a/P_b		→	→			b
5					X	X			A_a/A_b or $(nL)_a/(nL)_b$	$L(n)$ ↓↑	→	↑			c
6			X		X	X			A_a/A_b or $(nL)_a/(nL)_b$	$L(n)$ ↓↑	→				
7		X			X	X			A_a/A_b or $(nL)_a/(nL)_b$	$L(n)$ ↓↓		→			
A		X	X	X	X	X			A_a/A_b or $(nL)_a/(nL)_b$	$L(\dot{m})n(\dot{m})$ ↓↑		→			d
B	A	X	X	X				X	Q_a/Q_b	$L(\dot{m})n(\dot{m})$ ↑↓				→	e,f
C	A	X			X		X		P_a/P_b	$L(\dot{m})n(\dot{m})$ ↓↑		→			

[a] $\Delta p \downarrow$ is alternate objective, $Q \uparrow$ a consequence.
[b] $\Delta p \downarrow$ is alternate objective, $P \downarrow$ a consequence.
[c] For low NTU, $A_a A_b$ comes out directly.
[d] $(A_a/A_b)_5$ at low NTU.
[e] $(Q_a/Q_b)_B = (Q_a/Q_b)_3$.
[f] $\Delta T_i \downarrow$ is alternate, for fixed Q.

Source: Adapted from [15]. Reprinted with permission.

along with the assumption that ΔT_{lm} remains constant, we get

$$\frac{A_a}{A_b} = \frac{h_{b,i}}{h_{a,i}} = \frac{n_a L_a}{n_b L_b} \qquad (8.14)$$

where n is the number of tubes and L is their length per pass.

The pumping power equation is

$$P_b = n_b V_b \frac{\pi D^2}{4} \left(f_b \frac{L_b}{D} \frac{\varrho V_b^2}{2g_c} \right) = P_a = n_a V_a \frac{\pi D^2}{4} \left(f_a \frac{L_a}{D} \frac{\varrho V_a^2}{2g_c} \right)$$

This gives

$$\frac{n_a L_a}{n_b L_b} = \frac{A_a}{A_b} = \frac{f_b}{f_a} \frac{V_b^3}{V_a^3} \qquad (8.15)$$

where f is the Moody friction factor and V is water velocity. We are assuming that the diameter D of the tube will be changed little by augmentation.

Combining Eqs. (8.14) and (8.15), we can write

$$\frac{A_a}{A_b} = \frac{n_a L_a}{n_b L_b} = \frac{\mathrm{Nu}_b}{\mathrm{Nu}_a} = \frac{f_b}{f_a} \frac{\mathrm{Re}_b^3}{\mathrm{Re}_a^3} \qquad (8.16)$$

where Nu is the Nusselt number, $\mathrm{Nu} = hD/k$, and Re is the Reynolds number for tube flow, $\mathrm{Re} = VD/\upsilon$.

Relationships for the heat transfer behavior (Nu) and the pressure drop (f) for the augmented surface must be available for assessment of the feasibility of augmentation. For this case, we choose low-transverse ribs as the in-tube augmentation promoter. Figures 8.19 and 8.20 show the Nusselt number and friction factor behavior for tubes with low-transverse ribs in terms of Reynolds number. For simplicity, the McAdams equation, Eq. (5.14),

$$\frac{\mathrm{Nu}}{\mathrm{Pr}^{0.4}} = 0.023 \, \mathrm{Re}^{0.8}$$

is used to represent the baseline heat transfer coefficient.

The procedure is as follows. First pick a Reynolds number for the baseline case, say $\mathrm{Re}_b = 2.5 \times 10^4$. Then from the plots:

$$f_b = 0.0063 \qquad \frac{\mathrm{Nu}}{\mathrm{Pr}^{0.4}} = 76$$

$$f_a = 0.036 \qquad \frac{\mathrm{Nu}}{\mathrm{Pr}^{0.4}} = 215$$

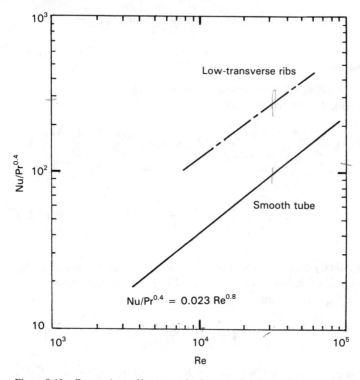

Figure 8.19 Comparison of heat transfer for smooth tubes and tubes with low-transverse ribs.

Equation (8.16) gives $A_a/A_b = 0.353$ and also allows computation of the augmented Re_a. To keep pumping power the same, Re_a must be lower than Re_b. Equation (8.16) gives $\text{Re}_a = 1.782 \times 10^4$ because f is almost constant in this range of Re. A lower Re requires more tubes to accommodate the flow rate, which must remain fixed so that ΔT_{lm} is constant. A constant flow rate yields

$$\varrho n_b \frac{\pi D^2}{4} V_b = \varrho n_a \frac{\pi D^2}{4} V_a$$

or

$$\frac{n_a}{n_b} = \frac{V_b}{V_a} = \frac{\text{Re}_b}{\text{Re}_a}$$

This relationship gives $n_a/n_b = 1.264$; i.e., the number of tubes must be increased by about 25%. However, the length can be reduced, because

$$\frac{A_a}{A_b} = \frac{n_a L_A}{n_b L_b}$$

or

$$\frac{L_a}{L_b} = \frac{n_b}{n_a} \frac{A_a}{A_b} = \frac{1}{1.264} (0.353) = 0.279$$

For this case, at the selected Reynolds number, the area can be reduced significantly, although the number of rows must be increased with augmentation. The required length decreases much more rapidly than the number of tubes increases.

In an actual case, the log-mean temperature difference could be allowed to vary and the resistance of the condensation process could be included in the analysis. This would yield a more realistic result. Nevertheless, the technique illustrated above is applicable to more practical cases.

8.4.3 Energy Savings through Augmentation

The preceding section dealt with a general procedure by which the desirability of augmentation could be determined. An example showed that the required area for an exchanger could be significantly reduced for a fixed duty. The savings were not in energy but rather in capital equipment costs.

Energy savings can be realized through more efficient heat transfer. To

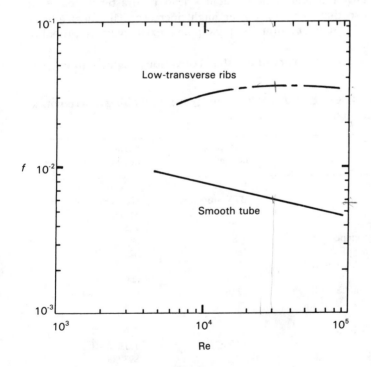

Figure 8.20 Comparison of friction factors for smooth tubes and tubes with low-transverse ribs.

examine this aspect of exchanger performance, we pick a heat duty and a fixed surface and see what effect augmentation would have on energy savings. Energy is not being "saved" in the exchanger itself; rather, more efficient heat transfer is allowing changes in the operating parameters such that less energy need be expended to accomplish the same duty. A good example is a heater where steam is condensed in a shell to heat fluid in the tubes. Heat exchange surface augmentation for this case leads to a reduction in steam pressure (and saturation temperature). Because it takes less energy to produce lower-pressure steam, both energy and dollar savings are produced.

An example of this is shown in Table 8.4. The augmented surface here is high-flux boiling surface on the tube side and fluted exterior condensing surfaces. The effect is to increase the heat transfer efficiency so that the steam pressure can be reduced from 400 to 125 psig. The estimated steam costs drop radically over this pressure range, leading to the energy/dollar savings.

In single-phase systems the positive effect of augmentation on energy conservation might be to decrease flow rate for one or both of the streams, thereby taking less energy for fluid circulation. Remember, however, that many heat transfer augmentation schemes lead to drastic increases in pressure drop so that pumping power might increase. The most advantageous effect of augmentation in single-phase exchangers is probably reduction in area.

Table 8.5 gives a summary of the status of various augmentation tech-

Table 8.4 Example of Energy Saving with High-Flux Tubing in Aromatics Reboiler Application

	Bare tube design	High-flux design
Heat duty, Btu/hr	10×10^6	10×10^6
Boiling fluid	BTX aromatic	BTX aromatic
Condensing fluid	Steam	Steam
Steam pressure, psig	400	125
Steam condensing temperature, °F	448	352
Boiling temperature, °F	325	325
Mean temperature difference, °F	123	27
Exchanger area, ft^2	6000	6000
Exchanger size (diameter × length)	52 in. × 240 in.	52 in. × 240 in.
Overall heat transfer coefficient, Btu/hr ft^2 °F	95	435
Steam cost, $/1000 lb	2.50	1.50
Energy cost, $/yr	1,924,320	1,030,600
Energy savings, $/yr		893,000

Source: Linde Division, Union Carbide Corp. [17].

Table 8.5   Applications of Heat Transfer Augmentation Techniques

	Commercial availability	Mode			Typical material	Performance potential
		Forced convection	Boiling	Condensation		
Inside tubes						
Metal coatings	Yes	—	2	—	Al,Cu,St	High
Integral fins	Yes	2	3	4	Al,Cu	High
Flutes	Yes	4	4	4	Al,Cu	Mod.
Integral roughness	Yes	2	3	4	Cu,St	High
Wire coil inserts	Yes	3	4	4	Any	Mod.
Displaced promoters	Yes	2	4	4	Any	Mod. (Lam.)
Twisted-tape inserts	Yes	2	3	4	Any	Mod.
Outside circular tubes						
Coatings						
Metal	Yes	—	2	4	Al,Cu,St	High (boil)
Nonmetal	No	—	4	4	Teflon	Mod.
Roughness (integral)	Yes	3	2	4	Al,Cu	High (boil)
Roughness (attached)	Yes	3	4	—	Any	Mod. (for conv.)
Axial fins	Yes	1	4	4	Al,St	High (for conv.)
Transverse fins						
Gases	Yes	1	—	—	Al,Cu,St	High
Liquids/two-phase	Yes	1	1	1	Any	High
Flutes						
Integral	Yes	—	—	2	Al,Cu	High
Nonintegral	Yes	—	—	4	Any	High
Plate-fin heat exchanger						
Metal coatings	Yes	—	3	—	Al	High
Surface roughness	Yes	4	3	4	Al	High (boil)
Configured or interrupted fins	Yes	1	2	2	Al,St	High
Flutes	No	—	—	4	Al	Mod.
Plate-type heat exchanger						
Metal coatings	No	—	4	—	St	Low
Surface roughness	No	4	4	4	St	Low
Configured channel	Yes	1	3	3	St	High (for conv.)
Use code						

1. Common use
2. Limited use
3. Some special cases
4. Essentially no use

Source: Bergles [15].

niques. Although the feasibility of augmentation has been demonstrated, application is far from universal. Table 8.6 shows that the technology of augmented tubular surface is progressing but that further work needs to be done. Even the techniques that have been commercialized need further development in some cases.

8.5 SUMMARY

This chapter provides background material and some selected details for the consideration of energy conservation measures for heat exchangers. Different types of exchangers are described, and their applications and advantages are briefly discussed. Only a cursory treatment of heat exchanger theory is presented.

Fouling is a major cause of deteriorated heat exchanger performance; thus, we call more attention to it here than in the traditional heat transfer text. Although little can be done to prevent fouling, awareness of the problem can lead to intelligent decisions regarding the design and maintenance of heat exchangers.

Several methods of heat transfer augmentation are discussed. A general scheme for deciding on the merits of augmentation is presented, followed by an example of the calculation procedure required to determine the consequences of augmentation, in this case for in-tube augmentation in a condenser. We conclude with regret that the primary effect of most augmentation is to make exchangers smaller, thus saving capital costs rather than energy. Of course, in business the bottom line is dollars, so this effect is quite desirable.

Table 8.6   Status of Augmentation Technology Development

Configuration	Single-phase forced convection	Boiling	Condensation
Inside tubes			
Coatings	—	3, 5	—
Roughness	2, 3, 5	1	1
Internal fins	1, 2, 3, 5	1, 5	1
Flutes	5	5	—
Insert devices	5	1, 5	1, 5
Outside tubes			
Metal coatings	—	3, 5	—
Nonmetal coatings	—	1	4
Roughness	1, 2	1, 2, 5	1
Extended surface	5	5	5
Flutes	—	—	5

Use code
  1. Basic performance data
  2. Design methods
  3. Manufacturing technology
  4. Heat exchanger application
  5. Commercialization

Source: Bergles [15].

NOMENCLATURE

A	heat exchanger area
C_R	capacity ratio, C_{min}/C_{max}, defined by Eq. (8.7)
c_p	heat capacity
C	thermal capacitance, $\dot{m}c_p$
D	diameter
f	friction factor
F	exchanger correction factor
g_c	gravitational constant
h	heat transfer coefficient
k	thermal conductivity
L	tube length
$\dot{m}$	mass flow rate
n	number of tubes
Nu	Nusselt number, hD/k
NTU,N	number of transfer units, UA/C_{min}, defined by Eq. (8.9)
P	pumping power
Pr	Prandtl number, $c_p\mu/k$
Q	heat transfer rate
r	radial
R	capacity ratio, defined by Eq. (8.5)
R_f	fouling resistance factor; see Eq. (8.11)
Re	Reynolds number, $\varrho VD/\mu$
T	temperature
U	overall heat transfer coefficient
V	velocity
x	linear distance
Δp	pressure drop
ΔT	temperature difference
ΔT_{lm}	log-mean temperature difference, defined by Eq. (8.2)
ε	heat exchanger effectiveness, defined by Eqs. (8.4) and (8.7)
ϕ_d	fouling deposition rate
ϕ_r	fouling removal rate
τ	time
β	fouling deposit removal rate constant
ϱ	density

Subscripts

1	inlet
2	outlet
a	augmented
b	baseline

c	cold, clean
f	fouled
h	hot
i	in, inner
lm	log-mean
o	out, outer
s	shell
T	total
t	tube

REFERENCES

1. A. C. Mueller, Heat Exchangers, in *Handbook of Heat Transfer*, W. M. Rohsenow and J. P. Hartnett (eds.), McGraw-Hill, New York, 1973.
2. N. Afgan and E. U. Schlünder, *Heat Exchangers: Design and Theory Sourcebook*, McGraw-Hill, New York, 1974.
3. R. K. Shah, Classification of Heat Exchangers, in *Heat Exchangers: Thermal-Hydraulic Fundamentals and Design*, S. Kakac, A. E. Bergles, and F. Mayinger (eds.), Hemisphere, Washington, D.C., 1981.
4. K. J. Bell, Preliminary Design of Shell and Tube Exchangers, in *Heat Exchangers: Thermal-Hydraulic Fundamentals and Design*, S. Kakac, A. E. Bergles, and F. Mayinger (eds.), Hemisphere, Washington, D.C., 1981.
5. J. H. Lienhard, *A Heat Transfer Textbook*, Prentice-Hall, Englewood Cliffs, N. J., 1981.
6. R. A. Bowman, A. E. Mueller, and W. M. Nagle, Mean Temperature Difference in Design, *Trans. ASME* 62:283, 1940.
7. N. Shamsundar, A Property of the Log-Mean Temperature Difference Correction Factor, *Mech. Eng. News.*, 19(3):14–15, 1982.
8. T. R. Bott, Fouling in Shell-and-Tube Heat Exchangers, in *Advances in Thermal and Mechanical Design of Shell-and-Tube Heat Exchangers*, Birnichill Institute, Glasgow, 1973.
9. D. Kern and R. Seaton, Surface Fouling—How to Calculate Limits, *Chem. Eng. Prog.* 55 (6):71, 1959.
10. J. G. Collier, Heat Exchanger Fouling and Corrosion, in *Heat Exchangers: Thermal-Hydraulic Fundamentals and Design*, S. Kakac, A. E. Bergles, and F. Mayinger (eds.), Hemisphere, Washington, D.C., 1981.
11. M. G. O'Callaghan, Fouling of Heat Transfer Equipment: Summary Review, in *Heat Exchangers: Thermal-Hydraulic Fundamentals and Design*, S. Kakac, A. E. Bergles, and F. Mayinger (eds.), Hemisphere, Washington, D.C., 1981.
12. J. de Deus and R. S. Pinhiero, Fouling of Heat Transfer Surfaces, in *Heat Exchangers: Thermal-Hydraulic Fundamentals and Design*, S. Kakac, A. E. Bergles, and F. Mayinger (eds.), Hemisphere, Washington, D.C., 1981.
13. A. Watkinson, L. Louis, and R. Brent, Scaling of Enhanced Heat Exchanger Tubes, *Can. J. Chem. Eng.* 52:558, 1974.
14. A. E. Bergles, Principles of Heat Transfer Augmentation. I. Single Phase Heat Transfer, in *Heat Exchangers: Thermal-Hydraulic Fundamentals and Design*, S. Kakac, A. E. Bergles, and F. Mayinger (eds.), Hemisphere, Washington, D.C., 1981.
15. A. E. Bergles, Applications of Heat Transfer Augmentation, in *Heat Exchangers: Thermal-Hydraulic Fundamentals and Design*, S. Kakac, A. E. Bergles, and F. Mayinger (eds.), Hemisphere, Washington, D.C., 1981.

16. J. G. Withers, and E. H. Young, Steam Condensing on Vertical Rows of Horizontal Corrugated and Plain Tubes, *Ind. Eng. Chem. Process Design Dev.*, 10:19–30, 1971.
17. Linde Division, Union Carbide Corp. *Technical Information-High Flux Tubing*, Tonawanda, N.Y., 1977.

PROBLEMS

8.1 A heat exchanger is to be made of 3/4-in., 14-gage copper tubing to furnish domestic hot water. The tube is to be immersed in the hot water heating boiler. The average boiler water temperature is 190°F and the domestic water is to be heated from 50 to 150°F. The heat transfer coefficient on the outside of the tube is 250 Btu/hr ft^2 °F. How many feet of tubing will be needed to heat 60 gallons of water per hour?

8.2 Crude oil flows at 2000 lb/hr through the inside of a double-pipe heat exchanger and is heated from 90 to 200°F. Kerosene initially at 450°F supplies the heat as it flows through the annular space. The overall heat transfer coefficient is 80 Btu/hr ft^2 °F. For safety, the minimum temperature difference between the oil and kerosene must be 20°F. Specific heats for oil and kerosene are 0.56 and 0.60 Btu/lb, respectively.

(a) Would the kerosene flow rate be greater for countercurrent or concurrent flow? (*Ans.*: concurrent)
(b) For which configuration would the heat transfer area be less? (*Ans.*: concurrent)

8.3 A counterflow double-pipe heat exchanger is employed to cool 8,000 lb/hr of a light oil (specific heat 0.4 Btu/lb °F) from 300 to 100°F. Water at 70°F is used for cooling and the exit water temperature is 100°F. The pipes are 1-in. and 2-in. schedule 40 steel. The film coefficients for water are 2,000 and 840 Btu/hr ft^2 °F, respectively, in the center pipe and in the annulus. Those for oil are 100 and 50 on the same basis.

(a) Should the oil or the water flow in the smaller pipe? Why?
(b) What length of heat exchanger is required?

8.4 A parallel-flow heat exchanger is used to cool 11,250 lb/hr of oil in a 1-in. outside diameter tube from 230 to 100°F by 20,000 lb/hr of water, which enters at 70°F. The c_p of oil is 0.45 Btu/lb°F while that of the water is about 1.0 Btu/lb°F. The overall coefficient is $U_o = 60$ Btu/hr ft^2 °F. Find the water outlet temperature and the number of 5-ft-long tubes required to perform the task.

8.5 Ammonia enters a two-shell-pass, four-tube-pass exchanger at 65°C with a mass flow rate of 0.8 kg/sec and $c_p = 4800$ J/kg °C and while passing through the tubes is cooled to 20°C by 1.5 kg/sec of water that enters at 10°C on the shell side. An overall coefficient of $U_o = 30$ W/m^2 °C has been determined. Find the required surface area and compare it to the area that a pure counterflow exchanger would need.

8.6 A heat exchanger with two shell passes and eight tube passes has 1.32 lb/sec of an oil, $c_p = 0.860$ Btu/lb °F, entering the shell at 248°F and leaving at 104°F. Water at 68°F enters the tubes with a mass flow rate of 1.1 lb/sec. The overall coefficient based on "clean" surfaces is $U_o = 53$ Btu/hr ft^2 °F and fouling factors of 0.005 and 0.0006 hr ft^2 °F/Btu are expected on the oil and water sides, respectively. The ratio of outside to inside tube diameters is 1:2. Compute the outside area needed for the exchanger to perform properly when fouled.

8.7 A process heater uses steam at 30 psia to heat liquid ammonia from 77 to 212°F. When the exchanger was first installed, tests showed that the overall coefficient was about 500 Btu/hr ft^2

°F. It is anticipated that after a few months of operation, fouling will cause U to fall to 300 Btu/hr ft^2 °F. Assuming that the heat exchanger must continue to meet its duty, i.e., to heat ammonia from 77 to 212°F, calculate the required increase in steam pressure.

8.8 A large power plant condenser is a one-shell-, two-tube-pass exchanger designed to condense steam at 1.0 psia ($T_{sat} = 38.7$°C, $h_{fg} = 2,407$ kJ/kg). Water enters the tube side at 21°C and exits at 30°C. The overall coefficient is 2,000 W/m^2 °C. What is the required heat transfer area for each kilogram of steam condensed per second? What is the required flow rate of water for each kilogram of steam condensed per second?

8.9 A chemical operation produces flue gas that is mostly air with a small amount of hazardous vapors. The flue gas is produced at 600°F at a flow rate of 80,000 ft^3/hr. Water at 70°F is used in a crossflow shell-and-tube heat exchanger to condense the vapors from the flue gas. Water flows in the tubes at 20,000 lb/hr. The overall coefficient is 10 Btu/hr ft^2 °F and the exchange area is 270 ft^2.

(a) What are the exit temperatures of the flue gas and water?
(b) Using the area from part (a), what would the overall coefficient have to be if the flue gas outlet must be 100°F?

8.10 Cooling pond water at 70°F is used to cool liquid ammonia at 150°F in a one-shell-pass, two-tube-pass heat exchanger. The ammonia has a flow rate of 80,000 lb/hr and is to be cooled to 100°F. Table 8.1 indicates that a U of 150 Btu/hr ft^2 °F is appropriate for this process. Estimate the required size of the exchanger if the maximum allowable water temperature for return to the cooling pond is 90°F.

8.11 Air enters a preheater at 70°F and leaves at 300°F; flue gases ($c_p = 0.25$ Btu/lb °F) enter at 600°F. Some 170,000 lb/hr of air is heated by 188,000 lb/hr of gas. The heat transfer coefficient U is 125 Btu/hr ft^2 °F. Find the temperature of the flue gases leaving the heater (°F). Compute the surface area of the tube (ft^2) if:

(a) The heater is arranged for parallel flow.
(b) The heater is arranged for counterflow.

8.12 A 6-pass, 36-tube shell-and-tube heat exchanger was designed to heat 643,000 lb/hr of a solution having a specific heat of 0.78 and a specific weight of 1.30 from 178 to 198°F. Steam at 20 psig condenses in the shell side. Solution velocity is 6 ft/sec, at which the film coefficient is 600 Btu/hr ft^2 °F. The tubes are 316 stainless steel, 1½ in. outer diameter, 16 gauge. Tube lengths are 12 ft. The steam-side coefficient is 1300 Btu/hr ft^2 °F. Calculate the effect on the temperature rise of the solution if fouling occurs, assuming the fouling resistance is internal and given by $R_i = 0.005$ hr ft^2 °F/Btu.

8.13 Condensing steam on the shell side of a shell-and-tube heat exchanger is used to heat 140,000 lb$_m$/hr of a chemical from 100 to 400°F. The specific heat of the chemical is 0.4 Btu/lb °F. The heat transfer area in the exchanger is 1,000 ft^2. The steam pressure required when the exchanger was first installed was 300 psia. After one year of operation, the steam pressure required had reached 400 psia. What is the fouling resistance for this heat exchanger after one year of operation?

8.14 A large steam condenser with smooth tubes is used to heat a chemical stream from 100 to 300°F. Tubes with low-transverse ribs, whose effects on friction factor and heat transfer are given by Figs. 8.19 and 8.20, are being considered as an energy conservation measure. In this case, the area, the heat duty, and the pumping power must remain constant, and the desired effect is to reduce the ΔT_{lm} so that the steam pressure might be reduced. The design Re for the smooth tube-side flow is 3.0×10^4. The steam pressure for the design with smooth tubes is 250 psia.

(a) Estimate the reduction in steam pressure if tubes with low-transverse ribs are used. Assume that h_{steam} remains unchanged although pressure will be changed.
(b) Find the required percentage of tubes that must be added to accommodate this change.

8.15 For the boiler that produces steam for the condensate heater of Problem 8.14, fuel is priced at $3 per million Btu's. The efficiency of the boiler is 85%. The condensate from the boiler is collected after further use at a pressure of 10 psia and returned to the boiler feed pump. For an $\dot{m}c_p$ = 140,000 Btu/hr °F for the chemical stream, calculate:

(a) The annual savings in energy realized by using tubes with low-transverse ribs, assuming that the steam leaves the boiler at saturated vapor conditions corresponding to the pressure in the boiler.
(b) The dollar savings for part (a).

NINE

HEAT RECOVERY

Many industrial processes reject heat to the surroundings at temperatures high enough above ambient and in sufficient quantities to make heat recovery economically attractive. This loss has become known as waste heat, since it has usually appeared uneconomical to reclaim it. If the waste heat is not recovered immediately, it quickly mixes with the atmosphere and becomes unavailable for energy conservation. In cases where large enough quantities of waste heat are rejected, thermal pollution can result.

Economics provides the major impetus for heat recovery. There are two ways in which recovered heat is economically advantageous. First, recovered waste heat can directly replace purchased energy and therefore reduce the energy cost per unit of industrial product. Second, capital equipment costs may be reduced when recovered heat replaces a piece of energy conversion equipment or substantially reduces the size requirements of the equipment. The replacement of a fired process heater with a device that uses condensate from another part of the process exemplifies the latter possibility.

There are sometimes additional advantages to recovering waste heat. In some cases, thermal pollution of the atmosphere can be reduced as a by-product of heat recovery. In other cases the cost of operating incineration equipment to decompose air pollutants can be reduced by recovering waste heat from the incineration exhaust gases.

The amount of waste heat rejected to the atmosphere each year in the United States is equivalent to the heating value of some 20 million barrels of oil per day. This could easily account for 40–45% of the total energy consumed in the United States. Of course, only a fraction of this energy can be practically recovered. Most of it is exhausted in concentrated streams at temperatures less than 200°F. It is difficult to recover energy economically at such low temperatures. The datum used to gauge the practicality of heat recovery is the state of the surrounding atmosphere. As a rule, the potential for heat recovery for streams at 200°F is referenced to 70°F.

Energy exhaust streams that are above 200°F are estimated to contain an energy equivalent of perhaps 15% of the total industrial consumption. Energy of this quality can be used for process or comfort heat, or even to generate electric power. The value of energy contained in streams above 200°F has been estimated to be about $12 billion at 1979 oil prices.

The value of waste heat exhausted from a small process industry can amount to several thousand dollars per month. Energy managers should be aware of waste heat streams, and the potential for heat recovery from these streams should be one of the primary checkpoints in an industrial energy audit. It has been estimated [1] that typical process industries could save 20% of their fuel requirements through heat recovery.

Economic incentives have led to increased use of heat recovery techniques. Many such techniques are well developed, and equipment is commercially available. Equipment costs are relatively low and the systems are reliable. Payback periods of 6 months are not uncommon, and these periods are rarely longer than 3 years. Waste heat recovery probably represents the largest energy conservation potential in the United States today.

9.1 CONSTRAINTS ON HEAT RECOVERY

Successful recovery of waste heat depends on five factors, as suggested in Fig. 9.1. The three quantities at the left in Fig. 9.1 must result from the basic processes of the plant. Obviously, steam at 250°F cannot be used directly to heat a process chemical to 400°F. Neither can a waste steam supply at 450°F that contains only 30,000 Btu/min be used to heat a stream to 400°F if it requires 300,000 Btu/min. Finally, it is futile to consider heat recovery if an appropriate use for the recovered heat is not available. The availability of equipment to recover and transfer the heat from one stream to another and the profitability of the venture are the final constraints on a potential heat recovery project.

The quantity and quality of the waste heat can be expressed as quantitative thermodynamic measures. Quantity is based on an energy balance for the system that is rejecting heat, as described by the first law of thermodynamics. The quantity of potentially recoverable energy flowing in a waste stream is

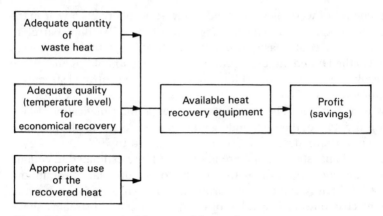

Figure 9.1 Factors required for successful waste heat recovery.

$$\Delta \dot{H} = \dot{m}(h - h_o) \tag{9.1}$$

where $h - h_o$ is the enthalpy per unit mass of the stream material referenced to the surroundings and $\dot{m}$ is the mass flow rate.

The quality of the waste heat is not defined by the first law but must be determined by applying the second law of thermodynamics. Quality is used in this context to indicate the potential of energy to do useful work, or its equivalent. Various versions of the second law are given in Chapter 5.

The property entropy is related to the second law and, to a certain degree, determines the quality of heat. As material containing heat is used in a series of processes, its entropy will increase. We can say that as a fixed amount of energy moves through a set of processes, its ability to do useful work steadily decreases. Therefore the quality of heat and its entropy level are intimately connected.

9.1.1 Availability

In accordance with thermodynamics a characteristic of a system called availability, b, can be defined. Availability can be used to quantify the ability of a system to do useful work; it is the maximum work that the system could perform in going from its existing state down to a state of equilibrium with its surroundings. In accordance with Eq. (5.17), the maximum reversible work for a stationary steady-state, steady-flow system is

$$w_{\text{rev}} = \left(h - T_o s + \frac{V^2}{2g_c} + z\,\frac{g}{g_c}\right) - \left(h_o - T_o s_o + \frac{V_o^2}{2g_c} + z_o\,\frac{g}{g_c}\right) \tag{9.2}$$

where the subscript o refers to the surroundings and the symbols are defined in Chapter 5. Setting w_{rev} equal to availability, and neglecting kinetic and potential energy changes, we write

$$b = (h - h_o) - T_o(s - s_o) \qquad (9.3)$$

When material is at the same state as its surroundings — i.e., $h = h_o$ and $s = s_o$ — its availability is zero.

Saturated vapor and liquid at the same pressure and temperature also have the same availability, although their states are clearly different. This can be shown by referencing the availability of the two states to the saturated liquid state. Then

$$b = (h - h_{ref}) - T_{ref}(s - s_{ref})$$

and

$$b_{vap} - b_{liq} = h_{fg} - T_{sat}\, s_{fg}$$

This is identically zero, because $h_{fg} = T_{sat}\, s_{fg}$ along the saturation line. As an example, select $T_{sat} = 200°F = 659.6$ R. Using the h_{fg} and s_{fg} values at 200°F from Table B.1 in Appendix B, we get

$$b_{vap} - b_{liq} = 977.9 - 659.6(1.4824) = 0.109 \text{ Btu/lb}_m$$

For all practical purposes, the difference in availability is zero; the discrepancy is caused by small uncertainties in properties. We conclude that steam is of no greater value to us than hot water at the same pressure and temperature as far as its ability to do work is concerned.

9.1.2 Irreversibility

Irreversibility is also a useful thermodynamic concept because it represents the degradation that occurs in a given process. It is defined as

$$I = W_{rev} - W_{cv}$$

where W_{cv} is the actual work performed by the control volume. Using the relationships given in Chapter 5 for a steady-flow system, we can write

$$I = \Sigma m_{out} T_o s_{out} - \Sigma m_{in} T_o s_{in} - Q_{cv} \qquad (9.4)$$

Irreversibility is similar to the concept of "lost work," that is, the conversion of energy during a process into a form that can no longer be converted into useful work. In some cases, irreversibility and loss of availability during a process are identical. For example, processes that involve only heat transfer are totally irreversible and the irreversibility balances the loss of availability for the process. Example 9.1 illustrates this concept.

Example 9.1   Availability of Combustion Products

An air preheater is used to preheat combustion air by cooling the products of combustion from a boiler furnace. The air flow rate is 90,000 lb$_m$/hr and the inlet air temperature is 80°F. Combustion products flow at the rate of 97,000 lb$_m$/hr and the products are cooled from 600 to 400°F. For the products, use the specific heat $c_p = 0.26$ Btu/lb$_m$ °F. Calculate the change in availability in Btu's per hour for the combustion products and the air. What is the irreversibility of the heat exchange process?

Solution: The availability per unit mass is

$$b = (h - h_o) - T_o(s - s_o)$$

The ambient state is $T_o = 80°F = 540$ R, and $p_o = p = 1$ atm. Then,

$$\Delta b = b_{out} - b_{in}$$
$$\Delta b = (h_{out} - h_{in}) - T_o(s_{out} - s_{in})$$

If the products and air behave as ideal gases, then in accordance with Chapter 5,

$$\Delta h = c_p \, \Delta T \qquad T_o \, \Delta s = T_o c_p \ln \frac{T_{out}}{T_{in}}$$

so,

$$\Delta b_p = 0.26(850 - 1{,}060) - 540(0.26) \ln (860/1{,}060)$$
$$= -22.64 \text{ Btu/lb}_m$$

and, if we define the extensive availability as $B = b \times$ mass,

$$\Delta \dot{B}_p = \dot{m}_p(\Delta b) = -2{,}196{,}000 \text{ Btu/hr}$$

The air outlet temperature is

$$T_{a,out} = T_{a,in} + \frac{(\dot{m}c_p)_p}{(\dot{m}c_p)_a} \Delta T_p$$

$$= 540 + \frac{97{,}000 \, (0.26)}{90{,}000 \, (0.24)} \, 200 = 773.5 \text{ R}$$

Therefore,

$$\Delta \dot{B}_a = 9{,}000 \, [0.24(773.5 - 540) - 540(0.24) \times \ln (773.5/540)]$$

$$= 852{,}000 \text{ Btu/hr}$$

The net loss of availability for the process is the sum of the air and the product streams.

$$\Delta \dot{B}_{net} = -1{,}344{,}000 \text{ Btu/hr}$$

The irreversibility of the product stream is

$$\dot{I}_p = T_o(s_{out} - s_{in})_p \dot{m}_p = T_o c_p \ln \left(\frac{T_{out}}{T_{in}} \right) \dot{m}_p$$

$$= -2{,}847{,}500 \text{ Btu/hr}$$

In a similar fashion,

$$\dot{I}_a = 4{,}195{,}500 \text{ Btu/hr}$$

The net irreversibility is the sum of these two,

$$\dot{I}_{net} = 1{,}344{,}000 \text{ Btu/hr}$$

The irreversibility matches the loss of availability, as we observed that it must.

We conclude from Example 9.1 that a heat transfer process consumes part of the available energy in a waste stream, although energy is conserved. When heat is transferred, the availability gain of one stream is not equal to the availability loss of the other stream. There is always a net loss.

9.1.3 Efficiency

Energy conversion devices can be characterized by their efficiency. *Efficiency*, based on the first law considerations given in Chapter 5, is

$$\eta_F = \frac{\text{actual energy change}}{\text{ideal energy change}} \tag{9.5}$$

This is the efficiency that is commonly used in energy conservation work. An efficiency based on the second law is more appropriate in that it reflects the quality of the energy that is being converted.

Moran [2] defines second-law efficiency as

$$\eta_s = \frac{\text{availability out in product}}{\text{availability in}} \tag{9.6}$$

or

$$\eta_s = 1 - \frac{\text{loss + destruction}}{\text{input}} \tag{9.7}$$

The parameter η_s accounts for the influence of the availability of energy in the various streams entering and exiting the device. Both losses and internal irreversibilities need to be dealt with to improve second-law efficiency. The analysis of specific devices, plants, or even classes of industries depends on the proper interpretation of Eq. (9.6). Equation (9.6) might also be used to evaluate the effects of steps taken to improve the efficiency of energy utilization, such as heat recovery.

All real processes are irreversible, so a second-law efficiency of 100% is unachievable. A realistic upper limit of η_s should be used as a goal. Some examples of second-law efficiencies follow.

Heating Water Electrically The second-law efficiency for heating of a fixed amount of water with resistance heating is

$$\eta_s = \frac{\Delta B}{W_e} \tag{9.8}$$

where W_e is electrical work. This is called a *task* efficiency; here the task is to heat water. Equation (9.8) can be evaluated to show that

$$\eta_s = \eta_F \left[1 - \frac{T_o}{\Delta T} \ln\left(\frac{T_i + \Delta T}{T_i}\right) \right] \tag{9.9}$$

where T_o is the surrounding temperature and $T_i + \Delta T$ is the initial temperature plus the temperature rise. For $T_o = 60°F = T_i$, and $\Delta T = 60°F$, $\eta_s = \eta_F/20$. We see that heating of water electrically is very inefficient.

Cycles

Power Cycles For the system shown in Fig. 9.2, the second-law efficiency

$$\eta_s = \frac{\eta_F}{1-(T_o/T_A)} \tag{9.10}$$

Equation (9.10) accounts for the irreversibility in the cycle as well as the loss of availability in the heat interaction at T_A. However, the loss of availability in the heat interaction at T_R is not charged as an energy loss. In other words, it is recoverable energy.

Refrigeration and Heat Pump Cycles For the refrigeration cycle shown in Fig. 9.3, the second-law efficiency for a refrigeration device intended to maintain a cold region is

$$\eta_h = \frac{- \int[1 - (T_o/T_a)]\delta Q_A}{W_{\text{cycle}}} \tag{9.11}$$

By defining a *coefficient of performance* for the refrigerator,

$$\text{COP} = \frac{Q_A}{W_{\text{cycle}}} \tag{9.12}$$

we can write,

$$\eta_s = \text{COP} \left\{ \frac{- \int[1 - T_o/T_a)]\delta Q_A}{Q_A} \right\} \tag{9.13}$$

Furthermore, for devices that extract heat at a constant temperature T_A,

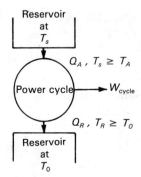

Figure 9.2 Power cycle with heat added at T_A and rejected at T_R.

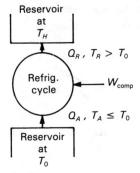

Figure 9.3 Cycle for transferring heat from a low- to a high-temperature region.

$$\eta_s = \text{COP} \left(\frac{T_o}{T_A} - 1 \right) \qquad (9.14)$$

A heat pump, on the other hand, is designed to maintain a hot region at T_H. The COP for this case is

$$(\text{COP})_{\text{H.P.}} = \frac{Q_R}{W_{\text{cycle}}} \qquad (9.15)$$

The second-law efficiency is

$$\eta_{s,\text{H.P.}} = \frac{\int [1 - (T_o/T_R)] \delta Q_R}{W_{\text{cycle}}} \qquad (9.16)$$

which simplifies to

$$\eta_{s,\text{H.P.}} = (\text{COP})_{\text{H.P.}} \left(1 - \frac{T_o}{T_R} \right) \qquad (9.17)$$

for a device that rejects heat at a fixed T_R.

Flow Devices   For devices that involve flow under steady conditions, the following second-law efficiencies are applicable:

Pumps, Compressors, and Turbines   Assuming negligible kinetic and potential energy changes, the second-law efficiency for pumps and compressors is

$$\eta_{s,\text{pump}} = \frac{b_e - b_i}{- \dot{W}/\dot{m}} = \eta_{s,\text{comp}} \qquad (9.18)$$

where b_e and b_i are the availabilities per unit mass of the exiting and entering fluid, respectively. For turbines,

$$\eta_{s,\text{turb}} = \frac{(\dot{W}/\dot{m})}{b_i - b_e} \qquad (9.19)$$

Throttling Devices   There is no work or heat output for these devices. For steady adiabatic flow and negligible kinetic and potential energy changes,

$$\eta_{s,thr} = \frac{b_e}{b_i} = 1 - \frac{\dot{I}/\dot{m}}{b_i} \tag{9.20}$$

In this case, Eq. (9.6) is interpreted in terms of output/input.

Heat Exchange with Mixing   Devices that involve mixing of two or more streams are commonly called *open* feedwater heaters. Figure 9.4 shows such a device. The appropriate expression is

$$\eta_s = \frac{(1 - y)(b_3 - b_2)}{y(b_1 - b_3)} \tag{9.21}$$

where $y = \dot{m}_1/\dot{m}_2$. Equation (9.21) is limited to temperatures above the environmental temperature.

Heat Exchange Without Mixing   For a heat exchanger with two streams, as shown in Fig. 9.5, Moran [2] defines

$$\eta_s = \frac{\dot{m}_c}{\dot{m}_h} \frac{b_4 - b_3}{b_1 - b_2} \tag{9.22}$$

Equation (9.22) is based on negligible kinetic and potential energy changes and steady adiabatic heat transfer. Furthermore, all temperatures are assumed to be higher than the temperature of the surroundings.

9.1.4 Temporal Availability

Another important issue in heat recovery is whether or not the waste heat is available when it is needed. Consider 425°F exhaust from an oven as a waste heat source to heat water for a washing process. Suppose the oven operates on the second plant shift while the washing process occurs during the first shift. Heat can be recovered only if operations are rescheduled or if the waste heat from the oven is collected and stored.

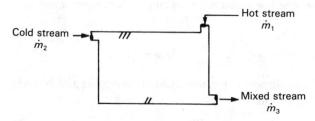

Figure 9.4 Open feedwater heater.

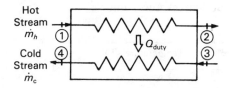

Figure 9.5 Closed heat exchanger.

9.2   WASTE HEAT CLASSIFICATION (omit)

Thermodynamic constraints on waste heat recovery depend strongly on the temperature of the waste heat source. Accordingly, the primary classification for waste heat is based on the source temperature. The classification is as follows:

High-temperature range	$1100°F \leq T \leq 3000°F$
Medium-temperature range	$400°F \leq T < 1100°F$
Low-temperature range	$80°F \leq T < 400°F$

These temperature ranges match a similar classification of commercial heat recovery equipment. Table 9.1 shows various sources for the three temperature ranges. High-temperature waste heat has a high availability of energy and can be used for material preheating and for work production in gas and steam turbines. This makes high-temperature waste heat recovery very economical. On the other hand, high temperatures can cause problems in that special materials and careful design may be required in the recovery equipment. This can lead to higher capital costs for the heat recovery equipment.

The use of waste heat can be either direct or indirect. Direct use means that the recovered heat is returned directly to the primary process. Combustion air preheating by flue gases is an example of direct use. Indirect or secondary use is the recovery of heat to be used elsewhere in the plant. A waste heat boiler recovering heat from a gas turbine exhaust and providing steam for a nearby process heater is an example of an indirect use.

We may further classify the utilization of waste heat as follows:

Direct utilization: No heat exchange equipment is required. An example is the use of heat recovered from cooling metal parts to preheat the incoming metal charge by simply directing the cooling air stream over the parts.

Recuperation: This involves the use of recuperative heat exchangers (see Chapter 8) in which the waste heat stream is separated from the stream to be heated.

Regeneration: This variation of recuperation involves the storage of waste heat in a medium and the subsequent recovery of the heat by the cool

Table 9.1 Classification of Waste Heat Sources According to Temperature Range

Type of device	Temperature (°F)
Sources in the high-temperature range	
Nickel refining furnace	2500–3000
Aluminum refining furnace	1200–1400
Zinc refining furnace	1400–2000
Copper refining furnace	1400–1500
Steel heating furnaces	1700–1900
Copper reverberatory furnace	1650–2000
Open hearth furnace	1200–1300
Cement kiln (dry process)	1150–1350
Glass melting furnace	1800–2800
Hydrogen plants	1200–1800
Solid waste incinerators	1200–1800
Fume incinerators	1200–2600
Sources in the medium-temperature range	
Steam boiler exhausts	450–900
Gas turbine exhausts	700–1000
Reciprocating engine exhausts	600–1100
Reciprocating engine exhausts (turbocharged)	450–700
Heat treating furnaces	800–1200
Drying and baking ovens	450–1100
Catalytic crackers	800–1200
Annealing furnace cooling systems	800–1200
Sources in the low-temperature range	
Process steam condensate	130–190
Cooling water from:	
Furnace doors	90–130
Bearings	90–190
Welding machines	90–190
Injection molding machines	90–190
Annealing furnaces	150–450
Forming dies	80–190
Air compressors	80–120
Pumps	80–190
Internal combustion engines	150–250
Air-conditioning and refrigeration condensers	90–110
Liquid still condensers	90–190
Drying, baking, and curing ovens	200–450
Hot-processed liquids	90–450
Hot-processed solids	200–450

Source: Kreider and McNiel [1].

stream passing over the medium. These systems usually use a storage medium that rotates between the hot and cold streams, and allow some mixing between the streams.

Waste heat boilers: These are recuperators that are used to produce process steam and/or hot water.

Energy cascading: In this technique the energy is used at its highest availability first and then in a serial fashion until its availability is so low as to be of no further economic value. This is the optimum way to use waste heat, but it requires adequate uses of the "cascading" waste heat. This topic will be discussed more fully later in this chapter.

Cogeneration: Plant site generation of steam for both electrical generation and process steam is called cogeneration. In essence, it is a specialized case of energy cascading. Chapter 12 gives a description of the techniques involved in cogeneration.

The temperature level of the waste heat source, the possible end use of the waste heat, and the unit that can be used to recover the heat can be roughly related as shown in Fig. 9.6. The bar plot shows the common end uses for waste heat from various sources along with heat recovery equipment that is used in the various temperature ranges. From Fig. 9.6 we see that a diesel exhaust could be used to preheat boiler feedwater or makeup, to preheat combustion air, or for the generation of steam. For preheating, several types of equipment could be used. For steam generation, a waste heat boiler would be the obvious choice.

There may be a wide variety of choices for heat recovery equipment. Sometimes the choice is obvious (as in the example of using a waste heat boiler for steam generation from a diesel exhaust), but more often it is not. The essential factors to be considered in order to make an optimal selection of the heat recovery device are quality or temperature of the waste heat, availability of the waste heat, chemical makeup of the waste heat stream (corrosive, toxic, pollutive, etc.), and temperature requirement of the end use.

9.3 PRACTICAL LIMITATIONS ON HEAT RECOVERY

In addition to the thermodynamic constraints on heat recovery, several practical limitations can arise: (1) low-temperature recovery can prove uneconomical; (2) dumping of waste heat, required when a waste heat load is interrupted, can be intolerable; and (3) temporal mismatch between load and source can make heat recovery impractical. In the following sections, we shall briefly discuss what can be done to accommodate these limitations.

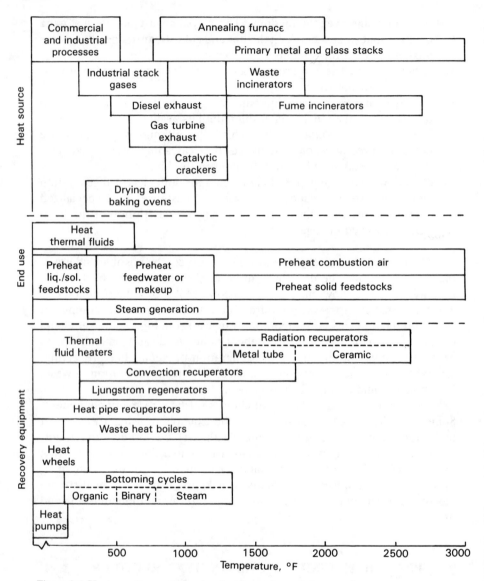

Figure 9.6 Heat recovery equipment and end uses for various waste heat sources.

9.3.1 The Heat Pump: Low-Temperature Recovery

The heat pump is gaining acceptance as a means of upgrading low-temperature waste heat to a temperature at which it can be used economically. The heat pump might be thought of as a reversed Rankine-cycle device with a few practical modifications, as shown in Fig. 9.7. The primary change, in contrast to a reversed Rankine cycle, is an expansion valve replacing the reversed pump; therefore, no expansion work can be reclaimed.

The first-law efficiency of the heat pump is the coefficient of performance:

$$\text{COP} = \eta_F = \frac{Q_H}{W_{\text{cycle}}} \tag{9.23}$$

where Q_H and W_{cycle} are shown in Fig. 9.7. Net work input is required for the

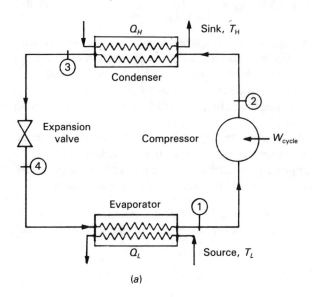

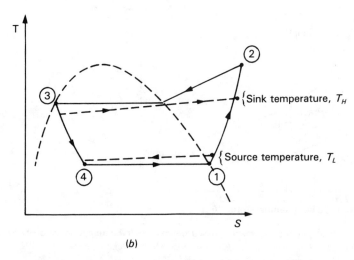

Figure 9.7 Heat pump cycle for heat recovery. (*a*) Typical heat pump components; (*b*) temperature-entropy plot for the heat pump refrigerant and the heat source and sink.

operation of the heat pump, but the work is converted to heat, and the net delivery of heat to the load, Q_H, is the total of Q_L and W_{cycle}. The COP of the heat pump should be greater than about 3 to be economically attractive because electric motors commonly used to drive the compressor generally have an efficiency[1] of about 33%.

The COP of a heat pump operating on the Carnot cycle is given by

$$\text{COP}_{\text{ideal}} = \frac{T_H}{T_H - T_L} \qquad (9.24)$$

where T_L is the temperature of the waste stream and T_H is the temperature required at the load. In practice, the sink and source temperatures do not remain constant; rather they behave as shown in Fig. 9.7b. The irreversibilities of the heat transfer processes, the pressure losses, and the compressor inefficiencies all cause the actual COP to be considerably less than the ideal COP.

Figure 9.8 shows the relation between the ideal COP and the actual COP for an industrial heat pump using water at 90°F as the heat source. The dashed line shows the approximate current limit on refrigerant temperature, 230°F. New refrigerants may allow the application of heat pumps up to 400°F.

Heat pumps have traditionally been powered by electric motors because motors are cheap and reliable. However, if the cost of energy is traced back

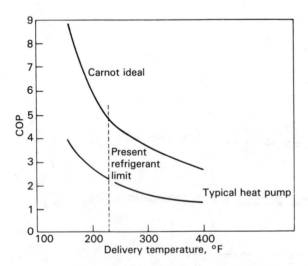

Figure 9.8 Typical heat pump performance, based on cold reservoir temperature of 90°F. (From M. H. Chiogioji, *Industrial Energy Conservation*, Marcel-Dekker, New York, 1979.)

[1] This efficiency is based on the heat content of fuel consumed by the power plant.

to the power plant, electric motors are not the optimal choice to power heat pumps. Other prime movers are under consideration for large industrial heat pumps. Reciprocating engines are the leading contenders at present. It must be emphasized, however, that the economic attractiveness of these prime movers depends strongly on the reclamation of energy from the engine exhausts. Figure 9.9 shows an example of a gasoline engine-driven heat pump system. This is a water-to-water heat pump, with heat being recovered from both the exhaust and engine cooling water and lubricating oil in the overall process. This additional heat recovery from the engine is necessary for the system to be economically feasible.

Open-Cycle Heat Pumps One means of extending the temperature limits of the heat pump cycle is to compress the process stream directly rather than using intermediate heat exchangers and an intermediate refrigerant. We call this an open-cycle heat pump because of the once-through nature of the process. Figure 9.10 shows two such devices: a simple vapor recompression device, and a device in which a vapor recompression process stream is used to evaporate a fluid at a lower temperature.

It is clear that these open systems have much more flexibility of application than do the conventional closed systems discussed before. Furthermore, because a separate working fluid (refrigerant) is not needed, the open-cycle heat pump is potentially less expensive in installed equipment costs than the closed heat pump.

The coefficient of performance for the open cycle is defined in basically the same way as for the closed-cycle heat pump, i.e.,

$$\text{COP}_C = \frac{Q_H}{W_{\text{comp}}}$$

If we visualize the system as a cycle operating between T_H and T_L, then the Carnot efficiency would be

$$\text{COP}_C = \frac{T_H}{T_H - T_L}$$

which would give the minimum work input required for a given heat load, Q_H. Common practice is to represent the *actual* COP for a recompression device as

$$\text{COP} = \eta_{\text{mech}} \, \text{COP}_C \qquad (9.25)$$

That is, the Carnot COP is lessened by a fraction, η_{mech}, reflecting the losses in the compression device. Operational testing yields the value of η_{mech}.

The actual recompression process can be accomplished by mechanical compressors or so-called thermocompressors, which operate on the *ejector* principle. For example, in Fig. 9.11 motive steam at 265 psig can be used to compress 2.5-psig steam up to 4.1 psig. Thermocompressors are inexpensive but do not operate efficiently from either a thermodynamic or an

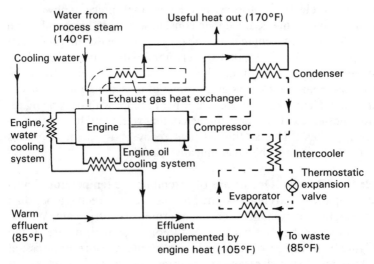

Figure 9.9 Heat pump driven by a gasoline engine with heat recovery from the engine. The dashed line indicates the heat pump fluid circuit. (From Reay [4].)

economic viewpoint. Mechanical compressors cost more initially but impose far less irreversibility on the process than do thermocompressors.

One way of determining the cost savings by using a heat pump is to analyze the performance on the basis of 1 million Btu's delivered to the high-temperature user. Then,

$$S/10^6 \text{ Btu} = D_u - D_s \frac{\text{COP} - 1}{\text{COP}} - \frac{D_p}{\text{COP}} \qquad (9.26)$$

where S is the dollar savings, D_u is the cost of energy at the high-temperature use, D_s is the cost of energy at the low-temperature source, and D_p is the cost of energy required to drive the compressor, all evaluated for 10^6 Btu of heat

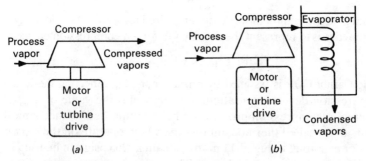

Figure 9.10 Open-cycle heat pumps. (a) Simple vapor recompression; (b) vapor recompression evaporation. (From Gilbert [4].)

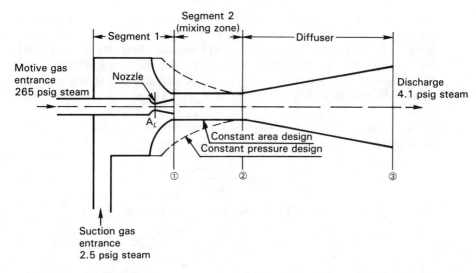

Figure 9.11 Cross section of an ejector used as a heat pump compressor. (From M. Altin, *Proceedings of the Fourth Annual Industrial Energy Conservation Technology Conference*, vol. 1, April 7–9, 1982, Houston, Texas.)

delivered to the high-temperature user. Example 9.2 illustrates this concept.

Example 9.2 Cost Savings with an Open-Cycle Heat Pump

Steam at 200°F is available by flashing of condensate collected during a process. There is a need for steam at 300°F at another point in the process. What is the COP for an open-cycle recompression heat pump whose total mechanical efficiency is 0.75? What is the saving per 10^6 Btu delivered at 300°F if D_u = \$5.50/10^6 Btu, D_s = \$1.50/10^6 Btu, and D_p = \$0.045/kWh?

Solution:

$$COP = 0.75 \ (760/100) = 5.70$$

Converting D_p gives:

$$D_p = \$13.19/10^6 \ \text{Btu}$$

and

$$S/10^6 \ \text{Btu} = 5.50 - 1.50(4.7/5.7) - 13.19/5.7$$
$$= \$1.95/10^6 \ \text{Btu}$$

This represents about a 35% savings compared to generating steam to serve the 300°F use directly.

Gilbert [3] indicates that mechanical single-stage compressors with pressure ratios of 2:1 yield an actual COP of about 15, which in general is cost-effective.

Table 9.2 shows typical measured COPs for open-cycle heat pumps with η_{mech} = 0.7. The COPs listed are based on the saturation temperatures at the given inlet and outlet pressures. The energy of superheat at the outlet pressure is available for transfer, however, the steam is often "desuperheated" by spraying liquid water into it, bringing it down to the saturation temperature. This has the advantage of increasing the available flow rate of the steam at the higher pressure. The COPs in Table 9.2, if based on the actual temperature following compression, would be on the order of 5 rather than 15.

Equation (9.26), if set to $S = 0$, would yield the break-even case, COP_0, for a given set of D_u, D_s, and D_p, the costs of energy:

$$COP_0 = \frac{D_p - D_s}{D_u - D_s} \tag{9.27}$$

The COP_0 for the values of D_u, D_s, and D_p in Example 9.2 is

$$COP_0 = \frac{13.19 - 1.50}{5.50 - 1.50} = 2.92$$

The no-savings COP is generally considered to be in the COP range 2.5 to 3.5. Clearly, to achieve a reasonable rate of return on invested capital, the COP must be considerably higher than COP_0.

9.3.2 Waste Heat Dumping

When waste heat is reclaimed from a process, some provision must be made to continue the heat recovery from the process if the load is changed or temporarily interrupted. Otherwise, the energy balance in the source process is upset, and this could lead to equipment damage or process shutdown. Thus, the design of the recovery operation must include provisions for dumping the waste heat.

In some cases dumping is easily achieved. Figure 9.12, for example, shows a system in which heat is recovered from an air-conditioning condenser to heat water. In a plant where large amounts of hot water are required and air-conditioning is required almost year-round, this is an attractive means of conserving energy and saving dollars. Regardless of the hot-water load, however, the condenser must reject its heat. This can be easily accomplished by directing the water from the hot-water circuit to the water main. This type of system is called an open system because no recirculation of the water is required.

Systems that use a recirculating fluid to recover heat require more care and safeguards than do open systems. As an example, take a system similar to that in Fig. 9.12, but with the recovered heat used for space heating. The condenser in this case might be for a refrigeration system rather than an air-conditioning system. Figure 9.13 shows a schematic diagram of the system. When the need for space heating lessens, there is no way to redirect the water flow since it is a closed loop. An effective way to dump waste heat

Compression		Horsepower required per 1000 lb/hr	Inlet enthalpy (Btu/lb)	Outlet enthalpy (Btu/lb)	Superheat energy (Btu/lb)	COP
Inlet p_1 (psia)	Outlet p_2 (psia)					
5	10	28.8	1131.1	1198.6	55.3	16.4
10	20	30.2	1143.3	1214.1	57.8	15.8
20	40	31.5	1156.3	1230.2	60.4	15.3
40	80	32.9	1169.8	1247.0	63.9	14.9
80	160	34.1	1183.1	1263.1	68.0	14.6

Source: Gilbert [3].

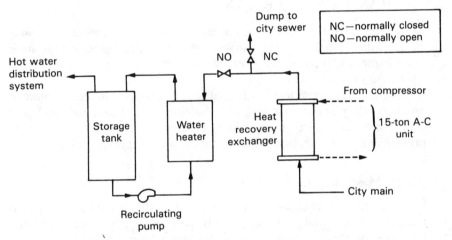

Figure 9.12 Schematic diagram of an open heat recovery system.

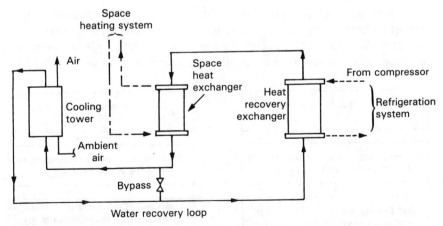

Figure 9.13 Schematic diagram of recirculating heat recovery loop with a cooling tower to dump waste heat.

is to place a cooling tower in the water circuit in series with the water-air heat exchanger. When the heat load diminishes, the cooling tower can dissipate the extra heat.

9.3.3 Heat Storage

Storage of recovered waste heat can overcome some temporal mismatch between load and sources. But storing vast quantities of heat is expensive because additional equipment is required. This can create space problems and can cause a potential heat recovery project to be uneconomical. It is often better to alter cycles to obtain a better match between load and source.

9.4 WASTE HEAT SURVEYS

In this section we elaborate somewhat on the identification and classification of the waste heat sources that might be found in a plant energy audit (see Chapter 2). The waste heat survey usually would proceed in the manner illustrated in Fig. 9.14.

Waste heat streams leaving the plant that are not part of the commercial product of the plant could include hot condensate and steam, flue gases, and heated air as well as streams containing combustible fuels such as paper by-products and sawdust, organic compounds, oven vapors, and fumes.

9.4.1 Mass-Heat Balances

Following the identification of exiting waste heat streams, the next step is to characterize accurately the waste heat by performing a heat balance on the system. Figure 9.15 is a typical waste heat survey form showing the information that is generally required for a heat balance. Both fired units and heat exchangers can be accommodated on this form.

An accurate determination of the characteristics of waste heat streams is important because capital costs tend to be large in industrial heat recovery projects. The potential for energy savings is correspondingly high, however, so these ventures may be worthwhile if an accurate analysis so indicates.

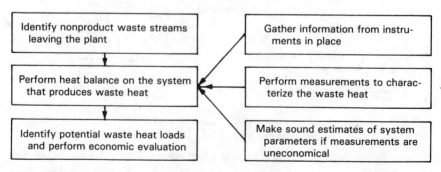

Figure 9.14 Typical sequence of events for a waste heat survey.

Because of the need for accuracy, this stage of the survey may be relatively costly. It may be necessary to install metering equipment to give accurate data on unit operating parameters.

Generally speaking, flow rates, temperature and pressure levels, and operating histories are required to assess waste heat. Properties of the materials in the streams, when they are needed, can usually be found tabulated in terms of pressure and temperature. The thermophysical properties of several materials are given in the Appendixes. In some cases additional measurements must be made to determine the exact composition of the stream; for example, the characterization of flue gas might require an Orsat analysis.

Example 9.3 illustrates the technique of performing a heat balance for a piece of equipment.

Example 9.3 Heat Balance for Continuous Dryer

A continuous dryer produces 15 tons/day of a product containing 3% water. The feed material contains 33% water. The air for drying enters an air heater at 70°F and a relative humidity of 45%. It is heated to 300°F before entering the dryer. It exists with a relative humidity of 68%. Methane is the fuel. A volumetric flue gas analysis gives 83.9% N_2, 3.7% O_2, and 12.4% CO_2. We want to make a mass-energy balance for this system.

Solution: In the figure we split the flows into three streams for convenience. Several important quantities are missing, such as the flow rate of air through the dryer, the temperature T_3, and the fuel flow rate. However, they can be estimated with a fair degree of accuracy from the information given, as shown below.

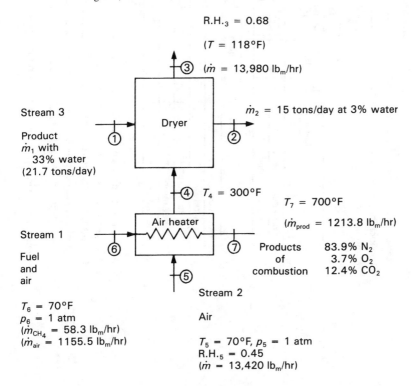

R.H.$_3$ = 0.68

$(T = 118°F)$

③ $(\dot{m} = 13,980 \; lb_m/hr)$

Stream 3

Dryer

$\dot{m}_2$ = 15 tons/day at 3% water

②

Product
$\dot{m}_1$ with
   33% water
(21.7 tons/day)

①

④ $T_4 = 300°F$

$T_7 = 700°F$

Air heater

$(\dot{m}_{prod} = 1213.8 \; lb_m/hr)$

Stream 1

⑥ ⑦ Products   83.9% N_2
Fuel   of   3.7% O_2
and   combustion   12.4% CO_2
air

⑤

Stream 2

$T_6 = 70°F$
$p_6 = 1 \; atm$
$(\dot{m}_{CH_4} = 58.3 \; lb_m/hr)$
$(\dot{m}_{air} = 1155.5 \; lb_m/hr)$

Air

$T_5 = 70°F, \; p_5 = 1 \; atm$
R.H.$_5 = 0.45$
$(\dot{m} = 13,420 \; lb_m/hr)$

WASTE HEAT SURVEY
SURVEY FORM FOR INDUSTRIAL PROCESS UNITS

Name of process unit _____ Inventory number _____

Location of process unit, plant name _____ Building _____

Manufacturer _____ Model _____ Serial number _____

		Firing rate	HHV	Temperature of			Flue Gas Composition, % volume				
	Name			Comb. air	Fuel	Stack	CO_2	O_2	CO	CH	N_2
Primary Fuel											
First alternative											
Second alternative											

	Flow path 1	Flow path 2	Flow path 3	Flow path 4
Fluid composition				
Flow rate				
Inlet temperature				
Outlet temperature				
Description				

Annual hours operation _____ Annual capacity factor (%) _____

Annual fuel consumption: primary fuel _____ First alternative _____ Second alternative _____

Present fuel cost: primary fuel _____ First alternative _____ Second alternative _____

Annual electrical energy consumption, kWh _____

Present electrical energy rate _____

Figure 9.15 Typical waste heat survey form for industrial processes.

Stream 3: Calculate the amount of water removed from the product.

$$\text{Dry product/day} = 30,000 \text{ lb}_m \times 0.97 = 29,100 \text{ lb}_m$$
$$\text{Wet feed/day} = 29,100/0.67 = 43,433 \text{ lb}_m/\text{day}$$
$$\text{Water removed/day} = 13,433 \text{ lb}_m$$

This must be absorbed by stream 2, the heated air.

Stream 2: At the inlet, the amount of water per pound mass of dry air can be found from the psychometric chart in Appendix B.

$$\omega_5 = 0.007 \text{ lb}_m/\text{lb}_m \text{ air}$$

Air is heated to T_4 in a constant-ω process. Therefore $\omega_4 = \omega_5$ and $T_4 = 300°F$. Neglect heat losses in the dryer. Then $h_3 = h_4$, and from the psychometric chart we read

$$T_3 = 118°F \qquad \omega_3 = 0.049 \text{ lb}_m \text{ water/lb}_m \text{ air}$$

A mass balance on stream 2 gives

$$\dot{m}_{\text{dry air}} = \frac{13,433 \text{ lb}_m \text{ water/day}}{\omega_3 - \omega_5} = 319,833 \text{ lb}_m \text{ air/day}$$

$$\dot{m}_5 = (1 + \omega_5) = 13,420 \text{ lb}_m/\text{hr} = \dot{m} \text{ dry air}$$

Stream 1: We use the information from stream 2 to analyze stream 1. The heat required to bring the air from 70 to 300°F is calculated as follows. Based on a mass-weighted average, the specific heat of the mixture of air and water vapor is

$$\bar{c}_p = 0.24 + 0.446(0.007) = 0.243 \text{ Btu/lb}_m \text{ air}$$

and

$$_4Q_5 = \dot{m}c_p \, \Delta T = (13,326)(0.243)(300 - 70)$$
$$= 744,811 \text{ Btu/hr}$$

The combustion equation based on the flue gas analysis is

$$CH_4 + 2.30(O_2 + 3.76 N_2) \rightarrow CO_2 + 0.298 O_2 + 2 H_2O + 6.76 N_2$$

The combustion process uses 115% theoretical air for which the air-fuel ratio is

$$AF = 10.948 \, \frac{\text{moles air}}{\text{moles } CH_4}$$

The first law for the combustion system gives

$$Q_{\text{air}} + \sum_R n_i \bar{h}_i = \sum_P n_e \bar{h}_e$$

where

$$\bar{h}_i = \bar{h}_f° + (\bar{h} - \bar{h}_{537})$$

where $\bar{h}_f°$ is the enthalpy of formation.

Then, neglecting the sensible enthalpy between 77° and 70°F,

$$q_{\text{air}} + (-32,211) = [-169,247 + (6,483)] + 0.298(4,600)$$

$$+ 2[-104,036 + (52,201)] + 6.766(4,407)$$

$$= -204,405 \text{ Btu/lb mole } CH_4$$

Therefore, the required number of moles per hour is

$$\dot{n}_{CH_4} = \frac{744,311}{204,405} = 3.64 \text{ lb mole/hr}$$

and

$$\dot{m}_{CH_4} = 58.3 \text{ lb}_m/\text{hr}$$

$$\dot{m}_{\text{comb. air}} = \dot{m}_{CH_4}(\text{AF}) = 58.3(19.82) = 1,155 \text{ lb}_m \text{ air/hr}$$

This completes the mass-energy balance on the system. The quantities in parentheses in the figure are those calculated to complete the balance.

Quite a bit of information was missing in Example 9.2, and it required thermodynamic analyses to determine the missing quantities. Realistically, data may be unattainable for some systems and these estimates must be made. Example 9.3 required the use of psychrometric charts, combustion calculations, and mass and energy balances. It also illustrates the systematic way in which energy balances are constructed. The next step in the analysis of this particular case would be to identify a need for 700°F combustion air on a continuous basis. An obvious application is to preheat the combustion air to save fuel.

9.5 HEAT RECOVERY EQUIPMENT

The device most used for heat recovery is the heat exchanger in one form or another. Exceptions are reversed pumps and turbines that are used to recover mechanical work from a relatively high pressure stream, and vapor and condensate collection systems, described in Chapter 7. The basic types of heat exchangers were described in Chapter 8; examples are regenerators and recuperators. Heat exchangers used for heat recovery are subject to the same problems in their operation as in other applications — fouling and flow-induced vibration, for example.

Heat exchangers are broadly classified according to the nature of the streams exchanging energy. The same is true for recovery exchangers. The classes are gas to gas, gas to liquid, and liquid to liquid.

9.5.1 Gas-to-Gas Recovery

The most common application for gas-to-gas heat recovery is preheating air with process flue gases. Other applications are the use of process waste heat for space conditioning and the use of the exhaust stream from an air-conditioning unit to preheat supply air. The commercial devices available to perform these functions are recuperators, regenerators, run-around loops, and heat pipe exchangers.

Recuperators Recuperators are closed heat exchangers. Several types are made, the simplest being a radiation recuperator, shown in Fig. 9.16. A

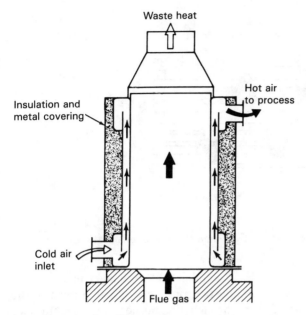

Figure 9.16 Metallic radiation recuperator for high-temperature recovery. (From Kreider and McNeil [1].)

radiation recuperator is designed for high-temperature application. The hot flue gases flow through the inner passage of this device while the air to be heated flows through an external annulus. When the air is used for combustion, preheating reduces the energy required from the fuel to heat the air to the combustion temperature.

Convective recuperators are necessary when radiation heat transfer is not effective. These exchangers can be either the tubular type or the plate type. Figure 9.17 shows a shell-and-tube convection recuperator. The operation of this exchanger can be predicted by the methods of Chapter 8.

In some situations a recuperator might have a radiation section where gas temperatures are high, followed by a convection section where the gases are cooler.

Figure 9.18 shows a typical plate-type recuperator. These units consist of thin plates that separate the two flows into a crossflow configuration. They come in packaged form for ease of installation. They have no moving parts, involve no cross-contamination of streams, and can be maintained with relative ease.

Recent developments in recuperator design include glass tube units for corrosive atmospheres, ceramic surfaces for high-temperature applications, and heat pipe exchangers.

The Recuperative Burner   This device, shown in Fig. 9.19, uses combustion products from the combustion zone to preheat combustion air. This is done

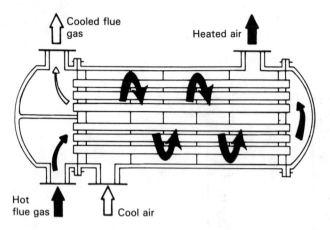

Figure 9.17 Shell-and-tube convection recuperator. (From Kreider and McNeil [1].)

by surrounding the burner with an annular heat exchanger in which the incoming air is heated by the combustion products. Usually, most of the combustion products are drawn through the burner, with the furnace pressure being controlled by the flow rate of the gases through the burner. Recuperative burners that cover a thermal input range of 100–900 kW are available. They can operate with up to 100% excess air.

Reay [4] cites fuel savings of 40 to 50% resulting from the use of recuperative burners. The applications include batch-steel reheating furnaces and intermittent kilns for ceramic firing.

Regenerators Regenerators operate in a transient manner, in that they contain elements that are alternately heated and cooled. The most common

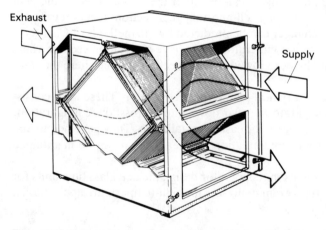

Figure 9.18 Plate-type convection recuperator. (From Reay [4].)

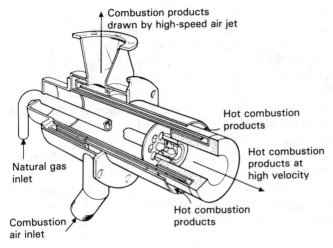

Figure 9.19 Recuperative burner for preheating combustion air. (From Reay [4].)

type of regenerator for an industrial application is the Ljungstrom rotary air-preheater. This type of heat recovery unit has been in use for over 50 years in large power plants and industrial combusion processes.

Figure 9.20 shows a rotary-type regenerator. An element of the unit is heated as it rotates through the flue gas stream. It then gives up its stored heat to the counterflowing air stream as the unit rotates full-circle. Similar smaller systems, called heat wheels, are used to reclaim exhaust heat from building air-conditioning systems.

The wheel matrix may contain a material as simple as wire mesh; a common mesh material is aluminum. Such materials are relatively inexpensive, and the heat transfer performance is good because the wire mesh presents a great deal of surface area for its volume. However, fouling of the matrix can be a problem and the pressure drop is high for densely packed meshes.

Pressure drop problems can be alleviated by substituting a corrugated metallic plate matrix for the wire mesh. Wheels of this type are called laminar-flow wheels because the flow is directed along passages, whereas motion through the wire matrix is chaotic. Fouling is not as prevalent for laminar-flow wheels, and when fouling occurs, cleaning is much easier.

Operating temperatures might range up to 1200–1300°F for conventional industrial applications. The upper limit is controlled by material behavior — primarily corrosion and thermal stress. Reay [4] reports that some progress has been made in developing ceramic wheel regenerators for application up to about 1800°F. In this temperature range, heat from metal and glass processing can be reclaimed to heat air for charge preheat or for combustion.

Run-Around Loops   If the heat recovery sink is not adjacent to the heat source, a run-around loop may be utilized. In this case there is no need for rerouting of duct work for high-volume gas flows. Rather, a closed liquid loop is used to transport the recovered heat from source to sink.

Figure 9.21 shows such a system applied in an industrial dryer system. Dryers usually have their inlet and exhaust streams separated by long distances, so they are especially conducive to the application of run-around loops. Heat transfer is accomplished at the two ends of the loop with conventional heat exchangers, usually finned-surface recuperators.

Run-around loops are used more frequently in HVAC applications than in industrial processes. They are usually designed for single-phase heat transfer (for the loop fluid) and consequently are limited in temperature application. Ethylene glycol solutions and high-temperature organic heat transfer fluids[2] such as Dowtherm may be used, but even the use of these

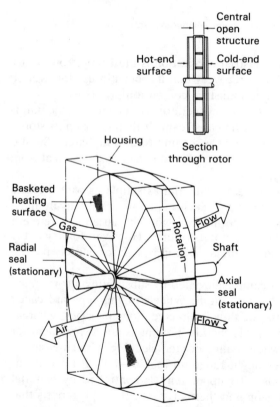

Figure 9.20 Rotary air preheater. (From *Steam: Its Generation and Use*, 38th ed. Copyright 1972 by Babcock and Wilcox, Barberton, Ohio. Reprinted with permission.)

[2] See Table 9.3 for a listing of organic heat transfer fluids.

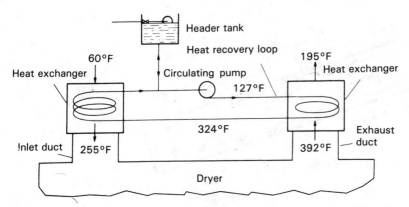

Figure 9.21 A run-around loop for heat recovery from an industrial dryer. (From Reay [4].)

fluids is limited to temperatures on the order of 500°F, and the fluids are subject to thermal degradation.

Heat Pipe Exchangers Heat pipe heat exchangers are a recent development for industrial application. They could be called recuperators because they operate steadily and the two streams are commonly separated. However, they are usually treated separately from recuperators because they use an array of heat pipes to transport the heat across the separating wall.

Figure 9.22 shows a typical layout for an air preheater heat recovery unit with heat pipes. The heat pipe operates as shown in Fig. 9.23. It is basically a closed evaporation-condensation loop. Because of the phase-change nature of the operation, large quantities of heat can be transported from the

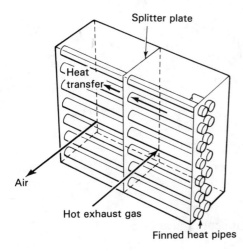

Figure 9.22 Typical layout for gas-to-gas heat pipe exchanger. (From Reay [4].)

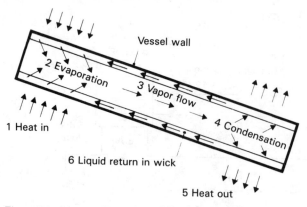

Figure 9.23 Schematic of heat pipe, showing principle of operation. (From Reay [4].)

heated to the cooled end with a very small temperature drop. Thus the heat pipe is almost an isothermal transporter of energy.

The layout of Fig. 9.22 suggests that the pressure drop should be low through the bundle of heat pipes. However, the gas-side heat transfer coefficient is low, and usually fins must be added in gas-to-gas application. This drives the pressure drop up while lowering the temperature difference required for heat transfer.

Heat pipe units have gained some industrial acceptance in relatively low temperature applications where the rotary regenerator, run-around loop, and plate-type recuperator are used. High-temperature units requiring ceramic surfaces and liquid-metal working fluids are not cost-competitive at present. However, research directed toward solving some of the temperature limitations on heat pipe performance is continuing.

Some of the processes in which heat pipe exchangers are used are listed below:

Air-dryer recuperation
Automotive paint-drying ovens
Biscuit and bread ovens
Brick-kiln heat recovery
Laundries
Paper dryers
Pharmaceuticals
Pollution control plus heat recovery
Spray-drying
Textile dryers
Welding-booth heat recovery

Chi [5] gives a comprehensive treatment of heat pipes.

9.5.2 Gas-to-Liquid Recovery

Most gas-to-liquid heat recovery equipment falls in the recuperator category. The usual arrangement is for the gas to be circulated over tubes through which liquid is flowing. In some cases the equipment operates with the gas flowing in the tubes; devices of this type are called fire-tube heat exchangers. The run-around loop described in the preceding section is actually a gas-to-liquid recovery system if the heat exchangers at either end of the loop are examined. But the overall effect is to transport energy between gaseous systems, so we include run-around loops in the gas-to-gas recovery category.

Two devices that are often used in gas-to-liquid recovery are economizers and waste heat boilers. Economizers are so prevalent in power plants that they are not always regarded as heat recovery equipment. But they have applications in other systems, so their inclusion in a list of heat recovery applications is warranted.

Heat can be recovered from incineration plants by using waste heat boilers. Incineration is intended to get rid of waste material by combustion, but some of the heat of combustion, which otherwise would be dumped into the atmosphere, can be recovered. The heat recovery is usually achieved with a waste heat boiler.

Waste Heat Boilers   Waste heat boilers are unfired generators of steam or other fluid vapors. An exhaust stream, such as that from an incinerator, a reciprocating engine or turbine, a glass or metal processing furnace, or a chemical process, can be the source of the waste heat. Chemical reactions do not occur in a waste heat boiler as a matter of course, but supplement firing is sometimes both feasible and advantageous. For example, gas turbine exhausts are rich in hot, unburned air so that fuel can be injected and burned without furnishing additional air to a waste heat boiler. Such a system can produce more steam than the unfired boiler.

Waste heat boilers can be classified according to purpose as follows:

Total heat recovery: The cooled gas is vented directly to the atmosphere. The boiler is used to recover as much of the waste heat as possible. Boilers that are mounted on metal-processing furnaces, glass-melting furnaces, and reciprocating engine and turbine exhausts are examples of this class of heat recovery units.
Partial heat recovery: The gas is cooled down to a temperature suitable for use in a following process. Sulfur combustion plants and hydrogen sulfide plants use this type of heat recovery.
Process-control heat recovery: Excess energy is removed from the process stream to prevent further chemical reaction. The waste heat boiler reclaims a portion of the heat available while providing stability for the

preceding process. A good example of this technique is a cracked-gas cooler in an ethylene plant.

Waste heat boilers have the following desirable characteristics:

Units are compact because the rates of heat transfer during boiling are high. Their smaller size makes the units lower in capital costs than gas-to-gas units.

Higher-temperature exhaust streams can be accommodated because the more efficient heat transfer maintains lower tube temperatures.

The duty can be rapidly varied by changing the steam-side pressure. Precise gas and water flow rate controls are not needed.

There are some disadvantages of waste heat boilers. Costly water treatment may be needed if "clean" steam is required, and this will drive up both capital and operating costs. The waste heat boiler might not be capable of cooling the exhaust to the desired temperature. In that case additional heat recovery equipment would be required downstream of the waste heat boiler.

Types of Waste Heat Boilers   Waste heat boilers may be based on either forced or natural circulation. These types may be of the water-tube or fire-tube design. Fire-tube units usually have the shell-and-tube configuration.

A forced-circulation water-tube waste heat boiler is shown in Fig. 9.24. Forced circulation increases the heat transfer rate on the water side so that smaller and lighter tubes can be used. Smaller tubes can withstand greater stress so they can be constructed of carbon steel. They are easy to form and to weld into place. The smaller water quantities that result decrease

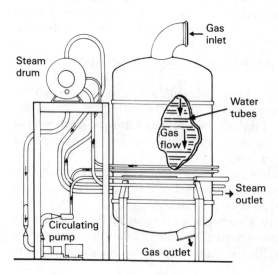

Figure 9.24 Forced-circulation waste heat boiler.

pumping costs, and start-up and shutdown times are decreased, lowering maintenance costs. Boyen [6] gives a thorough discussion of the relative merits of forced- and natural-circulation waste heat boilers.

Figure 9.25 shows a natural-circulation waste heat boiler. These units require larger tubes because of their lower heat transfer rates. They can respond rapidly to varying loads because the density difference between the downcomer and the riser creates a self-adjusting circulation mechanism. The main advantage of a natural-circulation boiler is the ruggedness and simplicity of its large thick-walled tubes. Furthermore, no pumps are required since the flow is self-generating. Natural-circulation boilers are larger than forced-circulation units of the same capacity.

Economizers Gas-to-liquid economizers are heat exchangers that produce a heated liquid stream, usually without vapor production. Economizers are almost always of the water-tube design. External finned tubes are usually required because the gas-side heat transfer coefficient is low compared to the water-side coefficient. Steel tubes and fins are common.

Power Plant Economizers Economizers in power plants operate at relatively high water pressure because the water is pumped through the economizer up to the boiler pressure. Large thick-walled tubes are required; tubes with 2-in. outer diameter are not uncommon. Economizer tubes can operate at temperatures up to 1500°F, although most applications are below this level.

Economizers and air preheaters take advantage of relatively low exhaust temperatures. In both of these devices, reducing the exhaust temperature down to the water vapor dew point causes condensation. The condensate can combine with flue gas components, notably sulfur, to form weak acidic solutions that corrode the economizer tubes. The temperature of the exhaust must remain above the dew point. Therefore, the feedwater is not heated to its economic potential and the magnitude of heat recovery diminishes. An 11°F rise in boiler feedwater in the economizer can save about 1% of the boiler fuel costs for a given steam output.

Figure 9.26 shows how condensation can be prevented in an economizer by recirculated a portion of the makeup feedwater back to the economizer, effectively preheating the water flow. Figure 9.26 shows about 30% of the feedwater being recirculated through the economizer. The increased flow makes the heat transfer more effective and the economizer can be smaller. The recirculation increases the feedwater inlet temperature, thereby preventing the temperature of the flue gas from dropping to the dew point.

Other Economizer Applications Economizers are used to recover heat from suitable exhausts for process water heating and for heating of organic fluids in thermal fluid heaters.

Figure 9.27 shows an economizer used to preheat water injected into a

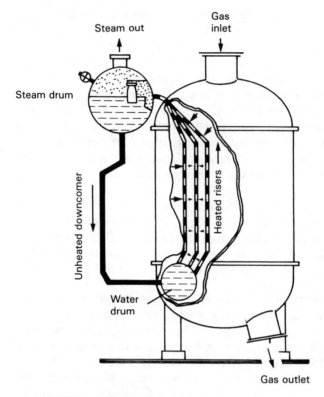

Figure 9.25 Natural-circulation waste heat boiler.

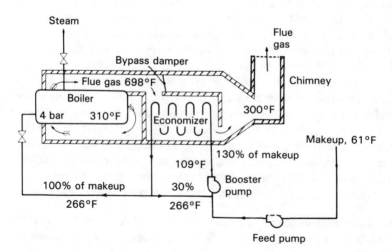

Figure 9.26 Economizer for preheating makeup water, with recirculation to prevent condensation.

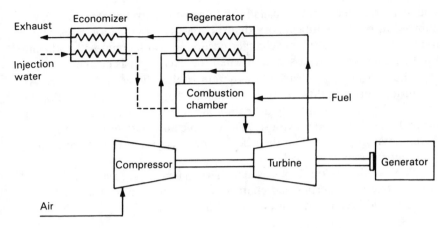

Figure 9.27 Economizer used to preheat injection water for a gas turbine.

gas turbine combustion chamber. The injection of water reduces the air to be compressed and also controls the maximum chamber temperature. Preheating the water reduces the sensible energy loss from the combustion zone to the water. The economizer is effective at lower temperatures than a gas-to-gas regenerator, so it is an economical addition to the gas turbine cycle.

Thermal fluid heaters use organic fluids, with higher boiling points than moderately pressurized water, as the heat recovery agent. Lower pressure requirements make this system more economical than pressurized-water or steam systems. Figure 9.28 shows a fluid heater on a gas turbine exhaust.

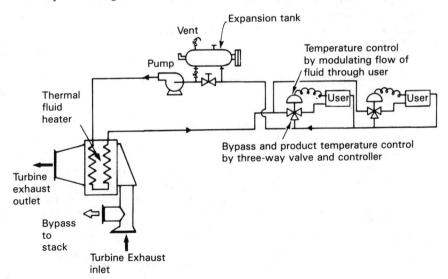

Figure 9.28 Thermal fluid heater economizer matched to a gas turbine exhaust.

The heat recovery unit is usually a finned-tube exchanger with liquid pumped through the tubes. An exhaust bypass is provided for cases where fouling of the finned-tube exchanger occurs, as can happen with light-oil-fired turbines. The bypass is used when the exchanger is cleaned.

The thermal fluid distribution system is also shown in Fig. 9.28. An expansion tank and flow controls are required at the point of usage of the recovered heat. Thermal fluid heaters are specialized forms of the run-around loop described previously. The organic fluids can suffer degradation at temperatures lower than their boiling points, so that fluid treatment or replacement may be necessary.

Some of the organic fluids used in thermal heaters are listed in Table 9.3 with pertinent properties and characteristics. The manufacturers of these fluids should be consulted if additional data are needed.

9.5.3 Liquid-to-Liquid Recovery

Liquid-to-liquid heat recovery units are compact because the density of both flow streams is high, and liquid-surface heat transfer coefficients are usually high. Compact recuperators are often used for liquid-to-liquid heat recovery.

Shell-and-tube and plate heat exchangers are most often used for liquid-to-liquid recovery. Both have compact configurations and involve complete separation of the liquid streams.

Liquid-to-liquid exchangers are used to recover heat from boiler "blowdown." Blowdown is the removal of boiler water that has accumulated a percentage of dissolved solids which is too high for normal operation. The water with impurities is forced from the boiler and replaced with clean makeup water. In cases where the blowdown is continuous the heat in the blowdown water can be partially recovered in a liquid-to-liquid exchanger. This practice is discussed in greater detail in Chapter 6.

The following example illustrates liquid-to-liquid heat recovery and its effects on a system.

Example 9.4 Liquid-to-Liquid Heat Recovery

Mammoth Industries maintains a cafeteria for its workers that requires a great deal of hot water for washing dishes. The existing system is shown in the accompanying figure. Water at 60°F is heated by condensing steam at 300 psia to 190 °F in a "calorifier." Dirty water at 160 °F is exhausted to the drain. It is proposed to add a small shell-and-tube exchanger to the system to recover part of the heat in the dishwater for preheating the clean water. The overall heat transfer coefficient for the exchanger is 350 Btu/hr ft^2 °F. It is desired to cool the dishwater to about 80 °F in the preheater. The clean water flows through the tubes and the dishwater through the shell of the single-pass exchanger. Calculate the required exchanger area, the outlet water temperature, and the reduction in the amount of steam required in the calorifier if evaporation reduces the amount of dirty water flowing from the dishwasher by 5%.

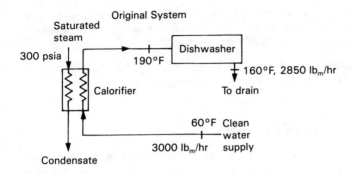

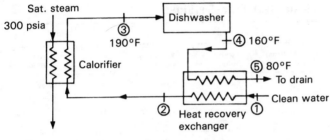

Solution: Without heat recovery,

$$\dot{m}_{cond} h_{fg} = (\dot{m}c_p)_{H_2O} \, \Delta T_{cal}$$

$$\dot{m}_{cond} = \frac{(3,000)(0.947)(130)}{800} = 481.1 \text{ lb}_m/\text{hr}$$

With heat recovery,

$$(\dot{m}c_p)_{dw} \, \Delta T_{dw} = (\dot{m}c_p)_{H_2O} \, \Delta T_{H_2O}$$

$$\Delta T_{H_2O} = \frac{(0.95)(0.977)(80)}{0.998} = 76.0 \,^{\circ}\text{F}$$

$$T_2 = 76 + 60 = 136 \,^{\circ}\text{F}$$

$$T_{lm} = \frac{(160 - 136) - (80 - 60)}{\ln (24/20)} = 21.9 \,^{\circ}\text{F}$$

$$A = \frac{(3,000)(0.997)(76)}{(350)(21.9)} = 29.7 \text{ ft}^2$$

Now we analyze the calorifier:

$$Q = (\dot{m}c_p)_{H_2O} \, \Delta T = \dot{m}_{cond} h_{fg}$$

$$\dot{m}_{cond} = \frac{(3,000)(0.997)(54)}{809} = 199.6 \text{ lb}_m/\text{hr}$$

The steam rate is reduced by 58.5% by heat recovery if the calorifier operates the same way at the two different inlet water temperatures. The reader should also be aware that this problem can be worked by using the NTU-ε charts of Chapter 8.

Table 9.4 summarizes the various types of heat recovery exchangers, their temperature application ranges, and other characteristics of their performance.

9.5.4 Mechanical Work Recovery

Energy can be recovered from high-pressure and/or high-flow-rate process streams. This does not technically fall into the heat recovery category;

Table 9.3 Properties of Some Heat Transfer Fluids

	Mineral oils				Diphenyl diph. oxide (Dowtherm A)	Ethylene glycol (Union Carbide)
Property	Humble Therm 500	Mobil Therm 600	Tidewater Avalon 90	Sunoco Circo XXX		
Chemical stability	Stable	Stable	Stable	Stable	Stable, but water can be dangerous	Stable
Oxidative	Up to 150 °F w/air	Up to 150 °F w/air	Up to 150 °F w/air	Up to 150 °F w/air	Stable	Low oxidation
Thermal stability	High resistance to decomposition. Deposits carbon	High resistance to decomposition. Deposits carbon	Slow decomposition starts at 575 °F, more at 750, rapid 850	Similar to tidewater Avalon 90	No appreciable decomposition below 650 °F Steady decomposition 650 +	Thermally stable to 400 °F
Maximum temperature						
Film	680 °F	625 °F	620 °F	590 °F	805 °F	400 °F
Bulk	615 °F	560 °F	555 °F	525 °F	700 °F	325 °F
Safety						
Fire resistance	None—burns	None—burns	None—burns	None—burns	None—burns	None—burns
Explosive	Negligible	Negligible	Negligible	Negligible	Explosive mists possible	None
Density, lb/gal						
70 °F	7.1	7.8	7.4	7.8	8.82	9.30
300 °F	6.5	7.3	6.7	7.3	7.98	
600 °F	5.5	6.4	5.9	6.5	6.60	Not used
700 °F	Not used	Not used	Not used	Not used	6.03	
Specific heat, Btu/lb$_m$ °F						
70 °F	0.48	0.38	0.43	0.43	0.379	0.625
300 °F	0.58	0.485	0.55	0.55	0.458	0.710
600 °F	0.72	0.624	0.70	0.70	0.560	Not used
700 °F	Not used	Not used	Not used	Not used	0.591	
Therm. cond., Btu/ft^2 hr °F/ft						
70 °F	0.078	0.070		0.070	0.081	0.167
300 °F	0.072	0.065		0.066	0.072	
600 °F	0.065	0.059		0.060	0.061	Not used
700 °F	Not used	Not used	Not used	Not used	0.057	
Pour point	+ 15 °F	+ 20 °F	+ 15F	+ 25 °F	Freezes at 53.6 °F	Freezes at − 60 °F
Viscosity, Cp						
70 °F		190	4,000	10,000	4.5	20.0
300 °F	1.9	2.4	7.4	7.0	0.6	1.0
600 °F	0.25		1.0	1.0	0.35	Not used
700 °F	Not used	Not used	Not used	Not used	0.30	
Compatible with materials	Ferrous and nonferrous. Oil-resistant gaskets and packing	Avoid copper and copper alloys. Oil-resistant gaskets and packing	Similar to other mineral oils	Similar to other mineral oils	Carbon steel, cast iron. Aluminum foil packing	Carbon steel, cast iron

Table 9.3 Properties of Some Heat Transfer Fluids (*Continued*)

Property	Polyalkylene glycols (Union Carbide)				Aromatic base fluids (Monsanto Chemical Co.)				
	UCON HTF-30	UCON HTF-14	UCON HTF-10	UCON HTF-L20	Therminol 44	Therminol 60	Therminol 66	Therminol 55	Therminol 88
Chemical stability	Stable	Stable	Stable	Stable	Stable	Stable	Stable	Stable	Stable
	Water-soluble	Water-soluble	Water-soluble						
Oxidative	Up to 150 °F in air	Up to 150 °F in air	Up to 150 °F in air	Up to 150 °F in air	Up to 150 °F in air	Up to 150 °F in air	Up to 150 °F in air	Up to 150 °F in air	High resistance
Thermal stability	Thermally stable up to 500 °F	Thermally stable up to 500 °F	Thermally stable up to 500 °F	Thermally stable up to 500 °F	Thermally stable up to 425 °F	Thermally stable up to 600 °F	Thermally stable up to 650 °F	Thermally stable up to 575 °F	Thermally stable up to 800 °F
Maximum temperature									
Film	565 °F	565 °F	470 °F	565 °F	475 °F	635 °F	705 °F	635 °F	850 °F
Bulk	500 °F	500 °F	400 °F	500 °F	425 °F	600 °F	650 °F	600 °F	800 °F
Safety									
Fire resistance	Burns	Burns	Burns	Burns	Burns	Burns	Burns	Burns	Burns
Explosive	Negligible	Negligible	Negligible	Negligible	Negligible	Negligible	Negligible	Negligible	Negligible
Density, lb/g									
70 °F	8.97	8.63	8.55	8.26	7.75	8.27	8.43	7.40	9.40
300 °F	8.22	7.84	7.76	7.46	6.85	7.60	7.59	6.69	8.42
500 °F	7.50	7.01	7.05	6.76	Not used	6.95	6.75	6.07	7.67
600 °F	Not used	Not used	Not used	Not used	Not used	6.65	6.42	5.76	6.58, 800 °F
Specific heat, Btu/lb$_m$ °F									
70 °F	0.44	0.44	0.44	0.44	0.46	0.383	0.365	0.460	
300 °F	0.54	0.54	0.54	0.54	0.54	0.495	0.480	0.572	0.467
500 °F	0.64	0.64	0.64	0.64	Not used	0.593	0.580	0.670	0.525
600 °F	Not used	Not used	Not used	Not used	Not used	0.643	0.630	0.718	0.613, 800 °F
Therm. cond., Btu/ft^2 hr °F/ft									
70 °F	0.121	0.115		0.099	0.083	0.076	0.071	0.0790	
300 °F	0.099	0.095		0.096	0.071	0.0705	0.067	0.0724	0.0712
500 °F				0.092	Not used	0.0654	0.064	0.0661	0.0686
600 °F	Not used	Not used	Not used	Not used	Not used	0.0630	0.062	0.0630	0.0608, 800 °F
Pour point	0 °F	−35 °F	−45 °F	−40 °F	−80 °F	−90 °F	−18 °F	−40 °F	Melts at 293 °F
Viscosity Cp									
70 °F	320.0	200.0	90.0	200.0	6.20	9.03	142	387	
300 °F	5.4	5.4	3.8	5.4	0.44	0.87	1.55	1.69	1.57
500 °F	1.85	1.85	1.5	1.85	Not used	0.38	0.45	0.58	0.55
600 °F	Not used	Not used	Not used	Not used	Not used	0.29	0.34	0.42	0.23, 800 °F
Compatible with materials	All common metals including copper	All common metals including copper	All common metals including copper	All common metals including copper	All common metals	All common metals	All common metals	All common metals	All common metals

Source: Boyen [6].

nonetheless, significant energy can be recovered in special circumstances, as in gas liquefication plants. Rotary devices placed in the process stream are usually used for work recovery. For gas streams, the term *turboexpander* is often used to describe the turbine through which the gas stream is expanded. The *liquid recovery turbine* serves a similar function for liquid flows.

Turboexpanders These expansion turbines convert the internal energy of a gas or vapor stream into mechanical work. Turboexpanders can cool a gas down to where condensation occurs. They are common in low-pressure air separation plants. The inlet air pressure for this application is about 75 psia.

Holm and Swearingen [7] report that power is not generally recovered from processes where less than 100 hp can be reclaimed. However, for power recoveries of 100 hp or more, it is economical to drive electrical generators or compressors. Turboexpanders recover far more energy in

Table 9.4 Operation and Application Characteristics of Industrial Heat Exchangers

Commercial heat transfer equipment	Low temperature sub-zero–250°F	Intermediate temperature 250–1200°F	High temperature 1200–2000°F	Large temperature differentials permitted	Packaged units available	Can be retrofit	No cross-contamination	Compact size	Gas-to-gas heat exchange	Gas-to-liquid heat exchanger	Liquid-to-liquid heat exchanger	Corrosive gases permitted with special construction
							Specifications for waste recovery unit					
Radiation recuperator			•	•	a	•	•		•			•
Convection recuperator		•	•	•	•	•	•		•			•
Metallic heat wheel	•	•			•	•	b	•	•			•
Ceramic heat wheel			•	•	•	•		•	•			•
Passive regenerator		•			•		•		•			•
Finned-tube heat exchanger		•		•	•	•	•	•		•		c
Tube shell-and-tube exchanger		•		•	•	•	•	•		•	•	•
Waste heat boilers	•	•		•	•	•	•		•	•		c
Heat pipes	•	•		d	•	•	•	•	•			•

[a] Off-the-shelf items available in small capacities only.
[b] With a purge section added, cross-contamination can be limited to less than 1% by mass.
[c] Can be constructed of corrosion-resistant materials, but consider possible extensive damage to equipment caused by leaks or tube ruptures.
[d] Allowable temperatures and temperature differential controlled by the equilibrium properties of the internal fluid.

Source: Kreider and McNiel [1].

hydrocarbon processing than in air separation, although they are not as frequently used in this application. The reason is that larger units can be used in hydrocarbon processing; for example, 10,000-hp turbines are available. They are usually directly connected to a process compressor as shown in Fig. 9.29.

Turboexpanders can operate in the range 2,000 – 3,000 psia with inlet temperatures up to about 1000°F. Enthalpy drops of 40–50 Btu/lb per expansion stage are typical. Rotor tip speeds of 100 ft/sec are allowable.

Turboexpander performance can be calculated by using classical thermodynamic relationships. Any basic thermodynamics text provides the information required for analysis; for example, see reference [8]. Example 9.5 shows such an analysis.

Holm and Swearingen [7] describe recent improvements in seals, bearings, flow controls, and other operational components that have limited the use of turboexpanders in the past.

Example 9.5   Power Recovery by a Turboexpander

Air at 800 psig and 250°F is exhausted from a certain process at the rate of 30 ft^3/min. The plant compressed air system operates at 100 psig and the air must be in the range 50–100°F. It is proposed to expand the processed air through a turboexpander into the plant compressed air system, saving both air and energy. The turboexpander efficiency is 80%. Calculate the following:

(a) The temperature of the turbine exhaust if expanded to 100 psig
(b) The heat required to bring the processed air to the temperature required for the plant air systems
(c) The shaft horsepower of the turbine

Solution: First, calculate the mass flow rate:

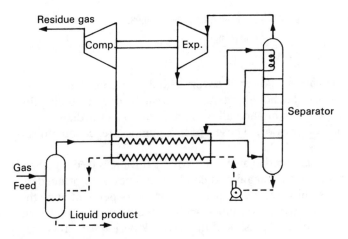

Figure 9.29 Turboexpander applied to the separation of propane and heavier hydrocarbons from a natural gas stream.

$$p = 814.7 \text{ psia} \qquad T = 710 \text{ R}$$

$$R = \frac{\bar{R}}{M} = \frac{1545}{28.95} = 53.34 \text{ ft lb}_f/\text{lb}_m \text{ R}$$

$$\rho = \frac{p}{RT} = \frac{(814.7)(144)}{(53.34)(710)} = 3.10 \text{ lb}_m/\text{ft}^3$$

$$\dot{m} = (30)(60)(3.10) = 5580 \text{ lb}_m/\text{hr}$$

Treat the air as an ideal gas:

$$\frac{p_2}{p_1} = \frac{114.7}{814.7} = 0.140 \qquad T_1 = 710 \text{ R}$$

$$pv^k = \text{const.} \qquad\qquad k = \frac{c_v}{c_p} = 1.4 \text{ (air)}$$

giving

$$\frac{T_2}{T_1} = \left(\frac{p_2}{p_1}\right)^{(k-1)/k} = (0.14)^{0.286} = 0.520$$

$$T_2 = 405 \text{ R} = -55 °F$$

also,

$$h = h(T) \text{ only}$$

$$\Delta h = c_p(T_1 - T_2) = 0.241(250 + 55) = 73.5$$

$$h_1 = 169.97 \text{ Btu/lb}_m \qquad \text{from Table B.3}$$

$$\eta = \frac{h_1 - h_2}{h_1 - h_2} = 0.80 \quad\Rightarrow\quad h_2 = 111.17$$

From Table B.3, $T_2 = 465 \text{ R} = 5 °F$. Now, heat required to bring the air to 50 °F is

$$Q = \dot{m}c_p \, \Delta T = \frac{5{,}580}{60}(0.241)(45) = 1{,}009 \text{ Btu/min}$$

Finally, for the shaft horsepower, pick a mechanical efficiency of 0.98.

$$\text{Horsepower} = \frac{0.98(5{,}580)(58.8)}{2{,}545} = 126.3 \text{ hp}$$

Liquid Recovery Turbines Power recovery should be considered when throttling is used to dissipate liquid pressure. This can be done by piping the liquid stream to a special pump and allowing it to run backward and drive another pump, or compressor. These special pumps are called power recovery turbines.

The horsepower recovered in a power recovery turbine is free. Furthermore, electric power lines or steam and condensate lines are not needed. The desirability of installing recovery turbines is largely dependent upon the operating costs of alternative power units, such as steam turbines or electric motors. As a rule, a project is feasible if the payback period for a recovery turbine is about 2 years.

Power recovery turbines are used to drive cooling-tower fans, reciprocating and rotary compressors, pumps, and on-line electric generators. With proper controls, power recovery turbines can be used with other drivers to allow for fluctuations in power recovery output. Evans [9] gives an excellent treatment of the operational characteristics of power recovery turbines.

9.6 ENERGY CASCADE SYSTEMS (skip)

The concept of heat recovery is part of a much broader concept known as energy cascading. Sternlicht [10] defines energy cascading as "matching the quality (temperature) of the available energy to the needs of the task." The principal idea of energy cascading is to use the exhausted heat from the highest-temperature process in the plant to drive a series of devices down to a temperature where heat recovery is no longer economical. Each successive device uses the preceding exhaust as part of its energy source.

Figure 9.30 shows a cascade energy system suitable for a plant where high temperatures are involved, in this case a steel rolling mill furnace. The furnace exhaust is used as input to a gas turbine, the turbine exhaust provides energy for a steam turbine, and finally the steam turbine exhaust is used as the heat source for an organic "bottoming" turbine. Waste heat is produced at a very low temperature, at which heat recovery is uneconomical. Recuperation of the rolling mill exhaust preheats furnace air, which

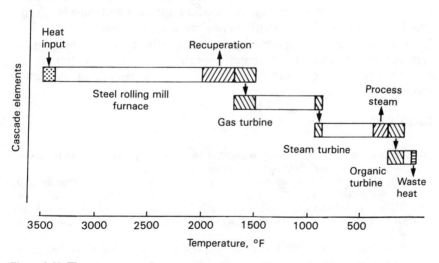

Figure 9.30 The energy cascading concept as illustrated by a steel rolling mill.

also reduces the energy requirement. Some process steam is extracted from the steam turbine rather than being used to produce power. By combining all these devices in a cascade so that the exhaust of one is the input to another, the full temperature-energy potential is exploited.

Most heat recovery systems are two-stage energy cascade systems. The combined generation of electricity and production of process steam, *cogeneration*, is a two-stage cascade system. Air preheaters and economizers are also part of two-stage cascade systems.

Bywaters [11] claims that the difficulty in implementing cascade systems is mostly due to industrial planning that has been based on cheap fuel. On-site electricity generation has decreased sharply in the past 20 years because industries could purchase electricity cheaply from central power station utilities. Industries had no incentives to install recovery equipment because of the low price of fuel. These factors led to plant designs in which all electricity is purchased, and even if cascading were feasible, there would be no available use for the generated power. Well-planned industrial parks, where several different types of companies and even a power utility might reside, could create a market for power generated by a cascade system.

9.7 SUMMARY

The potential for energy savings by heat recovery is great in many energy-intensive industries. There are many methods by which heat can be recovered, and most of them have been described in this chapter.

A waste heat source, of itself, is not a sufficient reason to consider heat recovery. An end use for the recovered heat must exist and must be physically located where the installation costs are not exorbitant.

The economics of heat recovery projects are the ultimate constraint on implementation. Thus, the technical aspects of heat recovery cannot be separated from the costs and potential dollar savings resulting from such a project.

REFERENCES

1. K. G. Kreider and M. B. McNiel, *Waste Heat Management Guidebook*, NBS Handbook 121, U.S. Government Printing Office, Washington, D.C., 1977.
2. M. J. Moran, *Availability Analysis: A Guide to Efficient Energy Use*, Prentice-Hall, Englewood Cliffs, N.J., 1982.
3. J. Gilbert, Heat Pump Strategies and Payoffs, in *Proceedings of the Fourth Annual Industrial Energy Conservation Technology Conference*, vol. 1, April 7–9, 1982, Houston, Texas, pp. 323–330.
4. D. A. Reay, *Heat Recovery Systems: A Directory of Equipment and Techniques,* Spon, London, 1979.
5. S. W. Chi, *Heat Pipe Theory and Practice,* Hemisphere, Washington, D.C., 1976.
6. J. L. Boyen, *Practical Heat Recovery*, Wiley, New York, 1975.

7. J. Holm and J.S. Swearingen, Turboexpanders for Energy Conservation, *Mech. Eng.*, 34–40, 1978.
8. B. D. Wood, *Applications of Thermodynamics*, 2d ed., Addison-Wesley, Reading, Mass., 1981.
9. F. L. Evans, Jr., Drivers, in *Equipment Design Handbook for Refineries and Chemical Plants,* 2d ed., vol. 1, Gulf Publishing Co., Houston, Texas, 1980.
10. B. Sternlicht, Capturing Energy from Industrial Waste, *Mech. Eng.*, 100 (8): 30–41, 1978.
11. R. Bywaters, Industrial Heat Recovery, notes for Industrial Energy Auditing Short Course, University of Houston, August 1980.

PROBLEMS

9.1 An air preheater of a gas turbine receives air at 540°F and 90 psia and heats it up to 1540°F. Compute the percent loss in available energy caused by a pressure drop of 20 psia in the heater. The sink temperature is 100°F.

9.2 A stream of liquid water at 200°F is mixed with another stream of liquid water at 80°F. Both streams have the same flow rate. Neglecting heat losses and kinetic and potential energy changes, determine the temperature after mixing and the irreversibility in Btu/lb_m. T_o is 70°F and p_o is 1 atm. What is the second-law efficiency for this process?

9.3 At the inlet of an air turbine operating at steady state p is 75 psia, T is 340°F, and V is 400 ft/sec. At the exit, p is 14.7 psia, T is 140°F, and V has been reduced to 100 ft/sec. Assuming ideal gas behavior, and using $T_o = 40°F$ and $P_o = 14.7$ psia, determine:

(a) the availability per pound mass of air at the inlet
(b) the irreversibility per pound mass for the process
(c) the work output in Btu/lb_m
(d) the second-law efficiency for the process

9.4 Water is to be heated electrically from 60 to 160°F. All the dissipated electrical energy is assumed to be transferred to the water. What is the second-law efficiency of the process if $T_o = 60°F$?

9.5 Calculate the first- and second-law efficiencies for a steam turbine that operates between an inlet condition of 1000°F, 800 psia and an outlet condition of 20 psia, $x = 1.0$. The ambient temperature is 70°F. Neglect heat losses to the surroundings.

9.6 Throttles are often used to reduce pressure (or to impede flow) in power and refrigeration devices. An isenthalpic process is commonly assumed, based on zero heat transfer and negligible kinetic energy changes. Calculate the irreversibility and the second-law efficiency of the throttling process occurring across an expansion value in a Freon-12 refrigeration system. The inlet condition is saturated liquid at 110°F and the outlet pressure is 50 psia. The ambient air is at 35°F.

9.7 Calculate the COP for an open-cycle heat pump (recompression device) that takes saturated steam at 5 psia and delivers it at 300°F. The mechanical efficiency of the compressor is 0.90. If the vapor were compressed isentropically, what would its exhaust pressure be?

9.8 Steam at 200°F can be formed by flashing of condensate collected from a process heater. Steam at 350°F is needed for an adjacent process. An open-cycle recompression system is being considered to upgrade the 200°F steam. The respective costs of energy are: for steam at 350°F, $D_u = \$6.00/10^6$ Btu; for 200°F steam, $D_s = \$1.50/10^6$ Btu; and for the energy to drive the compressor, $D_p = \$0.60/kWh$. Find the break-even COP for such a system. For a COP of 4, what is the saving per 10^6 Btu delivered at 350°F? What is the η_{mech} for such a system?

9.9 Thermal recompression is to be used in an evaporator designed to evaporate 14,000 lb/hr of water from an aqueous solution feed of 100 gal/min. The solution has a specific gravity of 1.05 and an incoming temperature of 185°F. The recompression ejector uses high-pressure motive steam at 120 psia, and has the capacity to entrain 1.0 lb of low-pressure vapor per pound of motive steam in the compression range equivalent to a temperature rise of 21°F. Find the mass flow rate of motive steam required, assuming that the c_p of the solution is 1.0 Btu/lb$_m$°F and that the solution has no boiling-point elevation. Neglect heat losses and subcooling of the condensate.

Recompression Ejector

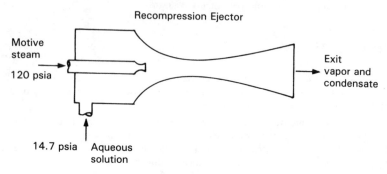

Motive
steam

120 psia

Exit
vapor and
condensate

14.7 psia   Aqueous
solution

9.10 A chemical plant employs a conventional sulfur burner in producing SO_2. Four hundred pounds per hour of crude sulfur (95% purity) are burned using 10% excess air. A waste heat boiler is used for cooling the gases, which leave the burner at 1650°F. The waste heat boiler produces saturated steam at 200 psia using 350°F feedwater and cools the gas mixture to 400°F. Assume 0.21 Btu/lb$_m$ °F for the specific heat of SO_2 and 0.28 for the specific heat of air and nitrogen. Assuming no SO_3 formation, no heat losses, and an overall heat transfer coefficient in the waste heat boiler of 15 Btu/hr ft^2 °F, calculate

(a) the steam production in pounds per hour
(b) the heat transfer surface required

9.11 We wish to recover heat from a dirty hot-water stream at 350°F by using it to preheat 10,000 lb$_m$/hr of distilled water from 70°F. The dirty water stream flow rate is also 10,000 lb$_m$/hr. What is the outlet distilled water temperature that could be *reasonably* obtained and the heat that could be recovered for:

(a) a counterflow heat exchanger?
(b) a cocurrent heat exchanger?
(c) a one-shell-pass, two-tube-pass exchanger with distilled water in the tubes?

9.12 A coal-fired boiler is to be equipped with an air preheater, and the following specifications are made:

steam output, avg. = 80,000 lb$_m$/hr
steam pressure = 3000 psia
steam temperature = 700°F
feedwater temperature = 212°F
cost of installed air preheater = $1 million
cost of coal = $85/ton

pounds of flue gas per pound of coal at 30% excess air = 13.8
overall efficiency when equipped with air preheater (at 30% excess air) = 83%
heating value of coal = 14,200 Btu/lb$_m$
flue gas temperature from boiler = 750°F
expected flue gas temperature from air preheater = 350°F

(a) What is the yearly fuel cost with the heater installed?
(b) What percentage of the fuel cost can be saved with the air preheater?
(c) What is the payback period for the preheater?

9.13 Dowtherm A is to be used as the heat transfer fluid in the run-around loop shown in Fig. 9.21. The inlet volumetric flow rate of atmospheric air is 50,000 CFM. For the conditions shown calculate:

(a) the required flow rate of Dowtherm A
(b) the Re$_D$ if a 3-in. inner diameter pipe is used to connect the two heat exchangers
(c) the pressure drop that the pump must offset if the 3-in. piping totals 80 ft and the heat exchangers, fittings, and valves add a pressure drop equal to that of the piping
(d) the minimum horsepower required for the pump

The properties of Dowtherm A are given in Table 9.2.

9.14 For the case of Problem 9.13, how much area for heat exchange is required if each end of the run-around loop is composed of fin-tube, single-pass cross-flow exchangers with the air side mixed. The overall coefficient for both exchangers is estimated to be 20 Btu/hr ft^2 °F.

9.15 A turbine receives air from a reservoir at 250 psia and 100°F. Exhaust is at 20 psia. The turbine operates steadily at an output of 50 hp with an isentropic efficiency of 70%. To improve plant efficiency, the exhaust from the turbine is to be used for refrigeration purposes. The refrigerated space is maintained at 0°F. How much refrigeration can be accomplished?

9.16 A certain process exhausts 10 CFM of air at 800 psig and 250°F. The plant service air system operates at 100 psig and the air must be in the range of 50–100°F. It is proposed that the process exhaust air be expanded through a back-pressure air turbine into the plant service air system, simultaneously saving the air and producing useful power.

(a) If the air turbine has an internal isentropic efficiency of 80%, what would the temperature of the turbine exhaust be in degrees Fahrenheit?
(b) How much heat would have to be added to the exhaust air in Btu's per hour in a heat exchanger to make it suitable for use in the plant service air system?
(c) Perhaps a better method would be to use a pressure-reducing (throttling) valve ahead of the turbine set at such a downstream pressure that the turbine exhaust temperature would come within the desired limits. What should be the downstream pressure setting of the reducing valve?
(d) What shaft horsepower might be available from the turbine in parts (a) and (c)?

9.17 Water at 100 psia and 500°F is available at the rate of 200,000 lb$_m$/hr. What is the maximum power in horsepower that could be extracted from this flow stream if it is expanded in a liquid recovery turbine, provided that cavitation does not occur in the turbine?

9.18 Engineers in a chemical plant are considering the addition of a waste heat boiler to a gas turbine exhaust. The boiler produces steam at 250 psia that is valued at 0.4¢/lb$_m$. The boiler costs $750,000. Production of steam in this way saves energy that would otherwise have to be expended in the plant boiler.

continuous
operat

The waste heat boiler produces 1,000 lb_m/hr of steam. The cost of invested money is 19%, the economic lifetime of the boiler is 20 years, and the salvage value is $150,000. Operating and maintenance costs are 3% of the first cost per year. The company uses straightline depreciation.

(a) Calculate the annualized costs of the project.
(b) What is the payback period for the boiler?

TEN

INDUSTRIAL INSULATION

The use of insulation for energy-efficient operation of hot and cold systems is almost mandatory in this era of expensive energy. Most insulations can reduce the heat loss or gain by 90% or more compared to that of bare surfaces. In fact, the amount of insulation that should be applied to a surface is usually based on an economic trade-off between energy savings and insulation costs.

Industrial insulation systems are emphasized in this chapter, although most of the ideas also apply to residential and commercial buildings. The physical mechanisms that contribute to the insulating effectiveness of various materials are categorized according to properties of importance. The factors that go into selecting an insulation system for a particular application are discussed, and techniques for selecting the appropriate thickness of insulation are the subject of the final section.

10.1 PHYSICS OF THERMAL INSULATION

The function of insulation is to reduce heat transfer, so the insulating material must have the appropriate characteristics to retard the transport of heat across it. Most insulation is a matrix of solid and gas so that heat transport across both the solid and gaseous components must be considered. In reality, the transport of energy across insulation can be by all of the modes of heat transfer described in Chapter 5 — conduction, convection, and radiation.

10.1.1 Conduction

Heat is conducted by molecular action. The "hotter" molecules are in more vigorous motion, and energy is transported to neighboring molecules by collision and/or vibration.

The effects of conduction in the solid part of an insulating matrix can be controlled in two ways. Either a material that is less conductive can be used or less of the solid material can be put into the matrix. Loosely spun fiberglass, in which the solid is in the form of small-diameter fibers and occupies only enough of the volume to give some structural rigidity, is an insulation whose conductivity depends on the amount of solid fibers present in the matrix.

The effects of conduction in the gas can be reduced by evacuation of the air from the matrix. Sealing of the system is required to maintain the vacuum and prevent air infiltration. Evacuation also reduces the effects of convection. Gases of lower thermal conductivity can be substituted for the air, but sealing the system is required. Some systems, such as low-temperature dewars, use evaluated insulating spaces, although it is costly to do so.

10.1.2 Convection

Convection occurs when the gas in the insulation matrix is set in motion because of temperature differences in the material. The rising of heated air and falling of cooled air gives rise to convection "cells" when the air is in an enclosure such as a porous insulation matrix. The cells can transport energy effectively if they are allowed to occupy relatively large pores in the matrix. It follows, then, that convection effects in insulation can be minimized by creating small cells across which the temperature difference is not large.

Convective effects can be eliminated by keeping the pore sizes in insulation small enough. Solid conduction usually increases when the cell size is decreased, so that the advantageous effect of reduced convection can be offset if care is not exercised.

Gas being forced through the insulation structure is clearly detrimental to the insulation's thermal performance, since the gas flow will directly carry energy with it. Fortunately, most insulation systems have some outer protective coating to prevent such direct infiltration.

10.1.3 Radiation

Thermal radiation contributes significantly to the energy transport across porous insulation. This is somewhat surprising, because radiation is usually thought to be important only in high-temperature applications. Radiation, however, is direct transfer of energy across the gas and can be of the same magnitude as solid and gas conduction, especially in insulation matrices, where the overall rates of heat transfer are very low. Evacuating the insulation matrix will not reduce radiation because it can be transmitted through a vacuum.

The radiation effect can be reduced by using materials with low emittance and by ensuring that the pore size of the matrix is small enough that radiation "paths" are broken into small distances. The effect is to cause radiation from the hotter wall of the pore to be absorbed by the opposite wall of the pore and reradiated at a lower temperature. Smaller pore sizes accentuate the temperature drop across a given thickness of insulation. In contrast, if the pore sizes are large, more energy can be transmitted because the net temperature difference for a given thickness of insulation is greater. However, increasing the density of insulation — and consequently decreasing pore size and reducing radiation — will lead to an increase in solid conduction.

It is clear from the discussion above that interactions between the various heat transfer mechanisms control the overall effectiveness of insulation. The net effect of the combined modes of heat transport is represented by a property called the *effective* thermal conductivity. The effective k is measured by imposing a temperature difference across a finite thickness of the insulation.

Figure 10.1 shows the effective k for glass fiber insulation, designed for use at 1000°F, as a function of density. On a microscopic level, the effective k

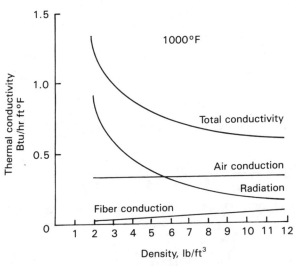

Figure 10.1 Thermal conductivity of glass fiber insulation showing the contribution of the heat transfer modes. From Greebler [1].

is composed of the conduction and radiation parts. Convection is negligible for this insulation because the fibers are closely packed, resulting in small cell sizes. As the density is increased the pore size goes down, but the solid content goes up, thus explaining the radiation and solid (fiber) conduction behavior.

10.1.4 Overall Heat Transfer

Engineers do not usually deal with the microscopic physics of heat transfer across layers of insulation. Rather, they use an effective thermal conductivity of the material that is based on the difference in temperature between the surface that is to be protected and the surroundings, and on the heat flux through the surface. The overall heat transfer coefficient, defined in Chapters 5 and 6, is normally based on an effective conductivity.

Another common way to describe the performance of a layer of insulation is to define its *resistance* to heat transfer. An analogy is made between heat flow and electric current flow so that the resulting model is a simple series thermal resistance circuit as shown in Fig. 10.2. For the system shown in Fig. 10.2,

$$Q = kA \frac{(T_w - T_{ins})}{L_{ins}} = hA(T_{ins} - T_{air}) \qquad (10.1)$$

By analogy with Ohm's law, we then define resistance, $\bar{R}_{ins}$ or $\bar{R}_{air}$, so that

$$Q = \frac{T_w - T_{ins}}{\bar{R}_{ins}} = \frac{T_{ins} - T_{air}}{\bar{R}_{air}} \qquad (10.2)$$

The resistances are then

$$\bar{R}_{ins} = \frac{L_{ins}}{k_{ins}A}$$

and

$$\bar{R}_{air} = \frac{1}{hA}$$

The outer insulation temperature, T_{ins}, is eliminated from Eq. (10.1) to give

$$Q = \frac{T_w - T_{air}}{(L_{ins}/k_{ins}A) + (1/hA)} \qquad (10.3)$$

Using the idea of an overall heat transfer coefficient, U, we write $U \equiv Q/A(T_w - T_{air})$, or

$$U = \frac{1}{(L_{ins}/k_{ins}) + (1/h)} \qquad (10.4)$$

The R Value If we call $C = k_{ins}/L_{ins}$ the *conductance* of a specified thick-

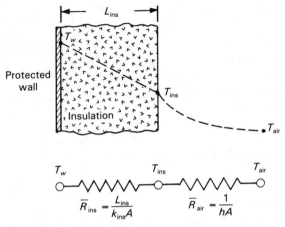

Figure 10.2 Thermal resistance circuit for plane insulation layer in contact with air.

ness of insulation, then the R value for the insulation is defined as

$$R \equiv \frac{1}{C} = \frac{L_{ins}}{k_{ins}} \qquad (10.5)$$

A 6-in.-thick blanket of fiberglass whose thermal conductivity is $k = 0.027$ Btu/hr ft °F has an R value of

$$R = \frac{0.5}{0.027} = 18.5 \text{ hr ft}^2 \text{ °F/Btu}$$

This particular combination of material and thickness is said to have an insulating effectiveness of R-19. The R system is commonly used by suppliers to categorize the effectiveness of insulation. The R value is calculated using L in feet and k in Btu's per hour per foot per degree Fahrenheit.

Many cylindrical systems require insulation for thermal protection; examples are pipes, tubes, vats, tanks, and heat exchangers. In accordance with Fig. 10.2, we write the heat transfer rate as

$$Q = \frac{A_o(T_w - T_{air})}{[r_o \ln (r_o/r_i)/k_{ins}] + 1/h} \qquad (10.6)$$

where r_o and r_i are the outer and inner radii of the insulation, A_o is the outer surface area of the insulation, and h is the heat transfer coefficient between the air and the insulation. In this case, U_o is

$$U_o = \frac{1}{[r_o \ln (r_o/r_i)/k_{ins}] + 1/h} \qquad (10.7)$$

and the R value for the insulation is

$$R = \frac{r_o \ln (r_o/r_i)}{k_{ins}} \qquad (10.8)$$

The factor $r_o \ln (r_o/r_i)$ is sometimes called the *equivalent thickness* of insulation. Use of the equivalent thickness makes it possible to write the planar and cylindrical heat transfer relations in the same form.

The R values for insulation are evaluated at the mean temperature of the fluid near the wall, $(T_w + T_{ins})/2$.

10.2 PROPERTIES AND CHARACTERISTICS OF INSULATION

The function of insulation is to retard heat transfer, so the material chosen for a given application would be insulation that has a lower k, all other factors being equal. Unfortunately, other factors are not always equal; for example, the operating temperature ranges for different insulating materials can differ greatly.

Certain properties and characteristics are useful in comparing different materials for their suitability for a particular insulation application. They are the temperature range over which the insulation must operate; the thermal conductivity; and other limiting factors, including compressive strength, fire hazard, cell structure, resistance to moisture penetration, and formability.

10.2.1 Temperature Range

The temperature range is one of the most restrictive characteristics of insulating materials. All insulation products have upper temperature limits at which they undergo serious — even damaging — changes. They may shrink and crack at high temperature. The higher the operating temperature, the more rapidly these changes occur. In these cases an upper limit of temperature is usually set below the point of rapid deterioration so that the material will maintain its performance for a reasonable service life.

Insulating materials that are made with organic binders undergo rapid degradation when a threshold temperature is reached at which the binder reacts chemically. These reactions are usually exothermic, and sometimes toxic or irritating fumes are given off. Consequently, the highest operating limit of temperature for these products is set well below the chemical reaction threshold temperature.

Lower temperature limits are not as restrictive as upper limits. However, some materials become brittle at low temperatures and are mechanically unsuitable for low-temperature applications. A far more frequent problem is vapor condensation in the regions of the insulation where the temperature is less than the dew point. Vapor barrier jackets are normally provided in these cases.

Table 10.1 is a listing of insulating materials broken into four temperature use ranges, from about 3600°F down to near absolute zero, −455°F. The high-temperature insulations are listed in ascending order of upper temperature limit.

The intermediate range includes insulation for steam systems, petrochemical processes, and thermal process equipment. Fiber-reinforced hydrous calcium silicate is perhaps the most versatile material in this group.

A variety of pipes, tanks, processing equipment, and architectural applications is covered by the low-temperature range. At temperatures below about 70°F, the principal design problem is insulation degradation caused by moisture penetration.

Finally, cryogenic insulation is used primarily to reduce boil-off of such inorganic liquids as oxygen, nitrogen, helium, and liquefied natural gas (LNG).

10.2.2 Thermal Conductivity

Once the temperature range is established and the insulating materials for a given application are identified, the effective thermal conductivity becomes the main property of interest. Figure 10.3 shows the thermal conductivity behavior as a function of temperature for several materials of practical interest. In general, k increases with temperature. The variation is almost linear, so that a mean temperature is appropriate for evaluating k.

Appendix D lists thermal conductivities for many of the more common insulating materials. These data are listed in terms of mean temperature of the tested sample.

Some materials have k values that change with time. Most notable are foam products in which a heavier-than-air gas is used to fill the cells. After manufacture, some of this gas diffuses out of the material, causing the k value to increase.

10.2.3 Other Limiting Factors

Several other characteristics can limit the applicability of various insulating materials for a particular case. Table 10.2 lists values of two of these characteristics — compressive strength and cell structure — for several insulating materials.

Compressive Strength Physical loading of insulation often occurs, sometimes by design and sometimes inadvertently by plant workers. Insulation supports and buried insulated lines involve a continuous loading of insulation that can be countered by good design.

The *compressive strength* of a system is specified as the approximate deformation that will occur under a given load. For example, in Table 10.2 for calcium silicate blocks, "100–250 at 5%" indicates a 5% volume defor-

Table 10.1   Temperature Use Range for Insulating Materials

High-temperature range (1000 to 3600°F)

MIN-K®[a]	Up to 1800°F
Mineral fiber	1000 to 1900°F
Calcium silicate	1200 to 2000°F
Metal foil systems	Up to 2500°F
Ceramic fibers	1600 to 2600°F
Ceramic refractories	2000 to 3000°F
Oxide fibers	2500 to 3000°F
Ceramic bricks	2000 to 3200°F
Carbon fiber	Up to 3500°F
Special metal foils	Up to 3500°F

Intermediate range (212 to 1000°F)

Calcium silicate
Fiberglass blankets
Rock wool (mineral fiber)
Foam glass
Diatomaceous silica
Glass fiber with binders
Asbestos fibers with binders
Magnesium carbonate
Vermiculite/perlite
Reflective metal sheets

Low-temperature range (−150 to +212°F)

Foam glass
Polyurethane foam
Urea-formaldehyde foam
Polystyrene foam (styrofoam)
Super-insulation
Fiberglass batts and blankets
Fiberglass rigid boards
Cellulose fiber
Aluminum foil/fiberglass
Rock wool (mineral fiber)
Rubber compounds

Cryogenic range (−455 to −150°F)

Vacuum wall enclosure
Multiple metal reflective foils
Opacified powders
Micro-glass spheres
Low-emissivity surfaces (silver-aluminum)
Combinations of the above known as super-insulation blankets
Polyurethane foam

[a]MIN-K® is composed primarily of fumed silica.

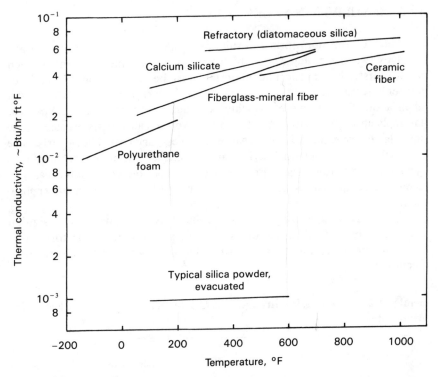

Figure 10.3 Effect of temperature on thermal conductivity for several low- to intermediate-range materials.

mation in the load range 100–250 psi. Deformations of 5 and 10% are the test results that are most commonly reported. Samples should be compared at the same deformation level, i.e., the stress required to deform the samples by the same percentages.

Cell Structure The cell structure of insulation controls the amount of moisture that the material will absorb. It also dictates the ease of vapor passage through the material. An open-cell structure is conducive to both vapor passage and moisture absorption. A closed-cell structure, however, is no guarantee that absorption will not occur. The thickness of cell walls and the type of material also play a role in the diffusion of moisture into the matrix. Vapor barriers are commonly used to prevent moisture penetration into low-temperature insulation, even for closed-cell materials.

Interested readers should consult Harrison [2] for details of fire hazard ratings and formability as well as for a comprehensive description of the physical and chemical makeup of many insulating materials.

10.3   INSULATION SELECTION

Selection of the appropriate insulation system from competing designs involves two basic decisions, which are not always independent of each other. First, the appropriate material must be selected on the basis of temperature range, thermal conductivity, and other factors that might limit application. Then the appropriate thickness must be determined for the particular application. A connection between the two decisions exists because of the interplay between thermal performance, thickness, and cost. With an expensive material of superior thermal performance, a smaller thickness is required than with a less expensive material of lower thermal performance. In this case, an economic comparison is needed to determine the best system for the application.

Figure 10.4 shows suggested steps that may be followed to design an economically optimal insulation system. Steps I and II are performed by using the properties and characteristics of insulating materials referred to in Section 10.2.

Table 10.2   Compressive Strength and Cell Structure for Selected Insulating Materials

Insulation type and form[a]	Temperature range (°F)	Compressive strength (psi) at % deformation	Cell structure (permeability and moisture absorption)
Calcium silicate blocks, shapes, and P/C	to 1500	100–250 at 5%	Open cell
Glass fiber blankets	to 1200	0.02–3.5 at 10%	Open cell
Glass fiber boards	to 1000		
Glass fiber pipe covering	to 850		
Mineral filter blocks and P/C	to 1900	1–18 at 10%	Open cell
Cellular glass blocks and P/C	−450 to 900	100 at 5%	Closed cell
Expanded perlite blocks, shapes, P/C	to 1500	90 at 5%	Open cell
Urethane foam blocks and P/C	(−100 to −450) to 225	16–75 at 10%	95% closed cell
Isocyanurate foam blocks and P/C	to 350	17–25 at 10%	93% closed cell
Phenolic foam P/C	−40 to 250	13–22 at 10%	Open cell
Elastomeric closed cell sheets and P/C	−40 to 220	40 at 10%	Closed cell
MIN-K blocks and blankets	to 1800	100–190 at 8%	Open cell
Ceramic fiber blankets	to 2600	0.5–1 at 10%	Open cell

[a]P/C, pipe covering.

Source: Harrison [2]. Copyright 1982 by John Wiley & Sons, New York. Reprinted with permission.

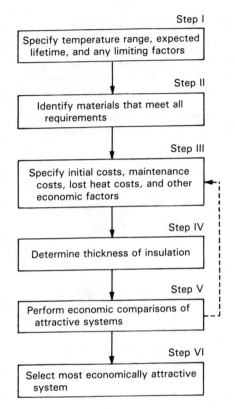

Step I
Specify temperature range, expected lifetime, and any limiting factors

Step II
Identify materials that meet all requirements

Step III
Specify initial costs, maintenance costs, lost heat costs, and other economic factors

Step IV
Determine thickness of insulation

Step V
Perform economic comparisons of attractive systems

Step VI
Select most economically attractive system

Figure 10.4 Procedure for economically selecting an insulation system.

Steps III to V bring the cost of the project into consideration. The dashed line from step V back to step III indicates that comparisons may eventually be required to reach the optimal economic decision, step VI.

The following example illustrates the use of the procedure sketched in Fig. 10.4 as well as some of the evaluative techniques.

Example 10.1   Selection of Insulation

Insulation is to be selected to cover a 6-in. pipe carrying steam at 600°F. The pipe is in a high-traffic area where loads can be placed upon it. Consequently, the compressive strength specification is that deformation must be limited to 5% under a 200-psi load. The heat transfer coefficient, h, is 200 Btu/hr ft^2 °F. A 90% reduction of the bare-pipe heat loss, 3500 Btu/hr ft^2 °F, is desired. Bids have been received for three possible insulations: fiberglass, calcium silicate, and diatomaceous silica. Which system should be selected? Use Fig. 10.4 as a guide for the selection process.

Solution: Insulation creates a large temperature difference between the pipe and the outer surface of the insulation. Consequently, if T_{ins} is unknown, it is commonly assumed to be equal to the air temperature. For this case,

$$T_m = \frac{T_w + T_{ins}}{2} = 335°F$$

Step I: The temperature range and compressive strength requirement are known. The expected lifetime is 15–30 years.

Step II: Figure 10.3 shows that all of the materials will function within the temperature range. But Table 10.2 shows that a fiberglass pipe covering does not meet the strength requirement. This leaves calcium silicate and diatomaceous silica for consideration. Information from the manufacturer of diatomaceous silica indicates that it will meet the compressive strength criterion.

Step III: The heat loss is the same for both systems, since a 90% reduction of heat loss is specified. The expected capital recovery rate is 22%. Other economic factors are tabulated below:

Factor	Calcium silicate	Diatomaceous silica
Lifetime	15 years	25 years
Annual costs:		
Taxes and insurance	0.02 × (first cost)	0.015 × (first cost)
Maintenance	First year, 0.03 of first cost, increasing 0.01 of first cost per year	Zero
First cost: (materials and installation)	$3.50 per inch of thickness/ linear foot	$3.75 per inch of thickness/ linear foot
Salvage value	None	None

Step IV: Determine thickness. In accordance with Eq. (10.7)

$$\frac{Q}{L} = \frac{2\pi r_{\text{ins}}(T_w - T_{\text{air}})}{[r_{\text{ins}} \ln (r_{\text{ins}}/r_p)/k_{\text{ins}}] + 1/h}$$

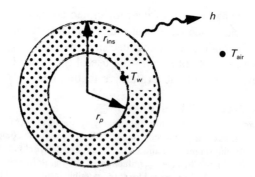

Rearranging, we get

$$\ln \frac{r_{\text{ins}}}{r_p} = k_{\text{ins}} \left(\frac{2\pi \, \Delta T}{Q/L} - \frac{1}{r_{\text{ins}} h} \right)$$

For most insulation applications, $R_{\text{ins}} \gg R_{\text{air}}$. Assuming that this is valid, we may neglect the term $1/r_{\text{ins}} h$. Then we get

$$\ln \frac{r_{ins}}{r_p} = k_{ins} \frac{2\pi(530\,°F)}{350 \text{ Btu/hr ft}} = 9.51 k_{ins} \text{ hr ft °F/Btu}$$

From Fig. 10.3

$$k_{c.s.} = 0.04167 \text{ Btu/hr ft °F}$$

and

$$k_{d.s.} = 0.0563 \text{ Btu/hr ft °F}$$
$$r_{ins\ c.s.} = 4.46 \text{ in.} \qquad \Delta r = 1.46 \text{ in.}$$
$$r_{ins\ d.s.} = 5.12 \text{ in.} \qquad \Delta r = 2.12 \text{ in.}$$

Checking $1/r_{ins}\,h$, we see that the assumption that it was negligible was appropriate.

$$\frac{1}{r_{ins}h} = \frac{1}{(4.46/12)/(200)} = 0.0135 \ll 9.51$$

Step V: Use the annualized cost technique (see Example 3.5).
Calcium silicate: $n = 15$, $i = 0.22$
AC = (tax and insurance) + (maintenance) + (first cost)$(A/P, i, n)$

$$(A/P, i, n) = \frac{i(1 + i)^n}{(1 + i)^n - 1} = 0.232$$

First cost: $3.50(1.46) = $5.11/linear foot

$$AC = 0.02(5.11) + 5.11 \underbrace{\frac{\displaystyle\sum_{n=1}^{15} (0.02 + 0.01n)}{15}}_{\text{maintenance costs}} + 5.11(0.232)$$

$$= \$1.80/\text{linear foot}$$

Diatomaceous silica: $n = 15$, $i = 0.22$
First cost: $3.75(2.12) = $7.95/linear foot

$$AC = 0.015(7.95) + (7.95)(0.232)$$

$$= \$1.96/\text{linear foot}$$

For $n = 25$ years, $i = 0.22$

$$(A/P, i, n) = 0.221$$

$$AC = 0.015(7.95) + 7.95(0.221) = \$1.88/\text{linear foot}$$

Step VI: The choice in this case is the calcium silicate, although it has a shorter lifetime and requires more maintenance.

10.3.1   Calculation of Insulation Thickness

The calculation of insulation thickness is based on Eq. (10.3) in the form

$$\frac{Q}{A} = \frac{T_w - T_{air}}{R_{ins} + R_{air}} \tag{10.9}$$

where the symbols are as defined previously. The equivalent thickness is required if Eq. (10.9) is used for cylindrical systems. Equation (10.9) may be solved for the heat loss for a given thickness of insulation or, in some cases, for the required thickness of insulation for a specified heat loss, as illustrated in Example 10.1.

The following example shows typical calculations for heat loss across a layer of insulation.

Example 10.2   Heat Loss Across a Fixed Insulation Thickness

A large rectangular steel tank contains 400°F pork fat, which is being rendered into oil. A 3-in.-thick blanket of 85% magnesia insulation is applied to the surfaces of the tank. What is the heat loss to 70°F air, per square foot of the tank? Assume that the average heat transfer coefficient to the air is 250 Btu/hr ft °F.

Solution: The mean temperature is approximately

$$T_m = \frac{T_w + T_{air}}{2} = 235°F$$

From Appendix D, k_{ins} = 0.043 Btu/hr ft °F. In accordance with Eq. (10.9),

$$\frac{Q}{A} = \frac{T_w - T_{air}}{R_{ins} + R_{air}} = \frac{(400 - 70)°F}{(3/12) \text{ ft}/0.043 \text{ Btu/hr ft °F} + 1/250 \text{ Btu/hr ft}^2 \text{ °F}}$$

$$= 56.72 \text{ Btu/hr ft}^2 \text{ °F}$$

Check T_{ins}:

$$\frac{Q}{A} = k_{ins} \frac{T_w - T_{ins}}{\Delta x}$$

$$T_{ins} = T_w \frac{(Q/A)(\Delta x)}{k_{ins}} \cong 70°F$$

Therefore, T_m = 235°F is appropriate.

The outer temperature of the insulation must be limited to safe values in cases of possible contact by plant personnel. The temperature accepted as safe for personnel contact is usually 140°F. If the outer insulation temperature is above 140°F, additional insulation must be added, or the proper thickness can be calculated for a surface temperature of 140°F. Example 10.3 illustrates this.

Example 10.3   Safe Outer Insulation Temperature

A horizontal 12-in. schedule 120 steel pipe carries superheated steam at 1000°F. What thickness of diatomaceous silica refractory material is required to reduce the outer temperature to 140°F?

Solution: We must use an energy balance to relate T_{ins} and D.

$$\frac{Q}{L} = \frac{\pi D(T_w - T_{ins})}{(D/2k_{ins})\ln(D/D_w)} = \pi Dh(T_{ins} - T_{air})$$

In this case h is a function of insulation diameter D. We use Eq. (5.72) from Chapter 5 for the free convection loss from the insulation:

$$h = \frac{k_{air}}{D}\left(0.60 + 0.387\left\{\frac{Ra_D}{[1 + (0.559/Pr)^{9/16}]^{16/9}}\right\}^{1/6}\right)^2$$

where

$$Ra = \frac{g\beta(T_{ins} - T_{air})\, D^3}{v\alpha}$$

for a horizontal cylinder. For this case β is $1/T_{air}$; other air properties are evaluated at $T_m = (140 + 70)/2 = 105°F$. From Appendix F we get

$$v = 0.658 \text{ ft}^2/\text{hr} \qquad Pr = 0.705$$

$$k = 0.0158 \text{ Btu/hr ft °F} \qquad \alpha = 0.941 \text{ ft}^2/\text{hr}$$

Using these, we obtain

$$Ra = 8.91 \times 10^7\, D^3$$

where D is in feet. Then

$$h = \frac{k_{air}}{D}(0.60 + 6.786\, D^{1/2})^2$$

and

$$\frac{1}{\ln(D/D_w)} = \frac{1}{2}\frac{k_{air}}{k_{ins}}\left(\frac{T_{ins} - T_{air}}{T_w - T_{ins}}\right)(0.60 + 6.786\, D^{1/2})^2$$

To solve for D, let us call

$$\frac{1}{\ln(D_1/D_w)} = fn_1$$

and

$$\frac{1}{2}\frac{k_{air}}{k_{ins}}\left(\frac{T_{ins} - T_{air}}{T_w - T_{ins}}\right)(0.60 + 6.786\, D^{1/2})^2 = fn_2$$

For the insulation, $T_m = (1000 + 140)/2 = 570$, which gives $k_{ins} = 0.06$ Btu/hr ft °F.

Then

$$fn_2 = \frac{1}{2}\left(\frac{0.0158}{0.06}\right)\left(\frac{70}{860}\right)(0.60 + 6.786\, D^{1/2})^2$$

$$= 0.0107(0.60 + 6.786\, D^{1/2})^2$$

Now we plot fn_1 and fn_2 against D as shown in the illustration.

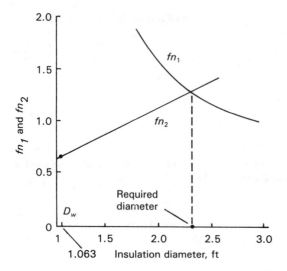

The required diameter is 2.32 ft. So the required thickness to keep the surface at 140°F is 7.54 in.

Condensation control is also possible if the temperature of the insulation is kept above the dew point, the temperature at which moisture condenses from ambient air. In this case the dew point is determined by the methods described in Section 5.1.7, and the thickness required to keep the temperature greater than the dew point at the outer edge of the insulation is calculated.

10.3.2 Economics of Insulation Selection

A project involving the application of insulation can be treated with the project investment analysis techniques described in Chapter 3. The construction of a cash flow diagram is the best way to visualize the projected investment. Initial costs, maintenance costs, reduced energy costs, the time value of money, and the expected rate of return on invested capital are all needed for a thorough economic analysis. Tax rates and credits should also be considered.

Insulation Costs The cost of insulation includes purchase price, installation and labor costs, and maintenance costs. All of these may vary from location to location and for different types of insulation.

Lost Heat Costs Reduced energy costs are related to the dollar value of the Btu's of heat that would be lost without insulation. The value of the lost heat should include the cost of the raw fuel that supplies it, taken along with the inefficiencies of the conversion process. For example, generation of steam in a natural gas boiler is accompanied by a conversion efficiency on the order of 80–90% at best. Chapter 7 includes a description of how to calculate the dollar value of steam.

The cost of the heat generation plant should also be included in the analysis. Saving energy allows a smaller plant to serve the same purpose. One way of including this effect in the analysis is to assign an incremental cost to an increase in plant capacity, stated as dollars per 1000 Btu's per hour. The use of this factor accounts for a reduction in required plant capacity by a well-insulated system.

10.4 ECONOMICAL THICKNESS OF INSULATION (skip)

The preceding sections dealt with thickness calculations for insulation and with economic factors that influence the selection of an insulation system. Conventional economic techniques can be used to determine which of a proposed set of alternatives is the most desirable. However, this does not necessarily reveal whether or not any of the proposed systems was based on an optimal thickness of insulation.

The economical thickness of insulation (ETI) is defined as the thickness of insulation for which the cost of the next increment of insulation is just balanced by increased energy savings over the life of the project. The situation is shown graphically in Fig. 10.5 as the thickness at which the total cost is a minimum.

McMillan [3] first developed the equations used for the calculation of economical thickness for both flat and cylindrical surfaces. McMillan's technique required a large number of charts to properly represent all the operating and financial variables. The procedure was quite cumbersome and lengthy.

A committee sponsored by NIMA[1] developed a manual for the determination of economical thicknesses, which was published in 1961 [4]. The laborious use of charts was shortened in this manual by the use of a smaller number of nomographs and charts. Still, the procedure is tedious and, in fact, has been supplanted in large part by the use of computer programs to solve the equations that fix the economical thickness.

The most recent set of nomographs was presented in a government document [5], accompanied by a 12-step procedure for determining economical thickness. The reader is urged to use computer programs devel-

---

[1]National Insulation Manufacturers Association; now known as TIMA, the Thermal Insulation Manufacturers Association.

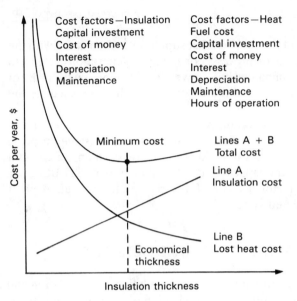

Cost factors—Insulation
Capital investment
Cost of money
Interest
Depreciation
Maintenance

Cost factors—Heat
Fuel cost
Capital investment
Cost of money
Interest
Depreciation
Maintenance
Hours of operation

Minimum cost

Lines A + B
Total cost

Line A
Insulation cost

Line B
Lost heat cost

Economical
thickness

Cost per year, $

Insulation thickness

Figure 10.5 Illustration of the economical thickness of insulation.

Table 10.3   Economical Thickness Worksheet

1. Calculate mean insulation temperature: T_m = _____ °F
   $T_m = (T_w + T_{air})/2$
2. Enter the thermal conductivity: k = _____ Btu in./hr ft^2 °F
3. Calculate the temperature difference:
   $(T_w - T_{air})$ ΔT = _____ °F
4. Enter annual hours of operation: _____ hr
5. Find D_s for flat surface or D_p for pipe D_s = _____
   using Fig. 10.6: D_p = _____
6. Determine B_3 using Figs. 10.7 and 10.8, i_3 = 0._____
   using i_3, money cost for insulation, n_1 = _____ years
   and n_1, insulation project life: B_3 = _____
7. Determine m_{c1}: m_{c1} = _____
8. Find Z_s for flat surface or Z_p for pipe Z_s = _____
   using Fig. 10.9: Z_p = _____
9. Calculate kR_s (R_s value of 0.7 is typical): kR_s = _____
10. Use Fig. 10.10 to determine ETI: w_1 = _____
11. If the economical thickness found in step 10 is within the single-layer
    range (corresponding to the single-layer slope, m_{c1} used in step 7), the
    thickness is correct. If the thickness is beyond the single-layer range,
    repeat the procedure from step 7 on, using the double-layer slope, m_{c2}: m_{c2} = _____
12. If the economical thickness with the double-layer slope is in the triple-
    layer range, repeat the procedure from step 7 on, using the triple-layer
    slope: m_{c3} = _____

Note: The ETI found with single-layer slope might fall in the double-layer realm and the subsequent
thickness found with a double-layer slope might fall in the single-layer realm; or the ETI found with
a double-layer slope might fall in the triple-layer realm and the subsequent thickness found with a
triple-layer slope might fall in the double-layer realm. Should either of these conditions occur, the
proper choice is the thickest single or thickest double layer, respectively.

oped under the sponsorship of TIMA, but he or she can gain insight into the procedures from the nomograph methods. Accordingly, the 12-step worksheet is shown in Table 10.3 and the nomographs and charts typically needed to complete the worksheet are illustrated in Figs. 10.6–10.10. Figure 10.10 is for the pipe size 16-in. outer diameter; similar plots for pipe sizes from 1/2- to 36-in. OD are given in [5]. A glossary of the terms appearing in the 12-step procedure is included at the end of the chapter.

The following example illustrates the nomograph procedure for determining economical thickness of insulation.

Example 10.4   Economical Thickness of Insulation   (will not be given)

Find the economical thickness of calcium silicate for a nominal 16-in. pipe carrying steam at 300°F. The cost of money is 20% and the value of energy is $3.50 per million Btu's. The expected lifetime for the insulation is 15 years. The following table gives the estimates for various installed insulation thickness.

Thickness, L	P, $/linear foot
2 in., single layer, L_1	P_1, 12.00
4 in., single layer, L_2	P_2, 17.50
5 in., double layer, L_3	P_3, 28.00
7 in., double layer, L_4	P_4, 37.00

Solution: To illustrate the procedure, we follow Table 10.3.

1. $T_m = (T_w + T_{air})/2 = 370/2 = 185°F$.
2. $k_m = 0.42$ Btu in./hr ft^2 °F; we use these units to conform to the ETI charts.
3. $\Delta T = 300 - 70 = 230°F$.
4. Annual hours of operation = 8760.
5. We now use Fig. 10.6 to find D_p. The dashed lines in the figure are the result of the parameters given in this example. The ΔT is connected to the thermal conductivity, $A \leftrightarrow A$. Likewise, the cost of energy and the hours of use are connected, $B \leftrightarrow B$. These two points are connected to give $D_p = 1.6$ along line C.
6. Figure 10.7 gives the compound interest factor $(1 + i)^n = 15.4$. It could also be quickly calculated; $i = 0.20, n = 15; (1.2)^{15} = 15.4$. Figure 10.8 yields the amortization factor for the insulation, B_3. This factor is equivalent to the capital recovery factor of Table 3.5, $(A/P, i, n)$. It could also be calculated without the use of Fig. 10.8. From Fig. 10.8, B_3 is 0.22.
7. The incremental cost of installed insulation, m_c, is given by

$$m_c = \frac{\Delta P}{\Delta L} \qquad m_{c1} = \frac{17.50 - 12.00}{4 - 2} = \$2.75 \text{ per inch/linear foot}$$

8. Figure 10.9 gives the Z_p factor. It is found by first connecting B_3 to D_p (A_3 to A_2) extended to line B. Then the m_{c1} value (A_1) is extended to the Z_p line, C, by passing through the B point just established. $Z_p = 2.8$.
9. $kR_s = 0.42(0.7) = 0.294$.

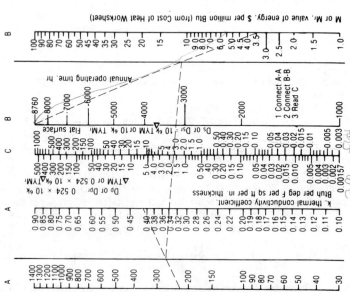

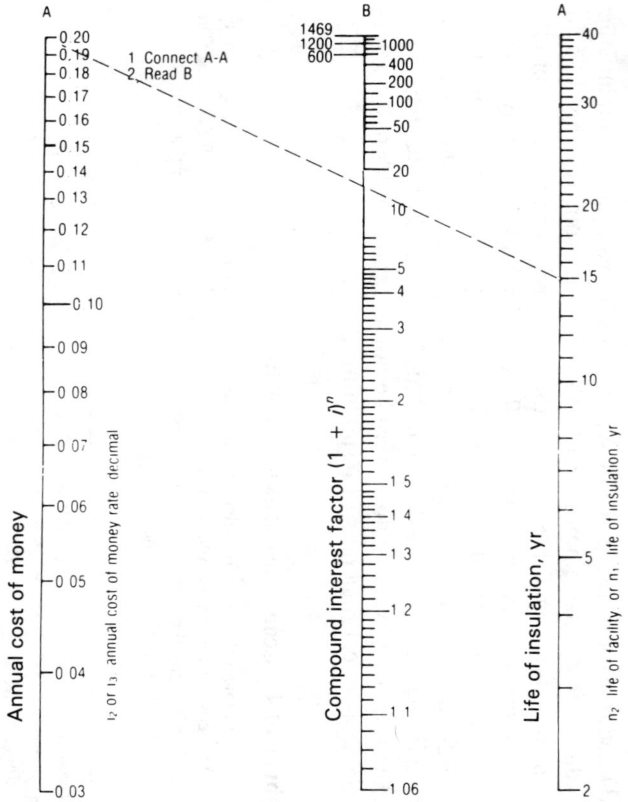

Figure 10.7 Compound interest factor for use with Fig. 10.8.

Figure 10.6 Annual cost of heat lost or gained.

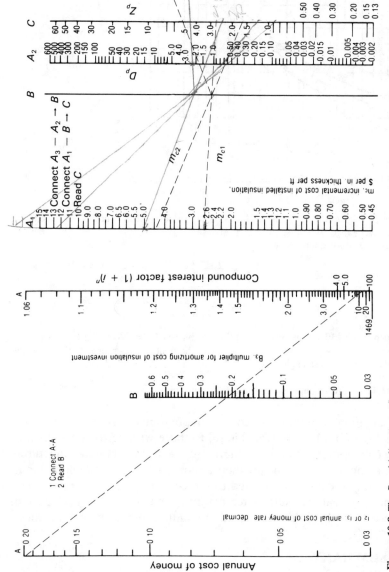

Figure 10.9 Z_p factor for round (pipe) surfaces; Z_p includes 10% annual maintenance charge for insulation.

Figure 10.8 The B multiplier $(1 + i)^n$, for amortizing an initial cost over a period of time with equal increments.

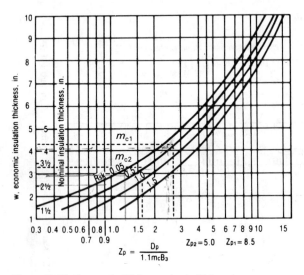

Figure 10.10 Economic thickness chart, 16-in. pipe.

10. Figure 10.10 gives the economic thickness, as shown, $w = 4.3$ in. But this lies outside the single-layer range. We must go to Step 11.
11. Calculate m_{c2}; for double layers

$$m_{c2} = \frac{P_4 - P_3}{L_4 - L_3}$$

$$= \frac{37 - 28}{2} = \$4.50$$

Return to Fig. 10.9. The m_{c2} dashed line establishes $D_p = 1.6$. Using Fig. 10.10 with $D_{p2} = 1.6$ gives $w_2 = 3.3$ in. This is less than the double-layer range. In this case, the choice is the thickest single layer.

The ETI for this example is 4 in. of calcium silicate.

As mentioned previously, computer programs developed under the sponsorship of TIMA are available [6] to those wishing to adapt them to their own computer systems. Furthermore, most insulation manufacturers will run a computer analysis for their customers through telephone input, with the output data being mailed to the customers as hard copy. These programs give heat loss, surface temperature, annual cost, payback period, and present worth of the heat saved in addition to economical thickness.

10.5 SUMMARY

Prevention of thermal losses from systems by applying insulating material is one of the first things an energy manager should investigate. Most surfaces in this era of expensive energy will already have some insulation, but it is worthwhile to examine the performance of such installations to see if additional (or more efficient) insulation might pay dividends.

We have presented some of the basic concepts of insulating materials in this chapter as well as brief discussions of some of the important properties other than thermal conductivity. The general procedure for selection of insulation is also presented.

GLOSSARY

B_3	annual amortization factor for the cost of insulation
D_p	annual cost of heat lost through 1 linear foot of pipe insulation
D_{pr}	annual cost of heat gain through 1 linear foot of pipe insulation
D_s	annual cost of heat lost through 1 ft^2 of flat surface insulation
D_{sr}	annual cost of heat gained through 1 ft^2 of flat surface insulation
i_3	cost-of-insulation rate or required rate of return on the last increment of insulation
M	average annual cost of energy, \$/10^6 Btu
m_{c1}	incremental cost of installed insulation, single layer, \$/in./ft
m_{c2}	incremental cost of installed insulation, double layer, \$/in./ft
m_{c3}	incremental cost of installed insulation, triple layer, \$/in./ft
n_1	term of insulation project, years
R_s	surface resistivity of the insulation system, hr °F ft^2/Btu
ΔT	$T_w - T_{air}$, difference between pipe or flat-surface temperature and ambient temperature, °F
T_{air}	average ambient temperature, °F (design temperature for application)
T_w	process temperature, °F (temperature of surface to be insulated)
w_1	economic thickness of insulation, single-layer application, in.
w_2	economic thickness of insulation, double-layer application, in.
w_3	economic thickness of insulation, triple-layer application, in.
Z	factor used in economic thickness determination, $Z = D/1.1m_cB_3$

REFERENCES

1. P. Greebler, Thermal Properties and Applications of High Temperature Aircraft Insulation, *Jet Propulsion,* American Rocket Society, 1954.
2. M. R. Harrison, Industrial Insulation, chapter 15 in *Energy Management Handbook,* W. C. Turner (ed.), Wiley, New York, 1982.
3. L. B. McMillan, Heat Transfer through Insulation in the Moderate and High Temperature Fields: A Statement of Existing Data, Paper 2034, American Society of Mechanical Engineers, New York, 1934.
4. Anonymous, *How to Determine Economic Thickness of Insulation,* NIMA Manual, 1961. See also *Handbook of Fundamentals,* p. 298, ASHRAE, New York, 1972.

5. Anonymous, *Economic Thickness of Industrial Insulation,* Conservation Paper 46, Federal Energy Administration, Washington, D.C., 1976. Stock No. 041-018-00115-0, U.S. Government Printing Office, Washington, D.C.
6. Anonymous, TIMA ETI Computer Program, Thermal Insulation Manufacturers Association, Mt. Kisco, N.Y.

PROBLEMS

10.1 A large cylindrical steel tank is used to collect steam condensate at 250 psia. The tank is 5 ft in diameter and 20 ft long. Calculate the reduction in heat loss in Btu's per hour if 2 in. of calcium silicate is installed on the tank. The surrounding air is still and at 70°F.

10.2 A nominal 5-in. schedule 40 wrought iron pipe is covered with 2 in. of 85% magnesia insulation. The pipe carries superheated steam at 750°F. The air temperature is 70°F. The inside and outside coefficients are 250 and 2.5 Btu/hr ft^2 °F, respectively. Determine the rate of heat loss per foot of pipe length, and the temperatures of the inside and outside pipe surfaces and the outside of the insulation.

10.3 Heavy crude oil at 200°F is pumped from a storage tank to a refinery holding vessel through a horizontal 8-in. nominal schedule 20 pipe. It is desired to apply enough fiberglass insulation so that the outer surface is safe for human contact. The average air velocity is 10 mph and the average air temperature is 70°F. What thickness of insulation is required? Round your answer off to the next highest 1/2-in. increment of thickness.

10.4 A 1-in. copper tube carries saturated liquid ammonia at 30 psia. How much polyurethane foam is required to reduce the heat gain of the ammonia by 90%? Air temperature is 70°F. Should a vapor barrier be used to prevent condensation?

10.5 Insulation is to be placed on a 1/2-in. OD line carrying liquid nitrogen vapor at –275°F. Estimate the thickness of 29.3 lb$_m$/ft^3 asbestos required to reduce the heat gain by 99%. The air temperature is 70°F.

10.6 Calculate the difference in installed costs between 2.5 and 3.0 in. of calcium silicate insulation on an 8-in. pipe 1000 ft long if the respective installed costs are: 2.5 in., $20.35/linear foot; 3.0 in., $22.95/linear foot. Assuming that energy is valued at $6/10^6 Btu and that the energy saved by adding the extra 0.5 in. of insulation is 20 Btu/hr per linear foot, how long would it take to pay for the extra insulation?

10.7 Rework Example 10.4 assuming that the cost of money is 14% and the value of energy is $4.50 per million Btu's. The other parameters remain the same.

10.8 A 6-in. pipe carrying steam at 650°F is to be covered with insulation. No loads will be placed on the insulation so that compressive strength is not a factor. A reduction of bare-pipe heat loss, 4000 Btu/hr ft, by 95% is desired. The outside heat transfer coefficient is estimated to be 175 Btu/hr ft^2 °F. Bids for installed costs have been received for two types of insulation: fiberglass and calcium silicate. When plotted as dollars per inch of insulation, the bids compare as shown in Fig. 10.8P. Table 10.8P gives other pertinent economic information. Which insulation should be chosen on the basis of economics? The insulation comes in 1/2-in. increments (2.0, 2.5, 3.0, etc.).

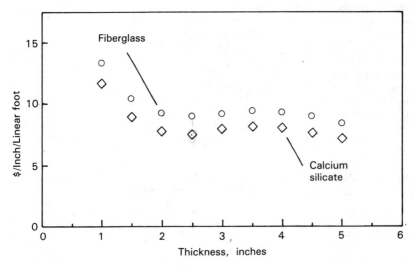

Figure 10.8P Comparative installed costs of two insulation systems.

Table 10.8P   Economic Factors

	Calcium silicate	Fiberglass
Lifetime	15 years	10 years
Annual costs:		
Taxes and insurance	0.02 × (installed costs)	0.025 × (installed costs)
Maintenance	First year, 0.03 × (installed costs), increasing by 0.01 of installed costs per year	First year, 0.04 × (installed costs), increasing by 0.01 of installed costs per year
Installed costs	See graph	See graph
Salvage value	Zero	Zero
Interest rate	20%	20%

10.9 A steel pipe of inside diameter 4 in. and wall thickness 0.25 in. is at 200°F. Two inches of asbestos insulation (ρ = 36.0 lb_m/ft^3) is added to the pipe. If the outer insulation surface is at 130°F, how much heat is lost from the pipe? If the coefficient at the outside of the insulation was 200 Btu/hr ft^2 °F, what would the temperature be at the outside of the insulation?

10.10 Find the economical thickness of insulation (ETI) for diatomaceous silica for a nominal 16-in. pipe carrying steam at 600°F. The cost of money is 22% and the value of energy is $4.50/$10^6$ Btu. The expected lifetime of the insulation is 20 years. Table 10.10P gives the cost estimates for installed insulation thickness.

Table 10.10P   Installed Insulation Costs

Thickness, L	$/linear foot, P
L_1 — 2 in., single layer	P_1 — \$30.00
L_2 — 3 in., single layer	P_2 — \$35.00
L_3 — 4 in., single layer	P_3 — \$55.00
L_4 — 6 in., double layer	P_4 — \$75.00
L_5 — 8 in., double layer	P_5 — \$110.00
L_6 — 9 in., triple layer	P_6 — \$125.00

10.11 Rework Problem 10.7, using the installed costs shown in Table 10.10P, but for a cost of money of 15%, a value of energy of \$5.50/10^6 Btu, and a lifetime of 15 years.

ELEVEN

ENERGY CONSERVATION IN INDUSTRIAL BUILDINGS

A significant fraction of the energy used in the world is consumed in the maintenance of buildings. In the United States, about 25% of the annual consumption of energy is for this purpose.

Energy consumption for industrial processes in a plant usually overshadows the energy used for space conditioning and lighting. Nonetheless, inefficient lighting and air-conditioning can cast doubts on the efforts of energy-conscious companies to conserve energy. Attention to this aspect of energy usage in industrial settings is warranted.

Adequate lighting must be provided for plant personnel, and the comfort levels of plant personnel must be maintained to allow maximum work efficiency. Certain temperature and humidity levels must be maintained for some machines operating in a plant. In heavy industries, such as foundries, glass plants, and stone and concrete plants, little attention is paid to comfort or to climate control for machines. However, even in heavy industries, ventilation is often required to remove fumes and debris for worker safety, and lighting of workstations might also be required. The basic energy needs of the plant enclosure must be accounted for in energy conservation studies.

Certain factors tend to set industrial plants and buildings apart from residential and commercial buildings. An industrial plant usually has streams of hot and cold fluids circulating throughout. These can be used to preheat air, to heat domestic hot water, to control humidity, to reheat cold air streams, and for many other functions. A commercial building does not have such systems, and additional equipment is usually required to perform these functions.

Both the art and the science of HVAC (heating, ventilating, and air-conditioning) are well developed. Design techniques and the required technical data are abundantly documented (e.g., see reference [1]). The relationship between HVAC design and increasing energy costs has been extensively studied since the 1974 oil embargo (e.g., see references [2–4]). Most of the publications cited here deal with commercial buildings, but many of the proposed strategies of energy conservation apply to industrial plants as well. Furthermore, the management of HVAC systems already in place with the goal of minimizing energy usage has received considerable attention [5]. Efficient use of lighting has also been addressed; significant advances have been made in this field [6].

This chapter emphasizes the special character of industrial plants rather than providing a comprehensive treatment of HVAC design. References to the copious literature in the HVAC field are made when appropriate. More in-depth treatments of the potential for energy savings in particular aspects of plant operation, such as ventilation and lighting, are given.

11.1 ENERGY LOADS IN BUILDINGS

HVAC systems maintain the interior of buildings at levels of temperature and humidity required for human comfort and in some cases for machine "comfort." Some machines, notably electronic computers, must be maintained at certain temperature levels to operate efficiently.

The HVAC system provides the necessary energy transfer from the building interior to offset the existing head loads for the building. The sketch in Fig. 11.1 shows the main components in the energy balance for an industrial building. The double-headed arrows indicate that heat must be added or taken away from the building, depending on the season, the size of the building, and other factors. In winter, heat is lost through the building structure and air infiltration causes a net heat loss. Heat may be required by the HVAC system to offset these heat losses. Heat is always generated by lighting, by building inhabitants, and by machines such as electric motors. Processes that operate at temperatures higher than building ambient temperatures also impose a heat gain.

Two types of heat load — sensible and latent heat — control the design of HVAC systems. Sensible loads are caused by temperature changes of the air moving through a building. They are described by the equation $\dot{Q} = (\dot{m}c_p\,\Delta T)_{\text{air}}$, where $\dot{m}$ is the air circulation rate, c_p is the heat capacity, and ΔT is the temperature rise of the air. The latent load owes its existence to

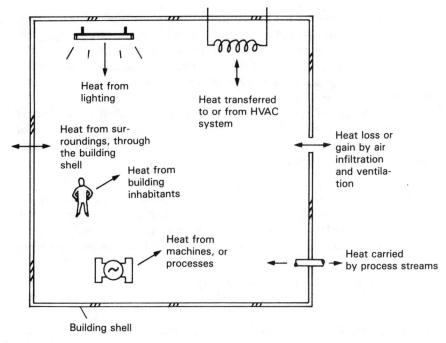

Figure 11.1 Diagram showing the components of a heat balance for an industrial building.

the moisture that is added to the air by the inhabitants' perspiration, process evaporation, and/or air infiltration. Latent loads are described by $\dot{Q} = \dot{m}h_{fg}$, where h_{fg} is the latent heat of the moisture and $\dot{m}$ is the moisture that must be removed from the air.

11.1.1 Estimation of Energy Usage

Annual energy usage in a building can be estimated in two ways, by the "top-down" or the "bottom-up" method. The bottom-up approach is the more theoretical of the two methods and can be performed by using the procedures detailed in the *ASHRAE Handbook—Systems* [7].[1] The bottom-up approach is based on the calculated heat loads and on the rated capacity of all energy-using equipment in the building. An inventory of the heat loads and equipment is made, and the total heat load is found by multiplying the loads by the hours of usage.

The top-down method is based on actual measured data obtained from utility bills, meter readings, or other sources. The fraction of the total load consumed by each component of the building is then estimated, or in some cases the loads of individual components are metered.

[1]ASHRAE, American Society of Heating, Refrigeration, and Air-Conditioning Engineers.

Results obtained by the bottom-up and top-down approaches seldom agree, because loads do not always correspond to nameplate ratings. The bottom-up approach usually overestimates the load and tends to mask the actual energy use of individual components. The top-down approach is preferred in energy auditing and conservation work because it is based on factual metered data.

The accuracy of either of the two methods described above is only 10–20% at best. There is little justification for attempting to calculate heat loads with greater accuracy. Time is better spent in attempting to identify opportunities for energy conservation in the plant.

Structural Heat Loads. The heat loads for structural elements of the building, such as walls, windows, roofs, and ceilings, are characterized in terms of so-called U factors. Students of heat transfer recognize the U factor as the overall heat transfer coefficient or thermal transmittance, given by

$$U = \frac{1}{\Sigma_j R_j} \tag{11.1}$$

where R_j is the thermal resistance[2] of the jth component. The heat load is

$$\dot{Q} = UA\,\Delta T \tag{11.2}$$

where A is the exposed area and ΔT is the temperature difference between inside and outside air.

Lighting Loads. Energy used for lighting is an additional heat load in a building. In winter this can be used to advantage if the heat is distributed to the parts of the plant building where it is needed. In summer it is another component of heat load that must be removed. Lighting is covered in Section 11.6.

Inhabitant Loads. Table 11.1 shows the rates of heat gain associated with people engaged in different levels of physical activity.

In industrial plants where few people are employed, this type of heat gain may be almost negligible. For example, a 2½-horsepower motor operating at 75% efficiency will generate about 1500 Btu/hr, which is the equivalent of one adult engaged in heavy work. In people-intensive, light manufacturing work, such as production-line assembly, the inhabitant heat gain is a much higher fraction of total heat gain than it is in heavy industry.

[2]See Section 10.1.4 for the definition of thermal resistance.

Table 11.1 Heat Gain from Adult Human Physical Activity

Activity	Total heat rate (Btu/hr)	Sensible heat (Btu/hr)	Latent heat (Btu/hr)
Seated at rest	350	195	155
Seated, light work	400	195	205
Standing, light work	450	200	250
Factory:			
Light bench work	750	220	530
Moderately heavy work	1000	300	700
Heavy work	1450	465	985

Source: Adapted from ASHRAE Guide and Data Book — Fundamentals 1977.

11.2 REDUCTION IN PLANT ENERGY USAGE

Until recently, concerted efforts toward energy conservation in industry were not economical. Fuel and electricity were cheap, and the economic emphasis in plant and process design was on minimizing costs for labor and raw materials. Now effective energy conservation programs geared to industries, large and small, result in large energy and dollar savings.

Chapter 4 dealt with the management of energy conservation programs and suggested ways in which programs might be organized and carried out. Successful programs must have the backing of top management. Furthermore, the engineer in charge must have a positive attitude toward energy conservation. Goals must be set, and records kept to show that the program is working. Education of all plant personnel to acquaint them with the need for and manner of conserving energy is imperative.

Reporting department-by-department reductions in energy usage, rather than a plantwide figure, creates pride among employees and brings peer pressure to bear on them. Individual metering for departments and/or large energy-consuming equipment provides feedback that is a valuable motivation for successful energy conservation.

An energy audit reveals the opportunities for energy conservation. Heat recovery techniques, thermal insulation, fine-tuning of combustion controls, and electrical load management can then be implemented for energy conservation. These topics are covered in other chapters in this text and are not discussed further here.

Some other aspects of plant operation tend to be overlooked, although their payback time for energy conservation projects might make them quite attractive. Systems such as lighting, ventilation, hot water, and compressed air are susceptible to reduction of energy use.

11.2.1 Short-Term Possibilities

Table 11.2 suggests ways in which energy conservation can be rapidly achieved for the above systems and others in a plant. From this list, it should be clear that the first axiom of industrial energy conservation is: *Shut it off, when you can.* Of course, there are situations in which more energy would be required to bring equipment back to its operating condition than would be saved by turning it off. Nonetheless, a willingness to turn things off is a good attribute for a plant energy conservation engineer.

11.2.2 Long-Term Possibilities

Implementation of short-term energy conservation opportunities such as those listed in Table 11.2 can pay immediate dividends. However, studies of energy consumption in the plant that might lead to long-term energy savings should not be ignored. Some of the possibilities are described below.

Table 11.2 Short-Term Energy Reduction Checklist

Lighting:	Turn lights off when not needed.
	Clean fixtures for better effectiveness.
	Remove lighting where not needed.
	Reduce parking lighting to the minimum required for personal safety and security.
	Remove advertising and decorative lighting.
Hot water:	Turn off parts washers when not needed.
	Keep covers on tanks, vats, etc.
	Insulate pipes, vats, etc.
	Reduce water temperature in restrooms.
	Use cold-water detergents for cleaning.
Ventilation:	Turn off ventilation and exhausts when not needed.
	Check process exhaust systems for proper amount of air being exhausted.
	Place shrouds around openings to furnaces, paint booths, and washers to minimize exhaust air usage.
	Check filters, dampers, etc. for proper operation.
Compressed air:	Analyze air-using equipment to determine the minimum air pressure possible.
	Fix all compressed air leaks.
	Never use compressed air for cooling.
	Turn off when not needed.
Miscellaneous:	Reduce heating in unoccupied areas.
	Shut off internal combustion equipment when not needed to reduce heat gain and reduce required ventilation.
	Repair faulty equipment, steam traps, valves, temperature controllers, etc.

11.2.3 Heating

The plant heating and air-conditioning system should be checked for proper design. Older systems may have been designed and installed in the era of cheap energy. Special attention should be paid to the method of delivery of the conditioned air. The conditioned air should be delivered to the space where people are working, rather than used to cool the entire plant.

A common problem during winter is stratification of air, with hot air collecting near the plant roof and cold air remaining near the floor. Mixing of the air can solve the stratification problem and can be done relatively cheaply. Vertical ductwork risers placed about one duct diameter from the floor can be used to move the cooler air to the upper regions of the plant, thus promoting air circulation and minimizing stratification. The general rule is that the plant volume should be circulated three times per hour for equalization of floor and ceiling temperature.

Replacing large unducted unit heaters with heating units with proper ductwork and controls for air delivery can also be economical. Clearly, however, such a replacement project has labor, equipment, and other associated costs. These additional costs must be weighed in relation to projected energy savings before such projects can be implemented.

Example 11.1 Redistribution of Air in a Plant

A large industrial plant, which has an open interior, operates with almost a 25°F temperature difference from floor to roof. To prevent this stratification, the plant engineer plans to install a set of 5-ft-diameter riser ducts equipped with axial fans to take air from near the floor and discharge it near the roof. The building is 50 ft high, 500 ft long, and 300 ft wide. The outlet air velocity is restricted to about 15 ft/sec. The risers are 42 ft high. How many ducts are needed to redistribute the flow? What is the power required?

Solution: The volume of the plant is

$$V = (50)(500)(300) = 7.5 \times 10^6 \text{ ft}^3$$

The total volumetric flow rate, Q, assuming a turnover rate of three times per hour, is

$$Q = (3)(7.5 \times 10^6) \frac{\text{ft}^3}{\text{hr}} \frac{\text{hr}}{60 \text{ min}} = 375{,}000 \text{ CFM}$$

For each duct,

$$Q_{\text{duct}} = A_{\text{duct}} \bar{V} = \frac{\pi}{4} (5)^2 (15)(60)$$

$$= 17{,}671 \text{ CFM}$$

The number of risers is

$$N_{\text{risers}} = \frac{375{,}000 \text{ CFM}}{17{,}671 \text{ CFM/riser}} = 21$$

The risers should be distributed around the plant for optimal air distribution.

The horsepower required to move the air through each riser is called the fluid horsepower (FHP) and is equal to

$$\text{FHP} = Q \, \Delta p$$

The Δp is found by using Eq. (5.77):

$$\Delta p = \frac{1}{2} f \frac{L}{D} \varrho V^2$$

The friction factor f is a function of the Reynolds number, Re:

$$\text{Re} = \frac{D \bar{V}}{\nu}$$

For air at 70°F, from Table B.3, $\nu = 0.589$ ft^2/hr and $\varrho = 0.075$ lb$_m$/ft^3. Thus,

$$\text{Re} = \frac{(5)(15)(3600)}{0.589} = 459{,}000$$

From Fig. 5.17 for a smooth duct, $f = 0.0135$. The pressure drop is

$$\Delta p = 0.0135 \left(\frac{42}{5} \right) \left(\frac{1}{2} \right) \frac{(0.075)(15)^2}{32.2} = 0.0297 \text{ lb}_f/\text{ft}^2$$

and

$$\text{FHP} = (17{,}671)(0.0297) = 525 \text{ ft lb}_f/\text{min}$$

Converting to horsepower, we get

$$\text{FHP} = 0.016 \text{ hp per riser}$$

In reality, more energy would be required because of fan and motor inefficiencies. Nonetheless, the amount of power consumed to redistribute the air is virtually negligible.

The heat wheel is another means of reducing heating costs. The heat wheel is a rotary regenerator designed to recover heat from exhaust gases. Figure 11.2 is a schematic diagram of a heat wheel used to preheat fresh air that is being taken into the building. The performance of heat wheels is

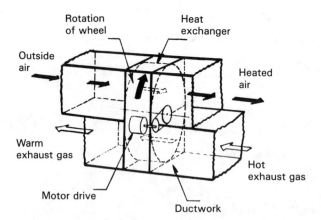

Figure 11.2 Heat recovery wheel used to condition air for a building. (From *Waste Heat Management Guidebook*, K. G. Kreider and M. B. McNeil, NBS Handbook 121, U.S. Government Printing Office, Washington, D.C., 1977.)

identical to that of industrial rotary regenerators, such as the air preheater discussed in Section 9.5.1.

11.2.4 Cooling

Another way to conserve energy is to turn off air-conditioning compressors at night and maintain the temperature with ventilating fans. The use of evaporative cooling rather than refrigeration is also worthy of study. Evaporative cooling costs remarkably less than refrigeration.

Another possibility for energy saving is cooling the roof with an intermittent water spray [8]. This may also extend the roof life because cycling of the roof material between extreme temperatures is reduced. In some cases, waste process water can be used for this purpose.

11.2.5 Ventilation

Energy can be saved in ventilation systems in a number of ways. The usual goal is to reduce the ventilation required in a particular situation. This usually results in a reduction in required fan power. However, for certain processes, government rules and regulations stipulate minimum ventilation rates. In these cases the aim in examining ventilation systems should be to reduce fan/blower motor power while maintaining the flow rate. Let us examine some possibilities.

Fan performance is briefly discussed in Chapter 5. Figure 11.3 shows typical centrifugal fan performance curves, plotted as total static pressure loss across the fan versus the volumetric flow rate in cubic feet per minute. The total static pressure, TSP, in inches of water, represents the total pressure diminished by the dynamic pressure, $\frac{1}{2}\varrho V^2$, generated by the fan. For a given fan operating at a fixed speed, a unique "fan curve" is generated.

Fan selection is performed by matching the fan curve to the "system curve," also shown in Fig. 11.3. The system curve is the sum of the pressure drops associated with friction losses, filter losses, and contraction and expansion losses in the ventilation duct system. The point where the fan curve and the system curve intersect is the operational point for the ventilation system. The required energy input to the fan at that point is the brake horsepower, BHP, which is also plotted on the performance curve.

Figure 11.3 shows several possibilities for reducing fan motor energy consumption, depending on the system changes that can be tolerated. The following three options will be discussed further:

1. Reduce the fan speed. This will result in lower flow rates, but a smaller motor can be used.
2. Alter the system curve to allow a smaller fan motor to be used for the same speed. This will also result in a lower flow rate.

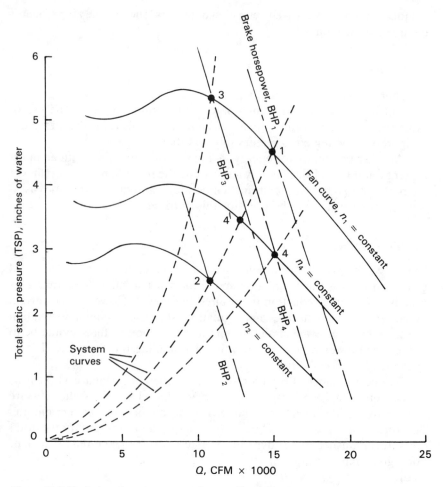

Figure 11.3 Typical performance curves for centrifugal fans.

3. Alter both the fan speed and the system curve to reduce the power required. This can be done while maintaining the same flow rate.

Fan-Speed Reduction. Determining the power savings through fan speed reduction for the same system curve is the most straightforward of the three options. We do this by using so-called fan laws, similar to the laws given for pumps and blowers in Chapter 5.

Fan Laws: Variation in fan speed with a constant air density, constant system curve.

$$Q \text{ (CFM) varies as fan speed} \qquad (11.3a)$$

$$\text{TSP (inches of water) varies as fan speed squared} \qquad (11.3b)$$

$$\text{Power (hp) varies as fan speed cubed} \qquad (11.3c)$$

To find the reduction in power caused by a reduction in fan speed, n, we write

$$\frac{Q_1}{Q_2} = \frac{n_1}{n_2} \qquad (11.4)$$

Also, according to the above fan laws,

$$\frac{\text{TSP}_1}{\text{TSP}_2} = \left(\frac{n_1}{n_2}\right)^2 \qquad (11.5)$$

$$\frac{\text{BHP}_1}{\text{BHP}_2} = \left(\frac{n_1}{n_2}\right)^3 \qquad (11.6)$$

The following examples illustrate this procedure.

Example 11.2 Power Savings by Fan Speed Reduction

A fan operating at $n_1 = 1,200$ RPM, $\text{TSP}_1 = 4.5$ in. of water delivers 15,000 CFM of air at 25 BHP. Assuming that the flow rate could be reduced to 11,000 CFM, find the new operating point on the performance curve.

Solution: In Fig. 11.3, move from point 1 to the new operating point, point 2, which lies along the curve for

$$n_2 = n_1 \frac{Q_2}{Q_1} = (1,200)\left(\frac{11,000}{15,000}\right) = 880 \text{ rpm}$$

The TSP at point 2, in accordance with Eq. (11.5), becomes

$$\text{TSP}_2 = (4.5)\left(\frac{880}{1,200}\right)^2 = 2.42 \text{ in. of water}$$

Finally, from Eq. (11.6) the power required is

$$\text{BHP}_2 = (25)\left(\frac{880}{1,200}\right)^3 = 9.86 \text{ hp}$$

About 15 hp can be saved for this reduction in flow rate.

System Curve Alteration. Energy savings are also possible for cases in which the fan/motor set characteristics are not altered. Point 3 in Fig. 11.3 illustrates what is required to reduce power consumption. The system curve has to be shifted upward and to the left so that the intersection of the fan curve and the system curve results in a lower BHP.

The fan laws of Eq. (11.3) do not apply to this case because the fan speed is constant. However, the performance is predictable by using other characteristics of the fan; for example, the static efficiency of a fan is

$$\eta_{st} = \frac{Q\,(\text{CFM}) \times \text{TSP}\,(\text{in. of water})}{6356\,\text{BHP}\,(\text{hp})} = \frac{\text{FHP}}{\text{BHP}} \qquad (11.6a)$$

where FHP is the horsepower imparted to the air. Example 11.3 illustrates the power reduction possible for system curve alteration.

Example 11.3 Power Savings for Constant Speed Fan

A ventilation system operating at 1,200 RPM, 15,000 CFM, and a total static pressure of 4.5 in. of water requires 25 BHP. At this operating condition, point 1 in Fig. 11.3, the static efficiency is 42.5%. Alteration of the system curve allows the fan to operate at 5.25 in. of water and 11,000 CFM, which is point 3 in Fig. 11.3.
The static efficiency for this case is 55%. What is the power reduction?

Solution: In accordance with Eq. (11.6a),

$$\text{BHP} = \frac{Q \times \text{TSP}}{6,356 \times \eta_{st}}$$

The power required at point 3 is

$$\text{BHP}_3 = \frac{(11,000)(5.25)}{6,356\,(0.55)} = 16.5\ \text{hp}$$

Almost 10 hp is saved in this case.

A note of caution is in order here. The fan should not be operated to the left of the maximum of the fan curve. In this region the flow is unsteady, resulting in reduced fan efficiency and, in some cases, excessive fan noise.

Fan Speed/System Curve Alteration. In cases where the flow rate cannot be reduced, power can be reduced by a combination of fan speed reduction and system curve alteration. Such a change is represented by moving from point 1 to point 4 in Fig. 11.3. The analysis is a combination of the predictions used for the two cases above. Example 11.4 shows the method.

Example 11.4 Power Savings at Constant CFM

The ventilating system operating at point 1 in Fig. 11.3 is to be operated at the same capacity, but at a TSP of 3.0 in. of water. What are the fan speed and BHP at the new point 4 if the respective efficiencies are η_{st1} = 42.5% and η_{st4} = 35.5%, respectively?

Solution: First, assuming operation along the same system curve to point 4', the new speed $n_{4'}$ is, in accordance with Eq. (11.5),

$$n_{4'} = n_1\,\frac{\text{TSP}_2}{\text{TSP}_1} = 1,200\,\sqrt{\frac{3}{4.5}}$$

$$= 980\ \text{RPM} = n_4$$

Now, moving along the 980 RPM fan curve to the new point 4, we write

$$\text{BHP}_4 = \frac{Q_4\,(\text{TSP}_4)}{6,356\,(\eta_{st4})} = \frac{(1,500)(3)}{6,356\,(0.355)}$$

$$= 19.9\ \text{hp}$$

In this case the power reduction is not as large as in Examples 11.2 and 11.3, but it is certainly significant.

The advent of the practical AC variable-speed motor allows for continuous changes of fan speed as the loads on the ventilation system change. Chapter 13 gives details of the AC variable-speed motor. A good control system linked to such a variable-speed drive can result in significant energy savings.

11.3 LIGHTING

The lighting systems in buildings built before the 1973–1974 oil embargo were designed without much regard to energy conservation. The main limiting design factor was initial cost; consequently, the energy manager now has many opportunities to improve the operating efficiency of the lighting system and, perhaps, to improve the quality of lighting as well.

Lighting systems are not different from other energy consumers. The energy they require is the power dissipation rate multiplied by the time of operation. Hence the total energy usage can be reduced by reducing the power consumption or shortening the time of usage. We shall concentrate on the reduction of power consumption in this section.

11.3.1 Lighting Systems

A lighting system is typically composed of lamps; circuitry, switches, and controls; and luminaires. Luminaires are the fixtures into which the lamps are placed; they usually include a reflecting surface to direct the light toward the space to be lighted. The characteristics of all of the components must be included when considering the efficiency of the lighting system.

Lighting systems operate according to the relationship

$$E = \frac{N(Lu)(Cu)(L_1)(L_2)}{A} \qquad (11.7)$$

where E is the illumination in foot-candles, N is the required number of lamps, Lu is the lumen output per lamp, Cu is the coefficient of utilization for the luminaire and other reflecting surfaces, L_1 is the lamp depreciation factor, L_2 is the luminaire dirt depreciation factor, and A is the area to be lighted.

The lamp depreciation factor L_1 takes into account the deterioration of lamp output with time. It is specified by the manufacturer on the basis of testing programs. The luminaire dirt factor, L_2, accounts for the effect of dirt accumulation on the luminaire.

The coefficient of utilization, Cu, indicates how efficiently the luminaire illuminates the plane of the task. It is related to the room geometry and reflectance, as well as the characteristics of the luminaire. The *IES Lighting Handbook* [9] contains Cu values for various types of luminaires and room characteristics. Manufacturers' literature usually also lists Cu values.

A lighting system designer has several options for a light source. We classify light sources into *incandescent, fluorescent,* and *high-intensity discharge* (HID) types. Table 11.3 lists light sources of these types along with comparative ratings of various aspects of their operation.

11.3.2 Types of Lighting

Incandescent Lamps. An electric current flowing through a thin wire filament causes it to heat to the point of incandescence and to emit light. The required circuitry is very simple, as shown in Fig. 11.4a. Incandescent lamps are the least efficient of all the commercially available light sources. This is shown in Table 11.3, where efficiency is given in lumens per watt.

A *lumen* is the amount of visible light flux which, falling normally on an area of 1 square foot, will produce an *illuminance* of 1 foot-candle. Illuminance, E, is defined by

$$E = \frac{I}{d^2} \tag{11.8}$$

where I is the luminous intensity of the light source and d is the distance from the source in feet. Illuminance has the units of foot-candles. Luminous intensity of various sources is standardized against platinum at 2033 K, which gives off 60 candles/cm^2. Luminous intensity is usually given in candlepower.

Incandescent lamps have shorter lives than light sources of other types. Their advantages are low initial and replacement costs, small physical size, and the fact that they can be dimmed to control the light output. Furthermore, no ballast is required in the circuits.

Fluorescent Lamps. Light is produced by a discharge arc passing through a gas in the tube. Ultraviolet rays are converted into visible light by interacting with a phosphor coating on the inside of the tube. A ballast is required in the circuit to regulate current and starting voltage, as shown in Fig. 11.4b.

Fluorescent lamps are more efficient and have longer lives than incandescent lamps. Higher costs, larger size, and the need for a ballast are their primary disadvantages.

High-Intensity Discharge Lamps. These are also discharge lamps and their operating characteristics are similar to those of fluorescent lamps. Figure 11.4c shows a circuit typical of a mercury vapor lamp system.

Table 11.3 Light Source Characteristics

Characteristic	Incandescent, including tungsten halogen	Fluorescent	High-intensity discharge			Low-pressure sodium
			Mercury vapor (self-ballasted)	Metal halide	High-pressure sodium (improved color)	
Wattage (lamp only)	15–1,500	15–219	40–1,000	175–1,000	70–1,000	35–180
Life^a (hr)	750–12,000	7,500–24,000	10,000–15,000	1,500–15,000	24,000 (10,000)	18,000
Efficacy^a (lumens/W) lamp only	15–25	55–100	50–60 (20–25)	80–100	75–140 (67–112)	Up to 180
Lumen maintenance	Fair to excellent	Fair to excellent	Very good (good)	Good	Excellent	Excellent
Color rendition	Excellent	Good to excellent	Poor to excellent	Very good	Fair (very good)	Poor
Light direction control	Very good to excellent	Fair	Very good	Very good	Very good	Fair
Source size	Compact	Extended	Compact	Compact	Compact	Extended
Relight time	Immediate	Immediate	3–10 min	10–20 min	Less than 1 min	Immediate
Comparative fixture cost	Low: simple fixtures	Moderate	Higher than incandescent and fluorescent	Generally higher than mercury	High	High
Comparative operating cost	High: short life and low efficacy	Lower than incandescent	Lower than incandescent	Lower than mercury	Lowest of HID types	Low
Auxiliary equipment needed	Not needed	Needed: medium cost	Needed: high cost	Needed: high cost	Needed: high cost	Needed: high cost

^{a}Life and efficacy ratings subject to revision. Check manufacturers' data for latest information.

Source: Energy Management Handbook, p. 406, W. C. Turner (ed.), Copyright 1982 by John Wiley & Sons, New York. Reprinted with permission.

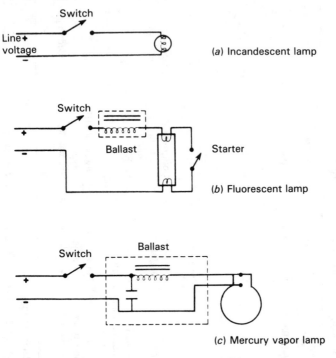

Figure 11.4 Typical lighting system wiring diagrams. (From Smith [5]. Copyright 1981 by Pergamon Press, Elmsford, N.Y.)

High-pressure sodium lamps are the most efficient of the three types of HID lamps. All HID lamps take a few minutes to relight after they have been extinguished; thus an emergency auxiliary lighting system might be required.

Low-Pressure Sodium Lamps. These lamps are usually classified separately from fluorescent sources, although both are low-intensity discharge lamps. This is done because there is a difference in the color of the light produced, which may limit its suitability for certain applications. Low-pressure sodium lamps emit a monochromatic yellow light that is not pleasing to the eye. However, they are the most efficient lamps in current use and have relatively long lifetimes. They have been extensively used in Europe.

11.3.3 Task Lighting

Task lighting is the use of lighting fixtures to light the task plane, rather than providing uniform, symmetrical lighting throughout the space. The Illuminating Engineering Society of North America recommends illuminance levels for various tasks that might be required in an industrial plant. Table

11.4 gives some suggested illuminance levels for various tasks. Consult the *IES Lighting Handbook* [9] for other types of spaces and tasks.

11.3.4 Reduction of Lighting Power Consumption

Two choices can be made in changing lighting systems: modification of the existing system, or replacement of one system with another.

Modification of Existing Systems. Several opportunities exist for power reduction in this category. One can:

- Remove lamps from fixtures. This is possible if a review of current accepted illuminance guidelines shows lower requirements than those for which the system was designed.
- Disconnect ballasts when lamps are removed. For systems other than incandescent ones, which have no ballasts, the ballast dissipates energy even when the lamp has been removed.
- Replace lamps with lower-wattage lamps.

Table 11.5 shows comparisons between incandescent lamps and various other types of lamps that could be substituted for them. It shows, for example, that one 100-watt incandescent lamp produces almost the same lumens as two 60-watt lamps although it has 16.7% less wattage.

The only discharge lamp that can be substituted for an incandescent lamp is the "self-ballasted" mercury vapor lamp described in Table 11.3. Standard mercury lamps are more efficient than self-ballasted mercury

Table 11.4 Suggested Illumination Levels

Types of space	Foot-candles
Industrial interiors:	
Manufacturing areas	
Ordinary tasks	50
Highly difficult tasks	200
Most difficult tasks	500–1000
Boiler rooms	10
Cable vaults	5
Corridors	5
Engine rooms	10
Garage areas (repairs)	70
Mechanical equipment room	10
Toilets	10
Exteriors:	
Parking	1–5
Building security	1–5

Table 11.5 Comparison of Incandescent Light Sources and Some Possible Light Source Substitutes

Lamp watts no. and unit	Light source[a]	Line watts	Life hours	Initial lumens	% of incandescent lumens	Lamp lumens per watt	Lumen[b] i or d		Watts[b] i or d		Cost per hour[c] (¢)
60	I	60	1,000	870	100	14.5					0.36
1-15	F	26	7,500	870	100	33	430	i	34	d	0.16
1-20	F	26	9,000	1,300	149	50	1,430	i	34	d	0.16
1-30	F (RS)	46	15,000	2,300	265	50	2,280	i	14	d	0.28
1-40	F (RS)	53	18,000	3,150	362	59			7	d	0.28
100	I	100	750	1,750	100	17.5					0.6
2-15	F (RS)	58	7,500	1,740	99	30	10	d	42	d	0.34
1-50	MV	63	16,000	1,575	90	25	185	d	37	d	0.38
200	I	200	750	4,010	100	20.1					1.20
2-30	F (RS)	75	15,000	4,600	115	61	590	i	125	d	0.46
1-55	F (SL)	82	12,000	4,500	112	55	490	i	118	d	0.50
1-60	F (HO)	90	12,000	4,300	107	48	290	i	110	d	0.54
1-100	MV	122	24,000	4,500	112	37	490	i	78	d	0.74
500	I	500	1,000	10,850	100	21.7					3.00
2-75	F (SL)	173	12,000	12,600	116	73	1,750	i	327	d	1.04
1-250	MV	285	24,000	13,000	120	46	2,150	i	215	d	1.72
1-400	MV	448	24,000	23,000	212	51	12,150	i	52	d	2.68
1-175	MH	210	7,500	14,000	133	67	3,150	i	290	d	1.26
1-250	MH	300	7,500	20,500	189	68	9,650	i	200	d	1.80
1-400	MH	460	15,000	34,000	313	74	23,150	i	40	d	2.76
1-150	HPS	200	15,000	16,000	147	80	5,150	i	300	d	1.20
1-250	HPS	310	15,000	25,500	235	82	14,650	i	190	d	1.86
1-400	HPS	475	20,000	50,000	460	105	39,150	i	25	d	2.86
750	I	750	1,000	17,040	100	22					4.50
2-110	F (HO)	243	12,000	18,600	109	77	1,560	i	507	d	1.46
1-400	MV	448	24,000	23,000	135	51	5,960	i	302	d	2.68
1-250	MH	300	7,500	20,500	120	68	3,460	i	450	d	1.80
1-400	MH	460	15,000	34,000	200	74	16,960	i	290	d	2.76
1-250	HPS	310	15,000	25,500	150	82	8,460	i	440	d	1.86
1-400	HPS	475	20,000	50,000	293	105	3,290	i	275	d	2.86

1000		I	1,000	1,000	23,740	100	23.7					6.00
	2-215	F (VHO)	440	9,000	29,000	122	66	5,260	i	d	560	2.64
	1-400	MV	448	24,000	23,000	97	51	740	i	d	552	2.68
	1-400	MH	460	15,000	34,000	143	74	10,260	i	d	540	2.76
	1-250	HPS	310	15,000	23,205	98	82	535	i	d	690	1.86
	1-400	HPS	475	20,000	50,000	211	105	26,260	i	d	525	2.86
1250		I (Q)	1,250	2,000	28,000	100	22.4					7.50
	2-215	F (VHO)	440	9,000	29,000	104	66	1,000	i	d	810	2.64
	2-250	MV	570	24,000	26,000	93	46	2,000	i	d	680	3.42
	1-400	MV	448	24,000	23,000	82	51	5,000	i	d	802	2.68
	2-400	MV	996	24,000	46,000	164	51	18,000	i	d	254	5.98
	1-1000	MV	1,130	24,000	63,000	225	58	35,000	i	d	120	6.78
	2-250	MH	600	7,500	41,000	146	68	13,000	i	d	650	3.60
	1-400	MH	460	15,000	34,000	121	74	6,000	i	d	790	2.76
	2-400	MH	920	15,000	68,000	243	74	40,000	i	d	330	5.52
	1-1000	MH	1,100	10,000	100,000	357	91	72,000	i	d	150	6.60
	2-150	HPS	400	12,000	32,000	114	80	4,000	i	d	850	2.40
	2-250	HPS	620	15,000	51,000	182	82	23,000	i	d	630	3.72
	1-400	HPS	475	20,000	50,000	179	105	22,000	i	d	775	2.86
	2-400	HPS	950	20,000	100,000	357	105	72,000	i	d	300	5.70
	1-1000	HPS	1,100	10,000	130,000	464	130	10,200	i	d	150	6.60
1500		I	1,500	1,000	34,400	100	22.9					9.00
1500		I (Q)	1,500	2,000	35,800	104	23.9					9.00
	2-215	F (VHO)	440	9,000	29,000	84	66	1,400	i	d	1,060	2.64
	2-400	MV	996	24,000	46,000	134	51	5,400	i	d	504	5.98
	1-1000	MV	1,130	24,000	63,000	183	58	11,600	i	d	370	6.78
	2-250	MH	600	7,500	41,000	119	68	28,600	i	d	900	3.60
	1-400	MH	460	15,000	34,000	99	74	6,600	i	d	1,040	2.76
	2-400	MH	920	15,000	68,000	198	74	4,000	i	d	580	5.52
	1-1000	MH	1,100	10,000	100,000	291	91	33,600	i	d	400	6.60
	2-150	HPS	400	12,000	32,000	93	80	65,600	i	d	1100	2.40
	2-250	HPS	620	15,000	51,000	148	82	2,400	i	d	880	3.72
	1-400	HPS	475	20,000	50,000	145	105	16,600	i	d	1,025	2.86
	2-400	HPS	950	20,000	100,000	291	105	16,500	i	d	550	5.70
	1-1000	HPS	1,100	10,000	130,000	378	130	95,600	i	d	400	6.60

[a] Types of light sources are abbreviated as follows: Incandescent (quartz), I (Q); fluorescent (F) rapid start (RS), slim line (SL), high output (HO), very high output (VHO); mercury vapor (MV); metal halide (MH); and high-pressure sodium (HPS).

[b] Abbreviations: i, increase; d, decrease (compared to incandescent).

[c] Based on a rate of $0.06/kWh.

Source: Smith [5]. Copyright 1981 by Pergamon Press. Reprinted with permission.

lamps, hence self-ballasted lamps are not suitable for new installations. Self-ballasted lamps also have a high initial cost. Thus, even though they are easy to install in an existing incandescent fixture, economics might justify a complete replacement with standard mercury lamps.

Replacement of Existing Systems. An economic analysis might indicate that complete replacement of a lighting system is desirable. In that case, more options for a particular type of lamp exist.

The decision to replace is based on the following steps:

1. Select a light source that has the highest efficiency, compatible with other system requirements such as life, color rendition, and other factors listed in Table 11.3.
2. Select the appropriate luminaire.
3. Estimate the required number of lamps per luminaire, and the number of luminaires.
4. Determine the initial and installation cost of the system and the energy savings compared to the old system.
5. Make an economic feasibility study.

The following example demonstrates this procedure.

Example 11.5 Lighting System Replacement

A small manufacturing facility uses 175 incandescent fixtures with 750-watt lamps as its existing system. The required lumens are 21,000 at the task plane and the corresponding power requirement is 131.25 kW. It is desired to replace this system with a more efficient one using the same number of fixtures. Electricity costs 4.75¢/kWh, and the lighting is used for 3,000 hours per year. Select a replacement system, calculate the annual savings, and find the return on investment.

Solution: Table 11.3 shows that low-pressure sodium lamps are the most efficient. We discard them, however, because of their poor color rendition. The next choice is metal halide.

Table 11.5 shows that a 250-watt metal halide lamp produces 120% of the lumens of a 750-watt incandescent system. Thus 175 250-watt metal halide lamps require 300 watts per fixture, corresponding to a required power of 52.5 kW, a saving of 78.75 kW.

The cost of the metal halide lamps is $39 per lamp; the fixtures cost $82 each, and the installation cost is $32 per fixture. The total cost is $26,775. The cost saving for 3000 hours of operation is

$$78.75(3,000)(0.0475) = \$11,222 \text{ per year}$$

The payback period is

$$p = \frac{26,775}{11,222} = 2.4 \text{ years}$$

and the return on investment is

$$\text{ROI} = \frac{11,222}{26,775} = 42\%$$

If fewer fixtures could be tolerated, the economics are even more favorable. The required fixtures for 21,000 lumens using 250-watt metal halide lamps are

$$N = 175 \left(\frac{17,040}{20,500}\right) \cong 145$$

assuming that the replacement luminaires are as efficient as the old ones.

The lighting designer is not constrained by existing equipment when designing a new installation, and can choose the most efficient available system that is compatible with a client's needs.

11.4 COMPRESSED AIR

Compressed air is often used for controlling and performing mechanical work because such systems have inexpensive and reliable components and installation and maintenance costs are relatively low. Until a few years ago, operating costs were also low because energy was so cheap. Even though the cost of energy has escalated, energy costs in compressed air systems do not totally dominate the system annual costs, as is the case in some other power delivery systems. Nonetheless, simple conservation measures can reduce operating costs, and it is worth checking what use of compressed air is being made.

A well-maintained compressed air system is a reliable energy distribution system with the advantages that leaks are not usually dangerous, ambient temperature has little effect, and there is no risk of pollution. Such systems usually continue to operate with little maintenance, although their effectiveness may fall off if they are neglected.

When shortages of system capacity occur, the usual remedial step is to simply add a second or larger compressor. This alleviates the problem, yet does not take into account potential energy conservation. In this section some aspects of the potential for energy savings in compressed air systems are discussed.

11.4.1 The Compressed Air System

Figure 11.5 shows a schematic of a well laid-out compressed air system. The motor-compressor provides the energy input to ambient air, raising it to the pressure required for its intended industrial use. A temperature rise during compression is an inevitable but unwanted side effect of the pressurization. Water or air is used as a medium for cooling the compressed air in the aftercooler. This cooling process condenses water from the air; the condensed water can be drained off to prevent corrosion of the system. The aftercooler itself presents an opportunity for heat recovery, and thus reduced energy costs, as shown in Fig. 11.5.

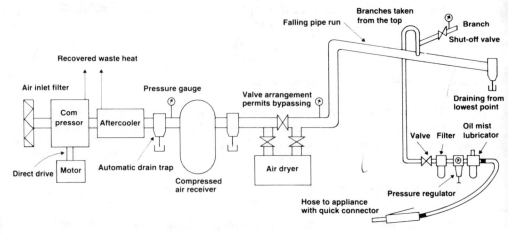

Figure 11.5 Schematic diagram of a typical industrial compressed air system. (Reprinted with permission from *Energy Technology,* Swedish Board for Technical Development, No. 4, 1982.)

The remainder of the system consists of a receiver for storing the compressed air and a piping system for distribution of the air to appropriate workstations. As shown in Fig. 11.5, additional drying of the air may be required before final distribution.

The distribution piping has frictional resistance to flow; thus, a pressure drop is induced between the receiver and the workstations. The temperature of the compressed air is usually near ambient temperature when it arrives at the workstation. For most industrial applications, the pressure at application is in the range 5–8 atmospheres absolute.

Although compressed air might be used for pneumatic controls, the bulk of it is used to drive machines such as presses, air hammers, and impact tools. Pneumatic cylinders or air motors are the devices usually used to produce work. These devices operate on the principle of a direct pressure force acting on a surface rather than an expansion through rotary blades. Thus, air remains basically at its supply pressure inside the machine and freely expands after it is exhausted.

Let us first examine how much work we might expect to deliver for such a system for a given work input at the compressor. To do this, a control volume is constructed that encompasses the entire system, as shown in Fig. 11.6. For the indicated pressure ratio of 8:1, the work required for loss-free (isentropic) compression is

$$w_{\text{comp}} = \int_1^2 v \, dp$$

The isentropic compression allows use of the relation $pv^k = c$, so that

$$w_{\text{comp}} = \frac{kR}{k-1} (T_2 - T_1)$$

where we also use the fact that air behaves as an ideal gas. The temperature ratio for this process is

$$\frac{T_2}{T_1} = \left(\frac{P_2}{P_1}\right)^{(k-1)/k}$$

Using $k = 1.4$ for air, T_2 is calculated to be

$$T_2 = 530(8)^{0.286} = 960 \text{ R}$$

Thus, the work required is

$$w_{comp} = \frac{1.4(53.34)}{0.4}(960 - 530)$$

$$= 80,277 \text{ ft-lb}_f/\text{lb}_m = 103.2 \text{ Btu/lb}_m$$

Based on an isentropic expansion, the temperature $T_4 = 292$ R, and the maximum possible work output for the motor is

$$w_{out} = \frac{1.4(53.34)}{0.4}(530 - 292) = 44,432 \text{ ft-lb}_f/\text{lb}_m = 57.1 \text{ Btu/lb}_m$$

At best, we can expect to extract only about 55% of the input energy as output work. In practice, this fraction is much reduced by air leakage, machine inefficiencies, and other factors. Electric motors can easily deliver over 90% efficiency, so the energy conservation engineer should keep systems other than compressed air in mind when considering energy delivery.

For compressed air systems already in place, what can be done to improve efficiency? The next section addresses that question.

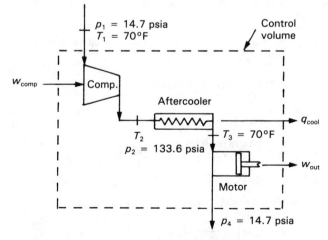

Figure 11.6 Control volume for determining the work output of a compressed air machine.

11.4.2 Efficiency Improvement

The following actions can be taken to improve compressed air system efficiency:

- Operate the compressor close to full load if possible.
- Install separate compressors if a number of different pressures are required in the plant.
- Recover waste heat from the compressor coolers.
- Use larger pipe diameters to reduce pressure drop.
- Minimize the air leakage from the system.

Compressors work most efficiently at full load. The demand for compressed air is seldom constant, however, so the load on a compressor can be quite variable. Thus, it is necessary to reduce compressor output either by throttling or by running at no load. The no-load power typically required for a reciprocating compressor is about 12% of full load, and that of a screw compressor is even higher. Efforts should be made to create a uniform load so that periods of low consumption are avoided.

The lower the pressure ratio across a compressor, the lower is the required power. Thus, if a number of different pressures are required, several compressors of different capacities might be more economical than one large compressor, which would require throttling at the various workstations.

Example 11.6 Comparison of Single and Multiple Compressors

A large manufacturing facility has the following compressed air needs: 100 CFM at 8 atm, 150 CFM at 6 atm, and 200 CFM at 2 atm, all at 70°F. A single compressor would have to compress all of the required air to 8 atm and then throttle it down to 6 and 2 atm. If individual compressors were installed, they would have to respond only to the exact needs at their output level. Which is the most energy-conservative?

Solution: Single compressor: First calculate the total mass flow rate required for all pressures.

8 atm: $p\dot{V} = \dot{m}RT$

$$\dot{m} = \frac{p\dot{V}}{RT} = \frac{(117.6)(100)(144)}{(53.34)(530)}$$

$$= 59.9 \text{ lb}_m/\text{min}$$

6 atm: $\dot{m} = \frac{6(14.7)(144)(150)}{(53.34)(530)} = 67.4 \text{ lb}_m/\text{min}$

2 atm: $\dot{m} = 29.95 \text{ lb}_m/\text{min}$

Therefore $\dot{m}_{total} = 157.25 \text{ lb}_m/\text{min}$ and the work required is

$$\dot{W} = \dot{m}w = \dot{m}\left[\frac{kR}{(k-1)}(T_{comp} - T_0)\right]$$

where T_{comp} = 960 R from our previous discussion. Thus

$$\dot{W} = 157.25 \left[\frac{1.4(53.34)}{0.4} (430) \right] = 1.262 \times 10^7 \text{ ft-lb}_f/\text{min}$$

Multiple compressors: Here we must analyze each system separately.

$$8 \text{ atm:} \quad \dot{W} = \frac{59.9}{157.25} (1.262 \times 10^7) = 4.81 \times 10^6 \text{ ft-lb}_f/\text{min}$$

$$6 \text{ atm:} \quad \frac{T_{comp}}{T_0} = \left(\frac{p_c}{p_0} \right)^{0.286} = (6)^{0.286} = 1.716$$

$$T_{comp} = 909 \text{ R}$$

$$\dot{W} = 67.4 \left[\frac{1.4}{0.4} (53.34)(379) \right] = 4.77 \times 10^6 \text{ ft-lb}_f/\text{min}$$

$$2 \text{ atm:} \quad T_{comp} = 646 \text{ R}$$

$$\dot{W} = 6.48 \times 10^5 \text{ ft-lb}_f/\text{min}$$

The total work rate is

$$\dot{W}_{total} = 1.023 \times 10^7 \text{ ft-lb}_f/\text{min}$$

The saving of 2.39×10^6 ft-lb$_f$/min is realized for the multiple-compressor system, which is about 19% of the single-compressor input.

The potential for waste heat recovery from compressor coolers is very good. Air-coolers usually raise the temperature of the cooling air by 25–20°C, while water-coolers might provide hot water above 90°C in some cases. Heat recovery by water is probably preferred because the recovered heat can be distributed in serial fashion at various lower temperatures. Figure 11.7 shows such a system that might be typical of a large plastics plant. Note that a cooling tower is available if the waste heat must be "dumped" in the case of reduced plant needs for the recovered heat.

A larger pipe diameter for a compressed air delivery system reduces the pressure drop and the operating costs, but increases capital costs. The design rule is to choose a pipe size that will give a maximum air velocity of about 20 ft/sec.

Leaky compressed air systems are commonplace in industrial plants. In fact, it is virtually impossible to operate a system involving hoses, valves, quick disconnections, and other heavily used components that does not leak. Industry experience indicates that 15 to 50% of the compressed air that is produced never reaches workstations. A well-organized preventive maintenance program to reduce leaks can result in significant savings.

11.5 SUMMARY

This chapter deals with types of energy consumption that are common to

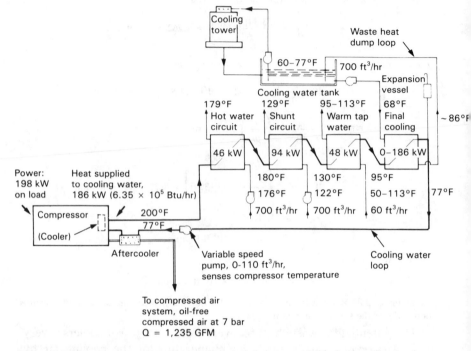

Figure 11.7 Heat recovery system for a compressed air system typical of a large plastics plant. (Reprinted with permission from *Energy Technology*, Swedish Board for Technical Development, No. 4, 1982.)

plants in most industries. The need to maintain tolerable temperature levels, provide adequate lighting, and adequately ventilate the work space is virtually the same for many different types of industries. Compressed air systems, which are used in many plants to deliver energy, also are discussed. Because of the widespread use of such systems, we associate them with the building itself rather than with any particular process in the building.

Most industrial plants need far less energy to maintain heating and cooling, lighting, and ventilation than is consumed in the machines and processes operating in the building. Energy savings are easily achieved for such systems, however, and they should not be overlooked when energy conservation opportunities are evaluated.

REFERENCES

1. American Society of Heating, Refrigeration and Air-Conditioning Engineers, *ASHRAE Handbook of Fundamentals*, New York, 1981.

2. R. W. Roose (ed.), *Handbook of Energy Conservation for Mechanical Systems in Buildings*, Van Nostrand Reinhold, New York, 1978.
3. A. Thumann, *Plant Engineers and Managers Guide to Energy Construction,* Van Nostrand Reinhold, New York, 1977.
4. Y. Y. Haimes ed., *Energy Auditing and Conservation: Methods, Management and Case Studies,* Hemisphere, Washington, D.C., 1980.
5. C. B. Smith, *Energy Management Principles: Applications, Benefits, Savings,* Pergamon, Elmsford, N.Y., 1981.
6. R. M. Harrold, Lighting, chapter 13 in *Energy Management Handbook*, W. Turner (ed.), Wiley, New York, 1982.
7. American Society of Heating, Refrigeration and Air-Conditioning Engineers, *ASHRAE Handbook—Systems*, New York, 1980.
8. J. I. Yellot, Roof Cooling with Intermittent Sprays, presented at the 73rd Annual Meeting of ASHRAE, Toronto, June 27, 1966.
9. Illumination Engineering Society of North America, *IES Lighting Handbook*, New York, 1981.

PROBLEMS

11.1 A light manufacturing plant building has 150 people engaged in light bench-assembly work, 25 people engaged in heavy work, and 10 in seated, supervisory positions. The plant has ten 5-hp electric motors operating at 95% efficiency and twenty-five 1-hp motors at 89%. Calculate the heat load in Btu's per hour due to human activity and energy dissipated by the electric motors.

11.2 From Table 11.5, make a list of lamp systems that will provide 100,000 lumens, e.g., one 1000-watt metal halide, three 400-watt metal halides and so forth. Rank the list according to ascending cost of operation.

11.3 A manufacturing company needs to illuminate part of its assembly line at 200 foot-candles. A region of 4 by 50 ft needs to be lighted. Luminaires whose coefficient of utilization is 0.15 are to be used, with two fluorescent lamps per luminaire. Factors L_1 and L_2 are 0.8 and 0.75, respectively. How many 75-watt lamps are needed? How many 110-watt lamps would be needed? How much total power is required for the two sizes of lamps?

11.4 An energy audit shows that lighting levels in a plant can be substantially reduced. The plant uses 210 110-watt individually ballasted fluorescent lamps in three-lamp luminaires. The ballast consumes one-tenth as much power as the lamp. It is estimated that the lighting level can be reduced so that one of the lamps in each luminaire can be removed. Calculate the energy savings if (a) only one lamp is removed from each luminaire and (b) both the lamp and the ballast are removed. Calculate the respective dollar savings per year if electricity costs 5.5¢/kWh, and the lamps are used 4400 hr per year.

11.5 A retrofit of the existing lighting system for a foundry building is under consideration as the result of an energy audit. The current system uses 50 fixtures with 200-watt incandescent lamps. The lights are used 5000 hr per year. The prime candidate for system retrofit is one with 45 luminaires using two 30-watt fluorescent F (RS) lamps. This proposed system entails the following costs per luminaire: initial cost of luminaire and lamps, $22.00; installation cost, $43.00; interest on investment, 0.20. The retrofit system is estimated to last for 30 years. Determine whether the retrofit should be implemented by using the data in Table 11.5P to determine the annual cost of ownership as described in Section 3.1.6.

Table 11.5P Annual Costs per Luminaire

	Existing system	Retrofit system
Maintenance	$150	$10
Lamp replacement	6/year	1/3 years
Lamp cost	$3/lamp	$7/lamp
Depreciation	—	Straight-line
Salvage value	None	None
Electricity	5¢/kWh	5¢/kWh

11.6 Points 1, 4', and 2 all lie on the same system curve in Fig. 11.3. At point 1, BHP = 25 hp and n = 1200 RPM. Calculate n_4', n_2, BHP_4', and BHP_2.

11.7 Calculate the static efficiency of a fan driven by a 10-hp motor connected to a duct system such that the operating point is 2.5 in. of water at 11,000 CFM. If the same system were used to drive the fan at 2.0 in. of water, delivering 12,000 CFM, what would its static efficiency be? Show that it is impossible to use this system to deliver 15,000 CFM at 4.5 in. of water.

11.8 A ventilation system that normally operates at point 4 in Fig. 11.3 is being redesigned to save energy. A study of the ventilation requirement shows that the flow rate can be reduced. Also, the duct system is to be altered so that a different system curve is accessible. The net effect is that the "new" operating point is point 2 in Fig. 11.3. Calculate n_2 and BHP_2 assuming that BHP_4 = 20 hp, n_4 = 1000 RPM, and the static efficiency increases by 10%.

11.9 Show that the static efficiency of a fan must remain constant along a given system curve as the fan speed is reduced.

11.10 A variable-speed motor is connected to a centrifugal fan so that when less air needs to be delivered the fan speed can be controlled rather than using dampers in the duct work. The variable-speed motor performs as shown in Table 11.10P. The system operates 50% of the time at point 1 and 50% at point 2 in Fig. 11.3. The BHP_1 is 20 hp and n_1 is 1200 RPM. If electricity costs 5.5¢/kWh, calculate the annual energy costs for the variable-speed system if it operates 4,500 hr/year. Compare this to the case where the motor must operate at point 3 for 50% of the time, i.e., with dampers in the system, and 50% of the time at point 1. The static efficiency of the fan at point 3 is 46%. What are the annual savings?

Table 11.10P Motor Characteristics

Speed (RPM)	Q (CFM)	Horsepower delivered	Efficiency
1,200	15,000	20	0.94
1,000	12,500	11.5	0.93
800	10,000	6.0	0.92

11.11 Certain axial fans operating at 1200 RPM are known to have the static head characteristics shown in Fig. 11.11P. They are being considered for a duct system that would require a static head of 4.0 in. of water to handle 4000 CFM of standard air.

(a) At what head, flow, and air horsepower would a single fan operate if connected to this duct system?

(b) At what total head, flow, and air horsepower would two fans in series operate if connected to this duct system?

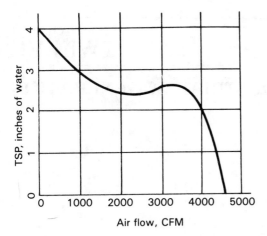

Air flow, CFM

Figure 11.11P Fan characteristics.

11.12 Calculate the temperature rise across an air compressor with a pressure ratio of 8:1. Intake air is at $p = 1$ atm and 70°F. How much intercooling between stages would be required for a flow rate of 10 CFM at the outlet if the compressor has a second stage with a pressure ratio of 6:1 and the desired outlet temperature is 425°F? What is the horsepower required for the two-stage compressor?

11.13 A large centrifugal air compressor has a total pressure ratio of 4. The inlet air is at 14.7 psia and 70°F. The air velocity in the inlet air duct is 350 ft/sec at a flow rate of 20 lb_m/sec. The velocity in the discharge duct is 300 ft/sec. The compressor adiabatic efficiency is 75%. Calculate:

(a) The static pressures and temperatures at the inlet and exit of the compressor
(b) The compressor static pressure ratio
(c) The horsepower needed to drive the compressor

11.14 Compressed air is to be delivered at 8 atm at 20 CFM through a distance of 150 ft. Calculate the respective diameters of two pipes — the first large enough to limit the air velocity to 20 ft/sec and a second pipe that requires a maximum velocity of 35 ft/sec. Also calculate the pressure drop through each pipe. Which pipe is desirable from an energy conservation point of view? Assume that the temperature of the air is 70°F.

11.15 Estimate the rate of air leakage in standard cubic feet per minute through a 0.010-in.-diameter pinhole in a pipe that contains air at 5 atm.

11.16 A compressor for a plant compressed air system delivers 200 CFM at 8 atm from inlet air at 70°F and 1 atm. Calculate the outlet air temperature for the compressor if its adiabatic efficiency is 80%. It is desired to recover energy in an aftercooler to heat water coming in at 80 °F. Find the size of a crossflow heat exchanger with the air mixed and the water unmixed required to heat 9.5 gal/min of water to 150°F. The overall coefficient for the exchanger is 20 Btu/hr ft^2 °F.

TWELVE

INDUSTRIAL COGENERATION

12.1 INTRODUCTION: THE COGENERATION CONCEPT

Cogeneration is the concurrent production of power (either mechanical or electrical) and process heat in a single integrated utility system. The concept is not a new one, but in recent years it has begun to receive renewed attention. At the turn of the century most industrial plants generated their own electricity, and in urban areas, such as New York City, electric utilities sold large amounts of low-pressure steam extracted from their generating turbines to factories and commercial buildings. In 1960, however, industry was generating only about 21% of its own electricity, and by 1968 this figure had declined to about 17% [1].

Several factors contributed to the decline of in-plant power generation and external distribution of utility steam. The economies of scale in power-generating equipment, the reliability of large interconnected power grids, and favorable rate regulations for the large utilities made the purchase of power by industry more economical in many cases than internal generation. At the same time, the difficulty of building and maintaining steam distribution systems and of recovering valuable condensate discouraged continued sale of steam by the utilities.

In the early 1970s, however, the economics of cogeneration began to shift. Rapid escalation in fuel prices focused attention on efficiency of energy use, while sharp increases in the capital cost of new central generating facilities caused the utilities to seek ways of delaying or avoiding new construction. Today, cogeneration is attractive both to industry and to many utilities, and it represents a major energy conservation opportunity in many industrial plants. The thermodynamic motivation for cogeneration can be clearly illustrated with an example.

Example 12.1

Consider the simple back-pressure turbine system shown in Fig. 12.1. Steam is generated at 100,000 lb/hr in a boiler of 80% efficiency at a pressure of 750 psia and a temperature of 700°F. An 80% efficient back-pressure turbine is used to generate electricity, while the 150-psia exhaust steam is passed on to a process unit, where it is condensed for process heat. The condensate from the process unit is returned to the boiler at 150 psia. We will determine the amount of power and process heat supplied by this integrated system, calculate the overall energy utilization efficiency, and compare this with the requirements of separate power- and heat-generating units.

From the steam tables for superheated steam at 750 psia and 700°F,

$$h_1 = 1342.5 \text{ Btu/lb}_m \quad \text{and} \quad s_1 = 1.5577 \text{ Btu/lb}_m \text{ °F}$$

If the steam expanded isentropically through the turbine, i.e., $s_1 = s_2$, it would exhaust in a saturated state with enthalpy h_{2s}. In fact, because the turbine is not 100% efficient, it will exhaust at a higher enthalpy. We will use the turbine efficiency to determine the actual exhaust state.

At 150 psia, the saturation properties of steam (from Table 7.2) are

$$h_f = 330.6 \text{ Btu/lb}_m \qquad h_g = 1194.1 \text{ Btu/lb}_m \qquad h_{fg} = 863.4 \text{ Btu/lb}_m$$

$$s_f = 0.5141 \text{ Btu/lb}_m \text{ °F} \qquad s_g = 1.5695 \text{ Btu/lb}_m \text{ °F}$$

Since we are considering isentropic expansion,

$$s_2 = s_1 = s_{f2} + x_{2s}(s_{fg2}) = 1.5577 \text{ Btu/lb}_m \text{ °F}$$

Solving for the quality under isentropic expansion conditions,

$$x_{2s} = 0.665$$

We can now determine the exhaust enthalpy for the isentropic case:

$$h_{2s} = h_f + x_{2s}(h_{fg}) = 1124.7 \text{ Btu/lb}_m$$

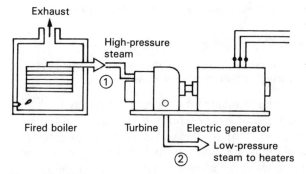

Figure 12.1 Back-pressure turbine cogeneration system.

The actual exhaust enthalpy is calculated from the definition of the turbine efficiency:

$$h_2 = h_1 - \eta_t(h_1 - h_{2s}) = 1168.26 \text{ Btu/lb}_m$$

The heat delivered to the process is, therefore,

$$\dot{Q}_p = (100,000 \text{ lb}_m/\text{hr}) (h_2 - h_{2f})$$

$$= 83.77 \text{ million Btu/hr}$$

The power generated is

$$P = (100,000 \text{ lb}_m/\text{hr}) (h_1 - h_2)$$

$$= 17.42 \text{ million Btu/hr} = 5.1 \text{ MW}$$

The fuel required by the boiler is

$$F = (100,000 \text{ lb}_m/\text{hr}) \, \frac{(h_1 - h_{2f})}{\eta_b}$$

$$= 126 \text{ million Btu/hr}$$

Thus, the overall energy utilization efficiency can be determined:

$$\eta_{oa} = \frac{P + \dot{Q}_p}{F} = \frac{17.42 + 83.77}{126} = 80\%$$

Now consider the fuel requirements to provide the same power and process heat by using conventional independent power and steam generation. A simplified schematic of a conventional power plant is shown in Fig. 12.2a. A typical efficiency for such a plant is on the order of 34%. This can also be expressed in terms of the plant heat rate (Btu's of fuel required per hour per kilowatt generated); a plant heat rate of about 10,000 Btu/kW hr is typical. The fuel required to generate 5.1 MW is thus

$$F_p = (5,100 \text{ kW}) (10,000 \text{ Btu/kW hr})$$

$$= 51 \text{ million Btu/hr}$$

Assume for this comparison that the process steam is generated in a boiler of 80% efficiency at the required process pressure of 150 psia (Fig. 12.2b). For our same process heat requirement of 83.77 million Btu/hr, the fuel to the boiler is

$$F_Q = \frac{83.77 \text{ million}}{0.80} = 104.7 \text{ million Btu/hr}$$

The total fuel required for the independent processes is the sum of that required for power plus that required for process heat:

$$F = F_p + F_Q = 155.7 \text{ million Btu/hr}$$

With cogeneration, then, we can save about 30 million Btu/hr in this particular application, or about 19% of the fuel required by the conventional method.

This example illustrates the rationale for cogeneration purely from an energy point of view. To be practical, a cogeneration system must, of course, be economically advantageous, that is, the added costs of equipment, maintenance, and operation must be satisfactorily balanced by savings in energy cost. Cogeneration does make economic sense in many industrial

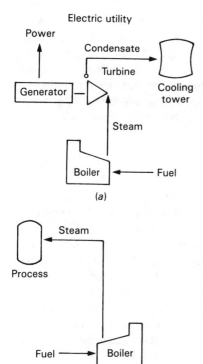

Figure 12.2 (*a*) Conventional utility electrical generation; (*b*) conventional industrial steam generation.

plants. The balance of heat and electricity requirements of the particular process largely determines the type of cogeneration system most suitable for a given situation. A number of alternatives are available; these will be discussed in the following section.

12.2 COGENERATION ALTERNATIVES

Several alternative schemes have been developed for concurrent generation of power and process heat, depending on the needs of the particular operation. The cogeneration system may be completely contained within a single plant or it may be jointly owned and operated by several different plants or by an industry/utility partnership. In this chapter we will be primarily concerned with in-plant cogeneration, the most common type of arrangement. Joint ownership operations will be discussed briefly.

Figure 12.3 shows a general schematic of different approaches to in-plant cogeneration, and in particular illustrates the distinction between topping and bottoming cycles. In a topping cycle, power generation takes place at the high-temperature end of the process. Fuel is burned to produce hot combustion gas, which is used either to generate power directly (as in a gas turbine or diesel engine) or to generate high-pressure, high-temperature steam for expansion through a steam turbine. The remaining energy in the gas or the low-pressure exhaust steam is used to provide heat to the process. Bottoming cycles provide process heat at high temperatures, as in steel, cement, or glass making, and utilize the waste heat to generate power. Several approaches to topping and bottoming cycles will be discussed in turn.

12.2.1 Steam Turbine Topping

The most common approach to topping is the steam turbine cycle shown in Figs. 12.4a and b. Steam is typically generated at 1000°F and 1,500 psia and expanded through either a simple back-pressure turbine (Fig. 12.4a) or an extraction turbine (Fig. 12.4b). The back-pressure turbine is simpler and less expensive, both initially and in operation. It is limited to operation at a single back-pressure (i.e., it cannot feed two headers at different pressures) and is somewhat less flexible in meeting varying plant steam and electric loads than an extraction turbine. The extraction turbine is more complex to install and control and hence is more expensive, but it can provide steam at various pressure levels, depending on where the extraction points are located in the turbine stages. If the particular plant typically has a high power demand and if process steam loads fluctuate significantly, a condensing extraction unit can be used in which the exhaust pressure is maintained

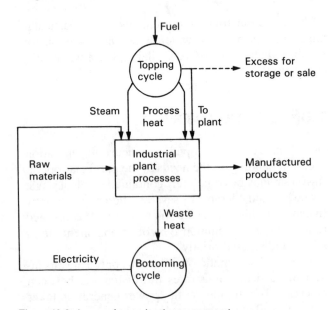

Figure 12.3 Approaches to in-plant cogeneration.

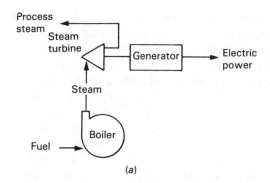

(a)

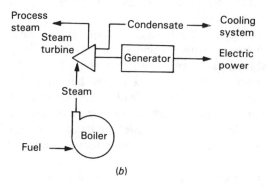

(b)

Figure 12.4 (*a*) Back-pressure turbine topping cycle; (*b*) extraction
turbine topping cycle.

at about 1 psia with a condenser. In this case some of the thermodynamic
advantage of cogeneration is lost, but operation is very flexible and can
adapt easily to varying plant energy needs.

In general, steam turbine topping is best suited to situations in which
process heat demand is high relative to electrical demand. This approach
also offers the significant advantage of fuel flexibility; that is, the boiler can
be fired with gas, oil, coal, or residual by-product fuels such as wood wastes
or refinery off-gas.

12.2.2 Gas Turbine Topping

Gas turbines (sometimes also referred to as combustion turbines) are often
used as alternative prime movers in cogeneration systems. Figure 12.5 shows
the general configuration used in these systems. The gas turbine consists of a
compressor section, which pressurizes atmospheric air to around 4–6 atm; a
combustor, which injects fuel and ignites the mixture, producing a gas
stream at a temperature of 1600–1700°F; and a turbine, which expands the
hot, high-pressure gas to produce power. In a simple gas turbine a significant amount of the turbine output is required to drive the compressor (on
the order of 40%), and the exhaust gas from the turbine is still quite hot,

typically 800–900°F. Thus, for power generation alone, gas turbines are not very efficient (characteristically 30% or less). When the hot exhaust gas can be used for process heating, however, the gas turbine can be quite attractive.

Gas turbine exhaust is relatively clean (compared to conventional boiler flue gas), as clean fuels are required in this type of combustor, and air flow rates are high (typically four to five times theoretical air). For this reason, the exhaust gas can be used directly for process heat in many processes, such as drying. If higher temperatures are required, additional fuel can be burned in the exhaust duct, since the gas still contains abundant oxygen.

Alternatively, the turbine exhaust can transfer its heat to a process fluid through a heat exchanger, or steam can be produced in a heat recovery boiler. Thus, the gas turbine system offers considerable flexibility in terms of the specific demands of the application.

Gas turbine topping is best suited to applications requiring a higher power/process heat ratio than steam turbine topping. In situations where this ratio is quite high, a combined gas/steam turbine topping cycle can be employed (Fig. 12.6). In this system, steam generated in the heat recovery boiler is utilized in a steam turbine to produce power and the turbine exhaust steam is utilized for process heat. Gas/steam turbine topping cycles usually use a fired heat recovery boiler, that is, additional fuel is burned in the gas turbine exhaust stream to allow steam generation at higher temperature and pressure.

While gas turbine cycles offer certain advantages as cited above, they also have limitations. Poor load-following ability is one of them. The output of a gas turbine cannot be modulated as easily as that of a steam turbine without seriously degrading its performance. Thus gas turbines operate best as base-loaded systems, i.e., where they can run at steady output all the time. This problem is significantly alleviated in the combined gas/steam turbine cycle. The gas turbine can be run at steady load, and load swings can be easily accommodated by controlling the fuel input to the fired heat recovery boiler and the steam flow to the back-pressure turbine. Gas

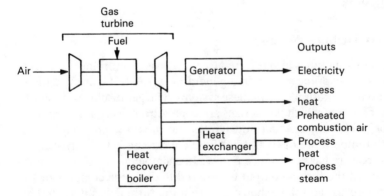

Figure 12.5 Gas turbine topping cycle.

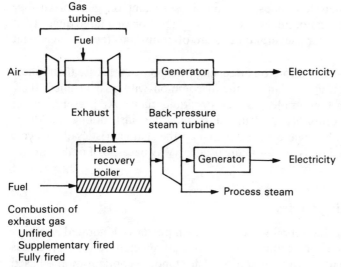

Figure 12.6 Gas/steam turbine topping cycle.

turbines are also presently limited to the use of clean fuels such as natural gas or light distillate fuel oils. This fuel inflexibility is currently the most serious impediment to wider use of gas turbine-based cogeneration. Research efforts are under way to develop techniques to allow gas turbines to run on heavy fuel oils and synthetic fuels.

12.2.3 Diesel Engine Topping Cycles

Diesel engine/generating units in capacities up to about 25 MW can be used as prime movers in cogeneration systems where a high power/heat ratio is needed. This type of system is shown schematically in Fig. 12.7. High-temperature exhaust gas from the engine is utilized in a heat recovery boiler for steam generation; cooling water from the engine jacket is used to preheat the boiler feedwater. Since, in contrast with the gas turbine, diesel exhaust is

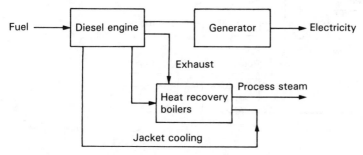

Figure 12.7 Diesel engine topping cycle.

nearly stoichiometric, it does not contain sufficient oxygen for further combustion, nor is it abundant enough for generation of a large quantity of steam. For this reason, several diesels are often used to feed a single heat recovery boiler.

The principal limitations on diesel engine topping cycles are their inherently high power/heat balance and, in common with gas turbines, their requirement for high-grade gaseous or liquid fuels. At present, diesel cogeneration is primarily used in commercial and institutional buildings where the energy balance is a good match for the characteristics of the cycle and where refined fuels are normally used anyway because of pollution and fuel-handling limitations.

12.2.4 Bottoming Cycles

An alternative to the topping cycle, in which power is generated at a high temperature, is the bottoming cycle, where heat rejected from the process is used as a source for power generation. The standard configuration is based on the Rankine cycle illustrated in Fig. 12.8. Waste heat from the process is used in a heat recovery boiler (vapor generator) to produce a vaporized working fluid. If the heat source is at a high enough temperature, water is used as the working fluid and the cycle is identical to that in a simple conventional steam power plant. Lower-temperature waste heat sources require the use of a low-boiling-point working fluid; organic refrigerants are presently used for this purpose.

Most of the systems currently used in industry utilize steam bottoming and are thus restricted to processes that inherently produce high-temperature waste-gas streams. Glassmaking plants, cement plants, and steel fabrication mills are examples. Organic Rankine cycle bottoming is a relatively new technology, which has been successfully applied in petroleum refineries and chemical plants, among others.

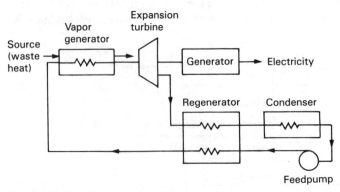

Figure 12.8 Rankine bottoming cycle.

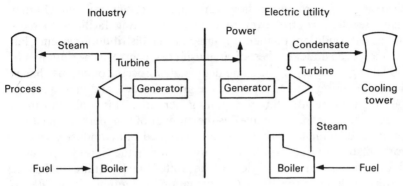

Figure 12.9 Electricity-export cogeneration.

12.2.5 Industry/Utility Cogeneration

The schemes discussed above are used primarily for in-plant cogeneration, although they also lend themselves to joint ownership by several plants having complementary power and heat demand requirements. Another cogeneration alternative is an interactive arrangement between an industrial plant or plants and the electric utility that serves them. This is becoming an increasingly attractive alternative, as favorable regulatory legislation coupled with the increasing cost of new utility plant construction has created a favorable climate for joint industry/utility ventures.

The most common arrangement is shown in Fig. 12.9. It is simply a conventional in-plant back-pressure steam cogeneration unit that provides the plant's own electricity needs and also produces some excess electricity for export into the utility grid. From the industry viewpoint, this arrangement is technically attractive because the utility essentially becomes an electricity "reservoir" that permits the turbine/generator unit to run at arbitrary load to satisfy process steam demands. If steam demand is high while plant electric demand is low, the necessary steam can be passed through the turbine and excess electricity sold to the utility. Since the industry/utility tie allows import as well as export of electricity, the plant buys electricity from the utility when the cogeneration system is running at a condition that does not allow it to fully meet the plant's power demand. This arrangement can also be economically attractive in that it ensures reliability of electric supply and permits the plant to benefit from some economy of scale by installing a larger unit than would be feasible if export of power were not allowed.

From the utility viewpoint, tying industry cogenerators into the grid is a mixed blessing. On the positive side, these additional sources of power can defer the need for new plant construction; industrial cogeneration plants can become reservoirs to draw on, particularly at times of heavy peak loading. This assumes that suitable coordination between the utility and the plants can be arranged. Coordination, however, can be a serious problem. Utilities can exercise close control over their own generating facilities to meet changing load conditions on their transmission and distribution systems, but they cannot, even under the best of circumstances, control a large number of individual cogenerators to the same degree. As process conditions change within each plant, electrical inputs to the grid will vary, and these must be compensated for by adjusting the utility generators. This problem can be expected to become more complex as the number of cogeneration facilities increases, and ultimately some sort of integrated utility/industry control network will be required. Another negative factor, in the view of some utilities, is overcapacity. In some regions of the United States, utilities built large generating facilities in the 1970s in anticipation of continued high rates of electrical demand growth, however, demand growth has moderated in recent years. Where generous capacity margins exist, it is not in the best interests of the utility to encourage new generating capacity in the form of cogeneration while large coal and nuclear plants operate at reduced load levels. This problem will gradually abate as overall electric demand growth catches up with capacity, and, in fact, some utilities that actively discouraged cogeneration in the middle to late 1970s have experienced a turnaround and are actively encouraging cogenerators in the 1980s.

The other form of utility/industry cogeneration is illustrated in Fig. 12.10. In this case the utility exports steam extracted from its generating turbines to the plant or plants. This type of system is desirable from the standpoint of fuel utilization, since the utility steam is normally generated with low-cost coal. Most industrial plants do not have sufficient steam demand to justify large coal-fired facilities and must generate their steam with gas or oil. It is also a more controllable system, primarily because a single electric generating plant is directly tied to a limited number of industrial plants. Steam exchange arrangements, however, are limited to plants in close proximity, since transmission of steam and condensate over long distances becomes too expensive.

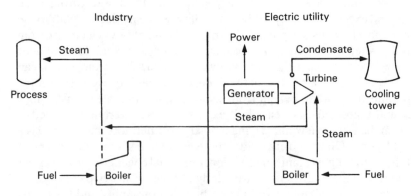

Figure 12.10 Steam-export cogeneration.

12.3 TECHNICAL AND ECONOMIC EVALUATION OF COGENERATION SYSTEMS

Determination of the technical and economic feasibility of a cogeneration system in a given application requires an analysis of the energy-using characteristics of the plant and compatibility with various types of cogeneration plants. In this section we will consider the systematic analysis of cogeneration systems.

12.3.1 Selection of Cogeneration System Configuration

The various types of topping and bottoming cycles used in industrial cogeneration and some of the advantages and disadvantages of each were discussed in Section 12.2. One of the principal determinants of the type of system to be selected is the balance of heat and power demands of the plant.

Figure 12.11 shows typical electrical and process steam capacities for steam turbine, gas turbine, and diesel engine topping cycles. Steam turbines are favored in situations where steam demand dominates the energy balance, and diesel engine cogeneration is better suited to electricity demand-dominated processes. Gas turbine cogeneration is best suited to processes that fall between these two extremes. A plant requiring 2 million pounds of process steam per hour and 10 MW of electrical power would probably best utilize a steam turbine cycle, while another plant requiring the same steam capacity but 50 MW of power would find a gas turbine system most suitable. At 20–30 MW, either system might prove feasible.

The feasibility of a Rankine bottoming cycle depends on the temperature of the waste heat source and the working fluid selected for the cycle. Figure 12.12 shows the inverse heat rate (kilowatt-hours of electricity produced per million Btu's of heat input) for various working fluids as a function of source gas temperature. As inverse heat rate goes down, indicating a decrease in overall efficiency of the cycle, the cost of the system to produce a given power output goes up, since more heat must be transferred from the source gas. While steam can produce satisfactory heat rates at high source gas temperatures, at temperatures below about 500-600°F, organic fluids are required.

12.3.2 Thermodynamic Evaluation of Cogeneration Systems

Decisions on implementation of cogeneration in an industrial plant are ultimately based on economic considerations. In energy-intensive industries, however, an important factor in the economic equation is the cost of fuel and electricity; as shown earlier in this chapter, cogeneration can affect significant overall energy savings. The detailed thermodynamic evaluation

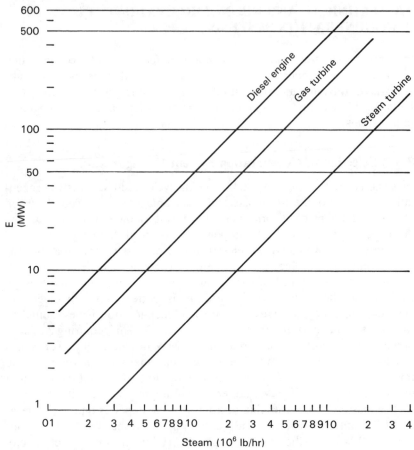

Figure 12.11 Application range for various topping cycles. (From Noyes [4].)

procedure for a cogeneration system depends on the type of prime mover used (i.e., steam turbine, gas turbine, or diesel engine) and how it is integrated into the process. In this section we will consider only the simple back-pressure steam turbine system. The general approach can be extrapolated to other types of systems as well.

Figure 12.13 shows the configuration to be considered. It is assumed that high-pressure steam is provided to the turbine through a high-pressure process header from the boiler. Power generated by the turbine provides for the plant electrical (or mechanical) demand, and any excess is sold to the utility grid. Steam exhausted from the turbine is delivered to the process (usually through a low-pressure steam header). In general, the plant electrical demand and steam demand will not be in balance; some excess or deficit will always exist in one demand or the other. Excess electricity is simply exported, but excess steam must be either exhausted to the atmosphere, a practice that is wasteful of both energy and valuable treated water, or condensed, in which case the latent heat is lost but the condensate recovered. If a deficit of process steam exists, it must be made up with low-pressure steam provided directly from a low-pressure boiler or through a pressure-reducing valve from the high-pressure header. A deficit in electricity is met by purchasing electricity from the grid.

Plant electricity and steam demands will, in general, vary with time, and thus the cogeneration system must follow a varying load profile. Figure

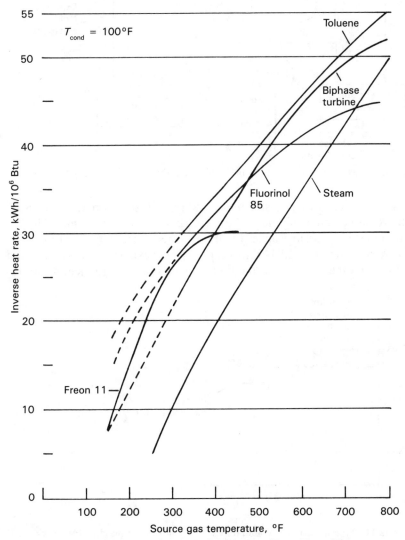

Figure 12.12 Performance of Rankine bottoming cycles. (From Noyes [4].)

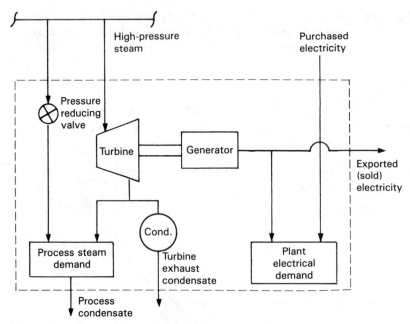

Figure 12.13 Schematic for analysis of back-pressure turbine cogeneration system.

12.14a illustrates a varying steam demand typical, say, of batch processing with steam-heated equipment. For a turbine unit of a given rating, a maximum steam-passing capacity will exist when the unit is operating at full load. As shown in Fig. 12.14a, this capacity may exceed the steam demand most of the time, and thus if the unit is being operated to meet steam re-

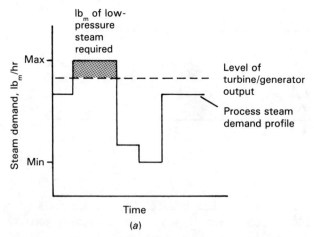

Figure 12.14 Typical plant energy-use profiles. (a) Plant steam demand profile showing T/G output.

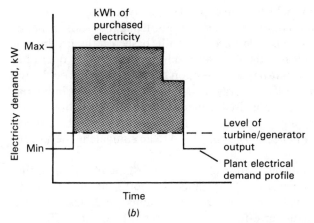

Figure 12.14 Typical plant energy-use profiles (*Continued*). (*b*) Plant electrical demand profile showing T/G output.

quirements, it will usually be running at partial load. During part of its operating cycle, however, steam demand will exceed capacity and additional process steam will have to be provided. Similarly, the plant will have a characteristic electricity demand profile, as shown in Fig. 12.14*b*. In the case shown, the cogeneration unit is rated well below usual electric demand requirements, and electricity must be purchased to supply the deficit.

The overall plant energy balance at any given time will thus depend on both the instantaneous steam and electric demands and the rating of the turbine/generator unit. Furthermore, it will depend on the operating strategy of the unit, that is, whether it is being operated to track steam demand or electricity demand. In general, it is advantageous to track steam demand and to operate the system so that no excess steam is ever passed down through the turbine. When electricity is being generated by the turbine, steam tracking is relatively straightforward, since the utility intertie serves to provide either a reservoir for excess electricity or a source of additional electricity to meet the plant deficit. When the turbine is directly coupled to a piece of mechanical equipment, such as a pump or compressor, operation is more constrained, since the equipment must operate at a particular input power level dictated by the needs of the process, somewhat equivalent to electricity demand tracking.

A procedure for evaluating energy consumption characteristics for a given unit follows.

1. Define the typical profiles of steam and electric (or mechanical) power demand for a selected time period. Ultimately, it will be desirable to convert all energy flows for the system to an annual basis. For detailed analysis, however, a shorter time period such as a day, week, or month must

be used; the time period selected will depend on the type of operation. In a chemical plant with large continuous process units, it might be satisfactory to define the plant operating profile for one month, based on the production mix anticipated for that month. A small food-processing plant operating batch cookers on a two-shift-per-day basis, on the other hand, would have to be defined in terms of daily energy demand profiles. In most industrial plants, process steam demands are defined in terms of mass flow (e.g., pounds per hour) of steam at a given saturation condition (usually a specified pressure). These will need to be converted to an equivalent energy rate (e.g., Btu's per hour).

    2. Define the turbine/generator efficiency as a function of operating load level. Turbine/generator (T/G) units do not generally maintain constant efficiency as load level is reduced. Most units are designed to give maximum or near-maximum efficiency at the rated load, and efficiency tends to drop off at lower load levels. Figure 12.15 shows a typical efficiency curve for small industrial turbine/generator units. Actual rated efficiency for a given unit will depend on the inlet and exhaust steam conditions as well as the unit's mechanical design; these figures can be obtained from turbine manufacturers. For each segment of the cogeneration system demand profile, the unit efficiency will be needed to determine actual steam flow rates and exhaust steam conditions.

    3. Select a rated (i.e., 100% load) power output or steam flow for the turbine/generator unit. In principle, a T/G unit of any arbitrary size could be used in a given plant, assuming that the local utility is willing to buy all the excess power generated. In practice, it is rarely economically desirable to produce either power or steam in amounts much in excess of the plant's own maximum demand. Thus the selected turbine/generator capacity will generally be somewhat below the maximum plant demand level. The primary

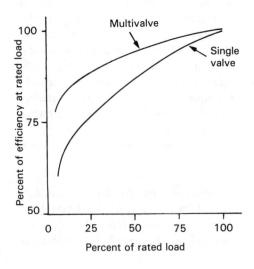

Figure 12.15 Turbine part-load performance.

objective of the overall engineering analysis of the cogeneration system is to determine the size of the unit that will give the best lifetime economic performance.

4. For each time increment, compute the overall energy balance for the cogeneration unit; use the procedure illustrated in Example 12.1, based first on a strategy of electric (or power) demand following and then on steam (or process heat) demand following. In each case, all inputs and outputs of the unit will be computed, including high-pressure steam to the turbine, additional process steam, process condensate, and electricity purchased or exported. Each of these energy flows has a value, and the net associated cash flow can be determined for each time increment.

5. Add the energy flows of each type for all the time increments and annualize the totals to determine an annual energy balance. Two results will be obtained, one for the steam load-following operating strategy and the other for the electric load-following strategy. The more desirable operating strategy from a thermodynamic viewpoint can thus be determined. Experience with cogeneration systems in industry tends to indicate that steam load following generally produces better overall efficiency and better economic results. This will depend, however, on the local situation vis-à-vis industrial fuel and electric rates.

6. Revise the analysis by considering turbine/generator units of various sizes to find the unit size that will give the best lifetime economic results.

Example 12.2

To illustrate the procedure described above, consider the case of a paper-products plant operating on two shifts per day, six days a week, 50 weeks per year. Operating profiles of electricity and process heat demand for the plant are shown in Figs. 12.16a and b. The profiles are assumed to be typical, and the cogeneration system is analyzed for one day's operation; costs are then extended to an annual basis.

We will assume that superheated steam at 800 psia and 1000°F is available from the boiler. Process heat is to be supplied at 160 psia, either by exhaust from the turbine or by direct let-down from the high-pressure steam header. The turbine/generator efficiency as a function of load is shown in Fig. 12.15. We will initially assume a turbine rated at 4,000 kW; at this load level, the rated efficiency is 83%.

Several thermodynamic properties will be useful in the analysis. From the steam tables:

State 1, high-pressure header: at 800 psia and 1000°F, $h_1 = 1511.4$ Btu/lb$_m$, $s_1 = 1.6807$ Btu/lb$_m$ °F

State 2, turbine exhaust: at $P_2 = 160$ psia, for an isentropic expansion through the turbine, $h_{2s} = 1299.3$ Btu/lb$_m$.

Process condensate: at $P_2 = 160$ psia, saturated liquid, $h_{2f} = 336.1$ Btu/lb$_m$.

Let us first determine the energy flows and costs if cogeneration is not used, that is, if all electricity is purchased and process heat is provided by direct let-down from the boiler. The results of this calculation are shown in Table 12.1. At each time step, the process heat and electric power demands are shown. These must be translated into total pounds of steam and kilowatt-hours of electricity required for one day. Kilowatt-hour demand is simply the power

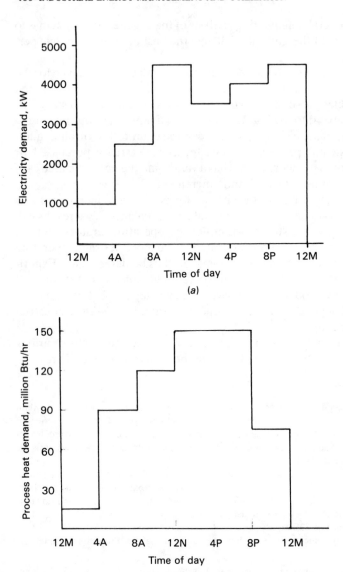

Figure 12.16 (*a*) Electricity demand; (*b*) heat demand.

demand multiplied by the number of hours (in this case, 4) in each time step. Steam flow required to meet the process heat demand for each time period is

$$\dot{m} \ (\text{lb}_\text{m}/\text{hr}) = \frac{\dot{Q}_p \ (\text{Btu/hr})}{(h_1 - h_{2f}) \ (\text{Btu/lb}_\text{m})}$$

and the total steam required is the flow rate multiplied by the number of hours. The kilowatt-hours and steam use in each time period are summed to give daily consumption values, and

Table 12.1 Energy Consumption and Cost; No Cogeneration

Time step	Process heat demand (10^6 Btu/hr)	Electricity demand (kW)	Direct process steam flow (10^3 #/hr)	Total steam (10^3 #)	Purchased electricity (kWh)
1	15	1000	12.76	51.05	4,000
2	90	2500	76.57	306.31	10,000
3	120	4500	102.10	408.41	18,000
4	150	3500	127.63	510.51	14,000
5	150	4000	127.63	510.51	16,000
6	75	4500	63.81	255.26	18,000
				2,042.05	80,000

Annual steam cost (1000's $)
2450

Total annual cost (1000's $)
4010

461

these are then annualized based on the number of days per year of plant operation. The cost figures shown in Table 12.1 assume a steam value of $4.00 per thousand pounds and a purchased electricity cost of $0.065/kWh. Condensate value is neglected in this example. Note that the analysis could easily accommodate time-varying electric rates by computing costs at each time step rather than aggregating them for an entire day.

Analysis of the cogeneration system for this case is summarized in Tables 12.2a and b and 12.3. For convenience in referring to the tables, the columns are designated A, B, etc. The cogeneration system may be operated in either an electric demand-tracking mode or a steam demand-tracking mode. For electric demand tracking, the turbine is operated to just meet the electric demand; if electric demand exceeds the turbine/generator capacity, the turbine is operated at rated load and the excess demand is met with purchased electricity. Note that in column D of Table 12.2a the system is at part load during time periods 1, 2, and 4, and fully meeting the plant demand. Column F shows the turbine/generator efficiency corresponding to operation at each load level. Given this efficiency, the actual exhaust steam enthalpy, h_2, can be computed:

$$h_2 = h_1 - \eta_t(h_1 - h_{2s})$$

It should be noted that turbine efficiencies are often given as functions of percent of rated steam flow rather than percent of rated electric load. In this case it will be necessary to iterate on the efficiency and exhaust enthalpy by determining the mass flow corresponding to the calculated value of h_2:

$$\dot{m}\ (\text{lb}_m/\text{hr}) = \frac{3413 \times P\ (\text{kW})}{(h_1 - h_2)\ (\text{Btu/lb}_m)}$$

A revised efficiency based on this mass flow can be determined, and new values of h_2 and η_t calculated until consistent values are obtained.

The amount of heat delivered to the process through the turbine, shown in column J, is computed from

$$\dot{Q}_t = \dot{m}\ (h_2 - h_{2f})$$

The balance of the process heat demand must be made up by direct let-down steam flow from the high-pressure header, column L. Note that during time steps 1 and 6, more heat is delivered by the turbine exhaust than is required by the process. This steam imbalance condition can and does occur, and excess steam must be condensed or vented. This is obviously wasteful, and in operation it would be desirable to operate the system in the steam-tracking mode during periods of low steam demand.

Table 12.2b shows the thermal analysis of the system operating in the steam-tracking mode. In this case the turbine is operated at a load level just sufficient to meet the heat demand of the process. If the process heat demand exceeds the maximum steam-passing capacity of the turbine at rated load, the excess demand is made up with let-down steam. In the steam demand-tracking mode, electricity in excess of the plant's internal demand may be generated, in which case it is sold to the utility grid.

Note that in the steam-tracking analysis it is always necessary to iterate on the turbine efficiency, since the quantity being tracked is itself directly tied to the efficiency through the exhaust enthalpy. At each time step, an efficiency is assumed and the value of h_2 determined as above. The mass flow required to meet the given process heat demand is then calculated:

$$\dot{m}\ (\text{lb}_m/\text{hr}) = \frac{\dot{Q}_p\ (\text{Btu/hr})}{(h_2 - h_{2f})\ (\text{Btu/lb}_m)}$$

and a new efficiency is determined based on this mass flow. Iteration is continued until consistent values of η_t and $\dot{m}$ are obtained. In this example the turbine operates at full load during all

Table 12.2 Thermal Analysis: 4000-kW Turbine/Generator Unit

A Time step	B Process heat demand (10^6 Btu/hr)	C Electric demand (kW)	D Capacity[a] (%)	E Electricity output (kW)	F Turbine efficiency	G True h_2 (Btu/#)	H h_2-h_{2f} (But/#)	I Turbine flow (10^3 #/hr)	J T/G process heat (10^6 Btu/hr)	K Additional process heat (10^6 Btu/hr)	L Direct process flow (10^3 #/hr)
(a) Electric demand tracking											
1	15	1000	25.00	1000	.65	1374	1037	24.76	25.68	0	0
2	90	2500	62.50	2500	.79	1345	1009	51.25	51.70	38.30	32.59
3	120	4500	100.00	4000	.83	1335	999	77.55	77.49	42.51	36.17
4	150	3500	87.50	3500	.83	1336	1000	68.27	68.29	81.71	69.52
5	150	4000	100.00	4000	.83	1335	999	77.55	77.49	72.51	61.69
6	75	4500	100.00	4000	.83	1335	999	77.55	77.49	0	0
(b) Steam demand tracking											
1	15	1000	19.36	774	.62	1379	1043	14.38	15.00	0	0
2	90	2500	100.00	4000	.83	1335	999	77.55	77.49	12.51	10.64
3	120	4500	100.00	4000	.83	1335	999	77.55	77.49	42.51	36.17
4	150	3500	100.00	4000	.83	1335	999	77.55	77.49	72.51	61.69
5	150	4000	100.00	4000	.83	1335	999	77.55	77.49	72.51	61.69
6	75	4500	96.78	3871	.83	1335	999	75.05	75.00	0	0

[a]Capacity is electric capacity in (a) and process heat capacity in (b).

time steps except 1 and 6; for these periods, the turbine operates at part load to meet the process heat requirements.

Table 12.3 shows the energy cost analysis for the system in the two assumed operating modes. Total steam use for each time period (turbine steam plus direct let-down steam) is shown in column N, purchased electricity in column O, and sold (exported) electricity in column P. Annualized total costs are summarized in Table 12.3, assuming steam at $4.00 per thousand pounds, purchased electricity at $0.065/kWh, and sold electricity at $0.03/kWh.

For operation in the electric demand-tracking mode, purchased electricity costs are minimized, since electricity is purchased only when demand exceeds the rated capacity of the turbine/generator unit. No electricity is exported to the grid in this mode.

In the steam-tracking mode, more electricity is purchased in this case, since during time steps 1 and 6 the unit is being operated at lower capacity than in the electric demand-tracking mode. Interestingly, though, steam tracking actually produces lower overall electric costs, since excess electricity is generated and sold during times of high steam demand (time steps 2 and 4)

Table 12.3 Cost Analysis: 4000-kW Turbine/Generator Unit

M Time step	N Total steam (10^3#)	O Purchased electricity (kWh)	P Sold electricity (kWh)
(a) Electric demand tracking			
1	99.02	0	0
2	335.34	0	0
3	454.87	2000	0
4	551.16	0	0
5	556.97	0	0
6	310.20	2000	0
Total	2,307.56	4000	0

Annual electricity revenue (thousand $): 0
Annual costs (thousand $)
  Electricity: 78.00
  Steam: 2769.08
  Net: 2847.08

(b) Steam demand tracking			
1	57.50	903	0
2	352.77	0	6000
3	454.87	2000	0
4	556.97	0	2000
5	556.97	0	0
6	300.20	2514	0
Total	2,279.28	5417	8000

Annual electricity revenue (thousand $): 72.00
Annual costs (thousand $)
  Electricity: 105.64
  Steam: 2735.14
  Net: 2768.78

and total steam costs are also slightly lower. Since this operating strategy never produces excess turbine exhaust steam, venting or condensing is not required at any time. In fact, analysis of many cogeneration systems has shown that, in general, steam tracking produces lower net energy cost than electric demand tracking, except where purchased electricity cost and buy-back price are very high compared to steam cost.

The annualized energy costs determined in the above analysis can now be used in an overall system economic analysis, as discussed in Chapter 3, by combining them with capital, labor, overhead, and other cost elements. Energy costs should be recalculated for each year of system life based on assumed escalation of electricity and steam prices, but the basic system thermal analysis need not be recomputed for each year unless significant changes in energy use profiles are anticipated.

A complete cogeneration analysis for a particular plant would require consideration of turbine/generator units of different sizes to meet the given steam and electric demands. For example, the case analyzed above was recalculated for an assumed turbine/generator rating of 2,000 kW instead of 4,000 kW; the results are summarized in Tables 12.4 and 12.5. With a smaller turbine unit less electricity can be generated, and hence, in either operating mode, purchased electricity costs are higher and revenues are zero. Although overall steam costs are somewhat lower, the net annual operating cost with the smaller unit is higher. Of course, the capital cost of the 2,000-kW unit is also lower; therefore, the decision about which unit is more economical cannot rest on energy costs alone, but must be based on a comprehensive analysis with discounted cash flow, present-worth, or other similar techniques.

12.3.3 Economic Evaluation

Procedures for economic evaluation of energy conservation systems were discussed in Chapter 3 and will not be repeated here. However, we will review the economic factors that are particularly pertinent to the evaluation of cogeneration systems.

To determine a bottom-line figure for the economic merit of a partic-ular system, it is necessary to compute annual cash flows for the system, considering all capital and operating costs, and sum these in a consistent way over the lifetime of the system. Since different alternative systems will produce different temporal cash flow profiles, the usual procedure is to apply a discount factor to each year's cash flow to normalize it to present value, and to take the present value of the total discounted cash flows as the common figure of merit for all systems. For cogeneration, the following cost elements are of special importance.

Capital Charges and Associated Financing Considerations. Cogeneration is a major capital investment, and the financing strategy can have a significant

Table 12.4 Thermal Analysis: 2000-kW Turbine/Generator Unit

A Time step	B Process heat demand (10^6 Btu/hr)	C Electric demand (kW)	D Capacity[a] (%)	E Electric output (kW)	F Turbine efficiency	G True h_2 (Btu/#)	H h_2-h_{2f} (Btu/#)	I Turbine flow (10^3 #/hr)	J T/G process heat (10^6 Btu/hr)	K Additional process heat (10^6 Btu/hr)	L Direct process flow (10^3 #/hr)
(a) Electric demand tracking											
1	15	1000	50.00	1000	.75	1352	1016	21.46	21.80	0	0
2	90	2500	100.00	2000	.83	1335	999	38.77	38.75	51.25	43.61
3	120	4500	100.00	2000	.83	1335	999	38.77	38.75	81.25	69.13
4	150	3500	100.00	2000	.83	1335	999	38.77	38.75	111.25	94.66
5	150	4000	100.00	2000	.83	1335	999	38.77	38.75	111.25	94.66
6	75	4500	100.00	2000	.83	1335	999	38.77	38.75	36.25	30.85
(b) Steam demand tracking											
1	15	1000	38.71	774	.71	1361	1025	14.64	15.00	0	0
2	90	2500	100.00	2000	.83	1335	999	38.77	38.75	51.25	43.61
3	120	4500	100.00	2000	.83	1335	999	38.77	38.75	81.25	69.13
4	150	3500	100.00	2000	.83	1335	999	38.77	38.75	111.25	94.66
5	150	4000	100.00	2000	.83	1335	999	38.77	38.75	111.25	94.66
6	75	4500	100.00	2000	.83	1335	999	38.77	38.75	36.25	30.85

[a]Capacity is electric capacity in (a) and process heat capacity in (b).

Table 12.5 Cost Analysis: 2000-kW Turbine/Generator Unit

M Time step	N Total steam (10^3#)	O Purchased electricity (kWh)	P Sold electricity (kWh)
(a) Electric demand tracking			
1	85.82	0	0
2	329.54	2,000	0
3	431.64	10,000	0
4	533.74	6,000	0
5	533.74	8,000	0
6	278.49	10,000	0
Total	2,192.97	36,000	0

Annual electricity revenue (thousand $): 0
Annual costs (thousand $)
  Electricity: 702.00
  Steam: 2,631.55
  Net: 3,333.55

(b) Steam demand tracking			
1	58.55	903	0
2	329.54	2,000	0
3	431.64	10,000	0
4	533.74	6,000	0
5	533.74	8,000	0
6	278.49	10,000	0
Total	2,165.70	36,903	0

Annual electricity revenue (thousand $): 0
Annual costs (thousand $)
  Electricity: 719.61
  Steam: 2,598.83
  Net: 3,318.44

influence on the economic desirability of the project. The installed cost of the system, including the cost of the turbine/generator unit and all associated piping and power transmission equipment, must be determined. If additional steam-generating capacity is needed, the fraction of the steam cost attributable to the cogeneration unit must also be computed. The financing interest rate and financing period must be specified; even if the project is to be financed internally, an internal payback period and cost of capital must be assumed.

Tax Considerations. As with all major corporate investments, tax consequences can play an important role in cogeneration economics. Cogeneration systems can, in some cases, qualify for special investment tax credits.

In addition, the tax rate for the particular company, the method of depreciation used, and the depreciation period must be specified to calculate tax effects associated with the system.

Energy Costs. With cogeneration projects, energy costs represent the largest component of annual cash flow. Present costs of steam and purchased electricity must be specified, as well as anticipated escalation rates of these costs. Similarly, the values and escalation rates of exported energy (electricity and possibly process condensate) must be specified. The value of electricity sold to the utility is an especially difficult number to pin down, since this value is set by public regulation, and regulatory policy on cogeneration is still evolving.

Other Operating Costs. The costs of maintenance and operating labor, while not large compared with those of energy, are not insignificant. Other cost elements that should be considered in the analysis include insurance, replacement parts, and operating overhead.

12.4 OTHER CONSIDERATIONS IN COGENERATION

In the preceding discussion the thermodynamic and (sometimes) economic benefits of cogeneration have been emphasized. The decision to invest in a cogeneration system, however, rests on a number of other factors, many of them negative [2]. Some of these factors will be discussed in this section.

12.4.1 Local Utility Attitudes

Cogeneration can be viewed as either an asset or a liability by a local utility, depending on its capacity margin. As discussed earlier, many utilities that added generating capacity slowly in the 1970s now welcome cogenerators as a source of new capacity which allows them to defer construction of expensive new plants. Others, with large underutilized coal and nuclear facilities, do not see cogeneration with scarce oil and gas as desirable, regardless of thermodynamic efficiency. Even though rates may be regulated, a cooperative attitude on the part of the utility is crucial to making a cogeneration project workable.

12.4.2 Regulatory Considerations

Fear of regulatory interference in business activities has been a major impediment to wider implementation of cogeneration in industry. In 1978 Congress passed the Public Utilities Regulatory Policies Act (PURPA) in an attempt to alleviate some of these impediments. PURPA exempts qualified cogenerators from many of the regulations that apply to public utilities, such as financial disclosure requirements and regulated allowable rate of return on electricity production. It also provides certain incentives in the form of fuel-use exemptions and tax credits to cogenerating companies. While PURPA enhances the attractiveness of cogeneration to industry, it does little to alleviate the concerns of the utility industry, and in fact is generally viewed negatively by the utilities, particularly in terms of rates to be paid to cogenerators for exported power. Several provisions of PURPA are being contested in the courts, and these proceedings are expected to continue for many years to come. Thus, it remains to be seen whether regulatory reform can overcome some of the inherent problems of close technical and financial interaction between regulated and nonregulated business partners.

12.4.3 Environmental Concerns

Detailed evaluations of cogeneration opportunities on a nationwide basis have indicated a potential for nearly 43,000 MW of generating capacity by the year 2000, or about 15% of the nation's total generating capacity [3]. This tremendous capacity, distributed throughout thousands of relatively small units, could represent a serious environmental control problem. In central utility generating plants, environmental controls are highly efficient and closely monitored. Emissions from smaller, decentralized facilities are more difficult to control and to monitor. Economy of scale in emission control equipment also makes it more expensive per unit of power generated to control emissions in a small cogeneration facility than in a central station.

12.5 CONCLUSION

In this chapter we have considered cogeneration of power and process heat primarily from a thermodynamic standpoint. Clearly, cogeneration makes thermodynamic sense. In analyzing cogeneration projects, however, it must be kept in mind that bottom-line decisions are based on economics and the overall business environment as well as energy considerations. The future outlook for cogeneration is positive, but a great deal of progress must be made in resolving regulatory and institutional problems before cogeneration can reach its full energy conservation potential.

REFERENCES

1. Stanford Research Institute, *Patterns of Energy Consumption in the United States*, Menlo Park, Calif., January 1972.
2. Mary Wayne, Plugging Cogenerators into the Grid, *EPRI J*, July/August 1981.

3. Industrial Cogeneration Potential: 1980–2000, *Cogeneration World*, March/April 1982.
4. Robert Noyes, Ed., *Cogeneration of Steam and Electric Power*, Noyes Data Corp., Park Ridge, N.J., 1978.

Additional Reading

D. J. DeRenzo, *Cogeneration Technology and Economics for the Process Industries*, Noyes Data Corp., Park Ridge, N.J., 1983.
D. B. Fisher and P. S. Schmidt, Analysis of Cogeneration Systems Using a Microcomputer, in *Proceedings of the 1983 Industrial Energy Conservation Technology Conference*, Houston, April 1983.

THIRTEEN

POWER CIRCUITS AND ELECTRICAL MACHINERY

13.1 INTRODUCTION

This chapter contains an overview of power circuit characteristics and conventional electric rotating machinery. Equivalent circuit models for machinery and for commonly encountered power circuit components are presented. The chapter establishes an analytical framework that can be used to treat practical application problems related to energy conservation. Details associated with the derivation or rigorous development of the equivalent circuit models are largely omitted, and the reader is referred to the references given at the end of the chapter for additional background material. Examples are presented to illustrate analytical techniques and to demonstrate the use of the equivalent circuit models. Representative data are furnished for use in ascertaining whether significant savings might be realized by particular courses of action.

13.2 FUNDAMENTAL CHARACTERISTICS OF POWER CIRCUITS

Figure 13.1 illustrates a simple power circuit consisting of a source, a load, and a line connecting the two. Analysis of the circuit can be accomplished by using well-known principles of circuit theory.

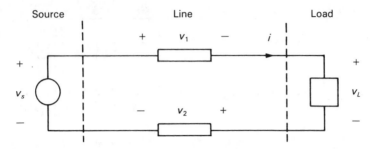

Figure 13.1 A simple power circuit.

Voltages v_1 and v_2 are voltages developed across the line because of the nonideal nature of the line. For example, if the circuit is a direct-current (dc) power circuit in which all voltages and currents are constant in time, Ohm's law ($e = ir$) requires that these voltages be equal to the product of the circuit current i and the resistance associated with each of the lines. It is customary to express constant dc circuit quantities with capital letters. Thus, if the circuit in Fig. 13.1 is taken to be dc: $v_s = V_s$, $v_1 = V_1$, $v_2 = V_2$, etc.

Line voltages V_1 and V_2 are obtained by applying Ohm's law:

$$V_1 = IR_1 \tag{13.1}$$

$$V_2 = IR_2 \tag{13.2}$$

The resistance of a conductor to the flow of dc current is directly proportional to conductor length and inversely proportional to conductor cross-sectional area:[1]

$$R = \varrho \frac{l}{A} \tag{13.3}$$

where ϱ is resistivity, l is circuit length, and A is conductor cross-sectional area. In most cases the two conductors required to complete the circuit are identical. A two-conductor cable with copper conductors of equal cross-sectional area insulated electrically from each other by the cable insulating material is an example. Thus, in most cases, since conductor lengths and cross-sectional areas are the same,

$$R_1 = R_2 = R \tag{13.4}$$

Kirchhoff's voltage law requires that the sum of the voltages around a closed path be zero. Thus,

$$-V_s + V_1 + V_L + V_2 = 0 \tag{13.5}$$

---

[1]For resistance to the flow of alternating current (ac) the expression can be modified to account for skin effect. Tables of resistance values for use in ac analysis generally are corrected to compensate for skin effect.

Substituting in Eq. (13.5) using Eqs. (13.1), (13.2), and (13.4) and rearranging,

$$V_L = V_s - 2RI$$

$$V_L = V_s - R_{eq}I$$

(13.6)

where $R_{eq} = 2R$.

Equation (13.6) suggests the somewhat simpler equivalent circuit shown in Fig. 13.2. In this circuit the resistance of the return line (bottom line) is assumed to be zero and the total circuit resistance $R_{eq} = 2R$ is "lumped" in the top line. The example illustrates the equivalent circuit concept, which is extremely useful in analysis. Essentially, the circuit correctly reflects terminal conditions in a relatively simple way. As a practical matter, it is generally easy to determine equivalent line resistance R_{eq} by referring to tabulated data giving resistance per unit length for standard cable sizes.

Referring to Fig. 13.2 or Eq. (13.6), it can be seen that voltage at the load V_L is equal to voltage at the source minus the voltage drop developed across the line. Loads are typically designed to operate satisfactorily only if their terminal voltage is maintained at or near the design value. Thus, an important function of a power circuit is to maintain load voltage at a relatively constant level over an anticipated range of load conditions. Power to the load for a power circuit is given by $P_L = V_L I$; variation of load power is reflected primarily in variation of load current, assuming that V_L remains relatively constant. The voltage drop across the line $V_s - V_L = IR_{eq}$ is the product of equivalent line resistance and the line current. The line is generally sized for a particular load so that this voltage drop is limited to 3–5% of the rated voltage at the maximum load current. Voltage regulation is another commonly used quantifier of a power circuit's ability to maintain load voltage over a range of load conditions. It is defined as

$$\text{Voltage regulation} = \frac{V_{NL} - V_{FL}}{V_{FL}}$$

(13.7)

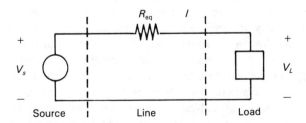

Figure 13.2 An equivalent circuit for a simple dc power circuit.

where V_{NL} is the no-load voltage at the load and V_{FL} is the full-load voltage at the load. Voltage regulation is often expressed as a percentage.

Transmission efficiency is defined as:

$$\eta = \frac{P_L}{P_s} \qquad (13.8)$$

where $P_L = V_L I$ is power consumed at the load and $P_s = V_s I$ is power supplied by the source. Power supplied by the source is equal to load power consumed plus losses in the line; therefore efficiency can also be expressed as

$$\eta = \frac{P_L}{P_L + \text{losses}} = \frac{P_s - \text{losses}}{P_s} \qquad (13.9)$$

where losses are the resistive line losses:

$$\text{Losses} = I^2 R_{\text{eq}} \qquad (13.10)$$

It is clear from Eqs. (13.6), (13.9), and (13.10) that voltage characteristics and transmission efficiency are improved by decreasing line resistance. This typically involves increasing conductor cross-sectional area, which increases cable cost. A compromise must usually be made between initial capital outlay for cable and anticipated cost of losses over the life of the system.

A well-designed source maintains a nearly constant source voltage V_s over the anticipated range of current. For example, dc generators are capable of performing this task and can yield very flat voltage versus current characteristics.

The above discussion of a simple dc power circuit clearly shows that improved voltage characteristics and transmission efficiency would result if the same power could be transmitted at a higher voltage and correspondingly lower current. In general, going to a higher voltage level would decrease transmission losses and improve voltage regulation but would require higher insulation levels in the various system components. If very high voltage levels are desired, it may be necessary to reduce voltage to a safe user level at the load. Voltage levels are restricted to relatively low values in dc rotating machinery primarily because of commutation considerations. The ability of the ac power transformer to efficiently step voltages up or down and the economically attractive features of the alternating-current induction machine have resulted in the widespread use of ac for power transmission. Power transmission by dc is attractive for relatively long-distance transmission (generally several hundred miles or more) and in situations where an asynchronous tie is required. The subject of dc power transmission is outside the scope of this text and will not be pursued further. nearly all industrial plant power circuits are ac, with dc circuits used primarily where variable-speed operation requires the versatility of the dc machine. Other applications requiring dedicated use of dc are electroplating and welding.

A simple single-phase ac power circuit is shown in Fig. 13.3. Analysis of ac circuits requires some additional techniques to effectively deal with variables that are sinusoidal functions of time. It should be noted, however, that the basic structure of the circuit shown in Fig. 13.3 does not differ significantly from the structure of the simple dc power circuit shown in Fig. 13.1. The previous remarks about voltage regulation, voltage drop, and transmission efficiency apply for ac power circuits. In particular, increasing cable size improves transmission efficiency and voltage characteristics in ac circuits as well as in dc circuits. It will be shown that other techniques can improve these characteristics as well in ac power circuits.

A final remark is in order to achieve a proper perspective with respect to the economics of energy conservation. Although increasing conductor cross-sectional area typically improves transmission efficiency in either dc or ac power circuits, economic considerations require a broader viewpoint. A deliberately simplified example illustrates this point. Consider a load consisting entirely of incandescent lights. The load may be treated as resistive in nature. Assume that the entire lighting load operates at a voltage of 95% of the rated voltage. That is, a 5% voltage drop is assumed in the line. If all wiring were resized to reduce the voltage drop to 1% (perhaps at a considerable cost) the voltage at the load would be 99% of rated if the voltage at the source remained the same. The lighting load is resistive; thus, the power consumed by the load will increase as the square of the voltage. The load power consumed would increase by roughly 8.6% $[(0.99/0.95)^2 = 1.086]$. Assuming that the lights operated satisfactorily to start with, the 8.6% increase in load power consumption might be regarded as a loss or at least an unwanted benefit. Furthermore, the 8.6% loss could easily exceed the percentage reduction in transmission losses.

The problem of unwanted increases in load power consumption can be eliminated by properly adjusting (reducing in this case) the voltage at the source, following the change in cable size. The capability to adjust source voltage where necessary is assumed in the following sections. The effects of voltage on other types of loads will be examined more completely in subsequent sections of this chapter.

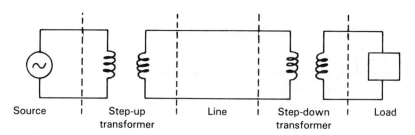

| Source | Step-up transformer | Line | Step-down transformer | Load |

Figure 13.3 A simple ac power circuit.

13.3 AC CIRCUIT ANALYSIS

An ac voltage waveform is illustrated in Fig. 13.4. The waveform is sinusoidal in nature, having a radian frequency of $\omega = 2\pi f$, where f is in hertz (cycles per second). In the United States, frequency is always 60 Hz and ω is thus approximately 377 rad/s. The peak value or crest value of the waveform is $\sqrt{2}\, V$, where V is the effective or root mean square (rms) value of the voltage. Most ac voltmeters and ammeters are calibrated to give readings in rms values, and nameplate data for various devices always reflect rms values for currents and voltages. For example, many single-phase ac power circuits are 120-V circuits, which implies 120 V rms. Power relationships for ac circuits are somewhat simpler if currents and voltages are expressed as rms values rather than peak values.

Power circuits may generally be analyzed by assuming linearity. This implies that in the steady state, all voltages and currents are assumed to be sinusoidal waveforms at 60 Hz. Harmonics may be present, but their effects are often negligible. The assumption of steady-state conditions further implies that short-duration electrical transients (generally decaying exponentials) following a switching action have decayed to negligible levels. Fortunately, such transients have little or no bearing on energy conservation considerations and the assumption of steady-state conditions is appropriate.

If all voltages and currents are assumed to be sinusoidal time functions at a known frequency, say 60 Hz, the amplitude and the phase angle of each such voltage or current completely characterize its waveform. A very simple yet powerful analytical technique involves representing such waveforms with complex constants. In essence, this permits solving ac circuits by simple algebraic manipulation where the solution of differential equations would otherwise be required. The approach is illustrated here by stating the necessary definitions and relationships and using them in the examples that follow. The reader is referred to the references [1–4] for a more rigorous development.

A phasor is a complex constant that represents a sinusoidal time

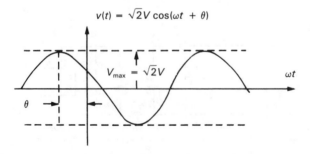

$$v(t) = \sqrt{2}V\cos(\omega t + \theta)$$

$$V_{max} = \sqrt{2}V$$

Figure 13.4 An ac voltage waveform.

function at a known frequency. Voltage and current time functions and their respective phasor representations are illustrated below.[2]

$$e(t) = \sqrt{2}\,E\,\cos(\omega t + \theta) <\!\!=\!\!> \underset{.}{E} = E\underline{/\theta} \qquad (13.11)$$

$$i(t) = \sqrt{2}\,I\,\cos(\omega t + \phi) <\!\!=\!\!> \underset{.}{I} = I\underline{/\phi} \qquad (13.12)$$

Note in Eqs. (13.11) and (13.12) the use of the symbol $<\!\!=\!\!>$ as opposed to a symbol denoting equality between a particular waveform and its phasor representation. Indeed, they are not equal. The phasors of Eqs. (13.11) and (13.12) are given in polar form. The significance of this is illustrated in Fig. 13.5, which shows each plotted in the complex plane.

In Fig. 13.5 the length of each phasor is taken as the rms value of the waveform it represents and the angle of each phasor is the angle it makes with respect to the real (horizontal) axis. The use of the rms value rather than the peak value is arbitrary but will be shown below to be a convenient choice. Because phasors are complex constants they can be expressed in a variety of ways, as illustrated in Table 13.1. The rectangular form is suggested by the geometry of Fig. 13.5. The exponential form follows from Euler's identity:

$$e^{j\theta} = \cos\theta + j\sin\theta \qquad (13.13)$$

The use of different forms facilitates the algebraic manipulations encountered and permits the development of important relationships. It is important to remember, however, that phasors are simply complex constants and are thus manipulated in accordance with the rules of complex algebra.

It can be shown that algebraic addition of sinusoidal time functions at the same frequency corresponds to addition of their respective phasor representations. That is, if

$$e_1(t) + e_2(t) + e_3(t) = 0$$

[2]A capital letter with a dot under the letter is used to denote a complex quantity.

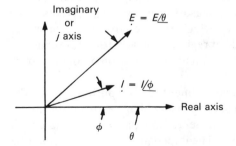

Figure 13.5 Phasors shown in the complex plane.

Table 13.1 Equivalent Phasor Forms

General		Polar		Rectangular		Exponential
$\underset{\cdot}{E}$	$=$	$E\underline{/\theta}$	$=$	$E \cos \theta + j \sin \theta$	$=$	$Ee^{j\theta}$
$\underset{\cdot}{I}$	$=$	$E\underline{/\phi}$	$=$	$I \cos \phi + j \sin \phi$	$=$	$Ie^{j\phi}$

where

$$e_1(t) = \sqrt{2}\, E_1 \cos(\omega t + \theta_1)$$
$$e_2(t) = \sqrt{2}\, E_2 \cos(\omega t + \theta_2)$$
$$e_3(t) = \sqrt{2}\, E_3 \cos(\omega t + \theta_3)$$

then

$$\underset{\cdot}{E}_1 + \underset{\cdot}{E}_2 + \underset{\cdot}{E}_3 = 0 \qquad (13.14)$$

where

$$\underset{\cdot}{E}_1 = E_1\underline{/\theta_1}$$
$$\underset{\cdot}{E}_2 = E_2\underline{/\theta_2}$$
$$\underset{\cdot}{E}_3 = E_3\underline{/\theta_3}$$

The addition of complex numbers suggested in Eq. (13.14) can be accomplished by adding the respective phasors in the manner of vectors, i.e., "tip to tail," or by resolving each into its real and imaginary parts and summing these parts separately. Thus, Eq. (13.14) implies:

$$E_1 \cos \theta_1 + E_2 \cos \theta_2 + E_3 \cos \theta_3 = 0$$

and

$$E_1 \sin \theta_1 + E_2 \sin \theta_2 + E_3 \sin \theta_3 = 0$$

It is clear from the discussion above that Kirchhoff's voltage law, which requires that the sum of the voltages around a closed path be zero, will also require that the phasors representing sinusoidal voltages in a particular loop sum to zero. In essence, phasors representing voltages must conform to Kirchhoff's voltage law. Similarly, phasors representing currents must conform to Kirchhoff's current law, which requires that the sum of the currents entering (or leaving) a node or region must be zero. These concepts are illustrated in Fig. 13.6, which shows an interconnection of four arbitrary circuit elements and a single source.

The impedance of a two-terminal element for the purpose of steady-state analysis is defined as the ratio of the phasor voltage across its terminals to the phasor current through the element. Impedances of basic circuit elements are illustrated in Table 13.2. The impedances shown in Table 13.2

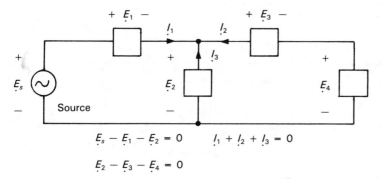

Figure 13.6 Kirchhoff's voltage and current laws in terms of phasor quantities for a two-loop circuit.

imply the phase angle relationships between voltage and current shown in Fig. 13.7.

Reactances of inductors and capacitors are defined in the following way:

$$X_L = \omega L \qquad (13.15)$$

$$X_C = -\frac{1}{\omega C} \qquad (13.16)$$

from which it follows that

$$\underset{L}{Z}_L = jX_L \qquad (13.17)$$

Table 13.2 Impedances of Basic Circuit Elements

Resistor	$+ \; \dot{E}_R \; -$ \quad $\xrightarrow{\quad}$ —/\/\/— \quad $\dot{I}_R \qquad R$	$\underset{R}{Z}_R = \dfrac{\dot{E}_R}{\dot{I}_R} = R$
Capacitor	$+ \; \dot{E}_C \; -$ \quad $\xrightarrow{\quad}$ —$\dashv\vdash$— \quad $\dot{I}_C \qquad C$	$\underset{C}{Z}_C = \dfrac{\dot{E}_C}{\dot{I}_C} = \dfrac{1}{j\omega C}$
Inductor	$+ \; \dot{E}_L \; -$ \quad $\xrightarrow{\quad}$ —mmmm— \quad $\dot{I}_L \qquad L$	$\underset{L}{Z}_L = \dfrac{\dot{E}_L}{\dot{I}_L} = j\omega L$

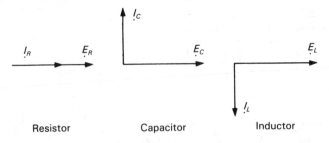

Resistor   Capacitor   Inductor

Figure 13.7 Phasor relationships for basic circuit elements.

and

$$Z_C = jX_C \qquad (13.18)$$

Data on inductors or capacitors are often furnished as appropriate reactances at 60 Hz rather than in the more fundamental units of henries for inductors, or farads for capacitors.

The foregoing concepts significantly simplify the analysis of steady-state ac circuits, as illustrated in the following example.

Example 13.1 Analysis of an AC Circuit

A 120-V circuit supplies a resistive load of 10 Ω by means of a feeder whose total resistance is 1 Ω and whose total inductive reactance is 0.5 Ω. If voltage at the source is 120 V rms, find the current in the circuit and the voltage at the 10-Ω load.

An equivalent circuit for the situation depicted is shown in the accompanying figure.

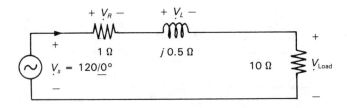

From Kirchhoff's voltage law:

$$V_s = V_R + V_L + V_{load}$$

where

$$V_R = (1)I$$
$$V_L = (j0.5)I$$
$$V_{load} = (10)I$$

Substituting:

$$V_s = (1)I + (j0.5)I + (10)I$$
$$= (11 + j0.5)I$$

The total equivalent impedance at the source is seen to be the sum of the line impedance and the load impedance. Current can be found as:

$$I = \frac{V_s}{11 + j0.5} = \frac{120\underline{/0^\circ}}{11.01\underline{/2.6^\circ}} = 10.9\underline{/-2.6^\circ}\,\text{A}$$

Voltage at the load is found by multiplying this current by load impedance. Thus

$$V_{\text{load}} = 10I = 109\,\underline{/-2.6^\circ}\,\text{V}$$

The voltage drop in the feeder is 11 V, which is roughly 9.2% of the 120-V source. This drop would be excessive in general. The power loss in the line can be shown to be excessive as well. In order to do this, some fundamental ac power relationships must be developed.

13.4 POWER IN AC CIRCUITS

The instantaneous power delivered to a two-terminal device is obtained by multiplying the instantaneous voltage across the terminals of the device by the instantaneous current through the device. An arbitrary two-terminal device with an assigned voltage polarity and current reference direction is illustrated in Fig. 13.8. The assignment of reference directions illustrated in Fig. 13.8 (current entering the terminal assigned the + or positive voltage polarity symbol) is in accordance with the passive sign convention. The product of instantaneous voltage $e(t)$ and instantaneous current $i(t)$, both assigned as illustrated in Fig. 13.8, yields instantaneous power $p(t)$ *consumed* by the device. That is, power consumed is given by

$$p(t) = e(t)i(t) \tag{13.19}$$

Equation (13.19) is fundamental and applies for any $e(t)$ and $i(t)$. For the special case of ac or sinusoidal time variation, assume ac voltage and current waveforms using the same notation introduced in Section 13.2:

$$e(t) = \sqrt{2}\,E\,\cos\,(\omega t + \theta)\quad\text{V} \tag{13.20}$$

$$i(t) = \sqrt{2}\,I\,\cos\,(\omega t + \phi)\quad\text{A} \tag{13.21}$$

Substitution of Eqs. (13.20) and (13.21) in Eq. (13.19) and application of a trigonometric identity yields:

$$\begin{aligned}
p(t) &= e(t)i(t)\\
&= (\sqrt{2}\,E\,\cos\,(\omega t + \theta)(\sqrt{2}\,I\,\cos\,(\omega t + \phi)\\
&= EI\,\cos\,(\theta - \phi) + EI\,\cos\,(2\omega t + \theta + \phi)\quad\text{W} \tag{13.22}
\end{aligned}$$

Figure 13.8 An arbitrary two-terminal device.

Examination of Eq. (13.22) reveals that instantaneous power $p(t)$ consists of a sum of two terms. The first is a constant and the second varies sinusoidally at twice the radian frequency ω. The sinusoidally varying term makes no contribution to average power, which is determined entirely by the constant term. Thus, average power P is given by:

$$P = \frac{1}{T} \int_0^T p(t)\,dt = EI \cos(\theta - \phi) \quad \text{W} \tag{13.23}$$

where T may be taken as the time period of a 60-Hz sinusoidal quantity. Note that E and I are rms or effective values of the respective waveforms. The quantity $\cos(\theta - \phi)$ is referred to as the power factor. It is easy to see that average power P for a particular set of rms voltage and current values at the terminals of a device will be maximized when the current and voltage are in phase, that is, when $\theta - \phi = 0$. This is referred to as operation at unity power factor in that $\cos(\theta - \phi) = 1.0$ if $\theta - \phi = 0$. Average power will be zero regardless of the magnitudes of E and I if current and voltage are 90° out of phase, that is, if $\theta - \phi = \pm 90°$. The phase angle difference between voltage and current, $\theta - \phi$, is referred to as the power factor angle. Specifically, it is defined as the angle by which the voltage waveform leads the current waveform. The power factor angle so defined is also the impedance angle for a passive load. For example, if the impedance of the two-terminal device illustrated in Fig. 13.8 is $Z_L = Z_L/\underline{\psi}$ it follows that

$$E = IZ_L \tag{13.24}$$

where E and I are the phasor representations for $e(t)$ and $i(t)$, respectively. Equating magnitudes on both sides of Eq. (13.24):

$$E = IZ_L \tag{13.25}$$

where all quantities in Eq. (13.25) are magnitudes of the respective complex numbers expressed in polar form. Equating angles on both sides of Eq. (13.24):

$$\theta = \phi + \psi \tag{13.26}$$

Thus, the power factor angle $\theta - \phi$ is easily seen to be the impedance angle ψ. By examining the impedances of basic circuit elements given in Table 13.2 and phase angle relationships illustrated in Fig. 13.7, it is seen that purely resistive loads have a power factor of unity while purely capacitive or inductive loads have power factors of zero and thus consume no average power.

Reactive power Q is defined as the product of rms voltage E, rms current I, and the sine of the power factor angle:

$$Q = EI \sin(\theta - \phi) \tag{13.27}$$

Reactive power is expressed in voltamperes, sometimes referred to as voltamperes-reactive (vars). If current I is resolved into components in phase with and 90° out of phase with (in quadrature with) voltage E, P as given by Eq. (13.23) is the product of "in-phase" components of E and I. Reactive power Q is the product of voltage magnitude and the component of current in quadrature. This is illustrated in Fig. 13.9.

Consideration of the impedances of basic circuit elements given in Table 13.2 and phase angle relationships illustrated in Fig. 13.7 shows that purely resistive loads consume average power P but do not consume reactive power Q. Inductive loads *consume* reactive power Q since for an inductive load $\sin(\theta - \phi) = +1$, but they consume no average power P. Capacitive loads *generate* reactive power Q since for a capacitive load $\sin(\theta - \phi) = -1$, but they consume no average power P. Generally, loads operate at power factors that are somewhat lagging (current lags voltage by some nonzero angle) due to inherent inductance present in the equivalent circuit for the load. Operation is thus generally at a power factor less than unity. For a given load power at a particular rms voltage, operation at a lagging power factor less than unity implies consumption of reactive power. Such a condition requires more current than would be required if the power factor were unity. An increased current requirement is generally regarded as undesirable because line losses are increased and line voltage drop is increased. Thus, reactive power Q consumed by a load, such as an induction machine, is a measure of the increased current requirement for a given operating condition. In addition to increased line losses and increased line voltage drops, reactive power consumption by a load implies a higher capacity requirement for the equipment supplying the load. For example, a transformer feeding a machine that produces 100 hp of usable shaft output power at a lagging power factor of 0.60 must have a higher voltampere rating than a transformer feeding a machine that produces the same usable shaft output power but that operates at unity power factor.

The relationships between average power P, reactive power Q, and apparent power, which is the product of rms voltage E and rms current I, are of fundamental importance in understanding ac circuit behavior. Some additional insights result if complex power S is defined in the following way:

$$\begin{aligned} S &= P + jQ \\ &= EI\cos(\theta - \phi) + jEI\sin(\theta - \phi) \end{aligned} \qquad (13.28)$$

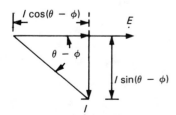

Figure 13.9 Resolution of current into components contributing exclusively to average power P or exclusively to reactive power Q.

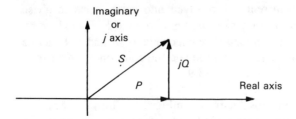

Figure 13.10 Complex power $\underset{\cdot}{S}$.

It is easy to show that

$$\underset{\cdot}{S} = \underset{\cdot}{E}\underset{\cdot}{I}^* \qquad (13.29)$$

where the asterisk denotes the complex conjugate of a complex number. If complex power is plotted in the complex plane, a triangular relationship between P, Q, and apparent power $S = |\underset{\cdot}{S}| = EI$ results. This is illustrated in Fig. 13.10. For a passive load whose impedance is $\underset{\cdot}{Z} = R + jX$:

$$\underset{\cdot}{S} = \underset{\cdot}{E}\underset{\cdot}{I}^*$$
$$= \underset{\cdot}{Z}\underset{\cdot}{I}\underset{\cdot}{I}^*$$
$$= I^2\underset{\cdot}{Z}$$
$$= I^2(R + jX)$$
$$= I^2R + jI^2X$$
$$= P + jQ \qquad (13.30)$$

Equation (13.30) shows that the complex power triangle is geometrically similar to the impedance triangle. This is illustrated in Fig. 13.11.

Figure 13.11 and the relationships given as Eqs. (13.28)–(13.30) neatly summarize much of what was discussed earlier. In particular, note that apparent power $S = |\underset{\cdot}{S}| = EI$ can be expressed as

$$S = \sqrt{P^2 + Q^2} \qquad (13.31)$$

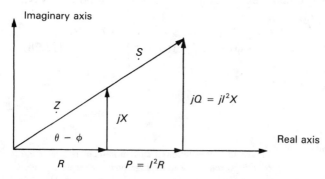

Figure 13.11 Geometric relationship between impedance triangle and complex power triangle.

Obviously, if a load consumes significant reactive power Q for a given average power P, it will require relatively high current, assuming voltage is fixed at a value near the rated value. Furthermore, assuming the load can be represented by an impedance, Fig. 13.11 shows that it is the reactive component of load impedance that gives rise to reactive power consumption.

Example 13.2 Analysis of an AC Motor Circuit

A single-phase 115-V motor develops its rated output shaft horsepower of 1 hp at a power factor of 0.82 when operated at rated voltage. Motor efficiency is 68%. The motor is fed by a two-conductor cable having a total cable resistance of 1.0 Ω and total cable inductive reactance of 0.5 Ω. Find the voltage regulation and transmission efficiency for the described condition. Repeat calculations if a capacitor is used to correct power factor to unity.

The input power to the motor is determined by the efficiency:

$$P_{in} = \frac{P_{out}}{\eta} = \frac{(1 \text{ hp}) (746 \text{ W/hp})}{0.68}$$

$$= 1097 \text{ W}$$

Motor current is determined from the fundamental average power relationship:

$$P = EI \cos \theta$$

where θ is the power factor angle. Thus

$$I = \frac{P}{E \cos \theta}$$

$$= \frac{1097 \text{ W}}{(0.82)(115 \text{ V})} = 11.63 \text{ A}$$

The voltage at the source end of the feeder is determined by adding the line voltage drop to the voltage at the motor.

$$V_s = V_m + IZ_l$$

where Z_l is the equivalent line impedance. For convenience, take the voltage at the motor to be the rated 115 V rms at a phase angle of zero. Then,

$$V_m = 115 \underline{/0°} \quad \text{and} \quad I = 11.63 \underline{/-\theta}$$

where

$$\theta = \cos^{-1} 0.82 = 34.92°$$

The minus sign appears in the expression for I since the power factor is assumed to be lagging (current lags voltage), which is typical for induction machines. The source voltage is:

$$V_s = 115 + (11.63 \underline{/-34.92°})(1.0 + j0.5)$$

$$= 115 + 13.00 \underline{/-8.35}$$

$$= 115 + j0 + 12.86 - j1.888$$

$$= 127.9 - j1.888$$

$$= 127.9 \underline{/-0.85°}$$

The line voltage drop is $127.9 - 115 = 12.89$ V. Voltage regulation is

$$\frac{V_{NL} - V_{FL}}{V_{FL}} = \frac{127.9 - 115}{115} = 0.112 \text{ or } 11.2\%$$

Cable real power losses are I^2R, where R is the cable resistance. Thus

$$\text{Losses} = I^2R$$

$$= (11.63)^2(1) = 135.3 \text{ W}$$

Transmission efficiency is

$$\frac{P_o}{P_{in}} = \frac{1097}{1097 + 135.3} = 0.890 \text{ or } 89.0\%$$

Voltage regulation is excessive and transmission efficiency is low. By placing a capacitor at the terminals of the motor in parallel with the motor, the magnitude of the line current can be reduced. The capacitor can be sized to draw a leading current that effectively supplies the reactive component of total required motor current. This is illustrated in the accompanying figure.

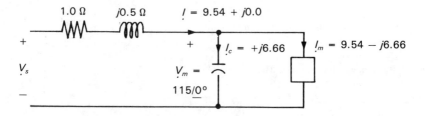

The figure shows that line current magnitude has been reduced from 11.63 A to 9.54 A. Note that average power P supplied to the motor-capacitor parallel combination is $P = (9.54)$ $(115) \cos 0° = 1097$ W. Since the capacitor consumes no average power, the entire 1097 W is still delivered to the motor. In essence, the terminal conditions of the motor remain the same, provided that V_s, the voltage at the source, is adjusted to take advantage of the improved situation. Recalculating voltage regulation and transmission efficiency:

$$V_s = 115 + 9.54(1.0 + j0.5)$$

$$= 124.5 + j4.77$$

$$= 124.6 \underline{/2.19°}$$

The line voltage drop is thus $124.6 - 115 = 9.6$ V. Voltage regulation is

$$\frac{V_{NL} - V_{FL}}{V_{FL}} = \frac{124.6 - 115}{115} = 0.083 \text{ or } 8.3\%$$

Transmission efficiency is

$$\frac{P_o}{P_{in}} = \frac{1097}{1097 + (9.54)^2(1)} = 0.923 \text{ or } 92.3\%$$

The situation could be further improved by increasing the conductor cross-sectional area.

A final remark is in order. Average power P and reactive power Q are conserved quantities. Thus, total real power required at the source can be found by simply summing the real power requirements of all the components supplied by the source. Similarly, total reactive power required at the source can be found by summing reactive power requirements of all the components supplied by the source. Capacitors are usually rated in vars or kilovars at a particular voltage rating because they are used to supply reactive power at a particular point in the system. In Example 13.2, correction to unity power factor would require selection of a capacitor with a reactive power rating exactly equal to the reactive power consumed by the motor. The fact that average power P and reactive power Q are conserved quantities provides a computational alternative for determining voltage at the source (and minimum voltampere ratings for devices "upstream" of the source). In Example 13.2, after application of the capacitor the total real power requirement at the source is

$$
\begin{array}{ll}
1097 \text{ W} & \text{motor} \\
0 \text{ W} & \text{capacitor} \\
\underline{(9.54)^2(1) = 91 \text{ W}} & \text{line } (I^2R) \\
1188 \text{ W} & \text{total}
\end{array}
$$

The total reactive power requirement is

$$
\begin{array}{ll}
(115)(11.63) \sin 34.92 = 765.6 \text{ vars} & \text{motor} \\
-765.6 \text{ vars} & \text{capacitor} \\
\underline{(9.54)^2(0.5) = 45.5 \text{ vars}} & \text{line } (I^2X) \\
45.5 \text{ vars} & \text{total}
\end{array}
$$

Thus, apparent power at the source is:

$$
V_s I = \sqrt{1188^2 + 45.5^2} = 1188.9
$$

and

$$
V_s = \frac{1188.9}{9.54} = 124.6 \text{ V}
$$

which is the same result obtained previously.

Example 13.2 illustrates the principle of power factor correction with capacitors. Additional considerations must be kept in mind, such as time variation of the motor load and a phenomenon known as motor self-excitation. Self-excitation limits the size of the capacitor that can be used. These issues will be addressed in the next chapter.

Generally speaking, capacitive compensation should be located as close as possible to the device consuming reactive power. This tends to minimize the current (and I^2R losses) in components serving the device. Overall economics may, however, dictate a centralized location for capacitor banks.

13.5 THREE-PHASE CIRCUITS

Nearly all ac circuits that transmit significant power are three-phase circuits. Single-phase ac circuits will necessarily have a double frequency component present in their instantaneous power waveform. This is evident from an examination of Eqs. (13.19)–(13.22). A three-phase circuit eliminates this double frequency component. It can be shown that instantaneous power in a three-phase circuit with properly balanced voltages and currents is constant with respect to time. Average power flow is consequently the same as instantaneous power in a three-phase circuit operating under balanced conditions.

Power is accordingly typically generated and transmitted on a three-phase basis. Large motors are typically three-phase machines because of improved performance characteristics. Single-phase loads can generally be powered by connecting them to one phase of a three-phase power circuit, but efforts should be made to distribute such loads evenly on the phases of the three-phase circuit in order to avoid operation with significant phase imbalance.

A typical break point between single-phase motors and three-phase motors might be at 1/3 hp. That is, in a particular plant it might be required that all motors of rating 1/3 hp or higher be three-phase motors, perhaps rated at 460 V. Smaller motors would be single-phase motors, perhaps rated at 120 V.

A three-phase circuit with a single source and single load is illustrated in Fig. 13.12. The circuit has both source and load Y-connected. Balanced three-phase operation requires that voltages (and currents) in each of the three phases be exactly the same in rms value and displaced in phase by $2\pi/3$ radians or 120°. Thus, for the circuit of Fig. 13.12 balanced operation would imply that:

$$V_{an} = E\,\underline{/\theta} \tag{13.32}$$

$$V_{bn} = E\,\underline{/\theta - 120°} \tag{13.33}$$

$$V_{cn} = E\,\underline{/\theta + 120°} \tag{13.34}$$

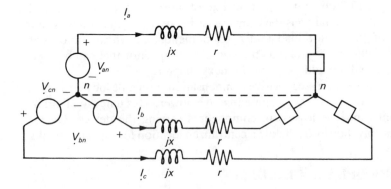

Figure 13.12 A three-phase circuit.

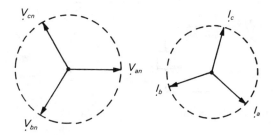

Figure 13.13 Phasor diagrams showing voltages and currents in a three-phase circuit assuming a balanced operating condition.

where V_{an} is the voltage of phase a with respect to the neutral point labeled n and V_{bn} and V_{cn} are similarly defined. It would also imply that

$$I_a = I \underline{/\phi} \tag{13.35}$$

$$I_b = I \underline{/\phi - 120^\circ} \tag{13.36}$$

$$I_c = I \underline{/\phi + 120^\circ} \tag{13.37}$$

Phasor diagrams illustrating the appropriate conditions are shown in Fig. 13.13.

The balanced condition defined by Eqs. (13.32)–(13.37) and illustrated in Fig. 13.13 generally results from using an inherently balanced source such as a three-phase generator and a three-phase device or three matched (equal) impedances as the load. The neutral return shown as a dashed line in Fig. 13.12 carries no current under balanced conditions since $I_a + I_b + I_c = 0$. Furthermore, the voltage at the center point of the Y in the load and at the center point of the Y in the source is exactly the same under balanced conditions. The neutral wire thus serves no purpose under balanced conditions and may be removed from the circuit. Addition of single-phase loads between line and ground (neutral) will require the presence of the neutral wire. A system operated with a neutral wire is sometimes referred to as a three-phase, four-wire system. Assuming that single-phase loads are properly distributed among the three phases, little error is introduced by assuming balanced conditions and analyzing a particular circuit accordingly. The assumption of balanced conditions permits analysis of a three-phase circuit on a single-phase basis. In essence, we analyze one phase and simply recognize that the behavior in the other two phases is the same, but phase-displaced in time by ±120°. An equivalent circuit for analysis is obtained by incorporating the neutral return (whether it physically exists or not) and considering one phase. A single-phase equivalent for the circuit of Fig. 13.12, assuming balanced operation, is shown in Fig. 13.14.

The reader should note the similarity between the equivalent circuit of

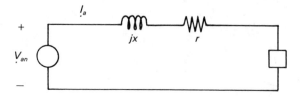

Figure 13.14 Single-phase equivalent circuit for the three-phase circuit of Fig. 13.12.

Fig. 13.14 and those that were introduced in previous examples. No new analytical techniques are required to deal with balanced three-phase circuits. It is necessary, however, to recognize certain conventions that are commonly used. Nameplate data and, in fact, virtually any data used to describe a three-phase device are always expressed as line data and as total three-phase power or voltamperes.

For example, if the circuit of Fig. 13.12 were described as a 460-V circuit this would imply an rms voltage between the lines (the neutral is not considered a line) of 460 V. This is sometimes referred to as the circuit line-to-line voltage. The relationship between line voltages and phase voltages, i.e., V_{an}, V_{bn}, and V_{an}, is found by applying Kirchhoff's voltage law. For example, the voltage of the phase a line with respect to the phase b line is given by

$$V_{ab} = V_{an} - V_{bn} \tag{13.38}$$

Similarly, for the other two phases:

$$V_{bc} = V_{bn} - V_{cn} \tag{13.39}$$

$$V_{ca} = V_{cn} - V_{an} \tag{13.40}$$

Equations (13.38)–(13.40) imply the relationship between phasors illustrated in Fig. 13.15.

Consideration of the geometry of Fig. 13.15 shows that the rms value of the line-to-line voltage is greater than the rms value of the line-to-neutral or phase voltage by a factor of $\sqrt{3}$. Using the subscript L to denote line quantities and the subscript P to denote phase quantities, a Y-connected source or load has the following relationships:

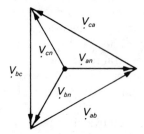

Figure 13.15 Relationship between line voltages in a three-phase circuit.

$$V_L = |V_{ab}| = V_{bc}| = |V_{ca}| \qquad (13.41)$$

$$V_P = |V_{an}| = V_{bn}| = |V_{cn}| \qquad (13.42)$$

$$V_L = \sqrt{3}\, V_P \qquad (13.43)$$

For a Y-connected source or load, phase currents are the same as the line currents:

$$I_L = I_P = |I_a| = |I_b| = |I_c| \qquad (13.44)$$

For a source or load connected in a delta configuration, as illustrated in Fig. 13.16, phase currents differ in rms value from line currents by a factor of $\sqrt{3}$. That is,

$$I_P = |I_{ab}| = I_{bc}| = |I_{ca}| \qquad (13.45)$$

$$I_L = |I_a| = |I_{ab} - I_{ca}| = \sqrt{3}\, I_P \qquad (13.46)$$

Line voltages and phase voltages are the same for a delta-connected source or load:

$$V_P = V_L = |V_{ab}| = V_{bc}| = |V_{ca}| \qquad (13.47)$$

All of the relationships between line and phase quantities follow directly from Kirchhoff's voltage and current laws.

Power specified for a three-phase circuit or the power rating of a three-phase device is always taken to be the total three-phase power. The total three-phase power is simply three times the power per phase:

$$P = 3P_P \qquad (13.48)$$

where

$$P_P = V_P I_P \cos(\theta - \phi) \qquad (13.49)$$

For a Y-connected device $V_L = \sqrt{3}\, V_P$ and $I_P = I_L$. Thus

$$P = 3P_P = \frac{3V_L}{\sqrt{3}} I_L \cos(\theta - \phi) = \sqrt{3}\, V_L I_L \cos(\theta - \phi)$$

For a delta-connected device $I_L = \sqrt{3}\, I_P$ and $V_L = V_P$. Thus

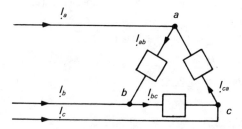

Figure 13.16 A delta connected load.

$$P = 3P_P = 3V_L \frac{I_L}{\sqrt{3}} \cos(\theta - \phi) = \sqrt{3} \, V_L I_L \cos(\theta - \phi) \qquad (13.50)$$

Note that the same three-phase power relationship applies regardless of the actual device connection. Three-phase wattmeters measure total three-phase power as given by Eq. (13.50).

Example 13.3 Analysis of a Three-Phase Induction Machine

A 100-hp, 460-V, 0.85-power factor induction machine has an efficiency of 92%. The machine is fed by a three-conductor cable having an equivalent impedance per phase of $0.15 + j0.06 \, \Omega$. Find the voltage, current, and power factor at the source end of the feeder if the induction machine operates at rated conditions.

Assume that the motor is Y-connected. On a single-phase basis the equivalent circuit situation depicted in the accompanying figure applies. For convenience, take the angle of the

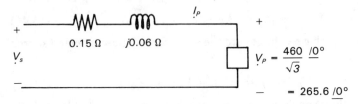

phase voltage at the motor to be $0°$. The electrical power into the motor is related to power out by the efficiency:

$$P_{in} = \frac{P_{out}}{0.92} = \frac{100 \text{ hp } (746 \text{ W/hp})}{0.92}$$

$$= 81.1(10^3) \text{ W or } 81.1 \text{ kW}$$

From Eq. (13.50):

$$P = \sqrt{3} \, V_L I_L \cos \theta$$

Thus

$$I_L = \frac{P}{\sqrt{3} \, V_L \cos \theta} = \frac{81.1(10^3)}{(\sqrt{3})(460)(0.85)} = 119.8 \text{ A}$$

Now, $I_p = I_L$ for a Y-connected device and $\theta = \cos^{-1} 0.85 = 31.8°$ is the angle by which phase voltage V_P leads phase current I_P. Thus

$$I_P = 119.8\underline{/-31.8°}$$

(It is assumed here that current lags the voltage in accordance with typical induction machine behavior. The power factor itself does not yield this information.) Solving the equivalent circuit for the source voltage V_s:

$$V_s = 265.6\underline{/0°} + (119.8\underline{/-31.8°})(0.15 + j0.06)$$

$$= 265.6 + 19.06 - j3.36$$

$$= 284.7 - j3.36 \cong 284.7\underline{/-0.68°}$$

Thus, at the source the line voltage is

$$V_{Ls} = \sqrt{3}\,(284.7) = 493.1 \text{ V}$$

The line current is

$$I_L = I_P = 119.8 \text{ A}$$

The power factor is

$$\text{p.f.} = \cos\left[-0.68° - (-31.8°)\right]$$

$$= \cos(31.12°) = 0.856$$

13.6 POWER CIRCUIT COMPONENTS AND ENERGY CONVERSION DEVICES

In this section important power system components are presented and circuit models for these components are developed.

13.6.1 Cables and Overhead Lines

From previous examples it can be seen that transmission efficiency, voltage regulation, and power factor correction calculations all require per phase line impedance data. Detailed descriptions of cable and overhead line characteristics are given in [1–3, 5, 6]. Positive sequence[3] data for cables or overhead lines can be determined from tables, and these data are appropriate for the type of analysis presented in previous examples. Appendix H Table H.1 lists positive sequence impedances for various three-conductor cable sizes, and Appendix H Table H.2 lists positive sequence data for some typical distribution class overhead line circuits. Note that resistance and reactance are given in the tables on a per-unit-length basis. Reactances of overhead lines vary with conductor spacing, and Appendix H Table H.2 gives correction factors for several different geometric mean spacings. The geometric mean spacing is defined as

$$\text{GMS} = \sqrt{D_1 D_2 D_3} \tag{13.51}$$

where D_1, D_2, and D_3 are the spacings illustrated in Fig. 13.17.

---

[3]Positive sequence refers to normal phase sequence. Negative and zero sequence quantities are of interest only when unbalanced three-phase conditions are present.

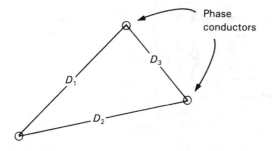

Figure 13.17 Illustration of spacings required to determine geometric mean spacing.

13.6.2 Transformers

An equivalent circuit for a power transformer is shown in Fig. 13.18. The equivalent circuit illustrated is essentially an "ideal transformer" (portion enclosed by the dashed line, modified to more accurately reflect actual transformer behavior. The primed variables V_1', I_1', V_2', and I_2' are related by the so-called ideal transformer relationships:

$$\frac{V_1'}{V_2'} = \frac{N_1'}{N_2'} \qquad (13.52)$$

$$\frac{I_1'}{I_2'} = \frac{N_2'}{N_1} \qquad (13.53)$$

where N_1 and N_2 are the numbers of turns on windings 1 and 2, respectively, of a single-phase transformer. The equivalent circuit represents one phase of a three-phase device under the assumption of balanced operation. Resistances r_1 and r_2 are associated with the actual windings themselves. Reactances x_1 and x_2 are leakage reactances. Ideally, the magnetic flux required for transformer action would confine itself to the iron core. In practice there is always some leakage flux, which gives rise to a voltage drop not accounted for by the ideal transformer. The leakage reactances effectively model this aspect of transformer behavior. The reactance x_m, which carries the current I_m, is the transformer magnetizing reactance. Magnetizing current I_m is the current required to establish the transformer flux. The resistance r_c, which carries the current I_c, is a resistance used to account for transformer core losses due to eddy currents and hysteresis effects. The current I_{ex} is the sum of I_m and I_c and is referred to as "exciting" current.

A well-designed power transformer approaches ideal transformer behavior. The exciting current I_{ex} might typically range from roughly 2 to 7% of the transformer rated current. The voltage developed across the

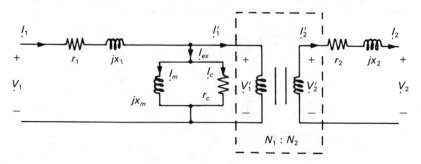

Figure 13.18 An equivalent circuit for a power transformer.

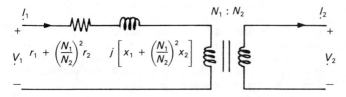

Figure 13.19 A simplified transformer equivalent circuit neglecting magnetizing impedance.

equivalent series impedance due to $r_1, r_2, x_1,$ and x_2 ranges from roughly 6 to 12% of rated voltage. These observations imply that the equivalent magnetizing impedance (the parallel combination of jx_m and r_c) will be relatively high and that the equivalent leakage impedance will be relatively small. Depending on the desired accuracy in a particular calculation, it is often possible to simplify the equivalent circuit of Fig. 13.18. For example, for voltage drop calculations, the magnetizing impedance can be neglected with little resulting error. Furthermore, from transformer theory (see [1–3]) impedances can be referred from one side of the transformer to the other by the turns ratio squared without affecting conditions at the transformer terminals. Thus, the simplified equivalent shown in Fig. 13.19 can be used for determining transformer terminal voltage characteristics.

When energy losses are of interest, both core losses and winding copper losses must be considered. Core losses are sometimes referred to as no-load losses because they occur regardless of whether the transformer secondary is loaded. They are also referred to as open-circuit losses. They vary approximately as the square of the transformer terminal voltage. Winding copper losses are sometimes referred to as load losses, and typically vary as the square of the transformer current. Since transformer voltage is maintained near the rated voltage, the load losses will vary approximately as the square of the actual transformer voltampere load. No-load losses in distribution class transformers range from roughly 20–50% of the load losses. Thus, significant savings can be realized by shutting down unloaded transformers, particularly larger transformers. Appendix H Table H.3, from reference [6], gives no-load losses and total losses for some typical distribution class transformers. Total losses are the sum of no-load and load losses when the transformer operates at rated conditions.

Transformer data are given by manufacturers in a variety of ways. Typically, either loss data, impedance data, or some combination of the two is available. Appendix H Table H.4 gives some typical transformer impedance data from [5] and [6] for the series impedance of various transformers.

Example 13.4 Analysis of a Three-Phase Transformer

A 500-kVA, three-phase transformer has Y-connected primary and secondary windings. The transformer has a rated line-to-line primary voltage of 12 kV and a rated line-to-line secondary voltage of 4.16 kV. No-load losses are 0.50%. The total series impedance of the transformer is 5% and the total series resistance is 1.1%. Transformer exciting current is 2% at rated voltage. Find an equivalent circuit for the transformer similar to that shown in Fig. 13.18. Show magnetizing impedance referred to the high-voltage side of the transformer.

Use of the per-unit system affords some degree of simplification. The system is described in considerable detail in references [1] and [2]. In this example, the equivalent circuit will be found with quantities expressed in terms of the more fundamental units of ohms, volts, and amperes. On a per-phase basis the rated voltamperes are given by

$$VA_{rated} = \frac{500}{3} = 166.7 \text{ kVA}$$

The high-side rated voltage is

$$V_1' = \frac{12}{\sqrt{3}} \text{ kV} = 6.93 \text{ kV}$$

The low-side rated voltage is

$$V_2' = \frac{4.16}{\sqrt{3}} \text{ kV} = 2.40 \text{ kV}$$

The turns ratio is

$$\frac{N_1}{N_2} = \frac{V_1'}{V_2'} = \frac{6.93}{2.40} = 2.88$$

The total equivalent series impedance is expressed as a percentage of the base impedance of the transformer. The base impedance is defined as transformer rated voltage divided by transformer rated current. Thus, the base impedance on the transformer high side is

$$\frac{V_1'}{I_1'} = \frac{V_1'}{VA_{rated}/V_1'} = \frac{V_1'^2}{VA_{rated}} = \frac{6.93^2 \times 1000}{166.7} = 288 \ \Omega$$

The total series impedance in ohms referred to the high side is

$$Z = (0.05)(288) = 14.4 \ \Omega$$

The total series resistance referred to the high side is

$$R = (0.011)(288) = 3.17 \ \Omega$$

Total series reactance is

$$X = \sqrt{Z^2 - R^2}$$
$$= \sqrt{14.4^2 - 3.17^2} = 14.05 \ \Omega$$

The series impedances of both windings are approximately equal when referred to the same side of the transformer. Thus, the series resistance R and the series reactance X can be halved and referred by the turns ratio squared to achieve the configuration of Fig. 13.18. That is,

$$r_1 = \frac{R}{2} = \frac{3.17}{2} = 1.59 \ \Omega$$

$$r_2 = \left(\frac{N_2}{N_1}\right)^2 \frac{R}{2} = \left(\frac{1}{2.88}\right)^2 \frac{3.17}{2} = 0.191 \ \Omega$$

$$x_1 = \frac{X}{2} = \frac{14.05}{2} = 7.025 \ \Omega$$

$$x_2 = \left(\frac{N_2}{N_1}\right)^2 \frac{X}{2} = \left(\frac{1}{2.88}\right)^2 \frac{14.05}{2} = 0.847 \ \Omega$$

The core loss resistance r_c can be found from the no-load losses and rated voltage. On a single-phase basis the no-load losses are

$$(0.005)(166.7 \text{ kVA}) = 833.5 \text{ W}$$

The voltage drop across the series impedance will be small, particularly under the no-load condition. Assuming that the voltage developed across the magnetizing impedance is rated voltage:

$$r_c = \frac{V_1'^2}{P_{NL}} = \frac{[6.93(10^3)]^2}{833.5} = 57.6 \text{ k}\Omega$$

The magnetizing reactance can be found from the exciting current, which is known to be 2% at rated voltage. On the high side:

$$I_{ex} = 0.02 I_1' = 0.02 \ \frac{VA_{rated}}{V_1'} = 0.02 \left(\frac{166.7}{6.93}\right) = 0.48 \text{ A}$$

Since $I_{ex} = I_m + I_c$ and I_m and I_c differ in phase by 90°:

$$I_{ex} = \sqrt{I_m^2 + I_c^2}$$

Thus

$$I_m = \sqrt{I_{ex}^2 - I_c^2}$$

$$= \sqrt{0.48^2 - \left(\frac{6.93}{57.6}\right)^2} = 0.465 \text{ A}$$

and

$$x_m = \frac{V_1'}{I_m} = \frac{6.93 \text{ kV}}{0.465 \text{ A}} = 14.9 \text{ k}\Omega$$

and the equivalent circuit is as shown in the accompanying figure.

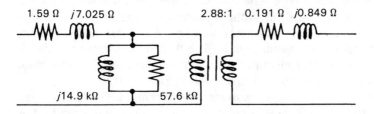

1.59 Ω   j7.025 Ω   2.88:1   0.191 Ω   j0.849 Ω

j14.9 kΩ   57.6 kΩ

If the equivalent circuit had been determined with all quantities expressed in per-unit, the ideal transformer portion of the equivalent circuit would not be required. The equivalent circuit with impedances expressed in per-unit is shown in the next figure. The per-unit system has some obvious advantages. The initial data were expressed in per-unit on the transformer rating as base. In particular, note that when expressed in per-unit:

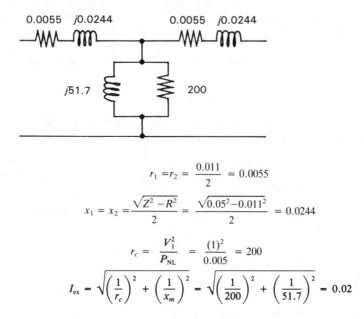

$$r_1 = r_2 = \frac{0.011}{2} = 0.0055$$

$$x_1 = x_2 = \frac{\sqrt{Z^2 - R^2}}{2} = \frac{\sqrt{0.05^2 - 0.011^2}}{2} = 0.0244$$

$$r_c = \frac{V_1^2}{P_{\text{NL}}} = \frac{(1)^2}{0.005} = 200$$

$$I_{\text{ex}} = \sqrt{\left(\frac{1}{r_c}\right)^2 + \left(\frac{1}{x_m}\right)^2} = \sqrt{\left(\frac{1}{200}\right)^2 + \left(\frac{1}{51.7}\right)^2} = 0.02$$

In what follows, the more fundamental units of ohms, amperes, and volts will be used. It should be recognized, however, that much of the available data is expressed in per-unit and working exclusively in the in per-unit system can have computational advantages.

13.6.3 The Induction Machine

The three-phase induction machine is widely regarded as the workhorse of industry insofar as electrical machinery is concerned. It is by far the most popular choice as a constant-speed electric drive and, unless specific application requirements preclude its use, it is generally the most economical choice of basic machine type. The simplicity of the machine is its principal advantage. In larger sizes the stator is made up of sheet steel punchings assembled in a stack. The punchings are aligned and fastened with bolts running parallel to the machine shaft. The three-phase stator winding is placed in slots around the inner periphery of the stator assembly. The stator of a large squirrel cage motor is illustrated in Fig. 13.20.

The rotor of the machine is also made up of punchings. The squirrel cage rotor configuration is the most popular. Instead of a winding comprised of turns of a coil, the conductors of the rotor of a squirrel cage machine are bars aligned parallel to the axis of the machine. These bars are equally spaced around the periphery of the rotor and typically are brazed to short-circuiting end rings on the ends of the rotor. A rotor for a large squirrel cage motor is illustrated in Fig. 13.21.

Figure 13.20 Stator construction of a large squirrel cage motor. (From "Industrial Electrical Systems" by the editors of Power Magazine, McGraw-Hill Book Company, copyright © 1967. Reprinted with permission.)

The type of construction illustrated in Figs. 13.20 and 13.21 yields a relatively simple configuration that is rugged, easy to maintain, and relatively low in cost. There are no brushes or commutation devices since the basic machine action is by induction. Figure 13.22 illustrates conceptually the basic mechanism of induction machine action. In Fig. 13.22*a*, the three-phase stator winding is shown Y-connected. The rotor bars effectively yield an equivalent polyphase winding with terminations short-circuited as illustrated. Figure 13.22*b* shows an end view of the machine with fundamental components of the various magnetic flux density waves represented by rotating space vectors. The vector Φ_s represents the armature winding flux wave and it is established by the three phase currents flowing in the armature

Figure 13.21 Rotor of a large squirrel cage machine. (From "Industrial Electrical Systems" by the editors of Power Magazine, McGraw-Hill Book Company, copyright © 1967. Reprinted with permission.)

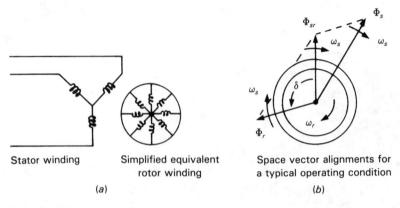

Stator winding   Simplified equivalent   Space vector alignments for
                rotor winding   a typical operating condition

(a)   *(b)*

Figure 13.22 Induction machine operation.

(stator) winding. It rotates at synchronous speed ω_s, which is determined by the frequency of the three-phase system to which the motor is connected (generally 60 Hz) and the number of poles the machine is wound for.[4] Strictly speaking, the fundamental component flux distribution suggested by Fig. 13.22*b* is correct only for a two-pole machine. For larger numbers of poles the pattern repeats itself in one pass around the periphery a number of times which is equal to the number of pole pairs the armature is wound for. In spite of this, the two-pole configuration illustrated in Fig. 13.22 correctly explains the essence of induction machine action for any number of poles. Synchronous speed ω_s in radians per second is given by

$$\omega_s = \frac{2\pi f}{p/2} \tag{13.54}$$

where p is the number of poles and f is the frequency of the three-phase supply in hertz. Synchronous speed in revolutions per minute, assuming a 60-Hz supply, is given by

$$n_s = \frac{3600}{p/2} \tag{13.55}$$

The vector Φ_r represents the rotor winding flux wave and it is established by the polyphase currents induced in the rotor conductors. The frequency of the induced rotor currents is $\omega_s - \omega_r$ because of relative speed between the armature winding flux wave and the rotor itself. Thus, Φ_r rotates at speed $\omega_s - \omega_r$ with respect to the rotor and at speed $\omega_s - \omega_r + \omega_r = \omega_s$ with respect to the stator. The fixed constant between Φ_s and Φ_r during normal steady-state operation is fundamental to the development of a

---

[4]The stator of an induction machine may be wound in such a way as to develop a magnetic field having 2, 4, 6, etc., magnetic poles.

torque on the rotor which is constant in time. The vector Φ_{sr} is simply the vector sum of Φ_s and Φ_r as illustrated in Fig. 13.22. It is important because it determines the required armature terminal voltage. Conversely, a specified armature terminal rms voltage fixes the magnitude, speed of rotation, and relative angle of Φ_{sr}. The vectors Φ_s and Φ_r must then adjust themselves in such a way as to satisfy the required vector addition illustrated in Fig. 13.22.

Induction machine torque is proportional to the product of the magnitudes of Φ_{sr} and Φ_r and the sine of the angle between the two:

$$T \propto \Phi_{sr}\Phi_r \sin \delta \qquad (13.56)$$

The applied terminal voltage fixes Φ_{sr}. The vector Φ_r and the angle δ vary as a function of machine steady-state speed ω_r. A typical torque versus speed curve is shown in Fig. 13.23. This figure also relates torque to slip, which is defined as

$$s = \frac{\omega_s - \omega_r}{\omega_s} = \frac{n_s - n_r}{n_s} \qquad (13.57)$$

Note that rated torque is developed at a relatively low value of slip. Rated slip is typically 2 to 6%. Normal operation of the induction machine implies operation at slip values near rated slip. Over this slip range, speed varies from no-load speed (approximately synchronous speed) to rated speed (approximately 94–98% of synchronous speed). Operation at high slip is typically inefficient because of excessive rotor circuit heating. Thus, the induction machine is generally regarded as a fixed speed drive unless a variable-frequency source supplies the machine. The advent of efficient solid-state drives has provided variable-frequency capability that permits efficient variable-speed operation of the induction machine. This type of operation will be examined in greater detail in the next chapter.

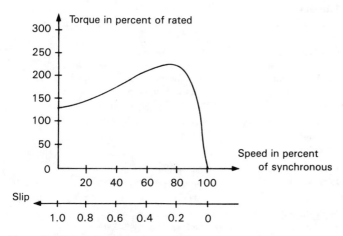

Figure 13.23 A typical induction machine torque speed curve.

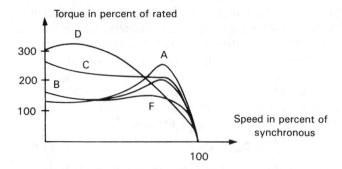

Figure 13.24 Squirrel cage induction machine torque speed characteristics as affected by rotor bar configuration.

Figure 13.24 illustrates the effects of varying the rotor bar geometry on the induction machine torque speed characteristics. Typical applications are shown in Table 13.3.

An equivalent circuit for steady-state operation of an induction machine is illustrated in Fig. 13.25. The circuit is similar to the equivalent circuit for a transformer. During normal steady-state operation assume balanced operation and analyze on a single-phase basis. Thus, V_p and I_p are armature phase voltage and current, respectively. The circuit element r_s is armature winding resistance, x_s is armature winding leakage reactance, x_m is magnetizing reactance, x_r is rotor circuit leakage reactance, and r_r is rotor circuit resistance. Slip s is as defined by Eq. (13.57). When expressed in per-unit, the circuit element values lie in a relatively narrow range. Stator

Table 13.3 Torque Characteristics of Various Induction Machine Design Classes

NEMA[a] class	Characteristics	Applications
A	Normal starting torque Normal starting current Low slip	Fans, blowers Rotary compressors Centrifugal pumps
B	Normal starting torque Low starting current Low slip	Generally same as A
C	High starting torque Low starting current	Reciprocating pumps and compressors Crushers
D	High starting torque High slip	Punch presses, shears, elevators

[a]National Electrical Manufacturers' Association.

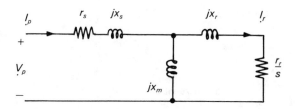

Figure 13.25 An induction machine steady-state equivalent circuit.

and rotor resistances r_s and r_r are roughly the same and approximately equal to rated slip. Leakage reactances typically range from 0.05 to 0.10 per unit. Magnetizing reactance ranges from roughly 3.0 to 7.0 per unit. To convert per-unit values to the more fundamental units of ohms, multiply by base impedance, which is $V_{P,rated}/I_{P,rated}$. Obviously, data for a specific machine are preferable if available. Other information may often be at hand that can be used to determine parameter values. For example, if the starting current at rated voltage is known (i.e., the locked rotor current), an approximate value for the leakage reactances can be found. If magnetizing reactance x_m and resistances r_r and r_s are neglected, locked rotor current occurs at slip $s = 1.0$ and is given by

$$I_{\text{locked rotor}} \approx \frac{V_p}{x_s + x_r} \approx \frac{V_p}{2x_s} \approx \frac{V_p}{2x_r} \qquad (13.58)$$

Equation (13.58) can be used to determine the reactances x_r and $x_s \approx x_r$.

The equivalent circuit can be used to compute quantities associated with steady-state operation. For a given value of slip s and terminal voltage V_p the circuit can be solved. Total power across the air gap for a three-phase machine is given by

$$P_g = 3I_r^2 \, \frac{r_r}{s} \qquad (13.59)$$

The factor of 3 is necessary when working in fundamental units, i.e., watts, amperes, and ohms. The mechanical power converted is

$$P_m = (1-s)P_g \qquad (13.60)$$

Torque is

$$T = \frac{P_m}{\omega_r} = \frac{P_g}{\omega_s} \qquad (13.61)$$

Rotor circuit losses are

$$P_r = sP_g \qquad (13.62)$$

Armature circuit losses are

$$P_a = 3I_p^2 r_s \qquad (13.63)$$

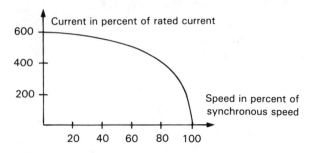

Figure 13.26 Induction machine current versus speed.

Some typical induction machine characteristics that can easily be determined from the equivalent circuit are illustrated in Figs. 13.26 and 13.27. Note that Fig. 13.27 reflects operation from no load to full load corresponding to the relatively narrow speed range of roughly synchronous speed down to rated speed.

A number of application considerations can be surmised from the foregoing discussion. Assuming a fixed frequency supply, the motor must operate at relatively low values of slip for efficient operation. Equation (13.62) is critical in this regard. Power factor tends to fall off at light load. Since reactive power consumption as a function of load is relatively flat, a capacitor bank switched with the motor can provide effective power factor correction over a relatively large range of loadings. The benefits of power factor correction will be examined more thoroughly in the next chapter. Current inrush during starting is several times the rated current. Motors must be sized to accelerate loads to the rated speed in a reasonable time, taking into account the reduced voltage that may occur because of the high starting current. An analysis of motor starting performance typically

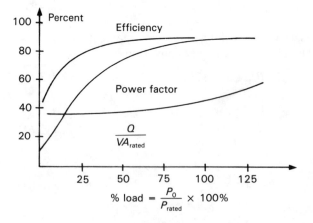

Figure 13.27 Induction machine efficiency, power factor, and reactive power versus load.

requires consideration of the impedance of the system to which the motor is connected and of the mechanical shaft load the motor is driving.

13.6.4 The Synchronous Machine

The synchronous machine, as its name implies, is strictly a fixed-speed machine. The field of the machine is established by supplying dc current to a winding on the rotor of the machine. The field so established has a fixed orientation with respect to the axis of the field winding. For the machine to develop a constant steady torque the rotor must turn in synchronism with the rotating field established by armature currents flowing in the stator windings. Figure 13.28 illustrates synchronous machine action for a machine that is operated as a motor.

In Fig. 13.28 the vector Φ_s represents the armature winding flux wave. This flux wave is established in exactly the same way as that in the induction machine, i.e., by balanced three-phase currents flowing in the phases of the armature winding. During normal steady-state operation the rotor turns at synchronous speed ω_s. The vector Φ_r represents the rotor winding flux wave and the vector Φ_{sr} represents the resultant of Φ_r and Φ_s. As in the case of the induction machine, a two-pole machine is illustrated in Fig. 13.28. A specified armature terminal rms voltage fixes the magnitude, speed of rotation, and relative angle of Φ_{sr}. The vectors Φ_r and Φ_s must adjust in such a way as to satisfy the required vector addition illustrated in the figure. Torque is proportioned to the product of Φ_{sr}, Φ_r, and the sine of the angle between the two.

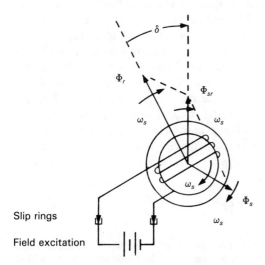

Figure 13.28 Synchronous machine operation (conceptual).

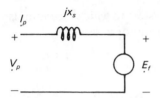

Figure 13.29 A synchronous machine steady-state equivalent circuit.

The synchronous machine has an important capability because of its externally supplied field. By varying the dc excitation level, the machine can be caused to operate at various power factors, both leading and lagging, without appreciably affecting its real power conversion. In other words, it can function as a variable source of reactive power. The ability to externally control the field of the machine is also important in establishing system voltage levels. For these reasons the synchronous machine is the primary electromechanical device used in power generation. As a motor it is not economically competitive with the induction machine except at relatively high horsepower ratings and relatively low rpm, where an induction machine may suffer by comparison because of its low power factor. In some cases the choice between an induction and a synchronous machine is not an easy one and a detailed study of the trade-offs is required.

An equivalent circuit for steady-state operation of a cylindrical rotor synchronous machine is illustrated in Fig. 13.29. Salient pole machines require modification of the model to account for the effects of saliency[5]; however, approximate behavior can be determined by using the equivalent circuit in Fig. 13.29. The references at the end of the chapter provide more detailed analysis of salient pole machines. Referring to Fig. 13.29, V_p and I_p are armature terminal voltage and current per phase, assuming steady-state balanced three-phase operation. The reactance x_s is synchronous reactance, which incorporates the effects of armature leakage flux and air gap flux due to armature current. It might typically be 1.5 if expressed in per-unit. In smaller machines armature resistance (not shown in Fig. 13.29) may be significant and should be added in series with synchronous reactance. The voltage E_f is the field-induced voltage. The rms value of this voltage induced in the armature of the machine is directly proportional to the field current. That is, $E_f = |E_f| \propto I_f$, where I_f is the dc current in the field winding on the rotor. The field excitation system controls the dc field current. The equivalent circuit shown in Fig. 13.29 implies a relatively simple relationship between quantities. For a motor:

$$V_p = E_f + jx_s I_p \qquad (13.64)$$

---

[5]Low-speed synchronous machines have rotor configurations suitable for accommodating a relatively large number of poles. These rotors are generally of the salient pole type, as opposed to the round or cylindrical configuration shown in Fig. 13.28.

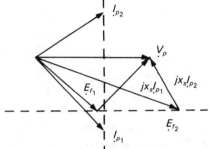

Figure 13.30 Effects of varying synchronous machine field current at constant shaft load.

The effects of varying the field at constant shaft load and constant terminal voltage are illustrated by the phasor diagrams shown in Fig. 13.30. A typical lagging power factor condition is illustrated with the variables subscripted "1". That is, I_{p_1} lags V_p when the field is adjusted to produce an induced voltage in the armature E_{f_1}. If the field excitation is increased, the magnitude of E_f must increase. Assuming that the armature terminal voltage V_p remains fixed and the shaft load on the motor does not change, a new steady-state condition with E_f increased to E_{f_2} results. This is illustrated by the variables subscripted "2". Note that current I_{p_2} leads the armature voltage V_p and that a leading power factor mode of operation has been obtained by increasing the field. The motor is said to be "overexcited," and it supplies reactive power to the system in this condition. The dashed vertical and horizontal lines in Fig. 13.30 result from the assumption of constant power. A family of V-shaped curves can be derived by using nothing more than the equivalent circuit shown in Fig. 13.29. Figure 13.31 illustrates such a set of curves.

It is not difficult to show that the real power converted per phase is given by

$$P = V_p I_p \cos \phi = \frac{E_f V_p}{x_s} \sin \delta \qquad (13.65)$$

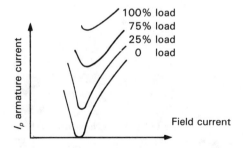

Figure 13.31 A family of V curves for a synchronous machine.

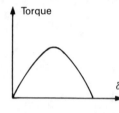

δ **Figure 13.32** Synchronous machine steady-state torque relationships.

where δ is the angle between E_f and V_p and is referred to as the steady-state power angle or steady-state torque angle. Torque per phase is given by

$$T = \frac{P}{\omega_m} = \frac{V_p E_f}{\omega_m x_s} \sin \delta \qquad (13.66)$$

where $\omega_m = (2/p)\,\omega_s$ is shaft speed in mechanical radians per second. Equation (13.66) yields the torque angle characteristic shown in Fig. 13.32.

Another important characteristic curve associated with the synchronous machine is its capability curve. A capability curve based on operation of the machine as a generator is shown in Fig. 13.33. This curve can be obtained from consideration of the equivalent circuit of Fig. 13.29. Operating points must fall within the area indicated by the hatching. Operation along curve *ab* is limited by field circuit heating. Along curve *bc*, operation is limited by armature circuit heating, with point *c* determined by stability considerations. By appropriately cooling the respective windings, it is possible to increase the region of allowable operation.

13.6.5 The DC Machine

The dc machine is perhaps the most versatile of all the conventional machine types. It performs efficiently over a wide range of speeds and is a logical choice for applications requiring variable speed and precise control.

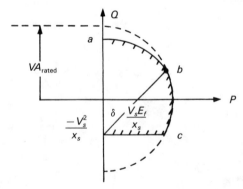

Figure 13.33 A synchronous machine capability curve.

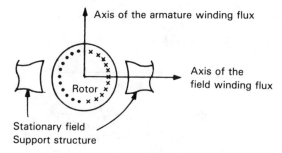

Axis of the armature winding flux

Axis of the
field winding flux

Rotor

Stationary field
Support structure

Figure 13.34 The basic dc machine.

Compared to the induction or synchronous machine, its configuration is relatively complex in that a brush/commutator system is required to supply the armature. Figure 13.34 illustrates dc machine action.

In Fig. 13.34 the field of the machine is on the stator. A dc current fed through the field winding establishes a stationary magnetic field. The armature winding is on the rotor. Ideally, the brush/commutator system distributes current to the armature conductors located in slots of the rotor in such a way that the pattern of currents illustrated in Fig. 13.34 is maintained regardless of the position or speed of the rotor. Thus, the axis of the field due to armature currents is stationary, and orthogonal to that of the field winding. The action so described seems simple, but a number of problems are inherent and must be overcome. The necessary armature current distribution requires that stationary brushes make sliding contact with armature commutator segments that rotate with the rotor. Sparking generally is present. Upper limits on the voltage between commutator segments essentially limit the maximum voltage across the terminals of the armature. The structure of the armature limits maximum speed. The machine requires more maintenance than the simple squirrel cage induction machine. Explosive environments may preclude its use entirely. In spite of these liabilities, the dc machine is regarded as an economically competitive choice as a variable-speed drive.

Torque-speed characteristics vary significantly depending on the manner in which the field and armature windings are excited. The basic types are illustrated in Fig. 13.35. A shunt field winding will typically have a large number of turns with winding material of small cross-sectional area. A series field winding is designed to carry currents at or near the rated armature current. This requires larger conductors and requires fewer turns on the windings. Torque-speed characteristics of the various types are illustrated in Fig. 13.36. Series motors develop relatively high starting torque for applications such as traction drives, cranes, and hoists. Speed regulation is excessive, as can be seen from the characteristics. Shunt

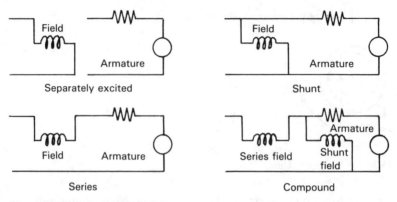

Figure 13.35 Basic dc machine types.

machines have a relatively flat curve, exhibiting little speed change in going from no load to full load. Compound machines can be designed to achieve characteristics in between those of a shunt or series machine. The separately excited machine fed by a solid-state drive is a very flexible package. A more or less typical system is illustrated in Fig. 13.37. The rectifier/control package essentially rectifies three-phase ac and feeds the field and armature circuits of the motor with dc. Armature circuit voltage is varied for speeds below the machine rated speed and field circuit voltage is varied for speeds above the rated speed. A speed range of 40 to 1 might be typical. The tachometer monitors speed for precise control. Speed can be set manually or can be controlled by the output of a process control device. Typically, such systems can incorporate, within the rectifier/control package, motor overload protection and loss-of-field protection. (Loss of field can produce dangerous motor overspeeds.) The rectifier/control package can also control armature current (and thus motor torque) effectively over the entire speed range. This eliminates systems that limit current at starting by inserting resistance into the armature circuit. These features allow the dc machine to be competitive even in light of current advances in adjustable-speed ac systems.

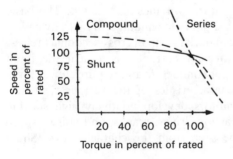

Figure 13.36 Typical torque-speed curves for dc machines.

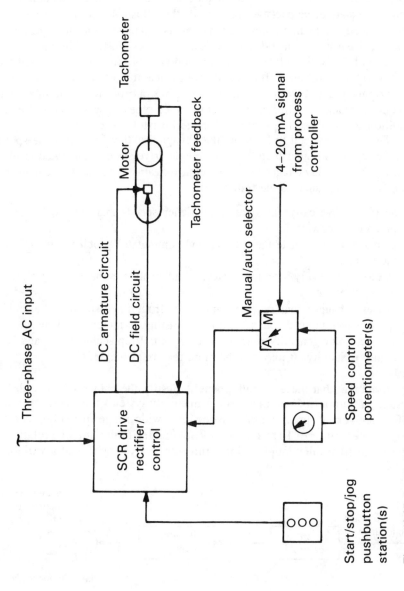

Figure 13.37 A typical dc machine drive/motor system.

511

13.6.6 Adjustable-Speed AC Drive Systems

The attractive features of the squirrel cage induction machine (simplicity, ruggedness, relative freedom from maintenance) and the development of efficient solid-state drive packages have made the adjustable-speed ac drive system competitive in many applications. In its simplest form an adjustable-speed ac drive system includes the basic components illustrated in Fig. 13.38.

The objective is to furnish a controlled variable-frequency three-phase supply to the motor itself. It is usually not sufficient to vary frequency alone, and voltage must be varied as frequency varies in order to maintain air gap flux in the machine. Thus, a more complicated type of control is required in addition to the basic requirement for the inverter system.

These factors tend to keep the initial capital outlay for adjustable-speed ac drive systems higher than those for the dc drive system illustrated in Fig. 13.37. These disadvantages must be weighed against potential benefits of the ac drive system, such as the following:

1. ac motors typically can operate at higher speeds than dc motors for a given horsepower.
2. ac machines eliminate the brushes and commutator problems inherent in the dc motor.
3. ac motors typically have lower weight and smaller frame sizes for a given horsepower rating.
4. Process changes that require conversion from fixed-speed to variable-speed drives favor the ac system if an existing ac motor can be used.
5. The ac motor generally requires a lower initial/replacement cost and has higher reliability. It may also be more readily available.

Assuming that these potential benefits justify the choice of an ac drive system, a number of manufacturers can supply systems that have very flat efficiency versus speed characteristics for a wide range of loading conditions. Speed ratios of 10 to 1 are easily obtained. Elimination of a relatively inefficient hydraulic coupling, or a throttling valve, by use of a variable-

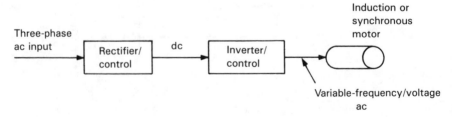

Figure 13.38 Components of an adjustable-speed ac drive system.

speed system can produce significant economic benefits. A specific case will be examined in the next chapter.

Inverter systems feeding induction machines generally do not exceed roughly 1500 hp. A somewhat different system, sometimes referred to as a synchroconverter, developed for use with synchronous machines, is more appropriate for higher horsepower requirements, say up to roughly 40,000 hp. The synchronous motor is preferable in the higher ratings because of its ability to simplify commutation requirements in the inverter. Bose [7] is an excellent source of more detailed information on ac drive systems. The first chapter of [7] provides a comparison table listing the trade-offs involved in the many choices of available ac drive systems.

NOMENCLATURE

v	instantaneous voltage
i	instantaneous current
V	dc voltage
V	root-mean-square or effective voltage
$\dot{V}$	phasor voltage
I	dc current
I	root-mean-square or effective current
$\dot{I}$	phasor current
VA	voltamperes or apparent power
R	resistance
P	real power or power
Q	reactive power
Z	impedance magnitude
$\dot{Z}$	complex impedance
L	inductance
C	capacitance
X	reactance
S	complex power
$\dot{S}$	magnitude of complex power or apparent power
p.f.	power factor
GMS	geometric mean spacing
D	distance
N	number of turns on a winding
n	speed (rpm)
p	number of poles
T	torque
T	time period
δ	power or torque angle

η efficiency
Φ rotating space vector
ρ resistivity
ω radian frequency

REFERENCES

1. William D. Stevenson, Jr., *Elements of Power Systems Analysis*, 3d ed., McGraw-Hill, New York, 1975.
2. Charles A. Gross, *Power System Analysis*, Wiley, New York, 1979.
3. B. M. Weedy, *Electric Power Systems*, 2d ed., Wiley, New York, 1972.
4. A. E. Fitzgerald, Charles Kingsley, Jr., and Alexander Kusko, *Electric Machinery*, 3d ed., McGraw-Hill, New York, 1971.
5. Westinghouse Electric Corporation, *Electric Transmission and Distribution Reference Book*, East Pittsburgh, Pa., 1964.
6. Westinghouse Electric Corporation, *Distribution Systems*, vol. 3, East Pittsburgh, Pa., 1965.
7. Bimal K. Bose, *Adjustable Speed AC Drive Systems*, IEEE Press, New York (selected reprint series), 1980.

PROBLEMS

13.1 A load consisting of a 10-Ω resistance in series with an inductive reactance of 5 Ω draws a current of 10 A rms. Determine load voltage.

13.2 For the data of Problem 13.1, determine load real power P, reactive power Q, and power factor.

13.3 A power feeder circuit supplying a particular load is known to have a voltage regulation of 6%. Assuming that the voltage at the load is 120 V rms, what voltage will be measured at the load end of the line if the load is disconnected?

13.4 A three-phase circuit has a rated voltage of 460 V. If three 20-Ω resistances are connected in a Y arrangement in the circuit, what is the total three-phase power delivered to the resistances?

13.5 Repeat Problem 13.4, but with the three resistances connected in a delta arrangement.

13.6 Nameplate data for an induction machine show a rated speed of 1730 rpm. What is the rated value of slip for the machine? If the machine is a 15-hp machine, estimate the rotor circuit I^2R losses and express the answer in kilowatts (kW).

13.7 A 2-kVA single-phase transformer has a rated primary voltage of 120 V and rated secondary voltage of 240 V. Determine rated primary and secondary currents.

13.8 If, for the transformer of Problem 13.7, rated voltage is applied to the primary winding and the secondary is left an open circuit, the no-load power consumption of the transformer is measured as 76 W. Find a value for the core loss resistance r_c, assuming that the resistance is referred to the secondary winding.

13.9 A 1000-hp synchronous motor has an efficiency of 95%. The machine is designed for operation at 4160 V and has a rated power factor of 0.9 leading. Determine the reactive compensation in kvar that the machine can supply to the system at rated conditions.

13.10 A 460-V, 0.85 power factor, 100-hp squirrel cage induction machine is known to have an efficiency of 92% when operated at rated conditions. Find voltage regulation and cable transmission efficiency if the machine is fed by a 1000-ft, 2/0 three-conductor cable.

13.11 Measurements made on a three-phase induction machine while it is running under load give the following results: clamp-on ammeter, 121 A; voltmeter (line voltage), 465 V; and three-phase wattmeter, 83.9 kW. At what power factor is the motor operating? What reactive power is the motor consuming? Determine the amount of capacitive reactive compensation required (in kvar) to correct the power factor to 0.95.

13.12 A 1000-kVA, three-phase transformer has Y-connected primary and secondary windings. The transformer primary is rated at 4160 V and the secondary is rated at 460 V. The total series impedance of the transformer is 6% and series resistance is 2%. Find the voltage regulation for an assumed unity power factor load of 1000 kW.

13.13 Consider the transformer described in Problem 13.12. If, in addition to the data given in Problem 13.12, the transformer no-load losses are known to be 0.45%, find the transformer efficiency if the load is 1000 kW at unity power factor.

13.14 A three-phase, 460-V, 100-hp induction machine has the following parameters:

rated slip	3.5%
r_s	0.075 Ω
r_r	0.081 Ω
x_s	0.325 Ω
x_r	0.350 Ω
x_m	15.0 Ω

Find the starting inrush current in amperes and the current during operation at rated slip.

13.15 A three-phase, Y-connected, 1000-kVA, 13.8-kV synchronous generator has a synchronous reactance of 1.65 per unit. if the machine delivers 1000 kVA at a power factor of 0.90 leading, what is the excitation voltage E_f?

13.16 Consider the synchronous machine described in Problem 13.15. If the machine is a four-pole machine with an estimated efficiency of 92%, what torque in foot-pounds must the prime mover develop for the machine to operate at the condition described?

FOURTEEN

ELECTRICAL ENERGY CONSERVATION

To achieve a cost-effective electrical system, the components of the system must be chosen carefully and the system must be operated in such a way as to minimize costs. The economics rarely simplify to the point where minimization of energy losses alone is sufficient to guarantee a cost-effective system. Generally, more efficient components have higher initial cost, and this increased cost must be weighed against the future savings to be realized in terms of reduced energy charges. Furthermore, a device that may be highly efficient can be operated in ways that significantly reduce its life or that incur significant demand or power factor penalty charges.

In this chapter, major sources of losses in typical industrial electrical systems are identified. Trade-offs involved in component selection are then examined. Finally, practices that can significantly affect operating costs are described.

14.1 SOURCES OF LOSSES

Hickok [1] provides an excellent summary of typical losses that occur in commonly encountered electrical system components. Table 14.1 is excerpted from reference [1] and illustrates ranges of losses that might be anticipated. The energy loss values shown reflect energy loss as a percentage

Table 14.1 Energy Losses in Typical Electrical System Components

Component	Energy loss (%)
Outdoor circuit breakers (15–230 kV)	0.002–0.015
Medium-voltage switchgear (5–15 kV)	0.005–0.02
Low-voltage switchgear (<5 kV)	0.13–0.34
Generators	0.09–3.5
Current-limiting reactors	0.09–0.30
Transformers	0.40–1.90
Load break switches	0.003–0.025
Medium-voltage starters	0.02–0.15
Busway (480 V and below)	0.05–0.50
Motor control centers	0.01–0.40
Cable	1.00–4.00
Motors	
1–10 hp	14.0–35.0
10–200 hp	6.0–12.0
200–1500 hp	4.0– 7.0
1500 hp	2.3– 4.5
Rectifiers	3.0– 9.0
Static variable-speed drives	6.0–15.0
Capacitors	0.5–2.0 W loss/kvar
Lighting	15–130 Lumens output/input watt

Source: Adapted, with permission, from H. N. Hickok, Electrical Energy Losses in Power Systems, *IEEE Transactions on Industry Applications,* vol. IA-14, no. 5, pp. 373–387, Sept./ Oct. 1978, Copyright © 1978 IEEE.

of the total throughput energy of the device. Equivalently, they reflect power loss as a percentage of the device power rating (or var rating in the case of capacitors). The losses shown in Table 14.1 are predominantly ohmic ones (I^2r losses), resulting in heat dissipation. The values for static variable-speed drives include losses attributed to all components, including the drive motor itself, for operation at rated speed and load.

Table 14.1 permits identification of the components that are most likely to cause significant electrical losses. The following sections examine some of these components in greater detail.

14.1.1 Motor Losses

Overall motor efficiency typically improves as motor capacity increases (see Table 14.1). Assuming that motor efficiency does not fall off appreciably with decreases in motor load, oversizing a motor somewhat for an intended application can reduce losses, albeit at the expense of increased and perhaps never used motor capacity. To illustrate this, the squirrel cage induction machine, which is the most commonly used machine, will be examined. The equivalent circuit presented in Chapter 13 can be used to develop the curves shown in Fig. 13.26. A similar set of curves based on actual motor operating data is shown in Fig. 14.1. Note that reductions in load from a nominal load of 100% can be significant before any appreciable change in efficiency occurs. Power factor, however, tends to fall off somewhat faster with decreases in load, and this will typically increase losses in the circuits serving the motor. It will also contribute to an overall lower power factor for the system. Benefits of oversizing in terms of improved efficiency are offset somewhat by power factor considerations. The relatively flat curve for reactive power versus load suggests that capacitors installed at the terminals of the motor would supply motor reactive power requirements reasonably well over a wide range of load conditions. This is, in fact, the case, and capacitors located at the motor terminals and switched with the motor can yield effective power factor correction over a wide load range. A motor oversized by 10 to 20% and equipped with capacitors for power factor correction if necessary can be an attractive option.

An additional benefit is afforded by some degree of oversizing. Winding insulation life is significantly affected by operating temperature. This is illustrated in Fig. 14.2, where the decrease in insulation life with temperature is apparent for all classes of insulation. A rough rule of thumb is that insulation life is halved for every 10°C rise in operating temperature. A motor that operates relatively lightly loaded will operate at a lower temperature. Operating temperature will be roughly proportional to the heat that must be dissipated, which in turn is roughly proportional to machine line current squared. These currents and the associated winding resistance losses can be evaluated at various loads by examining the equivalent circuit model

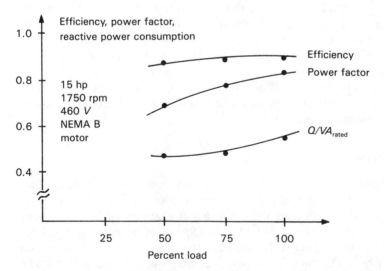

Figure 14.1 Induction machine efficiency, power factor, and reactive power versus load.

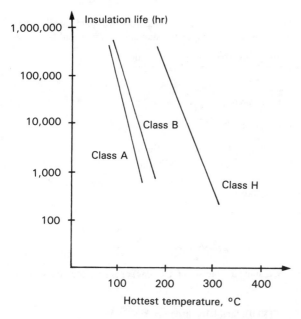

Figure 14.2 Motor insulation life versus temperature.

for the machine presented in Chapter 13. As a practical matter, efficiency, power factor, and line current data are generally readily available from manufacturers for motors of standard frame size at full, three-quarters, and half load. Thus, trade-offs involved in oversizing can be analyzed in considerable detail, assuming that up-to-date cost data are available for motors and for energy.

High-efficiency motors feature an improvement in efficiency ranging from roughly 0.5 to 1.5% at a cost increase of about 15 to 25%. The efficiency improvement typically comes from increased use of laminated steel, increased core length, use of more copper in the stator windings and rotor bars, and more rigid quality control measures. The trade-offs are illustrated in the following example.

Example 14.1 Comparison of a Standard Versus a High-Efficiency Motor

Determine annual operating cost savings and payback period for a high-efficiency 50-hp motor compared with a standard 50-hp motor. Assume a 1.5% efficiency increase at a price premium of 25% for the high-efficiency motor. The cost of the standard motor is $1,662.00 and its full-load efficiency is 91.5%. Assume that the cost of energy is $0.05/kWh.

The above data yield:

Motor	Efficiency	Cost
Standard	91.5	$1,662
High-efficiency	93.0	$2,078

In the absence of any data on actual load cycling, the comparison will be made on the basis of full-load operation. The hourly operating costs for the standard motor would be

$$C_s = \frac{(50 \text{ hp})(0.746 \text{ kW/hp})}{0.915} \times \frac{\$0.05}{\text{kWh}}$$

$$= \$2.038/\text{hr}$$

Similarly, for the high-efficiency motor:

$$C_h = \frac{(50 \text{ hp})(0.746 \text{ kW/hp})}{0.93} \times \frac{\$0.05}{\text{kWh}}$$

$$= \$2.005/\text{hr}$$

The hourly operating cost savings would be:

$$C_s - C_h = \$0.033/\text{hr}$$

To convert to annual cost savings, assume double-shift (16 hr/day) operation for 5 days a week, 52 weeks a year, i.e., 4160 hr/yr. Thus

$$C_s - C_h = (\$0.033/\text{hr})(4,160 \text{ hr/yr}) = \$137.28/\text{yr}$$

The payback period in years is obtained by dividing the price premium by the annual operating cost savings:

$$\frac{\$2,078 - \$1,662}{\$137.28/\text{yr}} = 3.03 \text{ yr}$$

The calculation can be improved by incorporating the techniques of Chapter 3, but the fundamental trade-off applies regardless.

14.1.2 Cable Losses

Cable losses are essentially ohmic (I^2r) losses, which must be dissipated as heat. As shown in Table 14.1, cable losses can be large compared with losses in other electrical system components. Generally, cables are sized on the basis of ampacity and voltage drop considerations. The ampacity for a given cable is the amount of current it can carry under some specified condition without overheating and seriously affecting (reducing) insulation life. Generally, the actual installation of the cable and ambient temperature are critical factors in determining its ampacity. Ampacities for cables are reduced in situations where heat dissipation is restricted. For example, derating factors specified by the National Electric Code for cables in trays increase as the percent fill of the cable tray increases.

Voltage drop considerations may dictate cable choice on longer cable runs, since total cable impedance increases directly with length. Regardless of whether ampacity or voltage drop considerations are used to determine

Table 14.2 Typical Cable Data for Three-Conductor Cable

Cable no.	Cable size, circular mils or (AWG)[a]	Ampacity (amps)	Resistance at 65°C (Ω/1,000 ft)	Cable cost ($/1,000 ft)
1	26,250 (6)	75	0.473	705
2	41,740 (4)	98	0.299	1,068
3	66,370 (2)	128	0.187	1,641
4	83,690 (1)	146	0.149	2,232
5	105,500 (1/0)	168	0.118	2,745
6	133,100 (2/0)	192	0.094	3,360
7	167,800 (3/0)	219	0.074	4,182
8	211,600 (4/0)	249	0.059	5,220
9	250,000	273	0.050	6,210
10	300,000	304	0.042	7,485
11	350,000	334	0.036	8,520
12	400,000	359	0.031	10,050
13	500,000	408	0.025	12,120
14	600,000	450	0.021	16,410
15	750,000	505	0.017	20,310

[a]AWG (American Wire Gage), a standard gauge used to identify the size of a conductor.

cable size for a given application, I^2r losses can always be reduced by choosing a larger cable with greater cross-sectional area, since cable resistance varies inversely with cross-sectional area. The situation is similar to that presented earlier in that losses can be reduced but at the expense of added capacity. Table 14.2 illustrates the trade-offs for 5-kV class three conductor cable. Ampacity and resistance values at 65°C are from reference [2]. Cable cost data are typical in 1980 dollars. For convenience, a cable numbering system is used in which increasing cable numbers correspond to increasing cable sizes. Ampacities in Table 14.2 are for duct bank installation with one conductor cable in a bank and an ambient earth temperature of 20°C. Correction factors for other conditions are cited in reference [2].

Example 14.2 Reduction of Cable Losses

Determine annual operating cost savings and payback period required if a load requiring a continuous current of 75 A is served by cable number 4 in Table 14.2 rather than cable number 1. Assume that the cost of energy is $0.05/kWh.

Note that cable number 1 is adequate on the basis of ampacity. Assuming that voltage drop considerations do not require oversizing, a choice of cable 4 is clearly conservative and will result in a relatively low cable operating temperature. Cable resistance, which varies more or less linearly with temperature, will also be reduced. These effects can be accounted for in an approximate way. The ampacities cited in Table 14.2 are based on some maximum permissible conductor temperature. Assume that steady-state operation of any cable at the value specified in Table 14.2 results in a conductor temperature of 85°C. The resistances cited in Table 14.2 are at 65°C and must be corrected for a temperature of 85°C. Resistance variation with temperature is often expressed as

$$\frac{r_2}{r_1} = \frac{T + t_2}{T + t_1}$$

where r_2 is resistance at temperature t_2 and r_1 is resistance at temperature t_1. The constant T is 241°C for hard-drawn copper of 97.3% conductivity [3]. Thus, resistance for conductor 4 at 85°C is given by

$$r_2 = r_1 \frac{(T + t_2)}{(T + t_1)}$$

$$= 0.149 \frac{(241 + 85)}{(241 + 65)} = 0.159 \ \Omega/1,000 \ \text{ft}$$

Heat dissipation is proportional to $I^2 r$. If a linear relationship between heat flux and temperature difference between the conductors and ambient is assumed, the heat flux q is given by

$$q = 3I^2 r = k(t - t_a)$$

where t is conductor temperature and t_a is ambient temperature. Solving for k for cable 4:

$$k = \frac{3I^2 r}{t - t_a} = \frac{3(146)^2(0.159)}{85 - 20} = 156.43 \ \frac{\text{W}}{1,000 \ \text{ft} \ °\text{C}}$$

Note that the constant k for cable 4 is simply determined on the basis of the ampacity of the cable (146 A) and the assumed temperature difference (85–20°C). If cable 4 is operated at 75 A its operating temperature t_2 and resistance r_2 are determined by solving simultaneously:

$$3I^2 r_2 = k(t_2 - t_a)$$

$$r_2 = r_1 \frac{(T + t_2)}{(T + t_1)}$$

where I is 75 A and t_1 is resistance at temperature t_1. Thus:

$$3(75)^2 r_2 = 156.43(t_2 - 20)$$

$$r_2 = 0.159 \frac{(241 + t_2)}{(241 + 85)}$$

Solving for r_2 and t_2:

$$t_2 = 34.47°\text{C}$$

$$r_2 = 0.134 \ \Omega/1,000 \ \text{ft}$$

These calculations illustrate in an approximate way the reduced operating temperature benefits afforded by oversizing.

The hourly operating costs per 1,000 ft with cable 1 would be

$$C_1 = 3I^2 r \ (\$0.05/\text{kWh})$$

$$= \frac{3(75)^2(0.504)(0.05)}{1,000} = \$0.4253/\text{hr}$$

Similarly, for cable 4

$$C_4 = \frac{3(75)^2(0.134)(0.05)}{1,000} = \$0.1131/\text{hr}$$

The hourly operating cost savings per 1,000 ft would then be

$$C_1 - C_4 = \$0.3122/\text{hr}$$

Assuming double-shift (16 hr/day) operation for 5 days a week, 52 weeks a year, i.e., 4,160 hr/yr, annual operating cost savings on a 1,000-ft basis are

$$C_1 - C_4 = \$0.3122/\text{hr} \times 4,160 \text{ hr/yr} = \$1298.75/\text{yr}$$

The payback period in years is obtained by dividing the price premium by the annual operating cost savings:

$$\frac{\$2,232 - \$705}{\$1,298.75/\text{yr}} = 1.18 \text{ yr}$$

Here again, as in Example 14.1, the calculation can be improved by incorporating the techniques of Chapter 3.

Dimachkieh and Brown [4] describe in detail a method for choosing cable sizes that minimizes total cable cost. A number of interesting results are presented for various assumptions about the cost of energy, anticipated cable life, and interest rate over the life of the cable. The procedure first requires determining a total cable cost for a particular cable at various operating current levels. For example, considering cable 15 in Table 14.2, a total cable cost per 1,000 ft is determined by summing the cable cost as shown in Table 14.2 and the present worth of the cost of losses over the anticipated life of the cable. Assumptions are required in order to do this. Table 14.3, excerpted from reference [4], illustrates results for the following assumptions:

1. 5 day/week, 16 hr/day operation.
2. Continuous operation at the assumed current level.
3. Operation at the ampacity levels shown in Table 14.2, resulting in a conductor temperature of 85°C.
4. An assumed cable life of 30 years.
5. An interest rate over the life of the cable of 12%.
6. Cost of energy starting at $0.05/kWh and tripling in 16 years, then staying level for the remaining life of the cable.

Note that as the assumed current level goes down the operating temper-

Table 14.3 Total Cable Cost per 1,000 ft for Cable 15[a]

Current	Annual losses (kWh/1,000 ft)	Total cable cost (thousands of dollars/1,000 ft)	Cable temperature (°C)
505	58,500	65.3	85.0
450	44,100	54.3	69.0
408	35,100	47.3	59.0
359	26,200	40.5	49.2
334	22,400	37.5	44.9
304	18,200	34.3	40.2
273	14,500	31.5	36.1
249	11,900	29.5	33.2
219	9,100	27.3	30.1
192	6,900	25.7	27.7
168	5,300	24.4	25.9
146	4,000	23.4	24.4
128	3,000	22.6	23.4
98	1,800	21.7	22.0
75	1,000	21.1	21.2

[a]750,000 circular mils.
Source: Dimachkieh and Brown [4].

ature approaches the assumed 20°C ambient, and the total cable cost per 1,000 ft approaches the cost of cable 15 as shown in Table 14.2. If similar tables are developed for the remaining 14 cable sizes, it will be seen that for any of the current levels shown in Tables 14.2 and 14.3 a cable size that minimizes total cable cost can be determined. This is illustrated for a current level of 98 A in Fig. 14.3, which is also excerpted from reference [4].

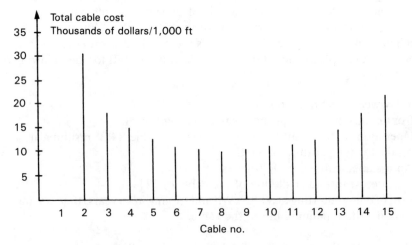

Figure 14.3 Total cable cost for a load current of 98 A versus cable size.

Figure 14.3 shows that cable 8 results in the lowest total cable cost for a current of 98 A. This implies choosing a cable six sizes larger than would be chosen on the basis of ampacity. Such results are clearly dependent on the assumptions one makes, and assumptions about the cost of energy are critical. Table 14.4 shows the results of varying the assumptions concerning the cost of energy. Total savings realized by oversizing are easily determined by following the procedure described above, and they are reflected in Table 14.4 as well.

The approach described above does not take into account higher installation costs associated with pulling and splicing larger cable. Cable costs reflected in Table 14.2 could be modified in an appropriate way to reflect these considerations. The savings associated with cable selection can be significant, which suggests that reasonably detailed comparison studies are justified.

14.1.3 Power Factor-Related Losses

Power factor correction was introduced briefly in Chapter 13 in Example 13.2. In essence, capacitors can supply the reactive component of current required by loads that operate with a lagging power factor. This effectively reduces the magnitude of the line current required to serve a particular load without reducing the load's actual power consumption. The reduced line current requirement results in lower I^2r losses and less voltage drop in the line serving the load. The capacity requirement for devices serving a particular load is also reduced. Capacitors effectively supply reactive power Q. Assuming that they can be located at or near a load that consumes reactive power, they can eliminate the requirement for generating and transmitting reactive power over significant distances.

The obvious choice for capacitor location is at the terminals of any load requiring reactive power. Induction machines consume reactive power as shown in Fig. 14.1. One approach is to locate capacitors at the terminals of an induction machine and switch them with the machine. This yields savings in all electrical equipment serving the machine, including the cable that supplies it. If power factor correction is to be applied for a number of machines, multiple capacitor bank installations are required. An alternative is to locate capacitors centrally, perhaps entirely at the substation serving a particular plant or in several load centers within the plant. The latter approaches are attractive in that installation is simplified and capacitors may be applied at higher voltages. The cost of capacitors decreases on a per-kilovar basis as the capacitor voltage rating increases. For example, capacitors for installation in 240-V systems cost roughly twice as much per kilovar as capacitors for installation in 480-V systems. A utility-imposed power factor penalty charge can be reduced or eliminated entirely by locating capacitors at the metering point, but such an approach will not reduce losses

Table 14.4 Effects of Changes in Assumptions about Cost of Energy

Current (amperes)	75	98	128	146	168	192	219	249	273	304	334	359	408	450	505
Cable choices on the basis of ampacity:															
Cable choice (cable no.)	1	2	3	4	5	6	7	8	9	10	11	12	13	14	15
Cable choices on the basis of minimum total cable cost:															
Cost of energy level at $0.02/kWh															
Cable choice (cable no.)	3	5	6	7	7	8	9	9	11	11	11	12	13	14	15
Total savings (thousand $ per 1,000 ft)	2.8	2.4	2.0	1.9	1.6	1.3	0.9	0.5	0.3	0.4	0.0	0.0	0.0	0.0	0.0
Cost of energy starting at $0.02/kWh and tripling in 16 years															
Cable choice (cable no.)	5	6	7	8	8	9	11	11	11	13	13	13	13	15	15
Total savings (thousand $ per 1,000 ft)	6.6	6.4	5.7	5.6	5.2	4.7	4.1	3.4	2.8	2.5	2.2	1.9	0.0	0.3	0.0
Cost of energy starting at $0.05/kWh and tripling in 16 years															
Cable choice (cable no.)	7	8	9	11	11	12	13	13	13	13	15	15	15	15	15
Total savings (thousand $ per 1,000 ft)	20.3	20.6	19.9	19.7	19.6	18.7	18.0	16.5	15.2	13.2	12.2	11.1	8.1	6.5	0.0

within the plant itself. It is important to remember that both real and reactive power are conserved quantities. In essence, if the real power requirements of individual electrical devices within a plant are added, the total real power required by the plant and metered at the plant substation is obtained. Similarly, if all of the reactive power requirements of electrical devices within a plant are added, the total reactive power required by the plant is obtained. This implies that a utility-imposed power factor penalty charge can be reduced either by centrally locating capacitors or by locating capacitors at points within the system where a requirement for reactive power exists. The most economical installation for a particular plant depends on the utility rate structure cited for the plant. It is desirable to obtain as much information as possible about not only current rate structures but also anticipated future rate structures before attempting to reduce costs by power factor correction measures.

Example 14.3 Power Factor Correction Calculations

A 250-hp, 460-V induction machine operates at 75% load on a double-shift schedule, i.e., 4,160 hr/yr. Its power factor at 75% load is 0.84. Its efficiency at 75% load is 0.95. The motor is served by 500 ft of 350 MCM[1] three-conductor cable. Power factor penalty charges are $0.20 per kilovar per month. Cost of energy is $0.05/kWh.

Determine the annual operating cost savings and payback period if capacitors are applied at the motor terminals in order to correct motor power factor. Assume that the cost of capacitors is $21.00/kvar installed.

Ideally, the motor power factor would be corrected to unity in order to minimize losses and capacity requirements. The possibility of motor self-excitation limits the amount of reactive compensation that can be applied at the terminals of a motor. This phenomenon occurs when a motor and its capacitors are disconnected from the system. During the coast-down period, the capacitors may supply the magnetizing current required by the machine and open-circuit terminal voltage may rise to roughly 150% of rated voltage. Furthermore, if the motor and its capacitors are reconnected to the system before the rotor comes to rest, severe transient torques can result. These considerations cause motor manufacturers to recommend maximum amounts of capacitive compensation to be applied for various motors. Generally, the maximum recommended compensation will correct power factor to roughly 95% at full load. In this example correction to 95% will be assumed and the compensation required will be determined at the operating condition of 75% load.

Operating cost savings come from reduced line losses and reduced power factor penalty charges. From Appendix H Table H.1 the resistance of 350 MCM three-conductor cable is seen to be 0.0378 Ω per 1,000 ft. Thus the 500-ft feeder will have $0.0378/2 = 0.0189$ Ω total resistance in each phase. Line current can be determined from the fundamental three-phase power relationship:

$$P = \sqrt{3}\, V_L I_L \cos \theta$$

where P is input power at the machine terminals, V_L is line-to-line voltage, I_L is line current, and $\cos \theta$ is the power factor. P is determined on the basis of the cited efficiency at 75% load. Thus

[1]MCM: 1,000 circular mils, where 1 circular mil is the area of a circle having a diameter of 1 mil or 0.001 in.

$$P = \frac{(0.75)(250 \text{ hp})(0.746 \text{ kW/hp})}{0.95} = 147 \text{ kW}$$

Line current at a power factor of 0.84 will be

$$I_L = \frac{P}{\sqrt{3} \, V_L \cos \theta} = \frac{147(10^3)}{\sqrt{3}(460)(0.84)} = 220 \text{ A}$$

Line losses at a power factor of 0.84 will be

$$3I_L^2 r = 3(220)^2(0.0189) = 2.74 \text{ kW}$$

Similarly, line current and line losses at the new power factor of 0.95 are determined by

$$I_L = \frac{147(10^3)}{\sqrt{3}(460)(0.95)} = 194 \text{ A}$$

$$3I_L^2 r = 3(194)^2(0.0189) = 2.13 \text{ kW}$$

Thus, hourly operating cost savings C_L due to reduced line losses will be

$$C_L = (2.74 - 2.13) \text{ kW} \times \frac{\$0.05}{\text{kWh}} = \$0.031/\text{hr}$$

Reactive compensation required to achieve correction from 0.84 to 0.95 power factor can be determined by using the fundamental three-phase reactive power relationship:

$$Q = \sqrt{3} \, V_L I_L \sin \theta$$

where θ is the power factor angle. At a power factor of 0.84 the motor consumes reactive power:

$$Q = \sqrt{3} \, (460)(220)(0.54) = 94.7 \text{ kvar}$$

At a power factor of 0.95 the combined load consisting of the motor and capacitors consumes reactive power:

$$Q = \sqrt{3} \, (460)(194)(0.31) = 47.9 \text{ kvar}$$

The difference between the two computed values of reactive power is the amount of reactive power that must be supplied by the capacitors:

$$94.7 - 47.9 = 46.8 \text{ kvar}$$

Rounding up, a requirement for 47 kvar of capacitive compensation is estimated. The price premium will be

$$47 \text{ kvar} \times \frac{\$21}{\text{kvar}} = \$987.00$$

The hourly operating cost savings C_{PF} due to penalty factor charge reduction will be

$$C_{PF} = 47 \text{ kvar} \left(\frac{\$0.20}{\text{kvar month}} \right) \left(\frac{0.00139 \text{ month}}{\text{hr}} \right) = \$0.013/\text{hr}$$

Note that the hourly operating cost savings C_L due to reduced line losses exceed the hourly operating cost savings C_{PF} due to reduction of penalty charges in this case. The importance of the utility rate structure in the calculation is obvious.

Total operating cost savings will be

$$C_L + C_{PF} = 0.031 + 0.013 = \$0.044/hr$$

On an annual basis, savings will be

$$(\$0.044/hr)(4160 \text{ hr/yr}) = \$183/yr$$

The payback period in years is obtained by dividing the price premium by the annual operating cost savings:

$$\frac{\$987}{\$183/yr} = 5.4 \text{ yr}$$

14.1.4 Transformer Losses

An equivalent circuit for a power transformer was presented in Section 13.6.2. The equivalent circuit (see Fig. 13.20) can be used for general transformer analysis. As discussed in Chapter 13, total transformer losses may be expressed as a sum of no-load and load losses. No-load losses are approximately proportional to transformer voltage squared. Because voltage is always at or near the rated voltage, no-load losses are generally assumed to be constant. Load losses vary approximately as the square of the current, which in turn is roughly proportional to the load. Some typical values for no-load and total losses are shown in Appendix H Table H.4.

Example 14.4 Transformer Loss Calculations

A 1,000-kVA transformer is to be operated at 90% capacity for double-shift operation, 5 days a week, 52 weeks a year, i.e., 4,160 hr/yr. During the remaining time, 4,576 hr/yr, it will operate at 8% capacity. Two alternative choices for the transformer are available as illustrated below:

	Transformer A	Transformer B
Price	$8,200	$8,800
No-load losses	2.8 kW	3.2 kW
Full-load losses	18 kW	15 kW

Cost of energy is $0.05/kWh. Determine annual operating cost savings realized by choosing transformer B and the payback period.

Losses must be determined for each transformer for operation at both 90% and 8% capacity. Load losses at 100% capacity are found by subtracting the transformer's no-load losses from its full-load losses. Load losses at reduced capacity are assumed to vary as the square of the load. Total losses at reduced capacity are found by adding load losses at reduced capacity to the no-load losses. Consider transformer A first.

$$\text{Load losses at 100\% load} = 18 - 2.8 = 15.200 \text{ kW}$$

$$\text{Load losses at 90\% load} = 15.2(0.90^2) = 12.312 \text{ kW}$$

$$\text{Load losses at 8\% load} = 15.2(0.08^2) = 0.097 \text{ kW}$$

$$\text{Total losses at 90\% load} = 12.312 + 2.8 = 15.112 \text{ kW}$$

$$\text{Total losses at 8\% load} = 0.097 + 2.8 = 2.897 \text{ kW}$$

For transformer B:

$$\text{Load losses at 100\% load} = 15 - 3.2 = 11.800 \text{ kW}$$

$$\text{Load losses at 90\% load} = (11.8)(0.90^2) = 9.558 \text{ kW}$$

$$\text{Load losses at 8\% load} = (11.8)(0.08^2) = 0.076 \text{ kW}$$

$$\text{Total losses at 90\% load} = 9.558 + 3.2 = 12.758 \text{ kW}$$

$$\text{Total losses at 8\% load} = 0.076 + 3.2 = 3.276 \text{ kW}$$

The annual cost to operate transformer A is

$$\frac{\$0.05}{\text{kWh}} \left[\left(15.112 \text{ kW}\right)\left(4{,}160 \frac{\text{hr}}{\text{yr}}\right) + \left(2.897 \text{ kW}\right)\left(4{,}576 \frac{\text{hr}}{\text{yr}}\right) \right] = \frac{\$3{,}806}{\text{yr}}$$

The annual cost to operate transformer B is

$$\frac{\$0.05}{\text{kWh}} \left[\left(12.758 \text{ kW}\right)\left(4{,}160 \frac{\text{hr}}{\text{yr}}\right) + \left(3.276 \text{ kW}\right)\left(4{,}576 \frac{\text{hr}}{\text{yr}}\right) \right] = \frac{\$3{,}403}{\text{yr}}$$

The annual cost savings realized by choosing transformer B is

$$3{,}806 - 3{,}403 = \$403/\text{yr}$$

The payback period is found by dividing the price premium $8,800 − $8,200 = $600 by the annual cost savings:

$$\frac{\$600}{\$403/\text{yr}} \cong 1.5 \text{ yr}$$

14.2 VARIABLE-SPEED DRIVES

Variable-speed electric drives can provide significant energy savings if they eliminate devices that are inherently inefficient at reduced speeds. Throttling valves and hydraulic couplings are examples of devices that have relatively low efficiency at reduced flow rate or speed. Large fans applied in such a way that they must provide a wide range of forced-draft flow rates are another example.

The dc machine has classically been regarded as the most versatile machine for operation over a wide speed range. The development of solid-state technology has made possible the efficient use of ac machines for variable-speed applications. This permits elimination of undesirable features associated with the dc machine such as high initial cost and brush/commutator maintenance problems. However, the solid-state drives used in conjunction with ac machines are typically more complex and require higher initial investments than drives used with dc machines. Other trade-offs between dc and ac variable-speed drives were discussed in Chapter 13. High energy costs can cause a variable-speed electric drive (ac or dc) to be an extremely attractive alternative when the application requires operation over a range of speeds.

Example 14.5

Determine annual operating cost savings if a throttling valve is eliminated by replacing a constant-speed centrifugal pump drive with a variable-speed drive. The two alternatives are illustrated in the accompanying block diagram.

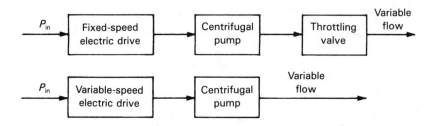

Potential savings are illustrated in the next figure, which shows hypothetical pump curves for the two situations. A curve representing the same system head is also shown for each of the two alternatives. The system head curve might typically reflect the sum of a static head and a head due to pipe friction. If a fixed-speed drive and throttling valve are used, the throttling valve head is added to the system head, yielding a total head curve as shown at the left in the figure. Steady-state solution points must lie on the pump characteristic curve that applies at

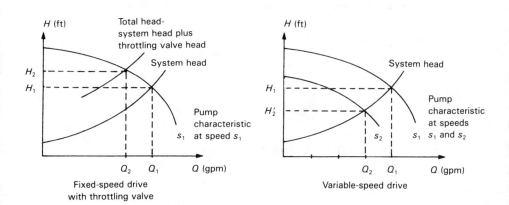

the fixed speed s_1. If a variable-speed drive is used, the pump characteristic changes for various steady-state operating speeds and solution points lie on the system head curve. Throttling head is eliminated. In either case, the pump shaft power (typically in brake horsepower, BHP) that the drive motor must supply is proportional to the product of head and flow rate divided by pump efficiency. Pump efficiency is a function of flow rate and typically falls off if flow is reduced at constant shaft speed. If the pump is driven at constant speed, the shaft power required at flow rates Q_1 and Q_2 is found by determining appropriate values of heads H_1 and H_2 from the pump curve at speed s_1. If the pump speed is reduced to effect flow reduction, the shaft power required at flow rates Q_1 and Q_2 is found by determining appropriate values of H_1 and $H_2' < H_2$ from the system head curve. Throttling losses at the reduced flow rate Q_2, which would be proportional to $Q_2(H_2 - H_2')$ with the fixed-speed drive, are eliminated with the variable-speed drive.

Typical pump test data can be used to develop characteristics similar to those illustrated in the next figure [5]. To determine savings the pump duty cycle must be considered.

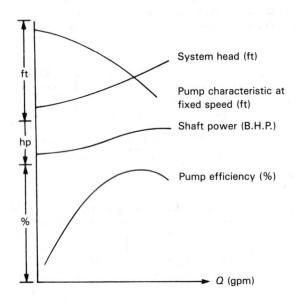

The following flow rates will be assumed to apply for the percentage of time indicated:

Flow rate, Q (thousand gpm)	Time, t (% of total time)
20	10
18	20
16	40
14	20
12	10

For the fixed-speed drive the following table can be established.

Flow rate, Q (thousand gpm)	Time, t (%)	Pump BHP, P_s (hp)	Drive motor efficiency, n (%)	Line power, P_{in} (kW)	Energy, E (kWh/hr)
20	10	733	94.0	582	58.2
18	20	721	94.0	572	114.4
16	40	708	93.5	565	226.0
14	20	692	93.0	555	111.0
12	10	665	93.0	533	53.3
					562.9

The values shown in the table are typical values, which are obtained as follows. The brake horsepower is read directly from the pump characteristic. Drive motor efficiency is line-to-shaft efficiency for the electric drive, such as might be determined from characteristics similar to those illustrated in Figs. 13.26 and 14.1. (Alternatively, it could be estimated by knowing motor efficiency at full, three-quarters, and half load.) For example, an 800-hp motor would operate at $665/800 \times 100\% = 83.1\%$ load at the minimum flow rate anticipated in this case, and this might correspond to a typical efficiency of 93%. Note that induction machine speed changes very little from no load to full load and may be regarded as essentially constant. Line power P_{in} is obtained by dividing pump brake horsepower P_s by drive motor efficiency n and converting to kilowatts. Energy E at a particular flow rate required per hour of operation is obtained by multiplying P_{in} by the time t. Thus, the total of 562.9 kWh per hour of operation can be used to determine energy costs for operation with the fixed-speed drive and throttling valve system.

A similar table can be developed for the variable-speed drive system.

Flow rate, Q (thousand gpm)	Time, t (%)	Pump BHP, P_s (hp)	Motor shaft speed, s (rpm)	Torque, T (ft-lb)	VS drive efficiency n (%)	Line power, P_{in} (kW)	Energy, E (kWh/hr)
20	10	733	900	4,278	85	643	64.3
18	20	632	857	3,873	82	575	115.0
16	40	522	804	3,410	80	487	194.8
14	20	428	752	2,988	78	409	81.8
12	10	348	702	2,604	75	346	34.6
							490.5

The values in this table are typical values, obtained as follows. The brake horsepower is determined by reading values of head from the system head curve corresponding to the various flow rates, taking the product of head and flow rate, dividing by pump efficiency, and converting to horsepower. For small speed changes pump efficiency remains relatively constant [6] and in this case a value corresponding to maximum efficiency for a typical pump characteristic at nominal speed was assumed (83%). The values of shaft speed s and torque T are required as input data for determination of the overall variable-speed drive line-to-shaft efficiency. Typically, these data would be furnished to a supplier of variable-speed drive systems so that the supplier could furnish the required drive system line-to-shaft efficiencies. The determination of speed s is not easy. If a family of pump H versus Q characteristics with speed s as parameter is available, it is possible to estimate speed from the curves at each of the desired flow rates, working with points along the system head curve. Given a pump characteristic (H vs. Q) at nominal pump speed, a family of curves can be developed by using approximations. In order to predict the curve for a new speed s, any point on the new curve has its head adjusted according to $H' = H(s'/s_{rated})^2$

and its Q adjusted according to $Q' = Q(s'/s_{\text{rated}})$, where Q and H are determined from points on the rated speed curve and H' and Q' locate points along the curve at speed s'. The procedure is described fully in [6]. A somewhat simpler approach is to assume that brake horsepower P_s is proportional to speed s to the third power ($P_s \propto s^3$), which is not unreasonable for a centrifugal pump [7]. Speed s in the table above was determined by using the latter approximation. Torque T is then easily determined from $T = (33{,}000 \text{ BHP})/2\pi s$ or, equivalently in this case, by assuming T proportional to speed squared ($T \propto s^2$). Efficiencies n are conservative (line-to-shaft) values for a variable-speed drive of this size. It is not difficult to find instances of higher efficiencies in manufacturers' data sheets. To determine variable-speed drive line-to-shaft efficiency at various speeds, the load torque versus speed characteristics are essential. Line power is obtained by dividing brake horsepower P_s by efficiency n and converting to kilowatts. Energy E is again obtained by multiplying P_{in} by t. The total of 490.5 kWh per hour of operation can be used to determine energy costs for operation with the variable-speed alternative. Assuming 4,160 hr/yr operation and an energy cost of \$0.05/kWh, annual operating cost savings would be

$$(562.9 - 490.5) \frac{\text{kWh}}{\text{hr}} \times 4{,}160 \frac{\text{hr}}{\text{yr}} \times \frac{\$0.05}{\text{kWh}} \cong \$15{,}060/\text{yr}$$

Alcock [7] describes additional considerations in the choice of variable-speed drives. Levers [8] gives a similar example based on a wound rotor motor with a static slip power recovery device used to control speed.

14.3 DEMAND CONTROL

An electric utility incurs costs in providing electrical service to its customers. These costs include, but are not limited to, the cost of fuel consumed in generating electrical energy and the cost of equipment required to generate, transmit, and distribute electrical energy. Consequently, customers are billed not only for total energy consumed in a given period, but also for their usage of utility-owned equipment required to serve them. The first charge is an energy charge. The second charge is a demand-related charge based on capacity requirements.

14.3.1 Energy Charges

Energy consumption E is obtained by integrating instantaneous power consumption $P(t)$ over a time interval of interest.

$$E = \int_0^T P(t) \, dt \tag{14.1}$$

Typically, a watt-hour meter or kilowatt-hour meter accomplishes this; the time period T is taken as 1 month and consumption is measured in kilowatt-hours (kWh). Many kilowatt-hour meters are electromechanical devices that total the number of revolutions of a disk whose speed of rotation is proportional to the instantaneous power consumption. This effectively integrates a customer's instantaneous power consumption in a manner consistent with Eq. (14.1).

14.3.2 Demand Charges

A customer's rate of energy consumption and power factor are used to determine the utility-owned capacity that must be available to serve the customer. Measurement of energy consumption alone would not yield this information. As an example, suppose one customer's load consists entirely of 1,000 hp of connected motor load that operates continuously. A second customer's load consists entirely of 2,000 hp of connected motor load that operates continuously but for only 12 hr a day. Assuming that the motors of both customers have the same efficiency, the two customers would consume the same total energy in a monthly period. It is clear that the equipment required to serve the second customer (transformers, feeders, etc.) must have twice the capacity of the equipment required to serve the first. In this simplified example, knowledge of each customer's maximum power consumption (which is the maximum rate of energy consumption) would be sufficient to determine the customer's installed capacity requirements.

Simple metering and recording of maximum power consumption is rarely adequate to properly assign charges related to capacity requirements. Devices such as transformers and feeders are sized by volt-ampere (VA) or kilovolt-ampere (kVA) requirements. A customer operating at a very low power factor would impose a requirement for installed capacity much greater than what would be arrived at by simply considering the customer's maximum power (kilowatt) consumption. Recall that power P is proportional to the product of volt-amperes and power factor. Thus, volt-ampere requirements are inversely proportional to power factor and increase if power factor decreases for a given power requirement. Utilities will frequently assess a power factor penalty charge for power factors significantly less than unity or will bill for reactive power consumption in volt-amperes reactive (vars) or kilovolt-amperes reactive (kvar).

Most installed equipment can handle short peaks in excess of the equipment kVA rating. Therefore, the prevalent practice is to charge customers for peak sustained power consumption over some reasonable but short time interval. Such charges are referred to as demand charges. (These charges may be supplemented with power factor penalty-related charges.) Power demand is defined as the average power consumption over a specified interval of time. Typical demand intervals range from 5 to 60 min. A 15-min interval is typical. More formally, demand over the ith time interval can be defined in the following way:

$$D_i = \frac{\int_{t_i}^{t_i+T_d} P(t)\, dt}{T_d} = \frac{E_i}{T_d} \qquad (14.2)$$

In Eq. (14.2) T_d is the demand interval. Note that the units of demand are the same as those of power, i.e., energy per unit time (perhaps expressed as kW). The concept is illustrated in Fig. 14.4.

Note that in Fig. 14.4, the area under the instantaneous power curve for the ith interval must be the same as the area under the straight-line segment indicating demand over the ith interval. That is:

$$D_i T_d = \int_{t_i}^{t_i + T_d} P(t) \, dt = E_i \qquad (14.3)$$

which follows directly from the definition of demand given by Eq. (14.2). The demand charge that a utility imposes is typically based on the highest demand occurring in the billing period, i.e., $\max\{D_i\}$ for all i contained in the billing period. Consideration of Eq. (14.2) shows that the maximum demand in a given period could be arrived at in a relatively simple way. A kilowatt-hour meter could be modified to move a pointer through an angle proportional to the energy E_i consumed in the ith interval. At the end of the interval, the pointer should be reset to zero with some means of preserving the angle excursion or, more specifically, the maximum angle excursion in all previous demand intervals in the period of interest. This can be accomplished by causing the pointer to move a friction-restrained second pointer, which remains at the maximum excursion until reset by utility personnel at the end of the billing period. This would accomplish the simple function of determining maximum demand in the billing period but would not yield

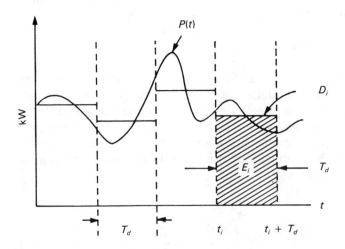

Figure 14.4 The definition of demand illustrated.

information that might be used to analyze and perhaps modify a customer's usage patterns. A recording device showing all D_i over a representative time period is essential for this, and such devices are readily available.

14.3.3 Determination of Potential Savings by Demand Control

From a customer's viewpoint, the objective is to minimize maximum demand, assuming that the utility is using maximum demand to establish the demand charge. Demand control is then nothing more than a technique for leveling out one's power consumption curve. It is clear from the definition of demand that it is possible to do this without affecting the total energy consumption in a given billing period. Figure 14.5 illustrates this.

In Fig. 14.5 it is seen that if a demand limit line is constructed that is greater than or equal to the average power consumption over the billing period, it is possible to "shave the peaks" and "fill in the valleys" in such a way that the demand limit line is never exceeded and energy requirements are still satisfied. In practical cases the demand limit will be greater than the average power requirement since it is rarely possible to operate with absolutely constant power consumption. The control required can be effected manually or automatically, assuming that sheddable loads are operating during times of peak consumption. The first step in assessing the merits of demand control schemes is to analyze the utility rate structure and the past history of demand. The concept of load factor is a particularly useful tool in

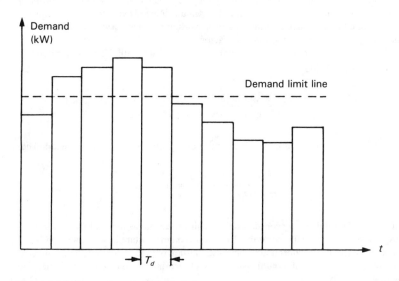

Figure 14.5 Demand control illustrated.

this type of analysis. The load factor for a given billing period is defined as the ratio of the average demand to the maximum demand. The load factor so defined is greater than zero and less than or equal to one. If the demand interval T_d is chosen so that there are n intervals in the billing period, it is clear from Eq. (14.2) that:

$$D_{avg} = \frac{\Sigma_i D_i}{n} = \frac{\Sigma_i E_i}{nT_d} \qquad (14.4)$$

Equation (14.4) shows that the average demand D_{avg} can be obtained by simply dividing the total kilowatt-hours of energy consumption in the billing period by the total hours in the billing period. Maximum demand and total kilowatt-hours are easily obtained from past billing statements, permitting determination of past performance in terms of load factor.

Example 14.6 Relation between Load Factor and Demand Charges

Demand charges and total energy charges are determined by the following rate schedule:

Fuel adjustment charge	$0.02872/kWh
Energy charge	$0.01152/kWh
Demand charge	$3.22/kW
Customer charge	$121.00/month

where demand charge is based on the highest demand recorded in a billing month.

Past records indicate that a total monthly charge of $10,659 occurred when kilowatt-hour consumption for the month was 220,968 kWh and maximum demand over the billing period was 511.5 kW. Assume that the same monthly energy is consumed but at various load factors and determine total charges as a function of load factor.

Start by determining the average demand for the recorded month. The average demand D_{avg} is obtained by applying Eq. (14.4):

$$D_{avg} = \frac{\Sigma_i E_i}{nT_d} = \frac{\text{total energy consumption}}{\text{total hours in billing period}} = \frac{220,968 \text{ kWh}}{(30)(24) \text{ hr}} = 306.9 \text{ kW}$$

The load factor for the recorded month is the ratio of D_{avg} to the maximum demand. Thus

$$LF = \frac{D_{avg}}{\max\{D_i\}} = \frac{306.9}{511.5} = 0.6$$

Now assume that the total energy consumption remains the same as the load factor varies. If total energy consumption remains constant, D_{avg} must remain constant in accordance with Eq. (14.4). Maximum demand is determined from the definition of load factor and all the information required to determine total monthly charges as a function of load factor is available. The following table is easily established.

Load factor	Energy consump-tion (kWh)	Average demand (kW)	Energy charge ($)	Customer charge ($)	Maximum demand (kW)	Demand charge ($)	Total charge ($)
1.0	220,968	306.9	8,891	121	306.9	988	10,000
0.9	220,968	306.9	8,891	121	341.0	1,098	10,110
0.8	220,968	306.9	8,891	121	383.6	1,235	10,247
0.7	220,968	306.9	8,891	121	438.4	1,412	10,424
0.6	220,968	306.9	8,891	121	511.5	1,647	10,659
0.5	220,968	306.9	8,891	121	613.8	1,976	10,988
0.4	220,968	306.9	8,891	121	767.3	2,471	11,483
0.3	220,968	306.9	8,891	121	1,023.0	3,294	12,306
0.2	220,968	306.9	8,891	121	1,534.5	4,941	13,953
0.1	220,968	306.9	8,891	121	3,069.0	9,882	18,894

The tabulated results are displayed graphically in the accompanying figure. It is easy to see that as the load factor approaches 1.0, total cost approaches $10,000 per month. A quick check of records for other previous months would permit a determination of load factor history.

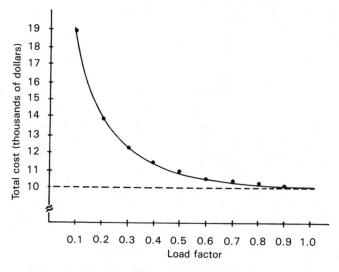

Total cost versus load factor for an assumed rate structure.

Potential benefits of demand control schemes could then be weighed against the cost of imple-mentation of various schemes. If load factor has been characteristically low, a small improve-ment in load factor significantly reduces costs, as seen from the figure. Obviously, the calcula-tions depend heavily on existing and future rate structures. It is also important to recognize that a utility may offer more than one rate structure to a large industrial user. The technique described in this example is helpful in ascertaining the most favorable rate structure.

14.3.4 Demand Control Considerations

The capacity of electrical equipment depends to a large extent on its ability to operate continuously without overheating. The operating temperature of a device is a complex function of the ambient temperature, the load the device supplies, the time duration of the load, and the initial operating temperature of the device when the load was applied. Demand metering is a relatively simple but inexact way of determining each customer's contribution to the heating of equipment required to serve the customer's load. Assuming that the demand interval is reasonably chosen, attempts to reduce the maximum demand which are not independent of the choice of time over which the maximum demand is measured will not be consistent with the principles underlying demand monitoring. Many utilities determine maximum demand by using a "sliding window" to ensure that demand charges are based on the maximum demand that occurred in *any* period of time equal to the demand interval. Demand control devices that function in such a way as to shed and restore nonessential loads in order to level out power consumption typically cost more if they accomplish true demand control that is sliding window-independent. This is sometimes referred to as instantaneous rate control, as opposed to ideal rate control or predictive control. The latter two types of control must be synchronized with the utility's demand interval. The various devices available for dedicated demand control are described in Palko [9].

Any practical demand control scheme requires identification and ranking of loads suitable for shedding. Production line machinery is generally excluded, although electric heating devices used in production may be amenable to some degree of demand control. Lights, air-conditioning, fans, and facility-heating equipment are loads generally considered suitable for shedding and restoring in a demand control scheme.

In general, motor life is reduced if a motor is subjected to repeated starting and stopping. Manufacturers typically specify a maximum permissible number of starts per hour for large motors to avoid overheating of the motors and resulting degradation of insulation life.

The above considerations can make the determination of an effective demand control scheme difficult, but a past history of operation at low load factor would suggest that there are significant potential savings.

NOMENCLATURE

I current
I_L line current
r electrical resistance per unit length
C cost per hour to operate
q heat flux
k heat transfer coefficient

Q reactive power
Q pump flow rate
P real power
V_L line-to-line voltage
H pump head
s pump shaft speed
P_s pump brake horsepower
E energy
n efficiency
n number of demand intervals
D demand
LF load factor
t time
t temperature ($^\circ$C)
T time period
T temperature constant
T torque
T_d demand time interval

REFERENCES

1. H. N. Hickok, Electrical Energy Losses in Power Systems, *IEEE Trans. Ind. Appl.* IA-14 (No. 5): 373–387, Sept./Oct. 1978.
2. *Westinghouse Electrical Transmission and Distribution Reference Book*, 4th ed., East Pittsburgh, Pa., 1964.
3. William D. Stevenson, Jr., *Elements of Power Systems Analysis*, 3d ed., McGraw-Hill, New York, 1975.
4. S. Dimachkieh and D. R. Brown, Choosing Power Cables on the Basis of Energy Economics, in *Proceedings of the 1980 Houston Conference on Industrial Energy Conservation Technology*, vol. 2, pp. 535–542, 1980.
5. H. H. Anderson, *Centrifugal Pumps*, Trade and Technical Press Ltd., Morden Surrey, England, 1972.
6. Austin H. Church, *Centrifugal Pumps and Blowers*, reprint, Krieger Publishing Co., Huntington, N.Y., 1972.
7. D. N. Alcock, The Equipment for and Economics of Variable Flow Well Pumping, *IEEE Trans. Ind. Appl.* IA-16 (No. 1): 144–153, Jan./Feb. 1980.
8. W. H. Levers, The Electrical Engineer's Challenge in Energy Conservation, *IEEE Trans. Ind. Appl.* IA-11: 392–404, July/Aug. 1975.
9. E. Palko, Saving Money Through Electric Power Demand Control, *Plant Eng.* 29 (No. 5): 58–63, March 6, 1975.

PROBLEMS

14.1 A three-phase, 460-V, 100-hp induction machine has an efficiency of 0.90 and power factor of 0.80 at half load. What reactive power in kilovars is required by the motor at this operating condition?

14.2 Measured input power for a three-phase induction machine is 52.5 kW when it drives a particular load. Line-to-line voltage is metered at 465 V and line current is metered at 86 A for the same power measurement. What power factor does the machine operate at and what reactive power is consumed for this conditon?

14.3 A standard 75-hp, three-phase induction machine has an efficiency of 92% at rated output. A high-efficiency motor of the same rating has an efficiency of 94%. At this output determine hourly energy savings in kilowatt-hours realized by choosing the high-efficiency motor.

14.4 Refer to Table 14.2 for cable data and determine hourly cable energy losses in kilowatt-hours for a 2,500-ft run of cable 4, assuming the cable operates at its ampacity limit. Assume cable temperature is 65°C.

14.5 Refer to Table 14.2 for cable data and estimate cable conductor temperature and electrical resistance for cable 5 if it carries a current of 150 A. Assume that operation at its ampacity limit results in a conductor temperature of 75°C. Conductors are hard-drawn copper of 97.3% conductivity.

14.6 A 350-hp, 4,160-V induction machine has a full-load efficiency of 94% and power factor at full load of 0.85. What reactive compensation is required to correct power factor to 0.95 at full load?

14.7 A 1,000-hp, 4,160-V induction machine has a full-load efficiency of 94% and power factor at full load of 0.82. Estimate the percentage reduction in feeder circuit losses at full load if motor power factor is corrected to 0.95 by connecting capacitors at the motor terminals.

14.8 A 1,500-kVA transformer has no-load losses of 0.273% and full-load losses of 0.98%. The weekly operating schedule for the transformer is as follows:

Time (hr)	Load (%)
48	8.5
40	85.0
40	75.0
40	50.0

Determine weekly energy losses in kilowatt-hours for the transformer.

14.9 Consider the tabulated data for Example 14.5. In comparing data for the fixed-speed and variable-speed drive motor options, what are the throttling losses in a 1-hr period at a flow rate of 16,000 gpm?

14.10 Suppose that data for the variable-speed drive option in Example 14.5 are to be extended to consider a flow rate of 10,000 gpm. If the pump BHP at the flow rate is 280 hp, what is the estimated motor shaft speed? Use the same assumptions as were used in the example.

14.11 Consider the tabulated data for Example 14.5. Suppose that instead of the duty cycle given for the example, the duty cycle was taken as: 16,000 gpm for 60% of the time and 14,000 gpm for 40% of the time. Determine hourly savings in kilowatt-hours realized by choosing the variable-speed drive option.

14.12 Consider the data tabulation for Example 14.6. What would total charges be for a load factor of 0.625?

14.13 A hypothetical curve representing daily power consumption versus time is given in the accompanying figure. What is the load factor for this curve?

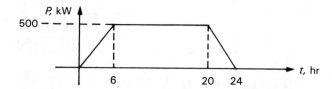

14.14 A certain industrial plant has an energy consumption of 250,000 kWh and a maximum demand of 850 kW occurring in the same month. What is the plant's load factor? How much load must be shed during the plant's peak consumption period in order to raise the load factor to 0.7?

14.15 The rate schedule cited for an industrial plant is as follows:

Fuel adjustment charge	$0.03217/kWh
Energy charge	$0.02161/kWh
Demand charge	$2.50/kW

The demand charge is based on the maximum demand occurring in a given month. The plant energy consumption on a monthly basis is 1,250,000 kWh. Determine monthly savings if the plant load factor is increased from 0.4 to 0.5.

FIFTEEN

INDUSTRIAL ENERGY USE PROFILES

15.1 INTRODUCTION

Up to now the material in this book has been presented in the following pattern. First, some history regarding energy consumption, energy resources, and the potential for energy conservation was presented. The basics of the energy audit process were then presented, followed by reviews of fundamental theory and procedures required to carry out audits, to decide whether energy conservation opportunities are technically and financially feasible, and to monitor implemented conservation projects. Finally, Chapters 6 through 14 dealt with specific issues related to energy conservation, ranging from the efficient production of steam and power to the effective use of electricity.

Many of the text examples and chapter problems have been designed to represent some narrow aspect of an energy conservation method. A concerted effort to generalize techniques to fit certain classes of industries has not been made.

In this chapter we discuss typical energy use profiles for several of the energy-intensive industries. The phrase *energy use profile* as used here includes the way in which energy is converted from one form to another, how much is converted, and what the potentials are for energy conservation for a particular industry, such as stone and concrete or food processing. Features of energy usage common to different industries as well as features that are unique to a given industry are pointed out.

In Table 1.1 the six highest energy-using industrial categories were listed. They are relisted here along with their Standard Industrial Classification (SIC) numbers: Foods and Kindred Products, SIC-20; Paper and Allied Products, SIC-26; Chemicals and Allied Products, SIC-28; Petroleum and Coal Products, SIC-29; Stone, Clay, and Glass Products, SIC-32; and Primary Metals Industries, SIC-33. These are the energy-intensive industries we will concentrate on in this chapter.

In the mid-1970s the U.S. Federal Energy Administration, in response to the Energy Policy and Conservation Act, developed energy efficiency improvement targets for the energy-intensive industries. In doing this, the FEA collected a wealth of data showing how energy had been used in various industries. These data were combined to create what were called Industry Composite Energy Use Profiles, which illustrate the end uses of purchased fuel and electricity for a given industry. When these data are combined with information of the type contained in Table 15.1, a general picture emerges of the energy conservation opportunities in various industries.

Table 15.1 Historical Industry Composite Energy Type Profile: Food and Kindred Products Industry, SIC-20

Fuel type	Units	1958	1962	1967	1971	1974
Distillate oil	percent	NA[a]	4.0	6.6	7.4	8.0
Residual oil	percent	NA	12.4	8.2	7.3	7.8
Total fuel oil	percent	22.4	16.4	14.8	14.7	15.8
Coal	percent	32.5	18.9	24.0	13.6	9.5
Coke and breeze	percent	0.2	0.2	0.2	0.2	0.2
Natural gas	percent	36.2	43.1	49.8	57.9	57.1
Other fuels	percent	1.7	3.3	1.9	1.9	4.2
Total purchased fuel	percent	93.0	91.9	90.7	88.3	86.8
Net electricity	percent	7.0	8.1	9.3	11.7	13.2
Total purchased fuel and net electricity	percent	100.0	100.0	100.0	100.0	100.0
Total purchased fuel and net electricity	10^{12} Btu	766.1	802.5	899.6	1,031.3	958.8

[a]Not available.

Sources: Census of Manufacturers, *Fuels and Electric Energy Consumed for 1963, 1967, and 1972;* unpublished Annual Survey of Manufacturers data for 1974.

15.2 THE FOODS INDUSTRY

The foods industry is a diverse one, including such extreme categories as fresh fruits and vegetables, where the main energy consumption is for transportation, and dehydrated food products, where the energy consumption is very high because of the need to remove moisture. Casper [1] in 1977 noted that there were about 28,000 individual food industry establishments in the United States. Forty-four percent of these firms had fewer than 20 employees, and most of them consisted of a single plant.

The foods industry accounted for about 7% of the industrial energy consumption in the United States in 1974. Table 15.1 shows the energy sources for the SIC-20 group for selected years up to 1974. The percentage of electricity usage rose dramatically from 7 to 13.2% in this period. The ratio of energy required to generate purchased electricity is about 3:1, so that electricity usage accounted for about 30% of the total energy required for foods in 1974.

Within the foods industry, beet sugar, wet corn milling, and meat packing are the largest energy-using industries. Table 15.2 shows that these

Table 15.2 Rank Order of the Component Industries in Food and Kindred Products Industry by Their Purchase of Fuels and Electric Energy, 1974

Industry	Rank order	Total purchased fuels and electrical energy (10^{12} Btu)	Cumulative total (10^{12} Btu)	Cumulative percent of total (%)
Food and Kindred Products		958.8	958.8	100.0
Priority Group I				
Beet sugar	1	81.5	81.5	8.5
Wet corn milling	2	78.8	160.4	16.7
Meat packing plants	3	97.1	237.5	24.8
Malt beverages	4	49.5	286.9	29.9
Canned fruits and vegetables	5	46.4	333.4	34.8
Fluid milk	6	43.7	377.0	39.3
Bread, cake, and related products	7	43.7	420.7	43.9
Soybean oil mills	8	43.3	464.0	48.4
Cane sugar	9	37.9	501.9	52.3
Frozen fruits and vegetables	10	31.4	533.3	55.6
Prepared feeds	11	31.4	564.7	58.9
Priority Group II				
Shortening and cooking oils	12	26.6	591.3	61.7
Food preparations	13	26.3	617.6	64.4

Table 15.2 Rank Order of the Component Industries in Food and Kindred Products Industry by Their Purchase of Fuels and Electric Energy, 1974 (*Continued*)

Industry	Rank order	Total purchased fuels and electrical energy (10^{12} Btu)	Cumulative total (10^{12} Btu)	Cumulative percent of total (%)
Priority Group II (*Continued*)				
Animal and marine fats and oils	14	23.9	641.5	66.9
Cheese, natural and processed	15	23.2	664.7	69.3
Condensed and evaporated milk	16	21.8	686.5	71.6
Canned specialties	17	21.2	707.6	73.8
Bottled and canned soft drinks	18	20.1	727.8	75.9
Poultry dressing plants	19	17.4	745.2	77.7
Distilled liquor	20	17.1	762.2	79.5
Sausage and prepared meats	21	15.7	777.9	81.1
Pet food	22	14.3	792.3	82.6
Raw cane sugar	23	12.6	804.9	84.0
Confectionery products	24	12.6	817.5	85.3
Frozen specialties	25	11.6	829.1	86.5
Flour and other grain mill products	26	11.6	840.7	87.7
Dehydrated food products	27	11.2	852.0	88.9
Cookies and crackers	28	10.9	862.9	90.0
Priority Group III				
Roasted coffee	29	10.2	873.1	91.1
Malt	30	9.2	882.3	92.0
Cereal preparations	31	8.2	890.5	92.9
Cottonseed oil mills	32	7.2	897.4	93.6
Pickles, dressings, and sauces	33	6.5	904.2	94.3
Ice cream and frozen desserts	34	6.1	910.3	94.9
Creamery butter	35	5.5	915.8	95.5
Chocolate and cocoa products	36	5.1	920.9	96.0
Flavoring extracts and syrups	37	5.1	926.0	96.6
Poultry and egg processing	38	4.8	930.8	97.1
Canned and cured seafood	39	4.8	935.6	97.6
Wines, brandy	40	4.1	939.7	98.0
Fresh or frozen package fish	41	4.1	943.8	98.4
Vegetable oil mills	42	3.4	947.2	98.8
Rice milling	43	2.7	949.9	99.1
Manufactured ice	44	2.7	952.6	99.4
Macaroni and spaghetti	45	2.4	955.0	99.6
Blended and prepared flour	46	2.0	957.1	99.8
Chewing gum	47	1.7	958.8	100.0

Source: Casper [1].

three industries alone accounted for about 25% of the energy consumed in 1974. Even when viewed in terms of Btu's required per dollar value of shipped product, beet sugar is highest, followed by wet corn milling. Consequently, we concentrate on one of these industries, beet sugar, in the following sections.

15.2.1 Energy Use Profile—Beet Sugar

Figure 15.1 is a detailed flow sheet for the beet sugar process. The dashed lines show energy-related flows through the process chain; the solid lines show the steps in the processing from raw material to refined product. Sugar beets are taken in and sugar, molasses, and pulp feed are produced. Fuels,

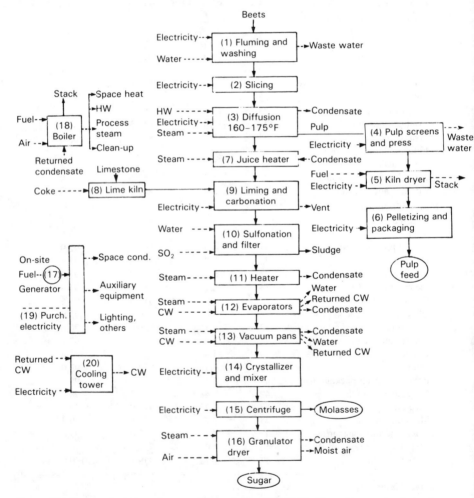

Figure 15.1 Process flow sheet for sugar beet refining (SIC-2063). (From Energy Analysis of 108 Industrial Processes [2].)

electricity, and coke make up the primary energy supply, with air and water providing the means for energy conversion and transport.

The energy balance shown in Table 15.3, together with Fig. 15.1, gives a complete picture of the composite beet sugar industry. Table 15.3 gives the energy input and the energy content of output products, both in Btu's per pound of beets; the temperature levels for each process; and the unrecoverable losses for the processes.

Selecting component 5, the kiln dryer, as an example, we see that beet pulp containing 10.3 Btu, 320 Btu of fuel, and 2 Btu of electricity are

Table 15.3 Beet Sugar Energy Balance

Unit operation	Material (In)	Energy (Btu/lb beets)	Temp. (°F)	Material (Out)	Energy (Btu/lb beets)	Unrecoverable losses	Temp. (°F)
1 Fluming and flashing	Beets	—	75	Beets	—	2	75
	Water	—	75	Water	—		75
	Electricity	2	—				
2 Slicing	Beets	—	75	Beets	—	6	75
	Electricity	6	—				
3 Diffusion	Beets	—	75	Pulp	31.3	7	160
	Hot Water	58	140	Juice	53.7		160
	Electricity	2	—	Condensate	3.2		180
	Steam	32	250				
4 Pulp screens and press	Pulp	31.3	160	Pulp	10.3	9	150
	Electricity	7	—	Water	21.0		150
5 Kiln dryer	Pulp	10.3	150	Dried pulp	3.0	30	200
	Fuel	320	—	Stack gas	299.3		400
	Electricity	2	—	and vapor			
6 Pelletizing and package	Dried pulp	3.0	200	Packed pulp	—	7	75
	Electricity	4	—				
7 Juice heater	Juice	53.7	160	Juice	92.3	15	185
	Steam	60	250	Condensate	6.4		100
8 Lime kiln	Limestone	—	75	Lime	1.5	5.8	250
	Coke	58	75	CO_2	50.7		600
9 Liming and carbonation	Juice	92.3	185	Juice	93.5	13.4	190
	Lime	1.5	250	CO_2	39.6		400
	CO_2	50.7	600				
	Electricity	2.0	—				

Table 15.3 Beet Sugar Energy Balance (*Continued*)

Unit operation	Material	In Energy (Btu/lb beets)	Temp. (°F)	Material	Out Energy (Btu/lb beets)	Unrecoverable losses	Temp. (°F)
10						8.0	
Sulfonation	Juice	93.5	190	Juice	84.0		150
and filter	Water	—	75	Sludge	1.5		150
	SO₂	—	75				
11						15	
Heater	Juice	84.0	150	Juice	158.5		200
	Steam	100	250	Condensate	10.5		180
12						42.5	
Evaporators	Juice	158.5	200	Juice	25		200
	Steam	390	250	Condensate	35		180
	Cold water	—	75	Cold water	421.6		100
				Water	25		100
13						25	
Vacuum pans	Juice	25	200	Juice	18		180
	Steam	100	280	Condensate	10.5		180
	Cold water	—	75	Cold water	70.5		95
				Water	1.0		95
14						15	
Crystallizer	Juice	18.0	180	Juice	8.0		130
and mixer	Electricity	5	—				
15						13.5	
Centrifuge	Juice	8.0	130	Sugar	1.0		100
	Electricity	8	—	Molasses	1.5		100
16							
Dryer	Sugar	1.0	100	Sugar	1.0		150
	Steam	5	250	Condensate	0.5		180
	Air	—	75	Moist air	3.5		220
17						8	
Electricity	Fuel	96	75	Electricity	30		—
generated	Cold water	—	75	Stack	25		450
				Cold water	33		95
18						80	
Boiler	Fuel	1011	75	Process	687		
	Condensate	66.1	180	Heat	31		
				Cleanup	77		
				Stack	202		
19							
Electricity	Generated	30	—	Process	44		
purchased	Purchased	18		Lighting and misc.	4		
20						29.2	
Cooling	Cold water	29.2	95	Cold water	—		75
tower	Electricity	5	—				

Source: Energy Analysis of 108 Industrial Processes [2].

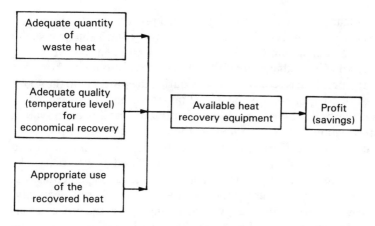

Figure 15.2 Factors required for successful waste heat recovery.

provided per pound of beet input. The dryer output is dried pulp containing 3.0 Btu at 200°F, and stack gases and vapors totaling 299.3 Btu at 400°F. Only 30 Btu of this total output is listed as unrecoverable. There is a clear opportunity to recover up to about 270 Btu per pound of beets at a temperature of 400°F. This represents about 80% of the energy input as fuel and electricity.

15.2.2 Energy Conservation Opportunities— Beet Sugar

Energy conservation opportunities depend on adequate recoverable energy, on the temperature of the process, and on an appropriate use of the recovered heat, as shown in Fig. 15.2. Many of the beet sugar processes in Table 15.3 involve a considerable energy flow but the temperature is too low for economical recovery. In Chapter 9 we pointed out the difficulty of recovering energy *economically* at temperatures below 200°F.

In keeping with the idea of cascading of energy from high to low temperatures, Table 15.4 shows the most attractive heat recovery opportunities

Table 15.4 Energy Conservation Targets for Beet Sugar Refining

Component	Temperature (°F)	Material	Energy level (Btu/lb beets)
Lime kiln (8)	600	CO_2	50.7
On-site generation (17)	450	Stack gas	25.0
Kiln dryer (5)	400	Stack gas and vapor	300.0
Lime carbonation (9)	400	CO_2	39.6
Boiler (18)	—	Process steam	687.0
		Stack gas	202.0

for the beet sugar process in descending order of process temperature. Any process that occurs below 200°F is excluded from the table. Table 15.4 shows the components that should be good targets for waste heat recovery.

The lime kiln CO_2 exhaust at 600°F is fed directly into the carbonation unit and its temperature is reduced to 400°F during the process. So the 400°F CO_2 exhaust of component 9 is the actual target for heat recovery.

Clearly, the boiler-house should be scrutinized first for conservation opportunities. About 69% of the fuel for the overall beet sugar process is provided to the boiler. Thirty percent goes directly to the lime kiln and dryer. Only about 1% of the input energy is purchased as electricity.

Figure 15.3, the energy profile for beet sugar, shows the percentages of input energy going for direct use to the boiler and for purchased electricity, as well as the ways in which the energy is eventually used in the process. This profile shows where most of the energy flows and helps to identify the process components where energy conservation might be fruitful. For example, we see that although improvement in lighting might be possible, it should have low priority because lighting represents only 0.07% of the total energy consumption.

The right-hand column of Figure 15.3 shows what strategies might be used to conserve energy in the major energy-using categories of a beet sugar plant. The potential for efficiency improvement is given in the column labeled "% end use efficiency improvement." To determine the total impact on the refining process, these figures must be weighted according to the percentages of intermediate uses and end uses shown on the left side of the table.

When this is done, the potential for energy use improvement is found to be somewhat greater than 20% for the beet sugar industry. The potential savings are related mostly to the boiler and, in fact, come mostly from dealing with process steam in an energy-conscious manner. Process steam consumes about 40% of the energy of the overall process.

In addition, more efficient pulp pressing can result in significant savings, because reducing the amount of moisture that remains in the pulp before drying also reduces the energy required in the dryer. This is an example of a low-temperature process that might be overlooked in an energy audit unless its effect on a subsequent process is accounted for.

Although other food processing industries might differ in varying degrees from the beet sugar industry, this method of using the energy use profile to identify targets for energy conservation in the beet sugar industry carries over to the other industries.

15.3 PRIMARY METALS

Energy use in the metals industry (SIC-33) is dominated by steel plants, which account for about 73% of the total energy used. Aluminum, copper,

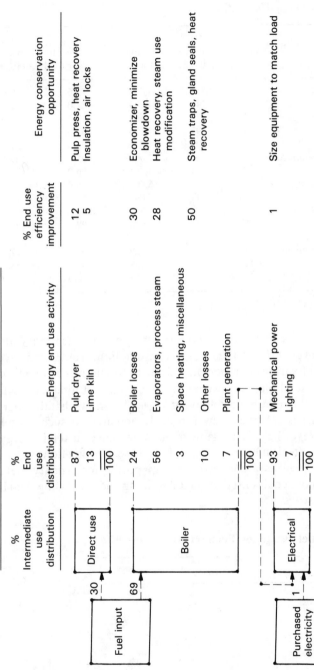

Figure 15.3 Energy use profile for beet sugar, showing intermediate use distribution and suggested ECOs. (Adapted from Casper [1].)

Table 15.5 Rank Order of the Component Industries in the Metals Industry in Terms of Energy Use, 1972

Component	Rank order	Total energy use (10^{12} Btu)	Cumulative total	Cumulative % of total	Normalized energy usage (10^6 Btu/ton)
Steel plants	1	3,753	3,753	72.6	29.9
Aluminum	2	741	4,494	87.0	112.6
Iron foundries	3	170	4,664	90.3	16.0
Copper	4	92	4,756	92.0	51.2
Ferroalloys	5	90	4,846	93.8	31.1
Other primary nonferrous	6	62	4,908	95.0	149.1
Steel foundries	7	53	4,961	96.0	28.0
Nonferrous foundries	8	44	5,005	96.9	34.1
Nonferrous processing	9	43	5,048	97.7	16.1
Miscellaneous metal products	10	42	5,090	98.5	26.1
Secondary nonferrous smelting and refining	11	39	5,129	99.3	17.7
Primary zinc	12	29	5,158	99.8	33.6
Primary lead	13	9	5,167	100.0	13.0

Source: Lownie et al. [4].

and iron foundries account for another 19%, so that these four industries account for over 90% of the energy consumption. Table 15.5 gives the contribution of various component industries to energy consumption.

Although steel plants consume the lion's share of total energy for the metals group, the right-hand column of Table 15.5 shows that the energy per ton of product is relatively low for steel compared to such products as aluminum. Nevertheless, because of its dominance in total energy consumption, the steel industry is singled out here for attention.

Table 15.6 shows a composite energy profile for primary metals for 1972. Much more coal is used for metal refining and processing than is used in other industries.

Table 15.6 Fuel Usage for Primary Metals in 1972

Fuel type	% of energy usage
Coal	18.7
Petroleum	15.3
Natural gas	35.0
Purchased electricity	17.8
Other fuels	13.2
Total	100.0

Source: Sittig [3].

15.3.1 Steel Industry Energy Use Profile

Figure 15.4 is the process flow sheet for steelmaking. The steel plant takes in iron ore and scrap steel, raw materials such as coal and limestone, and fuels and electricity and produces finished steel products, or in some cases intermediate products such as ingots, billets, and blooms.

The major energy-consuming processes are coking coal; agglomerating iron ore; ironmaking; steelmaking; soaking ingots; reheating blooms, billets, and slabs; and heat-treating/forging. Over 80% of the energy consumption is accounted for by these operations. Dashed lines in Fig. 15.4 represent energy flows; solid lines represent material flows through the process. Figure 15.4 is a process sheet for a specific type of steelmaking. Item 4, the basic oxygen furnace, might alternatively be an open-hearth furnace or an electric arc furnace. In addition, the casting, soaking unit, and primary rolling mill, steps 5 through 7, could be replaced by a continuous casting process. However, Fig. 15.4 serves reasonably well to illustrate the steekmaking operation.

Table 15.7 is the energy balance that accompanies Fig. 15.4. The energy levels in the table are normalized against a pound of finished products.

An exact energy accounting is difficult for steelmaking because of the diversity of the equipment that might be used in a particular plant. Published information such as the data in [3] is based on averages over the industry, and once a specific plant such as that of Fig. 15.4 is chosen, an energy balance based on such published data might be slightly in error. However, at the risk of incurring such errors and in the interest of simplifying the discussion, Table 15.7 has been prepared to accompany Fig. 15.4. It should be noted that this table represents a balance of energy for the specific processes in Fig. 15.4. Purchased electricity and fuel values would change for a different type of converting furnace and a different set of finishing processes.

Energy conservation opportunities abound in the steelmaking industry because of the high temperatures involved in many of the processes, and also because by-product gases can be burned in a furnace. Considerable heat recovery is practiced in the industry, but exploration of further conservation opportunities is warranted.

15.3.2 Energy Conservation Opportunities in Steelmaking

To rank order potential energy conservation opportunities (ECOs) in steelmaking would be tantamount to repeating Table 15.7. Alternatively, a composite energy use profile (Fig. 15.5) has been constructed showing the nature of energy consumption and giving some energy conservation opportunities. The "% end use distribution" values in Fig. 15.5 were found by tallying all the energy losses in each process, including heat, stack, and cooling air and/or water losses, as well as the energy contained in a product

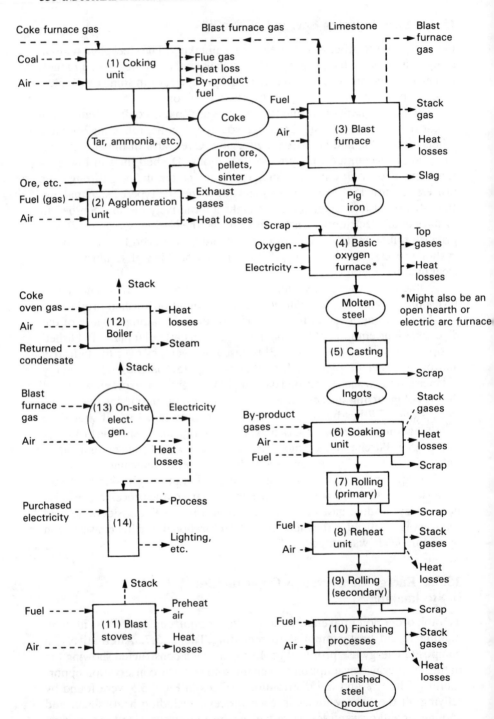

Figure 15.4 Process flow sheet for steelmaking.

Table 15.7 Energy Balance for Steelmaking

		In			Out		
Unit operation	Material	Energy (Btu/lb)[a]	Temp. (°F)	Material	Energy (Btu/lb)	Temp. (°F)	Unrecoverable (Btu/lb) heat losses
(1) Coking unit (2600°F)	Fuels	58.1	—	Coke	6783.4	75	
	Coal	9827.2	—	Coke breeze, etc.	958.7	75	
	Blast furnace gas	124.5	200	Coke oven gas	2207.0	85	489.4
	Coke oven gas	622.5	85	Cooling water, etc.	296.0	212	
	Electricity	20					
	Cooling water, etc.	83	75				
(2) Agglomeration unit (2500°F)	Coke breeze fines	348.7	75	Pellets, sinter, etc.	—	75	
	Ore, limestone etc.	—	75	Stack gas	374.1	250	
	Fuel	260.0	75	Cooling air	86.2	155	
	Electricity	40	—	Cooling air	116.9	95	77.5
(3) Blast furnace (2600°F)	Coke	6783.4	75	Pig iron and scrap	322.0	2550	
	Fuel	99.6	—	Slag	159.5	2550	
	Coke oven gas	20.8	85	Blast furnace gas	2726.6	200	1485.6
	Iron ore, pellets			Scrubber water and vapor	510.0	210	
	Sinter, limestone, etc.	—					
	Taps	29.1	75				
	Preheated air	761.0	2725				
	Endothermic reaction	−2490					
(4) Basic oxygen furnace (2840°F)	Oxygen	—	75	Molten steel	244	2800	
	Pig iron	181	2550	Stack	109	3000	
	Cooling and quench water	—	75	Slag	76	2800	
	Misc. materials	—	—	Cooling water	27	95	15
	Electricity	120	—				
	Exothermic reaction	170	—				
(5) Casting	Molten steel	383	2800	Ingots	107	1600	
	Electricity	300	—	Ingots	—	75	526
	Spray water	—	75	Waste spray	50	212	
(6) Soaking unit (2400°F)	Ingots	107	1600	Ingots	325	2400	
	Ingots	—	75	Stack	260	1200	574
	Coke oven gas	1042	85				
	Air	—	75				
	Electricity	10	—				
(7) Rolling mill	Ingots	325	2400	Billets	—	75	435
	Electricity	110	—	Scale	—	75	

Table 15.7 Energy Balance for Steel Making (*Continued*)

	In			Out			
Unit operation	Material	Energy (Btu/lb)[a]	Temp. (°F)	Material	Energy (Btu/lb)	Temp. (°F)	Unrecoverable (Btu/lb) heat losses
(8)	Billets	—	75	Billets	98	2400	
Reheat unit	Fuel	309	—	Stack loss	245		96
(2400°F)	Coke oven gas	130	—				
(9)	Billets	98	2400	Finished shape	—	75	
Rolling mill	Electricity	470	—	Scale	—	75	568
(10)	Finished	—	75	Finished	—	75	
Finishing	shapes			product			
processes	Fuel	960	—	Stack	650	870	
(≈650°F)	Steam	150	320	Cooling air	270	95	186
	Cooling air	—	75	Returned	14	180	
	Electricity	10	—	condensate			
(11)	Fuel	199.2	—	Preheat air	761.0	2725	
Blast	Oxygen	—	—	Stack	79	300	44
stoves	Blast furnace	684.8	85				
(2725°F)	gas						
	Air	—	75				
(12)	Fuel	294	—	Steam	357.7	320	
Boiler	Coke oven gas	256	85	Stack	196.3	500	28
	Returned	32	180				
	condensate						
(13)	Blast furnace	1180.7	200	Electricity	393	—	
On-site	gas			Stack	727.7	650	60
generation							
(14)	Purchased	738	—	Process	1080	—	
Electricity	Generated on	393	—	Lighting,	51	—	
	site			losses, etc.			

[a]Pound refers to finished product.

that is dissipated before the product enters the next process unit. Blast furnace gas, for example, is not considered as a loss for the furnace because it can be recovered and used elsewhere. Even under these conditions, the blast furnace has high energy losses and clearly is a prime candidate for energy conservation.

Some projected ECOs are listed at the right in Fig. 15.5, along with efficiency improvement estimates based on the total average energy consumption for steelmaking rather than on the consumption for an individual unit of the process. Some further explanation is given below.

Pellet Usage in Blast Furnaces. Using pellets in a blast furnace decreases the coke requirements and increases productivity. New and expanded pellet plants have had a significant impact on energy consumption in steelmaking,

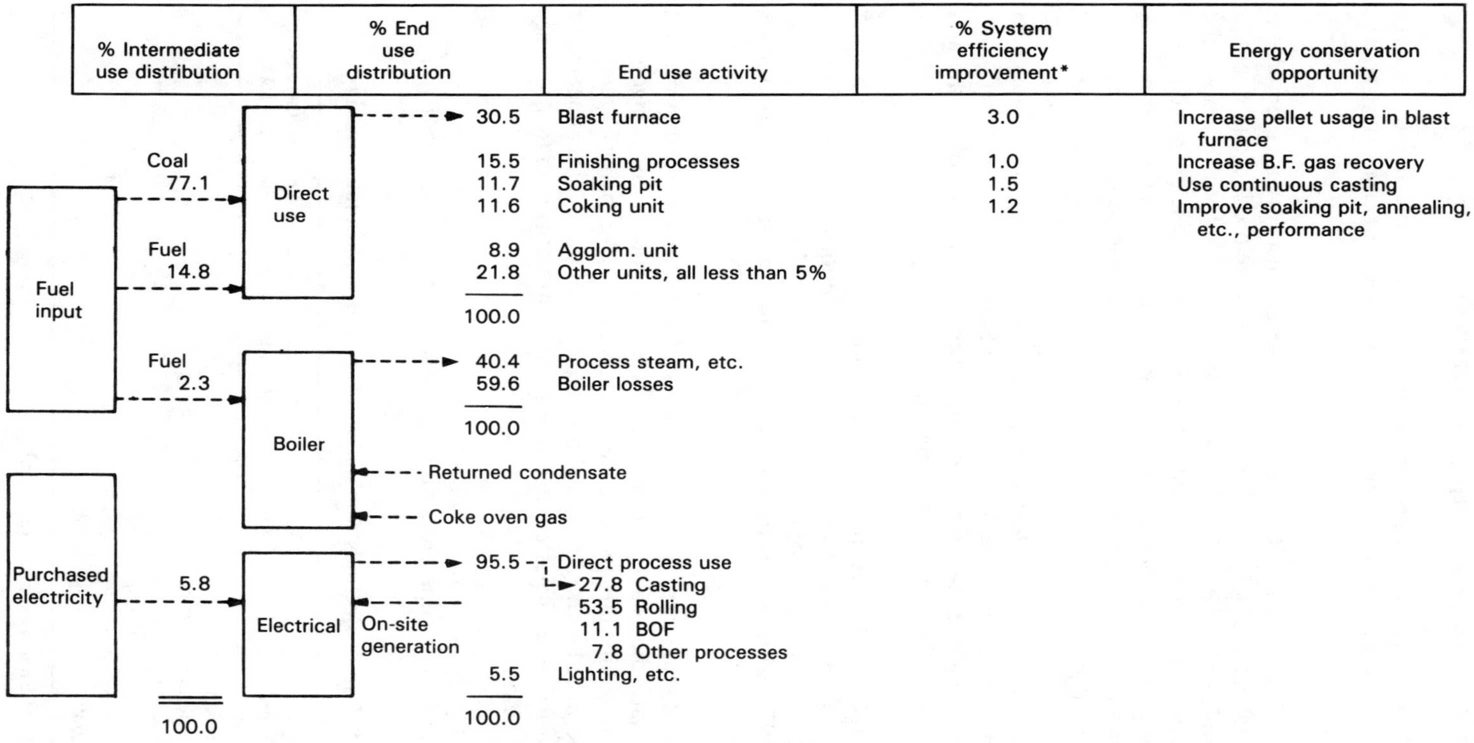

% Intermediate use distribution	% End use distribution	End use activity	% System efficiency improvement*	Energy conservation opportunity
Coal 77.1 → Fuel input; Fuel 14.8 → **Direct use**	30.5	Blast furnace	3.0	Increase pellet usage in blast furnace
	15.5	Finishing processes	1.0	Increase B.F. gas recovery
	11.7	Soaking pit	1.5	Use continuous casting
	11.6	Coking unit	1.2	Improve soaking pit, annealing, etc., performance
	8.9	Agglom. unit		
	21.8	Other units, all less than 5%		
	100.0			
Fuel 2.3 → **Boiler**	40.4	Process steam, etc.		
	59.6	Boiler losses		
	100.0			
Boiler: Returned condensate, Coke oven gas				
Purchased electricity 5.8 → **Electrical**	95.5 Direct process use	27.8 Casting / 53.5 Rolling / 11.1 BOF / 7.8 Other processes		
On-site generation	5.5	Lighting, etc.		
	100.0			

100.0

100.0

*Based on the total energy consumption for steelmaking.

Figure 15.5 Composite energy use profile for steelmaking.

559

and significant energy savings can be projected into the fugure. Energy savings come primarily from avoidance of the energy of converting coal to coke.

Recovery of Blast Furnace Gas. Blast furnace gas recovery is common but far from universal. About 65,000 cubic feet of gas are produced per net ton of hot metal, with a heating value of 95 Btu/ft^3. In 1972, for example, the estimated blast furnace gas produced was about 6×10^{12} ft^3, of which about 73% was recovered and burned. Increasing this recovery factor toward 90% would result in a very large improvement in steelmaking efficiency. Capital costs are relatively low for recovery equipment, so the 90% target should be attainable.

New and modern blast furnaces can also have a significant influence on steelmaking efficiency. The main effect will come from decreased coke consumption; indeed, a reduction of 50 lb of coke per net ton of hot metal is attainable in modernized blast furnaces.

Continuous Casting. Steel that is cast continuously requires about 3.5×10^6 Btu/ton less than ingot cast steel. The sooner continuous casting facilities are introduced, the more energy will be saved. Table 15.8 shows a projection made in 1979, with 1973 as the base year, for the conversion to continuous casting facilities.

Improved Soaking Pits. Table 15.8 shows that continuous casting will not totally replace the soaking pit in the foreseeable future. Steel ingots are reheated in soaking pits and again in reheat furnaces after being formed into slabs, bloom, or billets. After final rolling, much of the steel is annealed or heat-treated. Some suggested measures for saving energy in these processes are listed below:

- Improved reheat furnace control
- Installation of recuperators on reheat and annealing furnaces

Table 15.8 Use Projections for Soaking Pits and Continuous Casting

Year	Process	Output (%)	Weighted average (10^6 Btu/ton)	Improvement over base year (%)
1973	Soaking pits	93	1.7	—
	Continuous casting	7		
1980	Soaking pits	83	1.6	6
	Continuous casting	17		
1990	Soaking pits	58	1.3	24
	Continuous casting	42		

Source: Federal Energy Administration [6].

Table 15.9 Steelmaking Energy Use Projections

Year	Process	Percent of raw steel output	Energy use (10^6 Btu/ton raw steel)
1973	Open hearth	26.4	
	BOF	55.2	1.41
	Electric	18.4	
1980	Open hearth	15.6	
	BOF	63.1	0.88
	Electric	21.2	
1990	Open hearth	4.3	
	BOF	71.7	0.46
	Electric	23.9	

Source: Federal Energy Administration [6].

- Decreased leakage from furnaces
- Improved burner design and maintenance
- More effective use of refractory insulation in furnaces

Some steel companies have replaced fossil-fired reheat furnaces with induction heating machines for slabs. It is estimated that 1.3×10^6 Btu/ton could be saved in this way.

Basic oxygen furnaces (BOFs) and electric furnaces are replacing open-hearth furnaces, as shown in Table 15.9. Estimated energy reduction is also shown.

Recovery of BOF Off-Gas. A large volume of gas is generated during the oxygen-blowing period of a BOF cycle. The gas is rich in carbon monoxide, which could be recovered and burned to replace purchased fuel. Some steel plants in the United States recover part of the sensible energy from BOF off-gas by waste heat boilers. However, collection of the off-gas for combustion is rare, even though it represents an opportunity for significant energy savings. The savings potential is estimated as about 0.5×10^6 Btu per ton of steel. The capital costs of gas recovery equipment for this application are high compared to the potential savings. However, as fuel prices escalate, the economics of such recovery systems should become more attractive.

15.4 THE PAPER INDUSTRY

The paper industry (SIC-26) is the fifth largest industrial energy user in the United States, consuming about 11% of the total industrial energy per year. The American Paper Institute divides the industry into two major categories: paper and board, and pulp. The paper and board group is divided into six product groups: newsprint; printing, writing, and related paper types; packing and industrial converting; tissue; construction grades; and paperboard. The pulp industry is broken down into five process types: dissolving, sulfate, sulfate and soda, semichemical, and groundwood.

Pulp alone is produced in some mills and then shipped to paper mills for the final papermaking process. Recently, integrated pulp and paper mills have been built to take advantage of any possible capital cost or energy savings. Wastepaper may be recycled directly in a paper mill.

Much of the energy used in pulp and paper mills comes from waste or self-generated fuels, taking advantage of the combustibility of wood. For example, linerboard, the second highest consumer of energy in the paper industry, as shown in Table 15.10, has the lowest purchased energy requirement per ton because much of the energy used in its production is recovered energy rather than purchased energy.

Table 15.11 shows that the paper industry provides much of its primary energy through reclamation of by-product fuels. About 35% of the energy to make pulp and paper is self-generated. In contrast, in the metals industry virtually all primary energy must be purchased in one form or another.

15.4.1 Energy Use Profile—Writing Paper

Printing and writing paper top the list of energy consumers within the paper industry, both for total energy consumption and for energy required per ton of product. Thus, this is the logical choice for an energy profile.

Table 15.10 Rank Order of the Component Industries in Pulp and Paper, 1972

Component	Rank order	Total energy consumption (10^{12} Btu)	Cumulative total	Cumulative % of total	Normalized energy usage (10^6 Btu/ton)
Printing and writing paper	1	501	913	44.9	29.4
Linerboard	2	412	1434	70.5	17.7
Packaging, converting, and special paper	3	213	1126	55.3	23.4
Tissue	4	156	1282	63.0	29.0
SBS board	5	152	1719	84.5	26.0
Newsprint	6	146	501	24.6	27.1
Corrugating medium	7	139	1580	77.6	25.5
Recycled boxboard	8	121	1840	90.4	22.0
Construction paper and board	9	92	1932	94.9	18.0
Dissolving pulp	10	68	2000	98.3	20.8
Other combination board	11	35	2035	100.0	27.0

Source: Lownie et al. [4].

Table 15.11 Total Energy Usage for the Paper Industry, 1972

Fuel type	% of energy use
Coal	10.6
Refined oil products	20.1
Natural gas	18.9
Purchased electricity[a]	13.5
Purchased steam	0.7
Total purchased energy	63.8
Spent pulping liquor	24.7
Bark	4.2
Hogged fuels	2.1
Other	5.2
Total reclaimed energy	36.2
Total	100.0

[a]Purchased electricity valued at 10,500 Btu/kWh.
Source: Lownie et al. [4].

Figure 15.6 shows a process flow sheet typical of writing paper production in an integrated mill. It is laid out in a similar fashion as in the preceding sections: that is, solid lines indicate material flow and cross-hatched lines show where the energy goes. Table 15.12 is the energy balance sheet that accompanies Fig. 15.6. Most processes in papermaking occur at a relatively low temperature; exceptions are the lime kiln and the black liquor recovery furnace, where temperatures up to 1500°F may be encountered.

A composite energy use profile for the information of Table 15.12 is shown in Fig. 15.7. Recovered energy from wood chips, black liquor, and so forth is not included. Only purchased energy is represented. Most of the purchased fuel goes to the boiler to produce process steam. The boiler then represents a target of opportunity for energy conservation practices.

15.4.2 Papermaking Energy Conservation Opportunities

Several techniques for energy conservation are listed in Fig. 15.7, along with estimated percentages of energy efficiency improvement based on the total industry average of purchased energy for writing paper — i.e., 24.5×10^6 Btu/ton. Several of these techniques are discussed below.

Lime Kilns. In the caustic recovery process the lime kiln is the major user of energy, especially purchased fuel. The kiln is a rotary furnace for drying and calcining the lime mud. Both natural gas and fuel oil are used to fire lime kilns. There is a great opportunity to control excess air by using O_2 analyzers. Minimization of excess air in this way could save an estimated 0.5% of the fuel fired, corresponding to about 2,500 Btu per ton of paper products.

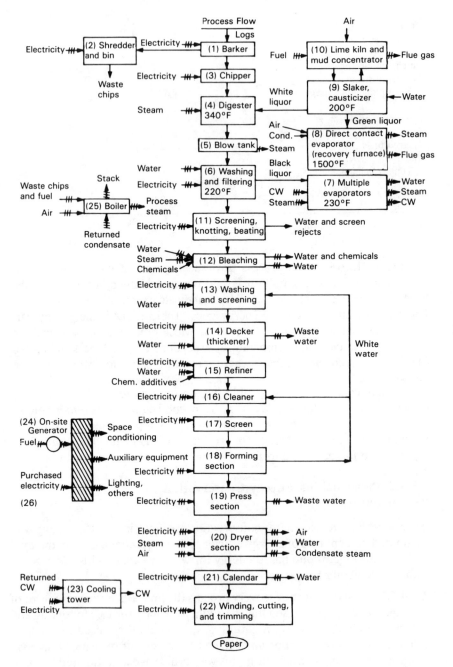

Figure 15.6 An integrated Fourdrinier paper mill.

564

Table 15.12 Energy Balance for the Integrated Paper Mill of Fig. 15.6

Unit operation	Material	In Energy (Btu/lb)[a]	Temp. (°F)	Material	Out Energy (Btu/lb)[a]	Unrecoverable losses	Temp. (°F)
1						35	
Barker	Logs	—	75	Wood	—		75
	Electricity	35	—	Bark	—		75
2						35	
Shredder	Bark	—	75	Waste chips	—		75
and bin	Electricity	35	—				
3						70	
Chipper	Wood	—	75	Wood chips	—		75
	Electricity	70	—				
4						184	
Digestor	Wood chips	—	75	Pulp mixture	2115		340
(340°F)	Steam	1905	340				
	White liquor	394	194				
5						44	
Blow	Pulp	2115	340	Pulp and liquor	855		220
tank	mixture			Vented steam	1215		220
6						172	
Washing and	Pulp and liquor	855	220	Black liquor	477		130
filters	Water	—	75	Pump and water	318		130
	Electricity	112	—				
7						242	
Multiple	Black liquor	477	130	Condensed vapor	834		200
evaporators	Steam	2030	292	Condensate	216		200
	Cold water	—	75	Black liquor	216		150
				Cold water	999		95
8						450	
Direct	Black liquor	180	150	Green liquor	635		1500
contact	Air	—	75	Steam	6300		400
evaporator	Return condensate	450	150	Stack	1795		300
	Exothermic reaction	8550	—				
9						188	
Slaker,	Green liquor	635	1500	White liquor	394		195
causticizor	Water	—	75	CaCO₃	200		109
	CaO	149	1200				
10						65	
Kiln	CaCO₃	200	109	CaO	149		1200
	Fuel	698	—	Flue gas	90		400
	Air	—	75				
	Endothermic reaction	− 594	—				
11						112	
Screening,	Electricity	225	—	Pulp	378		140
knotting,	Pulp and water	310	130	Water and rejects	45		140
and beating							
12						358	
Bleacher	Pulp	370	140	Water and	2105		100
	Cl₂	—	160	chemicals			
	Steam	2250	260	Condensate	270		210
	Water	—	75	Pulp and water	67		85
	Electricity	180	—				

Table 15.12 Energy Balance for the Integrated Paper Mill of Fig. 15.6 (*Continued*)

		In			Out		
Unit operation	Material	Energy (Btu/lb)[a]	Temp. (°F)	Material	Energy (Btu/lb)[a]	Unrecoverable losses	Temp. (°F)
13						52	
Washing and screening	Electricity	45	—	Pulp and water	60		78
	Water	—	75				
	Pulp and water	67	85				
	White water	—	75				
14						46	
Decker	Electricity	50	—	Pump and water	15		78
	Water	—	75	Waste water	66.6		78
	Pulp and water	78	78				
15						65	
Refining	Electricity	50	—	Pulp and water	—		75
	Water	—	75				
	Chemicals	—	75				
	Pulp and water	15	78				
16						43	
Cleaners	Electricity	43	—	Pulp and water	—		75
	White water	—	75				
	Pulp and water	—	75				
17						50	
Screens	Electricity	50	—	Pulp and water	—		75
	Pulp and water	—	75				
18						200	
Forming section	Pulp and water	—	75	Pulp and water	—		75
	Electricity	200	—	White water to cleaning	—		75
				White water to washing	—		75
19						250	
Press section	Pulp and water	—	75	Pulp and water	—		75
	Electricity	250	—	Waste water	—		75
20						450	
Drier section	Pulp, chemicals, water	—	75	Air and vapor	2651		240
				Condensate steam	464		220
	Steam	3500	250	Paper	35		150
	Air	—	75				
	Electricity	100	—				
21						78	
Calander	Electricity	43	—	Paper	—		75
	Water paper	35	150	Water	—		75
22						43	
Winding, cutting, trimming	Electricity	43	—	Paper	—		75
	Paper	—	75				
23						2150	
Cooling tower	Cold water	2120	95	Cold water	—		75
	Electricity	25	—				
24						125	
Electricity generation	Fuel	2475	—	Electricity	825		—
	Cold water	—	75	Stack	500		600
	Air	—	75	Cold water	1025		95

Table 15.12 Energy Balance for the Integrated Paper Mill of Fig. 15.6 (*Continued*)

| Unit operation | | In | | | Out | | |
	Material	Energy (Btu/lb)[a]	Temp. (°F)	Material	Energy (Btu/lb)[a]	Unrecoverable losses	Temp. (°F)
25						230	
Boiler	Fuel	4600	—	Process steam	3385		300
	Makeup	—	75	Auxiliary steam	300		300
	Air	—	75	Stack	685		450
26							
Electricity	Purchased electricity	1155	—	Process	1530		—
	Generated electricity	825	—	Lights and auxiliary equipment	450		—

[a]Pound refers to product output.

Flue Gas Waste Heat Recovery. Flue gas heat recovery [3], using economizers or air preheaters for boilers operating with flue gas temperatures greater than 450°F, would potentially save about 83,000 Btu per ton of paper production. This is based on a minimum allowable flue gas temperature of 350°F to prevent corrosion.

Boiler Combustion Controls. Boiler combustion control consists mainly of providing the proper amount of oxygen, carried by atmospheric air, to combine with the combustibles in the fuel. We pointed out in Chapter 6 that either too little or too much air will result in significant energy losses in the flue gas.

The tendency for boiler operators is to provide too much air, which drives heat losses up because of the sensible energy absorbed by the nitrogen. The best possible approach is to use automatic combustion controls along with flue gas analyzers. However, black liquor recovery boilers, in which the exact composition of the liquor is not constant over time, are generally not equipped with automatic controls. They are usually manually adjusted by using O_2 analyzers.

Implementation of boiler controls and instrumentation yields an estimate of 51,300 Btu/ton for the industry as a whole.

Boiler Feedwater Heating. Water provided to the boiler in a paper mill is made up of returned condensate, steam from the deaerator, and treated makeup water. Condensate usually provides 30 to 65% of the total water. The makeup stream is cold and is a good candidate for waste heat recovery from low-temperature processes in the plant. Waste streams that might be utilized for this purpose are digestor blow heat, dirty evaporator condensate, and boiler continuous blowdown.

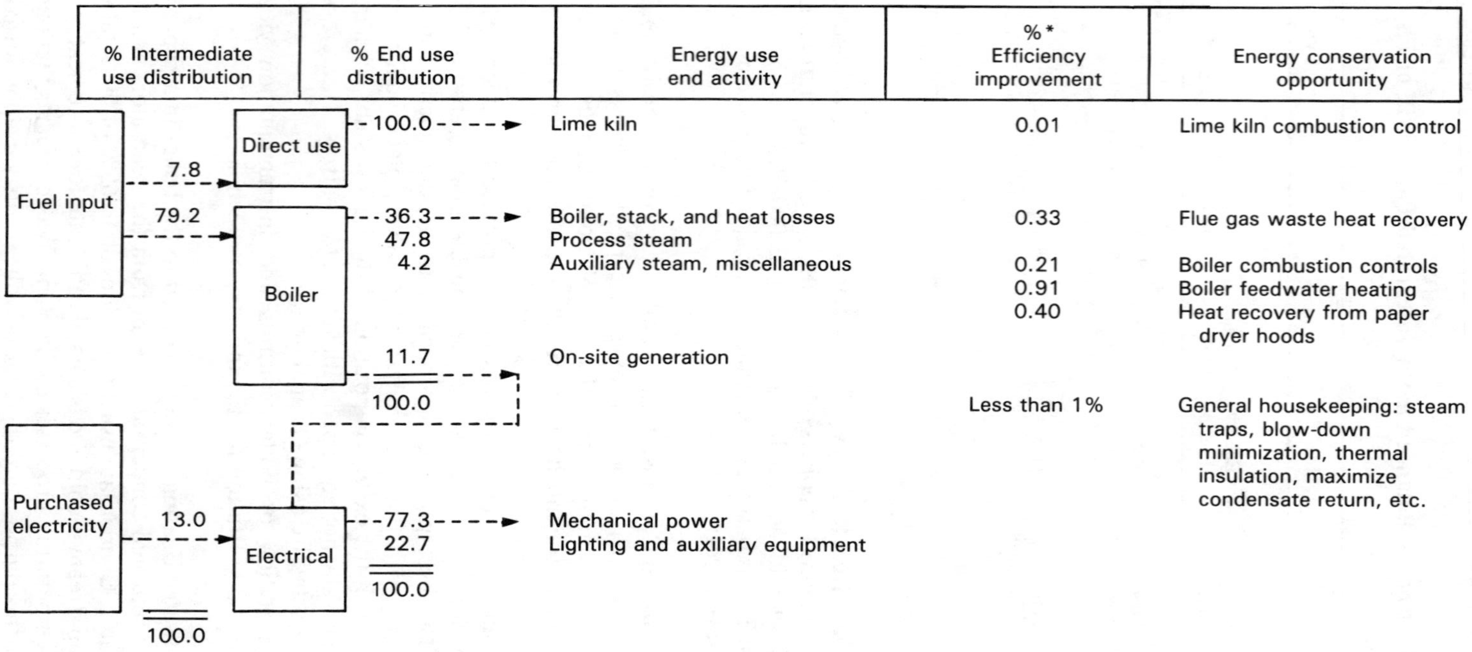
% Intermediate use distribution
% End use distribution
Energy use end activity
%* Efficiency improvement
Energy conservation opportunity
Fuel input
7.8
79.2
Direct use
Boiler
100.0
Lime kiln
0.01
Lime kiln combustion control
36.3
47.8
4.2
Boiler, stack, and heat losses
Process steam
Auxiliary steam, miscellaneous
0.33
Flue gas waste heat recovery
0.21
0.91
0.40
Boiler combustion controls
Boiler feedwater heating
Heat recovery from paper dryer hoods
11.7
100.0
On-site generation
Less than 1%
General housekeeping: steam traps, blow-down minimization, thermal insulation, maximize condensate return, etc.
Purchased electricity
13.0
Electrical
77.3
22.7
100.0
Mechanical power
Lighting and auxiliary equipment
100.0

*Based on total industry average of purchased energy per ton of writing paper.

Figure 15.7 Composite energy use profile for an integrated paper mill.

Based on the assumption that any makeup stream not heated can be heated to 100°F, an estimate of about 220,000 Btu/ton can be made for conserved energy.

Heat recovery from paper dryer hoods and various housekeeping procedures that do not require capital investment can also make a significant overall contribution to energy conservation in papermaking.

15.5 THE CHEMICALS INDUSTRY

The manufacture of chemicals involves many different technologies and is probably the most diverse industry of the top six energy-consuming industries in the United States. Chemical and allied products (SIC-28) include industrial organics; synthetic resins, rubber, and plastics; drugs; soap and detergents; paints and other surface coatings; inorganic industrial chemicals; agricultural chemicals; and miscellaneous chemical products.

The industry can be conveniently broken down by type of manufacturing process for the highest energy-consuming materials produced. Table 15.13 shows this breakdown for the nine highest energy-consuming processes, which account for about 48% of the total energy used by the chemicals industry. Ethylene/propylene production dominates the list.

Table 15.14 is the composite energy type profile for the chemical processes of Table 15.13. A unique feature of the industry is that a large fraction of the energy consumed is used in the form of fuels for feedstock, i.e., raw material for the process itself. Natural gas is the dominant fuel source, and in fact much of the feedstock, which accounts for 55% of the energy con-

Table 15.13 Rank Order of the Highest Energy-Consuming Processes in the Chemical Industry

Process	Rank order	Total energy use $(10^{12}$ kcal$)^a$	Cumulative total^b	Cumulative % of total
Ethylene/propylene	1	368	368	23.8
Ammonia	2	155	523	33.9
Chlorine/caustic soda	3	100	623	40.4
Ethylbenzene/styrene	4	35	658	42.6
Carbon black	5	35	693	44.9
Oxygen/nitrogen	6	20.5	713.5	46.2
Sodium carbonate	7	11.5	725.0	46.9
Cumene	8	10.6	735.6	47.7
Phenol/acetone	9	5.1	740.7	48.0

^{a}1 Btu = 3.968 kcal.
^{b}These processes account for 48% of the total energy consumed by the chemical industry.

**Table 15.14 Fuel Usage for Chemical Production in 1973
for Processes Listed in Table 15.13**

Fuel type	% of energy usage
Coal	2.6
Petroleum	1.5
Natural gas	30.8
Feedstock	55.9
Purchased electricity[a]	9.2
Total	100.0

[a]Purchased electricity evaluated at 2500 kcal/kWh.

sumption, is also natural gas. Very little coal or purchased petroleum products is used as fuel. Considerable gaseous by-products are produced in some processes and can be used as fuel. Ethylene processes, for example, produced 55×10^{12} kcal of fuel as by-product gases in 1973.

Ethylene processes are classified in the plastic materials industry (SIC-2821), which includes the subcategories shown in Table 15.15. Low-density polyethylene (LDPE) leads the list in both total energy consumption and energy cost per pound. Consequently, we single out the LDPE process for an energy use profile.

Table 15.15 Energy Use in the Plastic Materials Industry, 1972

	Estimated production, 1972 (10^9 lb)	Energy used, 1972	
		Total (10^{12} Btu)	(Btu/lb)
Low-density polyethylene	5.36	40.47	7550
Polyvinylchloride	5.12	30.82	6020
Polystyrene	4.89	11.0	2250
High-density polyethylene	2.30	9.13	3970
Polypropylene	1.73	5.09	2940
Phenolic	1.44	7.79	5411
Polyester	0.93	1.09	1176
Amino	0.93	5.03	5411
Acrylic	0.76	1.71	2250
Alkyds	0.63	0.74	1176
All other	1.83	5.65	3085
Total	25.92	118.52	
Average			4573

Source: Sittig [3].

15.5.1 Energy Use Profile—Low-Density Polyethylene

Figure 15.8 is a process flow sheet for an LDPE plant; it is set up in the same way as the flow sheets for other processes already discussed. The LDPE process is characterized by high pressures more than high temperatures. The reactor and several other components of the process require cooling, so heat recovery is possible. Table 15.16 is the corresponding energy balance for Fig. 15.8.

15.5.2 Low-Density Polyethylene Energy Conservation Opportunities

The composite energy use profile for LDPE is shown in Fig. 15.9. We have added the feedstock to the diagram in this case, because it represents a significant opportunity for heat recovery. The energy input of the raw feedstock here is estimated to be 22,400 Btu/pound of ethylene. Almost 10% of this is lost as heat of polymerization and is rejected eventually to cooling water in the cooling tower (step 16 in Fig. 15.8) through an intermediate heat exchanger (step 13 in Fig. 15.8). Low-pressure steam or electricity could be generated from this energy. The temperature of the coolant leaving the reactor is 450°F, so relatively large boilers, turbines, and condensers would be needed for a process recovery scheme. However, a system of reasonably high effectiveness is feasible. About 3% of the energy equivalent of the feedstock could be recovered as electricity in this way. This would result in significant energy savings.

A very large fraction of purchased energy goes to the boiler with attendant losses, which makes it advantageous to apply heat recovery to the boiler. Economizers, air preheaters, and combustion control with flue gas analyzers all provide energy conservation opportunities for the LDPE process. The boiler for this process has a stack temperature of 400°F, representing a moderate potential for heat recovery. Assuming a corrosion-safe stack gas temperature of 350°F, the efficiency improvement is about 1.25%, according to Fig. 6.21. Control of excess air from 15 down to 5% would give another 0.5% combustion efficiency improvement, based on Fig. 6.19.

15.6 PETROLEUM REFINING

The petroleum and coal products category (SIC-29) is dominated by petroleum refining and ranks third in energy consumption, behind chemicals and metals. The industry is broken down into five components: petroleum

Process Flow

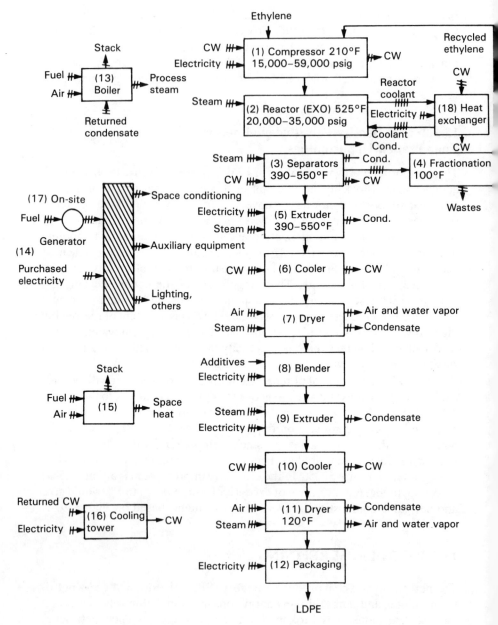

Figure 15.8 An LDPE plant.

Table 15.16 Energy Balance for LDPE

Unit operation	Material	In Energy (Btu/lb)[a]	In Temp. (°F)	Material	Out Energy (Btu/lb)[a]	Unrecoverable losses	Temp. (°F)
1 Compressor (210°F)	New ethylene	—	77	Compressed ethylene	463.7	84.3	210
	Recycled ethylene	65.9	100	Cold water	974.7		97
	Heat of vaporization	1316.8	—				
	Electricity	140.0	—				
	Cold water	—	77				
2 Reactor (525°F)	Compressed ethylene	463.7	210	Reacted ethylene	1300.6	119.2	450
	Steam	360.8	420	Condensate	42.9		220
	Coolant	7692.3	410	Reactor coolant	8616.3		450
	Exothermic reaction	1562.2	—				
3 Separator (500°F)	Reacted ethylene	1300.6	450	Separated ethylene	67.5	154.7	100
	Steam	240.7	490	Condensate	28.6		220
	Cold water	—	77	Cold water	1277.4		97
				Primary ethylene	12.6		100
4 Fractionation (100°F)	Separated ethylene	67.5	100	Recycled ethylene	65.9	0.16	100
				Waste	1.48		100
5 Extruder (500°F)	Primary ethylene	12.6	100	Extruded ethylene	177.6	63.05	400
	Electricity	72.9	—	Condensate	21.45		220
	Steam	176.6	295				
6 Cooler (77°F)	Extruded ethylene	177.6	400	Cooled ethylene	—	13.2	77
	Cold water	—	77	Cold water	164.4		97
7 Dryer (120°F)	Cooled ethylene	—	77	Dried ethylene	12.65	1.7	100
				Air and vapor	38.3		150
	Steam	59.8	370	Condensate	7.15		220
	Air	—	77				
8 Blender (77°F)	Dried ethylene	12.65	100	Blended ethylene	—	41.81	77
	Additives	—	77				
	Electricity	29.16	—				
9 Extruder (280°F)	Blended ethylene	—	77	Extended ethylene	111.6	97.6	280
	Steam	176.t	295	Condensate	21.45		220
	Electricity	54.7	—				

Table 15.16 Energy Balance for LDPE (*Continued*)

		In			Out		
Unit operation	Material	Energy (Btu/lb)[a]	Temp. (°F)	Material	Energy (Btu/lb)[a]	Unrecoverable losses	Temp. (°F)
10 Cooler (77°F)	Extruded ethylene	111.6	280	Cooled ethylene	—	14.6	77
	Cold water	—	77	Cold water	97.0		97
11 Dryer (120°F)	Cooled ethylene	—	77	Dried ethylene	7.15	7.2	90
	Steam	59.8	370	Air and vapor	38.3		200
	Air		77	Condensate	7.15		220
12 Packaging (77°F)	Dried ethylene	7.15	90	Ethylene	—	36.31	77
	Electricity	29.16	—				
13 Boiler	Fuel	1534.7	—	Process steam	483.4	53.5	375
	Air	—	77	Process steam	590.9		450
	Condensate	128.7	220	Stack	535.6		400
14 Electric	Purchased	433.5	—	Space conditioning	45.7		—
	Generated	22.1	—	Auxiliary equipment	22.7		—
				Lights	22.7		—
				Process	364.5		—
15 Space heating	Fuel	316.56	—	Space heat	253.2	5.76	—
	Air	—	77	Stack	57.6		350
16 Cooling tower	Cold water	3285	97	Cold water	—	3326.7	77
	Electricity	14.60	—				
	Cold water	26.96	140				
17 Electricity generation	Fuel	66.91	—	Electricity	22.1	4.46	—
	Cold water	—	77	Stack	13.39		—
				Cold water	26.96		140
18 Heat exchanger	Reactor coolant	8616.3	450	Coolant	7692.3	150	410
	Electricity	24.0		Cold water	774		95
	Cold water	—	75				

[a]Pound refers to product output.
Source: Energy Analysis of 108 Industrial Processes [2].

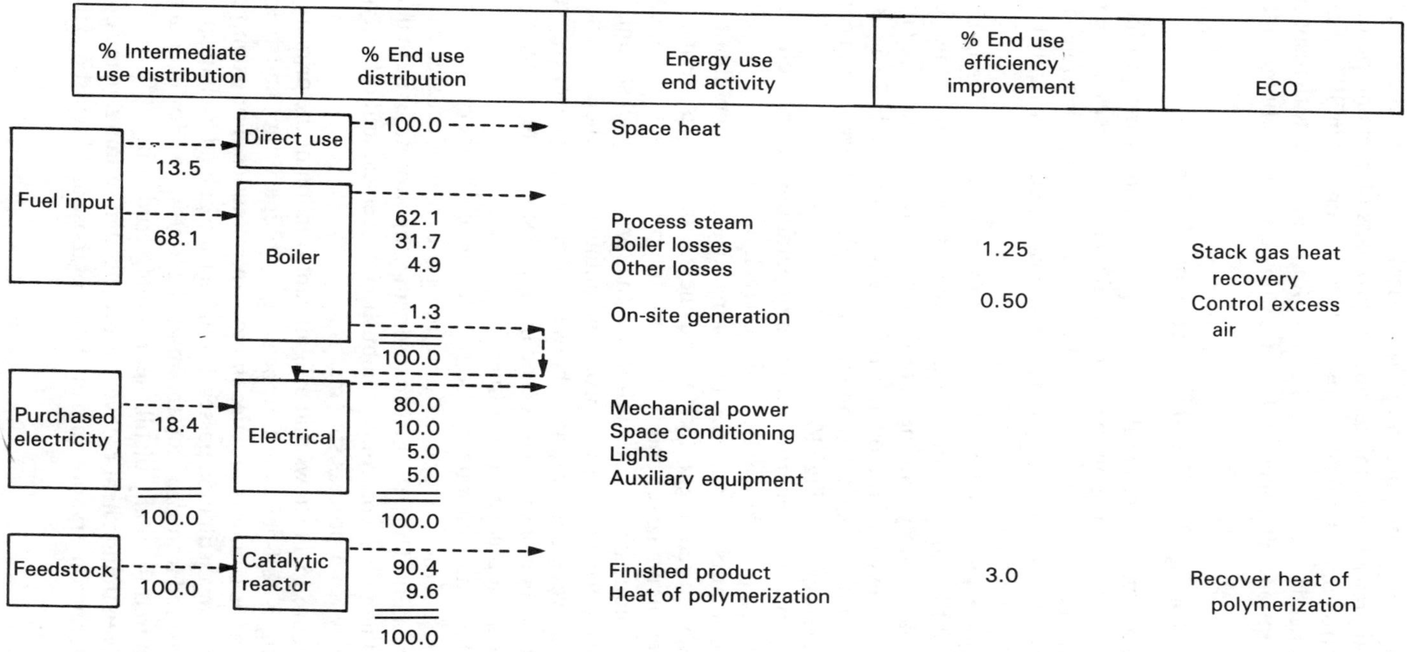

% Intermediate use distribution	% End use distribution	Energy use end activity	% End use efficiency improvement	ECO
	Direct use — 100.0	Space heat		
Fuel input 68.1	Boiler 62.1 / 31.7 / 4.9 = 100.0	Process steam / Boiler losses / Other losses	1.25 / 0.50	Stack gas heat recovery / Control excess air
	1.3 = 100.0	On-site generation		
Purchased electricity 18.4 = 100.0	Electrical 80.0 / 10.0 / 5.0 / 5.0 = 100.0	Mechanical power / Space conditioning / Lights / Auxiliary equipment		
Feedstock 100.0	Catalytic reactor 90.4 / 9.6 = 100.0	Finished product / Heat of polymerization	3.0	Recover heat of polymerization

Direct use 13.5

Figure 15.9 Composite energy use profile for an LDPE plant.

575

refining, paving mixtures and blocks, asphalt felts and coatings, lubricating oils and greases, and miscellaneous products of coal and petroleum.

Petroleum refining consumes 98% of the energy in SIC-29 so that there is little need to compare it to the other four categories with respect to energy consumption. We concentrate strictly on petroleum refining in this section.

15.6.1 Energy Use Profile—Petroleum Refining

Refineries vary considerably in complexity and in the product mix they produce. Depending on the product mix, certain processes are required to achieve the desired results. The many different combinations of processes that might be used in hundreds of refineries make it virtually impossible to describe the "average" refinery.

Figure 15.10 shows a flow sheet for petroleum refining that includes many of the required processes. All these processes might not be used in a particular refinery, but they are shown to illustrate how they interact with other processes in the overall refining scheme. Table 15.17 is an energy balance for a specific combination of the processes, indicated by the numbered units in Fig. 15.10.

The principal energy-consuming operations are crude oil distillation, desulfurization processes, catalytic cracking, naphtha reforming, alkylation, aromatics extraction, and coking. They consume about 80% of the energy required in the refining business. Natural gas, refinery-produced gas, petroleum coke, and fuel oil are the primary sources of energy.

A unique feature of a refinery is that practically all streams are energy streams, since hydrocarbon products contain stored chemical energy. Figure 15.10, however, differentiates between streams carrying energy required for refining operations (cross-hatched lines) and streams representing marketable hydrocarbon products (solid lines).

The composite energy use profile for this refinery, Fig. 15.11, shows that most of the energy used for refining goes to produce steam or directly to fired heaters. Thus, the main conservation opportunities lie in heat recovery from flue gases, control of combustion processes, and other types of heat recovery from process streams.

Figure 15.11 shows that a significant fraction of the primary energy for refining is self-generated, appearing as refinery fuel and fuel gas. Heat recovery is common practice in modern refineries. For example, crude oil undergoing distillation passes through a train of heat exchangers where heat is recovered from product streams going to storage. However, a fired heater is required to totally distill the incoming crude oil.

Even though heat recovery is practiced in refining, many opportunities for energy conservation remain. Some of those are discussed in the following section.

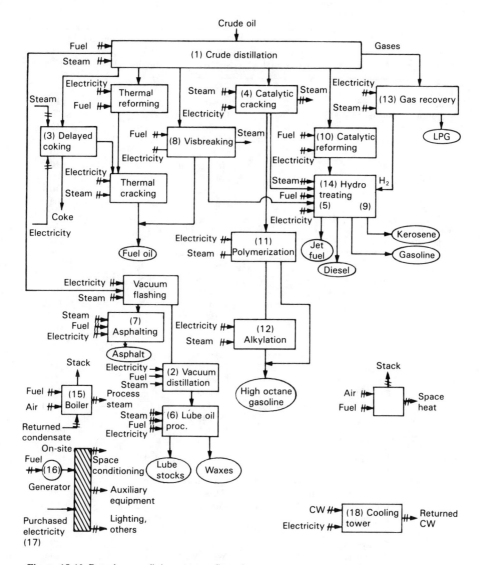

Figure 15.10 Petroleum refining process flow sheet.

Table 15.17 Energy Balance for Petroleum Refining

		In			Out		
Unit operation	Material	Energy (Btu/lb)^a	Temp. (°F)	Material	Energy (Btu/lb)^a	Unrecoverable losses	Temp. (°F)
1						14.1	
Crude	Crude oil	—	75	Diesel	1.3		110
distillation	Steam	50	422	Topped crude	110		670
(750°F)	Fuel	232	—	Wet gas	1.5		110
	Electricity	6	—	Naphtha	3.7		110
	Cold water	—	75	Kerosene	1.0		110
				Light gas oil	1.9		110
				Cold water	107		100
				Waste water	1.5		110
				Stack	46		900
2						6.1	
Vacuum	Topped crude	110	670	Lube distillate	1.6		150
distillation	Steam	61	422	Vacuum oil	5.6		150
(780°F)	Fuel	61	—	Asphalt stock	2.6		400
	Electricity	1	—	Residual oil	0.6		150
	Cold water	—	75	Residual oil	0.9		150
				Cold water	201.5		100
				Waste water	1.8		110
				Residual oil	0.1		150
				Stack	12.2		900
3						1.2	
Delayed	Residual oil	0.6	150	Fuels	—		100
coking	Steam	3.8	422	Gas oil	2.4		400
(950°F)	Fuel	20.7	—	Coke	0.2		200
	Electricity	0.5	—	Fuel gas	—		100
	Cold water	—	75	Cold water	17.7		100
				Waste water	—		100
				Stack	4.1		1000
4						7.1	
Catalytic	Light gas oil	1.9	110	Fuels	4.5		110
cracking	Vacuum oil	5.6	150	Fuel gas and air	—		100
(900°F)	Gas oil	2.4	400	Fuel oil	6.8		275
	Fuel	142	—	Fuels	0.8		200
	Electricity	11.5	—	Fuel oil	3.2		275
	Cold water	—	75	Hot steam	135.3		422
	Air	—	75	Cold water	5.7		100
	Catalyst	—	75	Off-gas	—		200
	Exothermic reaction	28.5	—	Waste water	0.1		200
				Stack	28.4		850
5						2.7	
Distillate	Fuel oil	6.8	275	Fuel oils	8.9		200
hydro-	Diesel	1.3	110	Fuel gas	—		100
forming	Steam	15.1	422	Waste water	0.4		100
(750°F)	Fuel	39.7	—	Cold water	39.9		100
	Electricity	4.1	—	Stack	7.9		850
	Cold water	—	75				
	Hydrogen	—	100				
	Endothermic reaction	−7.2	—				

Table 15.17 Energy Balance for Petroleum Refining (*Continued*)

Unit operation	Material	Energy (Btu/lb)[a]	Temp. (°F)	Material	Energy (Btu/lb)[a]	Unrecoverable losses	Temp. (°F)
6						5.6	
Lube oil	Lube distillate	1.6	150	Lube stocks	—		75
processing	Hydrogen	—	100	Waxes	—		75
	Steam	86.3	422	Fuel gas	—		100
	Fuel	26.5	—	Condensate	7.6		180
	Electricity	10.9	—	Stack	5.3		850
	Cold water	—	75	Cold water	106.8		100
7						1.0	
Asphalt	Asphalt stock	—	75	Asphalt	3.4		500
processing	Steam	6.6	422	Cold water	7.5		100
(500°F)	Fuel	13.3	—	Stack	2.7		600
	Electricity	0.5	—	Vapor	6.6		500
	Cold water	—	75				
	Inject air	—	75				
	Exothermic	0.8	—				
8						2.5	
Visbreaking	Residual oil	—	75	Fuel oil	—		100
(900°F)	Fuel	38.8	—	Steam	11.9		293
	Electricity	0.9	—	Stack	7.8		1000
	Cold water	—	75	Cold water	17.5		100
9						2.3	
Hydrogen	Naphtha	3.7	110	Desulfurized	2.7		100
treatment	Hydrogen	—	75	fuels			
				fuel gas	—		100
(800°F)	Steam	8.0	422	Stack	7.7		850
	Fuel	38.5	—	Condensate	0.7		180
	Electricity	2.5	—	Cold water	27.4		100
	Cold water	—	75				
	Endothermic	− 11.9	—				
10						10.9	
Catalytic	Desulfurized	2.7	100	Fuels	2.0		100
deforming	fuels			Fuels	0.4		100
(1000°F)	Fuel	218.0	—	Fuels	0.2		100
	Electricity	7.4	—	Fuel gas	0.1		100
	Cold water	—	75	Hydrogen	—		100
	Endothermic	−9.9	—	Cold water	161		100
				Stack	43.6		1100
11						0.4	
Isomerization	Fuels	0.2	100	Fuels	0.4		100
(400°F)	Fuels	0.2	100	Condensate	0.9		180
	Steam	6.9	422	Cold water	6.1		100
	Electricity	0.5	—				
	Cold water	—	75				
12						10	
Alkylation	Fuels	0.4	100	Fuels	—		70
(70°F)	Steam	2.1	422	Stack	39.6		400
	Fuel	198	—	Waste water	—		100
	Electricity	2.4	—	Cold water	155		100
	Cold water	—	75				
	Exothermic	1.7	—				

Table 15.17 Energy Balance for Petroleum Refining (*Continued*)

Unit operation	Material	In Energy (Btu/lb)^a	Temp. (°F)	Material	Out Energy (Btu/lb)^a	Unrecoverable losses	Temp. (°F)
13						1.4	
Gas	Wet gas	1.5	110	LPG	—		75
recovery	Steam	27.7	422	Fuels	—		100
(380°F)	Electricity	3.1	—	Refinery fuels	—		100
	Cold water	—	75	Cold water	28.5		100
				Condensate	2.4		180
14						0.7	
Hydrogen	Kerosene	1.0	110	Kerosene	1.0		110
treatment	Hydrogen	—	100	Fuel gas	—		100
(800°F)	Steam	1.5	422	Condensate	0.1		180
	Fuel	13.3	—	Stack	2.7		900
	Electricity	1.0	—	Cold water	9.0		100
	Cold water	—	75				
	Endothermic	−3.3	—				
15						16	
Boiler	Fuel gas	325.4	100	Process steam	121.8		422
	Makeup water	—	75	Stack	93.5		450
	Condensate	11.9	180	Misc. steam	104		422
				Space heat	2		422
16						2	
Electricity	Fuel	37.2	—	Electricity	12.4		—
generation	Cold water	—	75	Cold water	22.8		100
17							
Electric	Purchased electricity	72.5	—	Process electricity	52.3	—	
	Generated electricity	12.4		Miscellaneous electricity	5.9	—	
				Cool tower	26.7	—	
18						940	
Cooling	Cold water	913.3	100	Cold water	—		75
tower	Electricity	26.7	—				
19							
Fuels	Fuel gas	494	100	Boiler use	325.4		100
	Purchased fuels	710.4	—	Process	1041.8	—	
	Refinery fuels	200	100	Electricity generation	37.2	—	

^{a}Btu's per pound of input.

15.6.2 Conservation Opportunities in Refineries

The refining industry has a good record of conserving energy. The American Petroleum Institute reported 10.3% energy savings between 1972 and 1975. The industry-set goal of about 15% savings by 1980 was met; in fact, many refineries exceeded this goal considerably. Before becoming euphoric about this, however, we should realize that flows of energy through a refinery are huge and the potential for more efficient utilization of energy is also large.

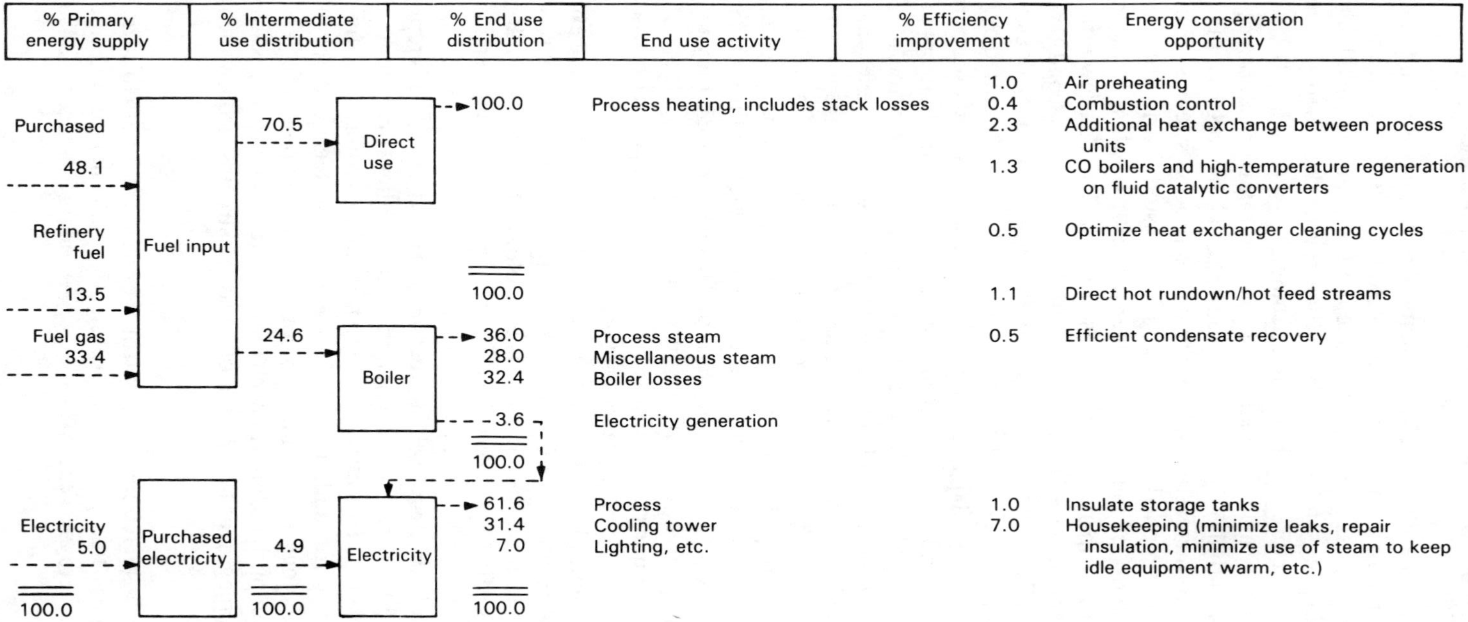

% Primary energy supply	% Intermediate use distribution		% End use distribution	End use activity	% Efficiency improvement	Energy conservation opportunity
Purchased 48.1	70.5	Direct use	100.0	Process heating, includes stack losses	1.0 0.4 2.3 1.3	Air preheating Combustion control Additional heat exchange between process units CO boilers and high-temperature regeneration on fluid catalytic converters
Refinery fuel 13.5			100.0		0.5	Optimize heat exchanger cleaning cycles
Fuel gas 33.4	24.6	Boiler	36.0 28.0 32.4 3.6	Process steam Miscellaneous steam Boiler losses Electricity generation	1.1 0.5	Direct hot rundown/hot feed streams Efficient condensate recovery
			100.0			
Electricity 5.0	Purchased electricity 4.9	Electricity	61.6 31.4 7.0	Process Cooling tower Lighting, etc.	1.0 7.0	Insulate storage tanks Housekeeping (minimize leaks, repair insulation, minimize use of steam to keep idle equipment warm, etc.)
100.0	100.0		100.0			

Figure 15.11 Composite energy use profile for a petroleum refinery.

Refineries use a great many heat exchangers, and chemical reactions abound. In such processes first-law analyses are not always indicative of the effectiveness of energy use. Second-law analyses, involving the concepts of availability and second-law efficiency as described in Chapter 9, give a better picture of what is happening.

An availability analysis of a typical catalytic reformer [7], for example, showed that only about 7% of the available energy supplied to the process went to increasing the energy of the process streams. Hence, 93% of the energy supplied to the process was wasted. Further examination showed that most of the available energy, 62%, was lost in the reactor furnace, where temperature differences are high. The first-law efficiency η_F for the reactor furnace is typically about 72%, but the second-law efficiency η_S is only 27%. The reason for the difference is that η_F does not discriminate between a Btu at 760°F in the flue gas and a Btu liberated by combustion that might be at 3500°F. Second-law efficiency values these energies differently.

Second-law efficiencies for refinery heat exchangers are typically low. For example, in the catalytic reformer study referred to above [7], about 16% of the available energy input was lost in heat exchangers. Most of this loss, 67%, was a direct loss resulting from heat transfer to air and cooling water. Clearly this practice should be replaced in favor of using process streams to pick up the rejected energy. Even if this were done, the available energy loss would be large unless temperature differences between interacting streams were minimized. This can be done only by increasing heat transfer area, thus increasing capital costs considerably. The refinery is a very complex plant, where both second-law and first-law analyses can show how energy might be conserved.

Figure 15.11 lists some practical means of increasing efficiency in petroleum refining. Percentage improvement factors are based on energy consumption for the industry as a whole, rather than on individual processes. These ECOs are discussed further in the following sections.

Additional Heat Exchange between Process Units. This represents by far the largest potential for energy savings other than general housekeeping measures. However, its implementation requires relatively large capital outlays for additional heat exchange surface and for installation. The technology is well known; commercial equipment is readily available to accomplish the goal of exchanging heat between process streams rather than dumping heat to air or cooling-water streams, from which it is unrecoverable.

Combustion Control. Control of excess air for both fired heaters and boilers reduces flue gas losses. Automatic controllers using input from flue gas analyzers are readily available. An energy saving of 0.4% is estimated.

Air Preheating. Air preheating can be accomplished by using stack gases or process streams. About 80% of the energy for refining is consumed by fired heaters. Stack gas temperatures typically vary from 400 to 1000°F, representing a clear potential for heat recovery.

Logistics can sometimes prevent retrofitting of fired heaters with air preheaters. However, virtually all new heater installations have air preheaters as an integral part. Air preheaters typically cost upward of $1 million, but rates of return of 30% or higher are commonplace.

CO Boilers. Fluid catalytic cracker regenerators typically operate near 1200°F. The waste gases are also rich in CO, a typical value being 9% CO by volume. Two actions are available: first, sensible energy can be recovered from the stream, and second, the CO can be burned to release energy, which can drive a boiler. Supplementary fuel might be required for combustion stability.

Carbon monoxide boilers are well accepted by industry. The catalytic cracker must be shut down, however, to install a CO boiler, so the economic analysis of the energy conservation opportunity must include the loss in production. Installation of a CO boiler can be carefully planned so that final tie-ins can be made during a routine maintenance shutdown.

Hot Rundown/Hot Feed Streams. Attention should be paid to directing hot streams produced by one unit directly to the following process, rather than storing the streams and then having to reheat them. For example, in many refineries, vacuum gas oil is cooled by water, then stored in a tank before being pumped to the catalytic cracker. The ECO here involves avoiding the cooldown-storage-reheat cycle by feeding the hot gas directly to the catalytic cracker from the crude unit.

Condensate Recovery. The principles of Chapter 7 can be applied to the collection of condensate, thus avoiding makeup losses. Also, the sensible heat in condensate can be usefully recovered when the condensate is returned to the boiler feedwater system. The rule of thumb is that for every 11°F rise in feedwater temperature, about 1% less fuel is required by the boiler to produce steam. In some plants it might be possible to recover condensate at pressures high enough to justify a flash drum from which steam can be withdrawn for use in other processes before being returned to the boiler.

Heat Exchanger Cleaning Cycles. The tendency for fouling of refinery exchangers is high because of the hydrocarbon streams and the high temperatures involved. A scheme for cleaning heat exchangers can pay big dividends.

It is often remarked that history repeats itself. Advantage can be taken of this fact by documenting the fouling behavior of exchangers in order to develop the optimum schedule for cleaning them. Availability calculations might also be helpful, if a threshold value for second-law efficiency can be determined from history and analysis. On-stream cleaning is a well accepted industrial practice; thus, the exchanger need not be shut down.

Insulation and good housekeeping practices can also be very fruitful in minimizing energy losses during refinery operation.

The major ECOs for petroleum refining have been discussed in this section. Other techniques are available, but they cannot all be discussed in this brief review. Sittig [3] gives ample information concerning potential ECOs for refining.

15.7 THE CEMENT INDUSTRY

The hydraulic cement industry is the biggest user of energy in the stone, clay, and glass classification (SIC-32). The hydraulic cement category includes portland, masonry, pozzolonic,[1] and natural cements.

Hydraulic cement is a powder formed by heating limestone, sand, clay, and other materials in a kiln. The kiln produces an amalgam of these substances, which is called *clinker*. The clinker is ground to form the powderized cement.

Two cement processing methods are commonly used: the wet and the dry method. These terms refer to the condition of the raw materials as they are ground and blended. The dry process uses dry materials, while the wet process uses a slurry of raw materials during grinding and blending. The other processes are essentially the same for both methods. A good description of the processes involved in cementmaking is given by Chiogioji [8].

15.7.1 Energy Use Profile—Portland Cement

Figure 15.12 is a flow sheet for one type of hydraulic cement—portland cement. The process is fairly simple, and can involve either the wet or the dry method. Both are shown to illustrate their relationship to the overall process.

Table 15.18 is an energy balance corresponding to Fig. 15.12. The principal energy consumer is the kiln in which calcining of the raw materials takes place. The calcining process yields the clinker; it involves an endothermic chemical reaction between the raw materials at high temperatures.

---

[1]Pozzolon is a material capable of reacting with lime in the presence of water to form cement-like compounds.

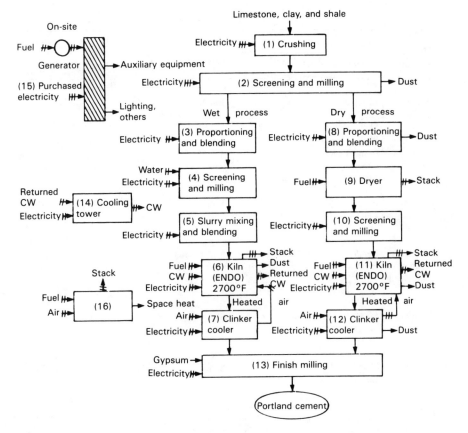

Figure 15.12 Process flow sheet for portland cement.

Kilns are long cylindrical furnaces that rotate slowly and are mounted at a slight incline to the horizontal. A typical kiln might be up to 750 ft long and 15–25 ft in diameter. The kiln rotates about one revolution per minute. The raw materials move gradually down the kiln because of its inclination, taking several hours to travel through it.

The kiln burner is located at the discharge end. Thus, the raw material becomes progressively hotter as it moves toward the discharge. At lower temperatures water vapor and CO_2 are driven off. In the higher-temperature sections of the kiln, where temperature is about 2700°F, a chemical reaction between lime and silica, aluminum, and iron takes place. The resulting clinker leaves the kiln as lumps and pellets.

In 1975 the wet cement process consumed about 8.2×10^6 Btu per ton of cement produced, compared to about 9.3×10^6 Btu per ton in 1950. Hence, some progress in conserving energy has been made over the years.

Table 15.18 Energy Balance for Portland Cement

Unit operation	Material	In Energy (Btu/lb)[a]	In Temp. (°F)	Material	Out Energy (Btu/lb)[a]	Out Unrecoverable losses	Out Temp. (°F)
1						75.6	
Crushing	Limestone	—	75	Batch	—		
(75°F)	Clay	—	75				
	Shale	—	75				
	Shell	—	75				
	Cement rock	—	75				
	Other	—	75				
	Electric	75.6	—				
2						2.1	
Screening	Batch	—	75	Batch	—		75
and milling	Electric	2.10	—	Dust	—		75
(75 °F)							
3						5.1	
Wet process	Batch	—	75	Mixed batch	—		75
Proportion	Electric	5.1	—				
and blend							
(75°F)							
4						2.1	
Screening	Mixed batch	—	75	Slurry	—		75
and milling	Water	—	75				
	Electric	2.1	—				
5						3.29	
Slurry mixing	Slurry	—	75	Slurry	—		75
and blend	Electric	3.29	—				
(75°F)							
6						97.96	
Wet kiln	Slurry	—	75	Clinker	243		2100
(700°F)	Cold water	—	75	Dust	—		—
	Fuel	1310	—	Stack	562		1100
	Electricity	12.96	—	Cold water	110		185
	Hot air from clinker	230	300				
	Cooler Endothermic	−540	—				
7						18.1	
Clinker	Clinker	145	2100	Clinker	2.4		100
cooler	Air	—	75	Hot air to kiln	230		300
(100°F)	Electric	7.5	—				
8						1.72	
Dry process	Batch	—	75	Mixed batch	—		75
Proportion	Electric	1.72	—	Dust	—		75
and blend							
(75°F)							
9						15	
Dryer	Batch	—	75	Dry mix	55		500
(325°F)	Fuel	270	—	Stack	200		725
10						1.78	
Screen and	Dry mix	55	500	Milled mix	55		500
mill (75°F)	Electric	1.78	—				

Table 15.18 Energy Balance for Portland Cement (*Continued*)

Unit operation	Material	In Energy (Btu/lb)[a]	In Temp. (°F)	Material	Out Energy (Btu/lb)[a]	Out Unrecoverable losses	Out Temp. (°F)
11						96.4	
Dry kiln	Milled mix	55	500	Clinker	145		2100
(2,700°F)	Cold water	—	75	Stack	458		1100
	Fuel	1050	—	Cold water	16		185
	Electric	5.42	—	Dust			
	Hot air from clinker Cooler	135	300				
	Endothermic	– 410	—				
12						17.5	
Clinker	Clinker	145	2100	Clinker	—		75
cooler	Air	—	75	Dust	—		—
(100°F)	Electric	7.5		Hot air to kiln	135		300
13						66.13	
Finish	Wet process	—	75	Hydraulic			
milling	Dry process	—	75	cement	—		75
(75°F)	Gypsum	—	75				
	Electric	66.13	—				
14							
Cooling	Cold water	246	95	Cold water	—	—	75
tower	Make up	—	75	Stack	246	—	—
15							
Electric	Purchased	190.8	—	Process	181	—	—
	Generated	9.2	—	Other	14	—	—
16							
Electric	Fuel	27.6	—	Electricity	9.2		—
generation				Stack	18.4		—

[a]Pound refers to product output.

Table 15.19 Fuel Consumption by Type for Hydraulic Cement

	Wet	Dry
Coal	30.4	42.6
Refined oil products	13.7	8.0
Natural gas	39.9	32.4
Purchased electricity	16.0	17.0
Total	100.0	100.0

Source: Gordian Associates [9].

Table 15.19 shows a composite energy type profile for the cement industry in 1975. Coal and natural gas are the principal sources of primary energy.

About 58% of the portland cement produced in the United States today is by the wet process [8]. In 1974 the wet process consumed 61% of the total energy used in cementmaking. Consequently, we choose the wet portland cement process for further discussion.

15.7.2 Energy Conservation Opportunities in Portland Cementmaking

Figure 15.13 is a composite energy use profile for wet portland cement. Thus, we selected components 3 through 7 in Fig. 15.12 for analysis rather than 8 through 12. The wet process consumes more energy per pound of product and is more commonly used.

Energy use is neatly apportioned between direct fuel used to fire the kiln and electric energy for the mechanical part of the various processes. Virtually no steam is used in cementmaking. Clearly, the target for energy conservation is the rotary kiln, which consumes 85.7% of the energy for the wet process of Fig. 15.12.

Many cement plants are old, and they are not very energy-efficient. "Wet" plants built before 1935 consume over 9×10^6 Btu per ton of cement, while plants built since 1965 consume slightly over 7×10^6 Btu per ton. A second-law analysis of a cement plant shows that the minimum required energy is about 1×10^6 Btu per ton. The modern Chichibu plant in Japan operates at about 3.0×10^6 Btu per ton, which is considerably better than virtually all of the plants currently operating in the United States.

One obvious way to reduce energy consumption is to convert from a wet to a dry process. The dry process is approximately 15% more efficient than the wet process.

Suspension Preheaters. A suspension preheater is a part of the kiln complex designed for more efficient heat transfer to the incoming raw feed. The raw particles are fed into the high-velocity combustion gas stream emerging from the rotary kiln, which "suspends" them in the gas flow. After picking up heat, the particles are collected, usually by cyclone separators, and then are fed into the rotary kiln for final calcining.

Typical suspension preheaters heat the feed to about 1400°F, so chemical reactions also occur and about 30–40% of the calcination is performed before the material goes into the rotary kiln. Up to four stages of heating might be used in a typical preheater.

Suspension preheaters require a tower beside the rotary kiln, because the process involves vertical movement of the particles and the combustion gases. However, the so-called long kiln, which uses conventional pre-

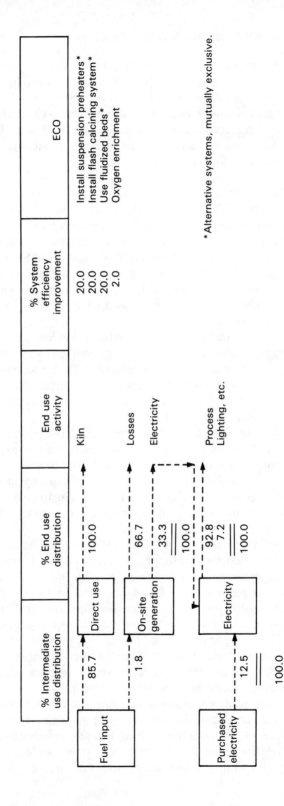

Figure 15.13 Composite energy use profile for portland cement (wet process).

heating, is shortened when a suspension preheater is installed. The suspension preheater tower usually occupies the space that the low-temperature sections of the long kiln would otherwise occupy.

The addition of a suspension preheater converts a plant from wet to dry, giving an additional boost to its efficiency. About 1.75×10^6 Btu per ton of cement can be saved by suspension air preheating, yielding an estimated 20% energy savings for the process.

Flash Calciner. This is a variation of suspension preheating that involves two stages of firing. The system is characterized by a flash calciner vessel located between the rotary kiln and the suspension preheater. Combustion gases passing from the rotary kiln and the raw material leaving the bottom of the third-stage suspension preheater discharge into the flash calcining vessel. Fuel is burned in the vessel to further calcine the preheated material.

The combustion gases and the kiln gases carry the material to the fourth-stage preheater, from which it is discharged to the rotary kiln for final clinker production.

In this system about half of the fuel is burned at low temperature with a low percentage of excess air. Considerable energy is saved in this way. About a 5% reduction in fuel is possible by using flash calcining rather than a four-stage suspension preheater. In addition, the production capacity for a kiln of a given size can be increased up to 2½ times compared to a typical four-stage suspension preheater kiln.

Flash calcining plants can also be large in size. The plant at Chichibu has a production capacity of 8,500 tons/day of clinker at a rate of 2.6×10^6 Btu per ton of clinker. In fact, smaller plants might not be able to economically justify conversion to such a system. Production capacity must be at least 2,500 tons/day for economic viability. It is interesting to note that in 1974 the average kiln capacity in the United States was less than 1,000 tons/day.

Fluidized-Bed Processing. There is some evidence that the rotary kiln might be replaced by a fluidized bed for production of portland cement clinker. In this system, raw material is introduced through the bottom of the bed of fluidized clinker particles. The bed is fluidized by hot combustion gases introduced by a distribution grid. The bed operates at 2400°F and heat transfer coefficients are very high; thus, raw material particles are rapidly brought to clinkering temperature. The clinker particles grow by amalgamation as more raw material enters the bed. Overflow outlets allow spillover of clinker. Clinker particles of sufficient size are separated, while smaller ones are returned to the bed for further growth.

Engineering studies indicate that savings of about 20% of the energy required for a dry long-kiln process are possible with a fluidized-bed system. Savings of 25% over the wet long-kiln process seem feasible. In fact, the energy requirements of a fluidized bed system are close to those of a short rotary kiln equipped with a suspension preheater.

The drawback of fluidized-bed technology is that it is somewhat new and untried in the cement industry. However, the technology has been proved in other fields, and fluidized beds should have a great deal of promise for energy conservation in the cement industry.

Oxygen Enrichment. Enriching combustion air with oxygen increases the capacity of an existing rotary kiln. Assuming that heat losses stay about the same with oxygen enrichment, clinker production goes up while the unit fuel consumption goes down. Tests have shown that oxygen enrichment can reduce fuel consumption in the kiln by 15%. Subtracting the energy used to produce the additional oxygen still yields substantial energy savings.

Other lesser improvements in efficiency might be gained by increasing kiln speed and by using more effective insulation to prevent heat loss. Increasing kiln rotary speed requires a lower tilt angle to keep the correct residence time for the material moving through the kiln. Older, slower kilns are inclined at 4–5%, while newer suspension preheater kilns are designed with 2.5–3% inclination.

15.8 SUMMARY

The six highest energy-consuming industries in the United States have been profiled in this chapter. One of the subindustries in each of the six major categories was selected for specific discussion to identify the principal energy conservation opportunities.

REFERENCES

1. M. E. Casper, *Energy-Savings Techniques for the Food Industry,* Noyes Data Corp., Park Ridge, N.J., 1977.
2. Energy Analysis of 108 Industrial Processes, Report to the Department of Energy under contract DOE E (11-1)2862, Drexel University, Philadelphia, June 1979.
3. M. Sittig, *Practical Techniques for Saving Energy in the Chemical, Petroleum and Metals Industries,* Noyes Data Corp., Park Ridge, N.J., 1977.
4. H. W., Lownie, Jr., et al., Draft Report on Development and Establishment of Energy Efficiency Improvement Targets for Primary Metals Industries, Federal Energy Administration, Washington, D.C., 1976.
5. M. Sittig, *Pulp and Paper Manufacture: Energy Conservation and Pollution Prevention,* Noyes Data Corp., Park Ridge, N.J., 1977.
6. *Project Independence Blueprint—Final Task Force Report—Energy Conservation in the Manufacturing Sector 1954–1990, Vol. 3,* Federal Energy Administration, Washington, D.C., Nov. 1974.
7. Gordian Associates, *The Data Base: The Potential for Energy Conservation in Nine Selected Industries, Vol. 2: Petroleum Refining,* Conservation Paper No. 10, Federal Energy Administration, Washington, D.C., June 1974.
8. M. H. Chiogioji, *Industrial Energy Conservation,* Marcel Dekker, New York, 1979.
9. Gordian Associates, *The Potential for Energy Conservation in Nine Selected Industries, Vol. 3: Cement,* U.S. Government Printing Office, Washington, D.C., 1975.

SOME CONVERSION FACTORS

Table A.1 Length

Centimeters, cm	Meters, m	Inches, in.	Feet, ft	Miles	Microns, μ
1	0.01	0.3937	0.03281	6.214×10^{-6}	10^4
100	1	39.37	3.281	6.214×10^{-4}	10^6
2.540	0.0254	1	0.08333	1.578×10^{-5}	2.54×10^4
30.48	0.3048	12	1	1.894×10^{-1}	0.3048×10^6
1.6093×10^5	1609.3	6.336×10^4	5280	1	1.6093×10^9
10^{-4}	10^{-6}	3.937×10^{-5}	3.281×10^{-6}	6.2139×10^{-10}	1
10^{-8}	10^{-10}	3.937×10^{-9}	3.281×10^{-10}	6.2139×10^{-14}	10^{-4}
9.4663×10^{17}	9.4663×10^{15}	3.727×10^{17}	3.1058×10^{16}	5.8822×10^{12}	9.4663×10^{21}

Table A.2 Mass

Grams, g_m	Kilograms, kg_m	Pounds, lb_m	Tons (short)	Tons (long)	Tons (metric)
1	0.001	2.2046×10^{-3}	11.102×10^{-6}	9.842×10^{-2}	10^{-6}
1.000	1	2.2046	0.001102	9.842×10^{-4}	10^{-3}
453.6	0.4536	1	5.0×10^{-4}	4.464×10^{-4}	4.536×10^{-4}
9.072×10^5	907.2	2.000	1	0.8929	0.9072
1.016×10^6	1.016	2.240	1.12	1	1.016
10^6	1.000	2.204 7	1.1023	0.9843	1
1.6598×10^{-24}	1.6598×10^{-27}	3.6593×10^{-27}	1.8297×10^{-30}	1.6337×10^{-30}	1.6598×10^{-30}

Table A.3 Flow

cm³/sec	ft³/min	U.S. gal/min	Imperial gal/min
1	0.002119	0.01585	0.01320
472.0	1	7.481	6.229
63.09	0.1337	1	0.8327
75.77	0.1605	1.201	1

Table A.4 Pressure

kg_f/cm²	lb_f/in.²	lb_f/ft²	cm Hg (0°C)	in. Hg (32°F)	in. H₂O (60°F)	atm	N/m²(Pa)
1	14.22	2,048	73.56	28.96	394.1	0.9678	2.022×10^{-4}
0.07031	1	144	5.171	2.036	27.71	0.06805	6896.6
4.882×10^{-4}	0.006944	1	0.03591	0.01414	0.1924	4.725×10^{-4}	47.87
0.01360	0.1934	27.85	1	0.3937	5.358	0.01316	1333.0
0.03453	0.4912	70.73	2.540	1	13.61	0.03342	3368.0
0.002538	0.03609	5.197	0.1866	0.07348	1	0.002456	249.1
1.033	14.70	2,116	76.0	29.92	407.2	1	1.013×10^{-5}
4.946×10^{3}	1.450×10^{-4}	2.089×10^{-2}	7.50×10^{-4}	2.953×10^{-4}	4.015×10^{-3}	9.869×10^{-6}	1

Table A.5 Energy

Ergs	Joules	kWh	g_m-cal	ft-lb_f	hp-hr	Btu
1	10^{-7}	2.778×10^{-14}	2.388×10^{-8}	7.376×10^{-8}	3.725×10^{-14}	9.478×10^{-11}
10^{7}	1	2.778×10^{-7}	0.2388	0.7376	3.725×10^{-7}	9.478×10^{-4}
3.6×10^{13}	3.6×10^{6}	1	8.598×10^{5}	2.655×10^{6}	1.341	3412
4.187×10^{7}	4.187	1.163×10^{-6}	1	3.088	1.56×10^{-6}	3.968×10^{-3}
1.356×10^{7}	1.356	3.766×10^{-7}	0.3238	1	5.051×10^{-7}	1.285×10^{-3}
2.685×10^{13}	2.685×10^{6}	0.7457	6.412×10^{5}	1.98×10^{6}	1	2545
1.055×10^{10}	1055	2.931×10^{-1}	252	778.2	3.93×10^{-4}	1

Table A.6 Heat Flux Density

watt/cm^2	cal/sec cm^2	Btu/hr ft^2
1	0.2388	3170.2
4.187	1	1.3272×10^4
3.155×10^{-4}	7.535×10^{-5}	1

Table A.7 Thermal Conductivity

watt/cm °C	cal/sec cm°C	Btu/hr ft°F	Btu in./hr ft^2 °F
1	0.2388	57.78	693.3
4.187	1	241.9	2903
0.01731	4.134×10^{-3}	1	12
1.441×10^{-3}	3.445×10^{-4}	0.08333	1

Table A.8 Viscosity

Centipoise	Poise	kg$_m$/sec m	lb$_m$/sec ft	lb$_m$/hr ft	lb$_f$ sec/ft^2
1	0.01	0.001	6.720×10^{-4}	2.419	2.089×10^{-5}
100	1	0.1	0.06720	241.9	2.089×10^{-6}
1,000	10	1	0.6720	2,419	0.02089
1,488	14.88	1.488	1	3,600	0.03108
0.4134	4.134×10^{-3}	4.134×10^{-4}	2.778×10^{-4}	1	8.634×10^{-6}
4.788×10^4	4.78.8	47.88	32.17	1.158×10^5	1

THERMODYNAMIC PROPERTIES

Table B.1 Saturated Steam: Temperature Table

Temp Fahr t	Abs Press. Lb per Sq In. p	Specific Volume			Enthalpy			Entropy			Temp Fahr t
		Sat. Liquid v_f	Evap v_{fg}	Sat. Vapor v_g	Sat. Liquid h_f	Evap h_{fg}	Sat. Vapor h_g	Sat. Liquid s_f	Evap s_{fg}	Sat. Vapor s_g	
32.0*	0.08859	0.016022	3304.7	3304.7	−0.0179	1075.5	1075.5	0.0000	2.1873	2.1873	32.0*
34.0	0.09600	0.016021	3061.9	3061.9	1.996	1074.4	1076.4	0.0041	2.1762	2.1802	34.0
36.0	0.10395	0.016020	2839.0	2839.0	4.008	1073.2	1077.2	0.0081	2.1651	2.1732	36.0
38.0	0.11249	0.016019	2634.1	2634.2	6.018	1072.1	1078.1	0.0122	2.1541	2.1663	38.0
40.0	0.12163	0.016019	2445.8	2445.8	8.027	1071.0	1079.0	0.0162	2.1432	2.1594	40.0
42.0	0.13143	0.016019	2272.4	2272.4	10.035	1069.8	1079.9	0.0202	2.1325	2.1527	42.0
44.0	0.14192	0.016019	2112.8	2112.8	12.041	1068.7	1080.7	0.0242	2.1217	2.1459	44.0
46.0	0.15314	0.016020	1965.7	1965.7	14.047	1067.6	1081.6	0.0282	2.1111	2.1393	46.0
48.0	0.16514	0.016021	1830.0	1830.0	16.051	1066.4	1082.5	0.0321	2.1006	2.1327	48.0
50.0	0.17796	0.016023	1704.8	1704.8	18.054	1065.3	1083.4	0.0361	2.0901	2.1262	50.0
52.0	0.19165	0.016024	1589.2	1589.2	20.057	1064.2	1084.2	0.0400	2.0798	2.1197	52.0
54.0	0.20625	0.016026	1482.4	1482.4	22.058	1063.1	1085.1	0.0439	2.0695	2.1134	54.0
56.0	0.22183	0.016028	1383.6	1383.6	24.059	1061.9	1086.0	0.0478	2.0593	2.1070	56.0
58.0	0.23843	0.016031	1292.2	1292.2	26.060	1060.8	1086.9	0.0516	2.0491	2.1008	58.0
60.0	0.25611	0.016033	1207.6	1207.6	28.060	1059.7	1087.7	0.0555	2.0391	2.0946	60.0
62.0	0.27494	0.016036	1129.2	1129.2	30.059	1058.5	1088.6	0.0593	2.0291	2.0885	62.0
64.0	0.29497	0.016039	1056.5	1056.5	32.058	1057.4	1089.5	0.0632	2.0192	2.0824	64.0
66.0	0.31626	0.016043	989.0	989.1	34.056	1056.3	1090.4	0.0670	2.0094	2.0764	66.0
68.0	0.33889	0.016046	926.5	926.5	36.054	1055.2	1091.2	0.0708	1.9996	2.0704	68.0
70.0	0.36292	0.016050	868.3	868.4	38.052	1054.0	1092.1	0.0745	1.9900	2.0645	70.0
72.0	0.38844	0.016054	814.3	814.3	40.049	1052.9	1093.0	0.0783	1.9804	2.0587	72.0
74.0	0.41550	0.016058	764.1	764.1	42.046	1051.8	1093.8	0.0821	1.9708	2.0529	74.0
76.0	0.44420	0.016063	717.4	717.4	44.043	1050.7	1094.7	0.0858	1.9614	2.0472	76.0
78.0	0.47461	0.016067	673.8	673.9	46.040	1049.5	1095.6	0.0895	1.9520	2.0415	78.0
80.0	0.50683	0.016072	633.3	633.3	48.037	1048.4	1096.4	0.0932	1.9426	2.0359	80.0
82.0	0.54093	0.016077	595.5	595.5	50.033	1047.3	1097.3	0.0969	1.9334	2.0303	82.0
84.0	0.57702	0.016082	560.3	560.3	52.029	1046.1	1098.2	0.1006	1.9242	2.0248	84.0
86.0	0.61518	0.016087	527.5	527.5	54.026	1045.0	1099.0	0.1043	1.9151	2.0193	86.0
88.0	0.65551	0.016093	496.8	496.8	56.022	1043.9	1099.9	0.1079	1.9060	2.0139	88.0
90.0	0.69813	0.016099	468.1	468.1	58.018	1042.7	1100.8	0.1115	1.8970	2.0086	90.0
92.0	0.74313	0.016105	441.3	441.3	60.014	1041.6	1101.6	0.1152	1.8881	2.0033	92.0
94.0	0.79062	0.016111	416.3	416.3	62.010	1040.5	1102.5	0.1188	1.8792	1.9980	94.0
96.0	0.84072	0.016117	392.8	392.9	64.006	1039.3	1103.3	0.1224	1.8704	1.9928	96.0
98.0	0.89356	0.016123	370.9	370.9	66.003	1038.2	1104.2	0.1260	1.8617	1.9876	98.0

Temp	P	v_f	v_{fg}	v_g	h_f	h_{fg}	h_g	s_f	s_{fg}	s_g	Temp
100.0	0.94924	0.016130	350.4	350.4	67.999	1037.1	1105.1	0.1295	1.8530	.9825	100.0
102.0	1.00789	0.016137	331.1	331.1	69.995	1035.9	1105.9	0.1331	1.8444	.9775	102.0
104.0	1.06965	0.016144	313.1	313.1	71.992	1034.8	1106.8	0.1366	1.8358	.9725	104.0
106.0	1.1347	0.016151	296.16	296.18	73.99	1033.6	1107.6	0.1402	1.8273	.9675	106.0
108.0	1.2030	0.016158	280.28	280.30	75.98	1032.5	1108.5	0.1437	1.8188	.9626	108.0
110.0	1.2750	0.016165	265.37	265.39	77.98	1031.4	1109.3	0.1472	1.8105	.9577	110.0
112.0	1.3505	0.016173	251.37	251.38	79.98	1030.2	1110.2	0.1507	1.8021	.9528	112.0
114.0	1.4299	0.016180	238.21	238.22	81.97	1029.1	1111.1	0.1542	1.7938	.9480	114.0
116.0	1.5133	0.016188	225.84	225.85	83.97	1027.9	1111.9	0.1577	1.7856	.9433	116.0
118.0	1.6009	0.016196	214.20	214.21	85.97	1026.8	1112.7	0.1611	1.7774	.9386	118.0
120.0	1.6927	0.016204	203.25	203.26	87.97	1025.6	1113.6	0.1646	1.7693	.9339	120.0
122.0	1.7891	0.016213	192.94	192.95	89.96	1024.5	1114.4	0.1680	1.7613	.9293	122.0
124.0	1.8901	0.016221	183.23	183.24	91.96	1023.3	1115.3	0.1715	1.7533	.9247	124.0
126.0	1.9959	0.016229	174.08	174.09	93.96	1022.2	1116.1	0.1749	1.7453	.9202	126.0
128.0	2.1068	0.016238	165.45	165.47	95.96	1021.0	1117.0	0.1783	1.7374	.9157	128.0
130.0	2.2230	0.016247	157.32	157.33	97.96	1019.8	1117.8	0.1817	1.7295	.9112	130.0
132.0	2.3445	0.016256	149.64	149.66	99.95	1018.7	1118.6	0.1851	1.7217	.9068	132.0
134.0	2.4717	0.016265	142.40	142.41	101.95	1017.5	1119.5	0.1884	1.7140	.9024	134.0
136.0	2.6047	0.016274	135.55	135.57	103.95	1016.4	1120.3	0.1918	1.7063	.8980	136.0
138.0	2.7438	0.016284	129.09	129.11	105.95	1015.2	1121.1	0.1951	1.6986	.8937	138.0
140.0	2.8892	0.016293	122.98	123.00	107.95	1014.0	1122.0	0.1985	1.6910	.8895	140.0
142.0	3.0411	0.016303	117.21	117.22	109.95	1012.9	1122.8	0.2018	1.6834	.8852	142.0
144.0	3.1997	0.016312	111.74	111.76	111.95	1011.7	1123.6	0.2051	1.6759	.8810	144.0
146.0	3.3653	0.016322	106.58	106.59	113.95	1010.5	1124.5	0.2084	1.6684	.8769	146.0
148.0	3.5381	0.016332	101.68	101.70	115.95	1009.3	1125.3	0.2117	1.6610	.8727	148.0
150.0	3.7184	0.016343	97.05	97.07	117.95	1008.2	1126.1	0.2150	1.6536	.8686	150.0
152.0	3.9065	0.016353	92.66	92.68	119.95	1007.0	1126.9	0.2183	1.6463	.8646	152.0
154.0	4.1025	0.016363	88.50	88.52	121.95	1005.8	1127.7	0.2216	1.6390	.8606	154.0
156.0	4.3068	0.016374	84.56	84.57	123.95	1004.6	1128.6	0.2248	1.6318	.8566	156.0
158.0	4.5197	0.016384	80.82	80.83	125.96	1003.4	1129.4	0.2281	1.6245	.8526	158.0
160.0	4.7414	0.016395	77.27	77.29	127.96	1002.2	1130.2	0.2313	1.6174	.8487	160.0
162.0	4.9722	0.016406	73.90	73.92	129.96	1001.0	1131.0	0.2345	1.6103	.8448	162.0
164.0	5.2124	0.016417	70.70	70.72	131.96	999.8	1131.8	0.2377	1.6032	.8409	164.0
166.0	5.4623	0.016428	67.67	67.68	133.97	998.6	1132.6	0.2409	1.5961	.8371	166.0
168.0	5.7223	0.016440	64.78	64.80	135.97	997.4	1133.4	0.2441	1.5892	.8333	168.0
170.0	5.9926	0.016451	62.04	62.06	137.97	996.2	1134.2	0.2473	1.5822	.8295	170.0
172.0	6.2736	0.016463	59.45	59.47	139.98	995.0	1135.0	0.2505	1.5753	.8258	172.0
174.0	6.5656	0.016474	56.95	56.97	141.98	993.8	1135.8	0.2537	1.5684	.8221	174.0
176.0	6.8690	0.016486	54.59	54.61	143.99	992.6	1136.6	0.2568	1.5616	.8184	176.0
178.0	7.1840	0.016498	52.35	52.36	145.99	991.4	1137.4	0.2600	1.5548	.8147	178.0

Table B.1 Saturated Steam: Temperature Table (*Continued*)

Temp Fahr t	Abs Press. Lb per Sq In. p	Specific Volume			Enthalpy			Entropy			Temp Fahr t
		Sat. Liquid v_f	Evap v_{fg}	Sat. Vapor v_g	Sat. Liquid h_f	Evap h_{fg}	Sat. Vapor h_g	Sat. Liquid s_f	Evap s_{fg}	Sat. Vapor s_g	
180.0	7.5110	0.016510	50.21	50.22	148.00	990.2	1138.2	0.2631	1.5480	1.8111	**180.0**
182.0	7.850	0.016522	48.172	48.189	150.01	989.0	1139.0	0.2662	1.5413	1.8075	**182.0**
184.0	8.203	0.016534	46.232	46.249	152.01	987.8	1139.8	0.2694	1.5346	1.8040	**184.0**
186.0	8.568	0.016547	44.383	44.400	154.02	986.5	1140.5	0.2725	1.5279	1.8004	**186.0**
188.0	8.947	0.016559	42.621	42.638	156.03	985.3	1141.3	0.2756	1.5213	1.7969	**188.0**
190.0	9.340	0.016572	40.941	40.957	158.04	984.1	1142.1	0.2787	1.5148	1.7934	**190.0**
192.0	9.747	0.016585	39.337	39.354	160.05	982.8	1142.9	0.2818	1.5082	1.7900	**192.0**
194.0	10.168	0.016598	37.808	37.824	162.05	981.6	1143.7	0.2848	1.5017	1.7865	**194.0**
196.0	10.605	0.016611	36.348	36.364	164.06	980.4	1144.4	0.2879	1.4952	1.7831	**196.0**
198.0	11.058	0.016624	34.954	34.970	166.08	979.1	1145.2	0.2910	1.4888	1.7798	**198.0**
200.0	11.526	0.016637	33.622	33.639	168.09	977.9	1146.0	0.2940	1.4824	1.7764	**200.0**
204.0	12.512	0.016664	31.135	31.151	172.11	975.4	1147.5	0.3001	1.4697	1.7698	**204.0**
208.0	13.568	0.016691	28.862	28.878	176.14	972.8	1149.0	0.3061	1.4571	1.7632	**208.0**
212.0	14.696	0.016719	26.782	26.799	180.17	970.3	1150.5	0.3121	1.4447	1.7568	**212.0**
216.0	15.901	0.016747	24.878	24.894	184.20	967.8	1152.0	0.3181	1.4323	1.7505	**216.0**
220.0	17.186	0.016775	23.131	23.148	188.23	965.2	1153.4	0.3241	1.4201	1.7442	**220.0**
224.0	18.556	0.016805	21.529	21.545	192.27	962.6	1154.9	0.3300	1.4081	1.7380	**224.0**
228.0	20.015	0.016834	20.056	20.073	196.31	960.0	1156.3	0.3359	1.3961	1.7320	**228.0**
232.0	21.567	0.016864	18.701	18.718	200.35	957.4	1157.8	0.3417	1.3842	1.7260	**232.0**
236.0	23.216	0.016895	17.454	17.471	204.40	954.8	1159.2	0.3476	1.3725	1.7201	**236.0**
240.0	24.968	0.016926	16.304	16.321	208.45	952.1	1160.6	0.3533	1.3609	1.7142	**240.0**
244.0	26.826	0.016958	15.243	15.260	212.50	949.5	1162.0	0.3591	1.3494	1.7085	**244.0**
248.0	28.796	0.016990	14.264	14.281	216.56	946.8	1163.4	0.3649	1.3379	1.7028	**248.0**
252.0	30.883	0.017022	13.358	13.375	220.62	944.1	1164.7	0.3706	1.3266	1.6972	**252.0**
256.0	33.091	0.017055	12.520	12.538	224.69	941.4	1166.1	0.3763	1.3154	1.6917	**256.0**
260.0	35.427	0.017089	11.745	11.762	228.76	938.6	1167.4	0.3819	1.3043	1.6862	**260.0**
264.0	37.894	0.017123	11.025	11.042	232.83	935.9	1168.7	0.3876	1.2933	1.6808	**264.0**
268.0	40.500	0.017157	10.358	10.375	236.91	933.1	1170.0	0.3932	1.2823	1.6755	**268.0**
272.0	43.249	0.017193	9.738	9.755	240.99	930.3	1171.3	0.3987	1.2715	1.6702	**272.0**
276.0	46.147	0.017228	9.162	9.180	245.08	927.5	1172.5	0.4043	1.2607	1.6650	**276.0**
280.0	49.200	0.017264	8.627	8.644	249.17	924.6	1173.8	0.4098	1.2501	1.6599	**280.0**
284.0	52.414	0.01730	8.1280	8.1453	253.3	921.7	1175.0	0.4154	1.2395	1.6548	**284.0**
288.0	55.795	0.01734	7.6634	7.6807	257.4	918.8	1176.2	0.4208	1.2290	1.6498	**288.0**
292.0	59.350	0.01738	7.2301	7.2475	261.5	915.9	1177.4	0.4263	1.2186	1.6449	**292.0**
296.0	63.084	0.01741	6.8259	6.8433	265.6	913.0	1178.6	0.4317	1.2082	1.6400	**296.0**

300.0	1.6351	1.1979	0.4372	1179.7	910.0	269.7
304.0	1.6303	1.1877	0.4426	1180.9	907.0	273.8
308.0	1.6256	1.1776	0.4479	1182.0	904.0	278.0
312.0	1.6209	1.1676	0.4533	1183.1	901.0	282.1
316.0	1.6162	1.1576	0.4586	1184.1	897.9	286.3
320.0	1.6116	1.1477	0.4640	1185.2	894.8	290.4
324.0	1.6071	1.1378	0.4692	1186.2	891.6	294.6
328.0	1.6025	1.1280	0.4745	1187.2	888.5	298.7
332.0	1.5981	1.1183	0.4798	1188.2	885.3	302.9
336.0	1.5936	1.1086	0.4850	1189.1	882.1	307.1
340.0	1.5892	1.0990	0.4902	1190.1	878.8	311.3
344.0	1.5849	1.0894	0.4954	1191.0	875.5	315.5
348.0	1.5806	1.0799	0.5006	1191.1	872.2	319.7
352.0	1.5763	1.0705	0.5058	1192.7	868.9	323.9
356.0	1.5721	1.0611	0.5110	1193.6	865.5	328.1
360.0	1.5678	1.0517	0.5161	1194.4	862.1	332.3
364.0	1.5637	1.0424	0.5212	1195.2	858.6	336.5
368.0	1.5595	1.0332	0.5263	1195.9	855.1	340.8
372.0	1.5554	1.0240	0.5314	1196.7	851.6	345.0
376.0	1.5513	1.0148	0.5365	1197.4	848.1	349.3
380.0	1.5473	1.0057	0.5416	1198.0	844.5	353.6
384.0	1.5432	0.9966	0.5466	1198.7	840.8	357.9
388.0	1.5392	0.9876	0.5516	1199.3	837.2	362.2
392.0	1.5352	0.9786	0.5567	1199.9	833.4	366.5
396.0	1.5313	0.9696	0.5617	1200.4	829.7	370.8
400.0	1.5274	0.9607	0.5667	1201.0	825.9	375.1
404.0	1.5234	0.9518	0.5717	1201.5	822.0	379.4
408.0	1.5195	0.9429	0.5766	1201.9	818.2	383.8
412.0	1.5157	0.9341	0.5816	1202.4	814.2	388.1
416.0	1.5118	0.9253	0.5866	1202.8	810.2	392.5
420.0	1.5080	0.9165	0.5915	1203.1	806.2	396.9
424.0	1.5042	0.9077	0.5964	1203.5	802.2	401.3
428.0	1.5004	0.8990	0.6014	1203.7	798.0	405.7
432.0	1.4966	0.8903	0.6063	1204.0	793.9	410.1
436.0	1.4928	0.8816	0.6112	1204.2	789.7	414.6
440.0	1.4890	0.8729	0.6161	1204.4	785.4	419.0
444.0	1.4853	0.8643	0.6210	1204.6	781.1	423.5
448.0	1.4815	0.8557	0.6259	1204.7	776.7	428.0
452.0	1.4778	0.8471	0.6308	1204.8	772.3	432.5
456.0	1.4741	0.8385	0.6356	1204.8	767.8	437.0

300.0	6.4658	6.4483	0.01745	67.005	**300.0**
304.0	6.1130	6.0955	0.01749	71.119	**304.0**
308.0	5.7830	5.7655	0.01753	75.433	**308.0**
312.0	5.4742	5.4566	0.01757	79.953	**312.0**
316.0	5.1849	5.1673	0.01761	84.688	**316.0**
320.0	4.9138	4.8961	0.01766	89.643	**320.0**
324.0	4.6595	4.6418	0.01770	94.826	**324.0**
328.0	4.4208	4.4030	0.01774	100.245	**328.0**
332.0	4.1966	4.1788	0.01779	105.907	**332.0**
336.0	3.9859	3.9681	0.01783	111.820	**336.0**
340.0	3.7878	3.7699	0.01787	117.992	**340.0**
344.0	3.6013	3.5834	0.01792	124.430	**344.0**
348.0	3.4258	3.4078	0.01797	131.142	**348.0**
352.0	3.2603	3.2423	0.01801	138.138	**352.0**
356.0	3.1044	3.0863	0.01806	145.424	**356.0**
360.0	2.9573	2.9392	0.01811	153.010	**360.0**
364.0	2.8184	2.8002	0.01816	160.903	**364.0**
368.0	2.6873	2.6691	0.01821	169.113	**368.0**
372.0	2.5633	2.5451	0.01826	177.648	**372.0**
376.0	2.4462	2.4279	0.01831	186.517	**376.0**
380.0	2.3353	2.3170	0.01836	195.729	**380.0**
384.0	2.2304	2.2120	0.01842	205.294	**384.0**
388.0	2.1311	2.1126	0.01847	215.220	**388.0**
392.0	2.0369	2.0184	0.01853	225.516	**392.0**
396.0	1.9477	1.9291	0.01858	236.193	**396.0**
400.0	1.8630	1.8444	0.01864	247.259	**400.0**
404.0	1.7827	1.7640	0.01870	258.725	**404.0**
408.0	1.7064	1.6877	0.01875	270.600	**408.0**
412.0	1.6340	1.6152	0.01881	282.894	**412.0**
416.0	1.5651	1.5463	0.01887	295.617	**416.0**
420.0	1.4997	1.4808	0.01894	308.780	**420.0**
424.0	1.4374	1.4184	0.01900	322.391	**424.0**
428.0	1.3782	1.3591	0.01906	336.463	**428.0**
432.0	1.32179	1.30266	0.01913	351.00	**432.0**
436.0	1.26806	1.24887	0.01919	366.03	**436.0**
440.0	1.21687	1.19761	0.01926	381.54	**440.0**
444.0	1.16806	1.14874	0.01933	397.56	**444.0**
448.0	1.12152	1.10212	0.01940	414.09	**448.0**
452.0	1.07711	1.05764	0.01947	431.14	**452.0**
456.0	1.03472	1.01518	0.01954	448.73	**456.0**

Table B.1 Saturated Steam: Temperature Table (Continued)

Temp Fahr t	Abs Press. Lb per Sq In. p	Specific Volume			Enthalpy			Entropy			Temp Fahr t
		Sat. Liquid v_f	Evap v_{fg}	Sat. Vapor v_g	Sat. Liquid h_f	Evap h_{fg}	Sat. Vapor h_g	Sat. Liquid s_f	Evap s_{fg}	Sat. Vapor s_g	
460.0	466.87	0.01961	0.97463	0.99424	441.5	763.2	1204.8	0.6405	0.8299	1.4704	460.0
464.0	485.56	0.01969	0.93588	0.95557	446.1	758.6	1204.7	0.6454	0.8213	1.4667	464.0
468.0	504.83	0.01976	0.89885	0.91862	450.7	754.0	1204.6	0.6502	0.8127	1.4629	468.0
472.0	524.67	0.01984	0.86345	0.88329	455.2	749.3	1204.5	0.6551	0.8042	1.4592	472.0
476.0	545.11	0.01992	0.82958	0.84950	459.9	744.5	1204.3	0.6599	0.7956	1.4555	476.0
480.0	566.15	0.02000	0.79716	0.81717	464.5	739.6	1204.1	0.6648	0.7871	1.4518	480.0
484.0	587.81	0.02009	0.76613	0.78622	469.1	734.7	1203.8	0.6696	0.7785	1.4481	484.0
488.0	610.10	0.02017	0.73641	0.75658	473.8	729.7	1203.5	0.6745	0.7700	1.4444	488.0
492.0	633.03	0.02026	0.70794	0.72820	478.5	724.6	1203.1	0.6793	0.7614	1.4407	492.0
496.0	656.61	0.02034	0.68065	0.70100	483.2	719.5	1202.7	0.6842	0.7528	1.4370	496.0
500.0	680.86	0.02043	0.65448	0.67492	487.9	714.3	1202.2	0.6890	0.7443	1.4333	500.0
504.0	705.78	0.02053	0.62938	0.64991	492.7	709.0	1201.7	0.6939	0.7357	1.4296	504.0
508.0	731.40	0.02062	0.60530	0.62592	497.5	703.7	1201.1	0.6987	0.7271	1.4258	508.0
512.0	757.72	0.02072	0.58218	0.60289	502.3	698.2	1200.5	0.7036	0.7185	1.4221	512.0
516.0	784.76	0.02081	0.55997	0.58079	507.1	692.7	1199.8	0.7085	0.7099	1.4183	516.0
520.0	812.53	0.02091	0.53864	0.55956	512.0	687.0	1199.0	0.7133	0.7013	1.4146	520.0
524.0	841.04	0.02102	0.51814	0.53916	516.9	681.3	1198.2	0.7182	0.6926	1.4108	524.0
528.0	870.31	0.02112	0.49843	0.51955	521.8	675.5	1197.3	0.7231	0.6839	1.4070	528.0
532.0	900.34	0.02123	0.47947	0.50070	526.8	669.6	1196.4	0.7280	0.6752	1.4032	532.0
536.0	931.17	0.02134	0.46123	0.48257	531.7	663.6	1195.4	0.7329	0.6665	1.3993	536.0
540.0	962.79	0.02146	0.44367	0.46513	536.8	657.5	1194.3	0.7378	0.6577	1.3954	540.0
544.0	995.22	0.02157	0.42677	0.44834	541.8	651.3	1193.1	0.7427	0.6489	1.3915	544.0
548.0	1028.49	0.02169	0.41048	0.43217	546.9	645.0	1191.9	0.7476	0.6400	1.3876	548.0
552.0	1062.59	0.02182	0.39479	0.41660	552.0	638.5	1190.6	0.7525	0.6311	1.3837	552.0
556.0	1097.55	0.02194	0.37966	0.40160	557.2	632.0	1189.2	0.7575	0.6222	1.3797	556.0
560.0	1133.38	0.02207	0.36507	0.38714	562.4	625.3	1187.7	0.7625	0.6132	1.3757	560.0
564.0	1170.10	0.02221	0.35099	0.37320	567.6	618.5	1186.1	0.7674	0.6041	1.3716	564.0
568.0	1207.72	0.02235	0.33741	0.35975	572.9	611.5	1184.5	0.7725	0.5950	1.3675	568.0
572.0	1246.26	0.02249	0.32429	0.34678	578.3	604.5	1182.7	0.7775	0.5859	1.3634	572.0
576.0	1285.74	0.02264	0.31162	0.33426	583.7	597.2	1180.9	0.7825	0.5766	1.3592	576.0
580.0	1326.17	0.02279	0.29937	0.32216	589.1	589.9	1179.0	0.7876	0.5673	1.3550	580.0
584.0	1367.7	0.02295	0.28753	0.31048	594.6	582.4	1176.9	0.7927	0.5580	1.3507	584.0
588.0	1410.0	0.02311	0.27608	0.29919	600.1	574.7	1174.8	0.7978	0.5485	1.3464	588.0
592.0	1453.3	0.02328	0.26499	0.28827	605.7	566.8	1172.6	0.8030	0.5390	1.3420	592.0
596.0	1497.8	0.02345	0.25425	0.27770	611.4	558.8	1170.2	0.8082	0.5293	1.3375	596.0

T	P	v_f	v_{fg}	v_g	h_f	h_{fg}	h_g	s_f	s_{fg}	s_g	T
600.0	1543.2	0.02364	0.24384	0.26747	617.1	550.6	1167.7	0.8134	0.5196	1.3330	600.0
604.0	1589.7	0.02382	0.23374	0.25757	622.9	542.2	1165.1	0.8187	0.5097	1.3284	604.0
608.0	1637.3	0.02402	0.22394	0.24796	628.8	533.6	1162.4	0.8240	0.4997	1.3238	608.0
612.0	1686.1	0.02422	0.21442	0.23865	634.8	524.7	1159.5	0.8294	0.4896	1.3190	612.0
616.6	1735.9	0.02444	0.20516	0.22960	640.8	515.6	1156.4	0.8348	0.4794	1.3141	616.0
620.0	1786.9	0.02466	0.19615	0.22081	646.9	506.3	1153.2	0.8403	0.4689	1.3092	620.0
624.0	1839.0	0.02489	0.18737	0.21226	653.1	496.6	1149.8	0.8458	0.4583	1.3041	624.0
628.0	1892.4	0.02514	0.17880	0.20394	659.5	486.7	1146.1	0.8514	0.4474	1.2988	628.0
632.0	1947.0	0.02539	0.17044	0.19583	665.9	476.4	1142.2	0.8571	0.4364	1.2934	632.0
636.0	2002.8	0.02566	0.16226	0.18792	672.4	465.7	1138.1	0.8628	0.4251	1.2879	636.0
640.0	2059.9	0.02595	0.15427	0.18021	679.1	454.6	1133.7	0.8686	0.4134	1.2821	640.0
644.0	2118.3	0.02625	0.14644	0.17269	685.9	443.1	1129.0	0.8746	0.4015	1.2761	644.0
648.0	2178.1	0.02657	0.13876	0.16534	692.9	431.1	1124.0	0.8806	0.3893	1.2699	648.0
652.0	2239.2	0.02691	0.13124	0.15816	700.0	418.7	1118.7	0.8868	0.3767	1.2634	652.0
656.0	2301.7	0.02728	0.12387	0.15115	707.4	405.7	1113.1	0.8931	0.3637	1.2567	656.0
660.0	2365.7	0.02768	0.11663	0.14431	714.9	392.1	1107.0	0.8995	0.3502	1.2498	660.0
664.0	2431.1	0.02811	0.10947	0.13757	722.9	377.7	1100.6	0.9064	0.3361	1.2425	664.0
668.0	2498.1	0.02858	0.10229	0.13087	731.5	362.1	1093.5	0.9137	0.3210	1.2347	668.0
672.0	2566.6	0.02911	0.09514	0.12424	740.2	345.7	1085.9	0.9212	0.3054	1.2266	672.0
676.0	2636.8	0.02970	0.08799	0.11769	749.2	328.5	1077.6	0.9287	0.2892	1.2179	676.0
680.0	2708.6	0.03037	0.08080	0.11117	758.5	310.1	1068.5	0.9365	0.2720	1.2086	680.0
684.0	2782.1	0.03114	0.07349	0.10463	768.2	290.2	1058.4	0.9447	0.2537	1.1984	684.0
688.0	2857.4	0.03204	0.06595	0.09799	778.8	268.2	1047.0	0.9535	0.2337	1.1872	688.0
692.0	2934.5	0.03313	0.05797	0.09110	790.5	243.1	1033.6	0.9634	0.2110	1.1744	692.0
696.0	3013.4	0.03455	0.04916	0.08371	804.4	212.8	1017.2	0.9749	0.1841	1.1591	696.0
700.0	3094.3	0.03662	0.03857	0.07519	822.4	172.7	995.2	0.9901	0.1490	1.1390	700.0
702.0	3135.5	0.03824	0.03173	0.06997	835.0	144.7	979.7	1.0006	0.1246	1.1252	702.0
704.0	3177.2	0.04108	0.02192	0.06300	854.2	102.0	956.2	1.0169	0.0876	1.1046	704.0
705.0	3198.3	0.04427	0.01304	0.05730	873.0	61.4	934.4	1.0329	0.0527	1.0856	705.0
705.47*	3208.2	0.05078	0.00000	0.05078	906.0	0.0	906.0	1.0612	0.0000	1.0612	705.47†

*The states shown are meta stable.

†Critical temperature.

v = specific volume, ft^3/lb_m

h = enthalpy, Btu/lb.

s = entropy, Btu/R/lb.

Source: Values reprinted, by permission, from *1967 ASME Steam Tables*, C. A. Meyer et al., 1968. Copyright © 1968 by the American Society of Mechanical Engineers.

Table B.2 Saturated Steam: Pressure Table

Abs Press. Lb/Sq In. p	Temp Fahr t	Specific Volume			Enthalpy			Entropy			Abs Press. Lb/Sq In. p
		Sat. Liquid v_f	Evap v_{fg}	Sat. Vapor v_g	Sat. Liquid h_f	Evap h_{fg}	Sat. Vapor h_g	Sat. Liquid s_f	Evap s_{fg}	Sat. Vapor s_g	
0.08865	32.018	0.016022	3302.4	3302.4	0.0003	1075.5	1075.5	0.0000	2.1872	2.1872	0.08865
0.25	59.323	0.016032	1235.5	1235.5	27.382	1060.1	1087.4	0.0542	2.0425	2.0967	0.25
0.50	79.586	0.016071	641.5	641.5	47.623	1048.6	1096.3	0.0925	1.9446	2.0370	0.50
1.0	101.74	0.016136	333.59	333.60	69.73	1036.1	1105.8	0.1326	1.8455	1.9781	1.0
5.0	162.24	0.016407	73.515	73.532	130.20	1000.9	1131.1	0.2349	1.6094	1.8443	5.0
10.0	193.21	0.016592	38.404	38.420	161.26	982.1	1143.3	0.2836	1.5043	1.7879	10.0
14.696	212.00	0.016719	26.782	26.799	180.17	970.3	1150.5	0.3121	1.4447	1.7568	14.696
15.0	213.03	0.016726	26.274	26.290	181.21	969.7	1150.9	0.3137	1.4415	1.7552	15.0
20.0	227.96	0.016834	20.070	20.087	196.27	960.1	1156.3	0.3358	1.3962	1.7320	20.0
30.0	250.34	0.017009	13.7266	13.7436	218.9	945.2	1164.1	0.3682	1.3313	1.6995	30.0
40.0	267.25	0.017151	10.4794	10.4965	236.1	933.6	1169.8	0.3921	1.2844	1.6765	40.0
50.0	281.02	0.017274	8.4967	8.5140	250.2	923.9	1174.1	0.4112	1.2474	1.6586	50.0
60.0	292.71	0.017383	7.1562	7.1736	262.2	915.4	1177.6	0.4273	1.2167	1.6440	60.0
70.0	302.93	0.017482	6.1875	6.2050	272.7	907.8	1180.6	0.4411	1.1905	1.6316	70.0
80.0	312.04	0.017573	5.4536	5.4711	282.1	900.9	1183.1	0.4534	1.1675	1.6208	80.0
90.0	320.28	0.017659	4.8779	4.8953	290.7	894.6	1185.3	0.4643	1.1470	1.6113	90.0
100.0	327.82	0.017740	4.4133	4.4310	298.5	888.6	1187.2	0.4743	1.1284	1.6027	100.0
110.0	334.79	0.01782	4.0306	4.0484	305.8	883.1	1188.9	0.4834	1.1115	1.5950	110.0
120.0	341.27	0.01789	3.7097	3.7275	312.6	877.8	1190.4	0.4919	1.0960	1.5879	120.0
130.0	347.33	0.01796	3.4364	3.4544	319.0	872.8	1191.7	0.4998	1.0815	1.5813	130.0
140.0	353.04	0.01803	3.2010	3.2190	325.0	868.0	1193.0	0.5071	1.0681	1.5752	140.0
150.0	358.43	0.01809	2.9958	3.0139	330.6	863.4	1194.1	0.5141	1.0554	1.5695	150.0
160.0	363.55	0.01815	2.8155	2.8336	336.1	859.0	1195.1	0.5206	1.0435	1.5641	160.0
170.0	368.42	0.01821	2.6556	2.6738	341.2	854.8	1196.0	0.5269	1.0322	1.5591	170.0
180.0	373.08	0.01827	2.5129	2.5312	346.2	850.7	1196.9	0.5328	1.0215	1.5543	180.0
190.0	377.53	0.01833	2.3847	2.4030	350.9	846.7	1197.6	0.5384	1.0113	1.5498	190.0
200.0	381.80	0.01839	2.2689	2.2873	355.5	842.8	1198.3	0.5438	1.0016	1.5454	200.0
210.0	385.91	0.01844	2.16373	2.18217	359.9	839.1	1199.0	0.5490	0.9923	1.5413	210.0
220.0	389.88	0.01850	2.06779	2.08629	364.2	835.4	1199.6	0.5540	0.9834	1.5374	220.0
230.0	393.70	0.01855	1.97991	1.99846	368.3	831.8	1200.1	0.5588	0.9748	1.5336	230.0
240.0	397.39	0.01860	1.89909	1.91769	372.3	828.4	1200.6	0.5634	0.9665	1.5299	240.0
250.0	400.97	0.01865	1.82452	1.84317	376.1	825.0	1201.1	0.5679	0.9585	1.5264	250.0
260.0	404.44	0.01870	1.75548	1.77418	379.9	821.6	1201.5	0.5722	0.9508	1.5230	260.0
270.0	407.80	0.01875	1.69137	1.71013	383.6	818.3	1201.9	0.5764	0.9433	1.5197	270.0
280.0	411.07	0.01880	1.63169	1.65049	387.1	815.1	1202.3	0.5805	0.9361	1.5166	280.0
290.0	414.25	0.01885	1.57597	1.59482	390.6	812.0	1202.6	0.5844	0.9291	1.5135	290.0
300.0	417.35	0.01889	1.52384	1.54274	394.0	808.9	1202.9	0.5882	0.9223	1.5105	300.0
350.0	431.73	0.01912	1.30642	1.32554	409.8	794.2	1204.0	0.6059	0.8909	1.4968	350.0
400.0	444.60	0.01934	1.14162	1.16095	424.2	780.4	1204.6	0.6217	0.8630	1.4847	400.0

450.0	456.28	0.01954	1.01224	1.03179	437.3	767.5	1204.8	0.6360	0.8378	1.4738	450.0
500.0	467.01	0.01975	0.90787	0.92762	449.5	755.1	1204.7	0.6490	0.8148	1.4639	500.0
550.0	476.94	0.01994	0.82183	0.84177	460.9	743.3	1204.3	0.6611	0.7936	1.4547	550.0
600.0	486.20	0.02013	0.74962	0.76975	471.7	732.0	1203.7	0.6723	0.7738	1.4461	600.0
650.0	494.89	0.02032	0.68811	0.70843	481.9	720.9	1202.8	0.6828	0.7552	1.4381	650.0
700.0	503.08	0.02050	0.63505	0.65556	491.6	710.2	1201.8	0.6928	0.7377	1.4304	700.0
750.0	510.84	0.02069	0.58880	0.60949	500.9	699.8	1200.7	0.7022	0.7210	1.4232	750.0
800.0	518.21	0.02087	0.54809	0.56896	509.8	689.6	1199.4	0.7111	0.7051	1.4163	800.0
850.0	525.24	0.02105	0.51197	0.53302	518.4	679.5	1198.0	0.7197	0.6899	1.4096	850.0
900.0	531.95	0.02123	0.47968	0.50091	526.7	669.7	1196.4	0.7279	0.6753	1.4032	900.0
950.0	538.39	0.02141	0.45064	0.47205	534.7	660.0	1194.7	0.7358	0.6612	1.3970	950.0
1000.0	544.58	0.02159	0.42436	0.44596	542.6	650.4	1192.9	0.7434	0.6476	1.3910	1000.0
1050.0	550.53	0.02177	0.40047	0.42224	550.1	640.9	1191.0	0.7507	0.6344	1.3851	1050.0
1100.0	556.28	0.02195	0.37863	0.40058	557.5	631.5	1189.1	0.7578	0.6216	1.3794	1100.0
1150.0	561.82	0.02214	0.35859	0.38073	564.8	622.2	1187.0	0.7647	0.6091	1.3738	1150.0
1200.0	567.19	0.02232	0.34013	0.36245	571.9	613.0	1184.8	0.7714	0.5969	1.3683	1200.0
1250.0	572.38	0.02250	0.32306	0.34556	578.8	603.8	1182.6	0.7780	0.5850	1.3630	1250.0
1300.0	577.42	0.02269	0.30722	0.32991	585.6	594.6	1180.2	0.7843	0.5733	1.3577	1300.0
1350.0	582.32	0.02288	0.29250	0.31537	592.3	585.4	1177.8	0.7906	0.5620	1.3525	1350.0
1400.0	587.07	0.02307	0.27871	0.30178	598.8	576.5	1175.3	0.7966	0.5507	1.3474	1400.0
1450.0	591.70	0.02327	0.26584	0.28911	605.3	567.4	1172.8	0.8026	0.5397	1.3423	1450.0
1500.0	596.20	0.02346	0.25372	0.27719	611.7	558.4	1170.1	0.8085	0.5288	1.3373	1500.0
1550.0	600.59	0.02366	0.24235	0.26601	618.0	549.4	1167.4	0.8142	0.5182	1.3324	1550.0
1600.0	604.87	0.02387	0.23159	0.25545	624.2	540.3	1164.5	0.8199	0.5076	1.3274	1600.0
1650.0	609.05	0.02407	0.22143	0.24551	630.4	531.3	1161.6	0.8254	0.4971	1.3225	1650.0
1700.0	613.13	0.02428	0.21178	0.23607	636.5	522.2	1158.6	0.8309	0.4867	1.3176	1700.0
1750.0	617.12	0.02450	0.20263	0.22713	642.5	513.1	1155.6	0.8363	0.4765	1.3128	1750.0
1800.0	621.02	0.02472	0.19390	0.21861	648.5	503.8	1152.3	0.8417	0.4662	1.3079	1800.0
1850.0	624.83	0.02495	0.18558	0.21052	654.5	494.6	1149.0	0.8470	0.4561	1.3030	1850.0
1900.0	628.56	0.02517	0.17761	0.20278	660.4	485.2	1145.6	0.8522	0.4459	1.2981	1900.0
1950.0	632.22	0.02541	0.16999	0.19540	666.3	475.8	1142.0	0.8574	0.4358	1.2931	1950.0
2000.0	635.80	0.02565	0.16266	0.18831	672.1	466.2	1138.3	0.8625	0.4256	1.2881	2000.0
2100.0	642.76	0.02615	0.14885	0.17501	683.8	446.7	1130.5	0.8727	0.4053	1.2780	2100.0
2200.0	649.45	0.02669	0.13603	0.16272	695.5	426.7	1122.2	0.8828	0.3848	1.2676	2200.0
2300.0	655.89	0.02727	0.12406	0.15133	707.2	406.0	1113.2	0.8929	0.3640	1.2569	2300.0
2400.0	662.11	0.02790	0.11287	0.14076	719.0	384.8	1103.7	0.9031	0.3430	1.2460	2400.0
2500.0	668.11	0.02859	0.10209	0.13068	731.7	361.6	1093.3	0.9139	0.3206	1.2345	2500.0
2600.0	673.91	0.02938	0.09172	0.12110	744.5	337.6	1082.0	0.9247	0.2977	1.2225	2600.0
2700.0	679.53	0.03029	0.08165	0.11194	757.3	312.3	1069.7	0.9356	0.2741	1.2097	2700.0
2800.0	684.96	0.03134	0.07171	0.10305	770.7	285.1	1055.8	0.9468	0.2491	1.1958	2800.0
2900.0	690.22	0.03262	0.06158	0.09420	785.1	254.7	1039.8	0.9588	0.2215	1.1803	2900.0
3000.0	695.33	0.03428	0.05073	0.08500	801.8	218.4	1020.3	0.9728	0.1891	1.1619	3000.0
3100.0	700.28	0.03681	0.03771	0.07452	824.0	169.3	993.3	0.9914	0.1460	1.1373	3100.0
3200.0	705.08	0.04472	0.01191	0.05663	875.5	56.1	931.6	1.0351	0.0482	1.0832	3200.0
3208.2*	705.47	0.05078	0.00000	0.05078	906.0	0.0	906.0	1.0612	0.0000	1.0612	3208.2*

†Critical pressure.

Source: Values reprinted, by permission, from *1967 ASME Steam Tables*, C. A. Meyer et al., 1968. Copyright © 1968 by the American Society of Mechanical Engineers.

Table B.3 Superheated Steam

Abs Press. Lb/Sq In. (Sat. Temp)		Sat. Water	Sat. Steam	Temperature – Degrees Fahrenheit													
				200	250	300	350	400	450	500	600	700	800	900	1000	1100	1200
1 (101.74)	Sh			98.26	148.26	198.26	248.26	298.26	348.26	398.26	498.26	598.26	698.26	798.26	898.26	998.26	1098.26
	v	0.01614	333.6	392.5	422.4	452.3	482.1	511.9	541.7	571.5	631.1	690.7	750.3	809.8	869.4	929.0	988.6
	h	69.73	1105.8	1150.2	1172.9	1195.7	1218.7	1241.8	1265.1	1288.6	1336.1	1384.5	1433.7	1483.8	1534.9	1586.8	1639.7
	s	0.1326	1.9781	2.0509	2.0841	2.1152	2.1445	2.1722	2.1985	2.2237	2.2708	2.3144	2.3551	2.3934	2.4296	2.4640	2.4969
5 (162.24)	Sh			37.76	87.76	137.76	187.76	237.76	287.76	337.76	437.76	537.76	637.76	737.76	837.76	937.76	1037.76
	v	0.01641	73.53	78.14	84.21	90.24	96.25	102.24	108.23	114.21	126.15	138.08	150.01	161.94	173.86	185.78	197.70
	h	130.20	1131.1	1148.6	1171.7	1194.8	1218.0	1241.3	1264.7	1288.2	1335.9	1384.3	1433.6	1483.7	1534.7	1586.7	1639.6
	s	0.2349	1.8443	1.8716	1.9054	1.9369	1.9664	1.9943	2.0208	2.0460	2.0932	2.1369	2.1776	2.2159	2.2521	2.2866	2.3194
10 (193.21)	Sh			6.79	56.79	106.79	156.79	206.79	256.79	306.79	406.79	506.79	606.79	706.79	806.79	906.79	1006.79
	v	0.01659	38.42	38.84	41.93	44.98	48.02	51.03	54.04	57.04	63.03	69.00	74.98	80.94	86.91	92.87	98.84
	h	161.26	1143.3	1146.6	1170.2	1193.7	1217.1	1240.6	1264.1	1287.8	1335.5	1384.0	1433.4	1483.5	1534.6	1586.6	1639.5
	s	0.2836	1.7879	1.7928	1.8273	1.8593	1.8892	1.9173	1.9439	1.9692	2.0166	2.0603	2.1011	2.1394	2.1757	2.2101	2.2430
14.696 (212.00)	Sh				38.00	88.00	138.00	188.00	238.00	288.00	388.00	488.00	588.00	688.00	788.00	888.00	988.00
	v	0.0167	26.799		28.42	30.52	32.60	34.67	36.72	38.77	42.86	46.93	51.00	55.06	59.13	63.19	67.25
	h	180.17	1150.5		1168.8	1192.6	1216.3	1239.9	1263.6	1287.4	1335.2	1383.8	1433.2	1483.4	1534.5	1586.5	1639.4
	s	0.3121	1.7568		1.7833	1.8158	1.8459	1.8743	1.9010	1.9265	1.9739	2.0177	2.0585	2.0969	2.1332	2.1676	2.2005
15 (213.03)	Sh				36.97	86.97	136.97	186.97	236.97	286.97	386.97	486.97	586.97	686.97	786.97	886.97	986.97
	v	0.01673	26.290		27.837	29.899	31.939	33.963	35.977	37.985	41.986	45.978	49.964	53.946	57.926	61.905	65.882
	h	181.21	1150.9		1168.7	1192.5	1216.2	1239.9	1263.6	1287.3	1335.2	1383.8	1433.2	1483.4	1534.5	1586.5	1639.4
	s	0.3137	1.7552		1.7809	1.8134	1.8437	1.8720	1.8988	1.9242	1.9717	2.0155	2.0563	2.0946	2.1309	2.1653	2.1982
20 (227.96)	Sh				22.04	72.04	122.04	172.04	222.04	272.04	372.04	472.04	572.04	672.04	772.04	872.04	972.04
	v	0.01683	20.087		20.788	22.356	23.900	25.428	26.946	28.457	31.466	34.465	37.458	40.447	43.435	46.420	49.405
	h	196.27	1156.3		1167.1	1191.4	1215.4	1239.2	1263.0	1286.9	1334.9	1383.5	1432.9	1483.2	1534.3	1586.3	1639.3
	s	0.3358	1.7320		1.7475	1.7805	1.8111	1.8397	1.8666	1.8921	1.9397	1.9836	2.0244	2.0628	2.0991	2.1336	2.1665
25 (240.07)	Sh				9.93	59.93	109.93	159.93	209.93	259.93	359.93	459.93	559.93	659.93	759.93	859.93	959.93
	v	0.01693	16.301		16.558	17.829	19.076	20.307	21.527	22.740	25.153	27.557	29.954	32.348	34.740	37.130	39.518
	h	208.52	1160.6		1165.6	1190.2	1214.5	1238.5	1262.5	1286.4	1334.6	1383.3	1432.7	1483.0	1534.2	1586.2	1639.2
	s	0.3535	1.7141		1.7212	1.7547	1.7856	1.8145	1.8415	1.8672	1.9149	1.9588	1.9997	2.0381	2.0744	2.1089	2.1418
30 (250.34)	Sh					49.66	99.66	149.66	199.66	249.66	349.66	449.66	549.66	649.66	749.66	849.66	949.66
	v	0.01701	13.744			14.810	15.859	16.892	17.914	18.929	20.945	22.951	24.952	26.949	28.943	30.936	32.927
	h	218.93	1164.1			1189.0	1213.6	1237.8	1261.9	1286.0	1334.2	1383.0	1432.5	1482.8	1534.0	1586.1	1639.0
	s	0.3682	1.6995			1.7334	1.7647	1.7937	1.8210	1.8467	1.8946	1.9386	1.9795	2.0179	2.0543	2.0888	2.1217

Abs Press, Lb/Sq In (Sat Temp)		Sat. Water	Sat. Steam												
35 (259.29)	Sh			40.71	90.71	140.71	190.71	240.71	340.71	440.71	540.71	640.71	740.71	840.71	940.71
	v	0.01708	11.896	12.654	13.562	14.453	15.334	16.207	17.939	19.662	21.379	23.092	24.803	26.512	28.220
	h	228.03	1167.1	1187.8	1212.7	1237.1	1261.3	1285.5	1333.9	1382.8	1432.3	1482.7	1533.9	1586.0	1638.9
	s	0.3809	1.6872	1.7152	1.7468	1.7761	1.8035	1.8294	1.8774	1.9214	1.9624	2.0009	2.0372	2.0717	2.1046
40 (267.25)	Sh			32.75	82.75	132.75	182.75	232.75	332.75	432.75	532.75	632.75	732.75	832.75	932.75
	v	0.01715	10.497	11.036	11.838	12.624	13.398	14.165	15.685	17.195	18.699	20.199	21.697	23.194	24.689
	h	236.14	1169.8	1186.6	1211.7	1236.4	1260.8	1285.0	1333.6	1382.5	1432.1	1482.5	1533.7	1585.8	1638.8
	s	0.3921	1.6765	1.6992	1.7312	1.7608	1.7883	1.8143	1.8624	1.9065	1.9476	1.9860	2.0224	2.0569	2.0899
45 (274.44)	Sh			25.56	75.56	125.56	175.56	225.56	325.56	425.56	525.56	625.56	725.56	825.56	925.56
	v	0.01721	9.399	9.777	10.497	11.201	11.892	12.577	13.932	15.276	16.614	17.950	19.282	20.613	21.943
	h	243.49	1172.1	1185.4	1210.4	1235.7	1260.2	1284.6	1333.3	1382.3	1431.9	1482.3	1533.6	1585.7	1638.7
	s	0.4021	1.6671	1.6849	1.7173	1.7471	1.7748	1.8010	1.8492	1.8934	1.9345	1.9730	2.0093	2.0439	2.0768
50 (281.02)	Sh			18.98	68.98	118.98	168.98	218.98	318.98	418.98	518.98	618.98	718.98	818.98	918.98
	v	0.01727	8.514	8.769	9.424	10.062	10.688	11.306	12.529	13.741	14.947	16.150	17.350	18.549	19.746
	h	250.21	1174.1	1184.1	1209.9	1234.9	1259.6	1284.1	1332.9	1382.0	1431.7	1482.2	1533.4	1585.6	1638.6
	s	0.4112	1.6586	1.6720	1.7048	1.7349	1.7628	1.7890	1.8374	1.8816	1.9227	1.9613	1.9977	2.0322	2.0652
55 (287.07)	Sh			12.93	62.93	112.93	162.93	212.93	312.93	412.93	512.93	612.93	712.93	812.93	912.93
	v	0.01733		7.945	8.546	9.130	9.702	10.267	11.381	12.485	13.583	14.677	15.769	16.859	17.948
	h	256.43		1182.9	1208.9	1234.2	1259.1	1283.6	1332.6	1381.8	1431.5	1482.0	1533.3	1585.5	1638.5
	s	0.4196		1.6601	1.6933	1.7237	1.7518	1.7781	1.8266	1.8710	1.9121	1.9507	1.987	2.022	2.055
60 (292.71)	Sh			7.29	57.29	107.29	157.29	207.29	307.29	407.29	507.29	607.29	707.29	807.29	907.29
	v	0.01738	7.174	7.257	7.815	8.354	8.881	9.400	10.425	11.438	12.446	13.450	14.452	15.452	16.450
	h	262.21	1177.6	1181.6	1208.0	1233.5	1258.5	1283.2	1332.3	1381.5	1431.3	1481.8	1533.2	1585.3	1638.4
	s	0.4273	1.6440	1.6492	1.6829	1.7134	1.7417	1.7681	1.8168	1.8612	1.9024	1.9410	1.9774	2.0120	2.0450
65 (297.98)	Sh			2.02	52.02	102.02	152.02	202.02	302.02	402.02	502.02	602.02	702.02	802.02	902.02
	v	0.01743	6.653	6.675	7.195	7.697	8.186	8.667	9.615	10.552	11.484	12.412	13.337	14.261	15.183
	h	267.63	1179.1	1180.3	1207.0	1232.7	1257.9	1282.7	1331.9	1381.3	1431.1	1481.6	1533.0	1585.2	1638.3
	s	0.4344	1.6375	1.6390	1.6731	1.7040	1.7324	1.7590	1.8077	1.8522	1.8935	1.9321	1.9685	2.0031	2.0361
70 (302.93)	Sh				47.07	97.07	147.07	197.07	297.07	397.07	497.07	597.07	697.07	797.07	897.07
	v	0.01748	6.205		6.664	7.133	7.590	8.039	8.922	9.793	10.659	11.522	12.382	13.240	14.097
	h	272.74	1180.6		1206.0	1232.0	1257.3	1282.2	1331.6	1381.0	1430.9	1481.5	1532.9	1585.1	1638.2
	s	0.4411	1.6316		1.6640	1.6951	1.7237	1.7504	1.7993	1.8439	1.8852	1.9238	1.9603	1.9949	2.0279
75 (307.61)	Sh				42.39	92.39	142.39	192.39	292.39	392.39	492.39	592.39	692.39	792.39	892.39
	v	0.01753	5.814		6.204	6.645	7.074	7.494	8.320	9.135	9.945	10.750	11.553	12.355	13.155
	h	277.56	1181.9		1205.0	1231.2	1256.7	1281.7	1331.3	1380.7	1430.7	1481.3	1532.7	1585.0	1638.1
	s	0.4474	1.6260		1.6554	1.6868	1.7156	1.7424	1.7915	1.8361	1.8774	1.9161	1.9526	1.9872	2.0202

Table B.3 Superheated Steam (Continued)

Abs Press. Lb/Sq In. (Sat. Temp)		Sat Water	Sat Steam	350	400	450	500	550	600	700	800	900	1000	1100	1200	1300	1400
								Temperature – Degrees Fahrenheit								
80 (312.04)	Sh			37.96	87.96	137.96	187.96	237.96	287.96	387.96	487.96	587.96	687.96	787.96	887.96	987.96	1087.96
	v	0.01757	5.471	5.801	6.218	6.622	7.018	7.408	7.794	8.560	9.319	10.075	10.829	11.581	12.331	13.081	13.829
	h	282.15	1183.1	1204.0	1230.5	1256.1	1281.3	1306.2	1330.9	1380.5	1430.5	1481.1	1532.6	1584.9	1638.0	1692.0	1746.8
	s	0.4534	1.6208	1.6473	1.6790	1.7080	1.7349	1.7602	1.7842	1.8289	1.8702	1.9089	1.9454	1.9800	2.0131	2.0446	2.0750
85 (316.26)	Sh			33.74	83.74	133.74	183.74	233.74	283.74	383.74	483.74	583.74	683.74	783.74	883.74	983.74	1083.74
	v	0.01762	5.167	5.445	5.840	6.223	6.597	6.966	7.330	8.052	8.768	9.480	10.190	10.898	11.604	12.310	13.014
	h	286.52	1184.2	1203.0	1229.7	1255.5	1280.8	1305.8	1330.6	1380.2	1430.3	1481.0	1532.4	1584.7	1637.9	1691.9	1746.8
	s	0.4590	1.6159	1.6396	1.6716	1.7008	1.7279	1.7532	1.7772	1.8220	1.8634	1.9021	1.9386	1.9733	2.0063	2.0379	2.0682
90 (320.28)	Sh			29.72	79.72	129.72	179.72	229.72	279.72	379.72	479.72	579.72	679.72	779.72	879.72	979.72	1079.72
	v	0.01766	4.895	5.128	5.505	5.869	6.223	6.572	6.917	7.600	8.277	8.950	9.621	10.290	10.958	11.625	12.290
	h	290.69	1185.3	1202.0	1228.9	1254.9	1280.3	1305.4	1330.2	1380.0	1430.1	1480.8	1532.3	1584.6	1637.8	1691.8	1746.7
	s	0.4643	1.6113	1.6323	1.6646	1.6940	1.7212	1.7467	1.7707	1.8156	1.8570	1.8957	1.9323	1.9669	2.0000	2.0316	2.0619
95 (324.13)	Sh			25.87	75.87	125.87	175.87	225.87	275.87	375.87	475.87	575.87	675.87	775.87	875.87	975.87	1075.87
	v	0.01770	4.651	4.845	5.205	5.551	5.889	6.221	6.548	7.196	7.838	8.477	9.113	9.747	10.380	11.012	11.643
	h	294.70	1186.2	1200.9	1228.1	1254.3	1279.8	1305.0	1329.9	1379.7	1429.9	1480.6	1532.1	1584.5	1637.7	1691.7	1746.6
	s	0.4694	1.6069	1.6253	1.6580	1.6876	1.7149	1.7404	1.7645	1.8094	1.8509	1.8897	1.9262	1.9609	1.9940	2.0256	2.0559
100 (327.82)	Sh			22.18	72.18	122.18	172.18	222.18	272.18	372.18	472.18	572.18	672.18	772.18	872.18	972.18	1072.18
	v	0.01774	4.431	4.590	4.935	5.266	5.588	5.904	6.216	6.833	7.443	8.050	8.655	9.258	9.860	10.460	11.060
	h	298.54	1187.2	1199.9	1227.4	1253.7	1279.3	1304.6	1329.6	1379.5	1429.7	1480.4	1532.0	1584.4	1637.6	1691.6	1746.5
	s	0.4743	1.6027	1.6187	1.6516	1.6814	1.7088	1.7344	1.7586	1.8036	1.8451	1.8839	1.9205	1.9552	1.9883	2.0199	2.0502
110 (334.79)	Sh			15.21	65.21	115.21	165.21	215.21	265.21	365.21	465.21	565.21	665.21	765.21	865.21	965.21	1065.21
	v	0.01782	4.048	4.149	4.468	4.772	5.068	5.357	5.642	6.205	6.761	7.314	7.865	8.413	8.961	9.507	10.053
	h	305.80	1188.9	1197.7	1225.8	1252.5	1278.3	1303.8	1328.9	1379.0	1429.2	1480.1	1531.7	1584.1	1637.4	1691.4	1746.4
	s	0.4834	1.5950	1.6061	1.6396	1.6698	1.6975	1.7233	1.7476	1.7928	1.8344	1.8732	1.9099	1.9446	1.9777	2.0093	2.0397
120 (341.27)	Sh			8.73	58.73	108.73	158.73	208.73	258.73	358.73	458.73	558.73	658.73	758.73	858.73	958.73	1058.73
	v	0.01789	3.7275	3.7815	4.0786	4.3610	4.6341	4.9009	5.1637	5.6813	6.1928	6.7006	7.2060	7.7096	8.2119	8.7130	9.2134
	h	312.58	1190.4	1195.6	1224.1	1251.2	1277.4	1302.9	1328.2	1378.4	1428.8	1479.8	1531.4	1583.9	1637.1	1691.3	1746.2
	s	0.4919	1.5879	1.5943	1.6286	1.6592	1.6872	1.7132	1.7376	1.7829	1.8246	1.8635	1.9001	1.9349	1.9680	1.9996	2.0300
130 (347.33)	Sh			2.67	52.67	102.67	152.67	202.67	252.67	352.67	452.67	552.67	652.67	752.67	852.67	952.67	1052.67
	v	0.01796	3.4544	3.4699	3.7489	4.0129	4.2672	4.5151	4.7589	5.2384	5.7118	6.1814	6.6486	7.1140	7.5781	8.0411	8.5033
	h	318.95	1191.7	1193.4	1222.5	1249.9	1276.4	1302.1	1327.5	1377.9	1428.4	1479.4	1531.1	1583.6	1636.9	1691.1	1746.1
	s	0.4998	1.5813	1.5833	1.6182	1.6493	1.6775	1.7037	1.7283	1.7737	1.8155	1.8545	1.8911	1.9259	1.9591	1.9907	2.0211

Superheated steam table (pressure in psia, saturation temperature in °F; columns headed by degrees of superheat, with v, h, s).

140 psia (353.04 °F)

	Sh	Sat. liq.	Sat. vap.	46.96	96.96	146.96	196.96	246.96	346.96	446.96	546.96	646.96	746.96	846.96	946.96	1046.96
v		0.01803	3.2190	3.4661	3.7143	3.9526	4.1844	4.4119	4.8588	5.2995	5.7364	6.1709	6.6036	7.0349	7.4652	7.8946
h		324.96	1193.0	1220.8	1248.7	1275.3	1301.3	1326.8	1377.4	1428.0	1479.1	1530.8	1583.4	1636.7	1690.9	1745.9
s		0.5071	1.5752	1.6085	1.6400	1.6686	1.6949	1.7196	1.7652	1.8071	1.8461	1.8828	1.9176	1.9508	1.9825	2.0129

150 psia (358.43 °F)

	Sh	Sat. liq.	Sat. vap.	41.57	91.57	141.57	191.57	241.57	341.57	441.57	541.57	641.57	741.57	841.57	941.57	1041.57
v		0.01809	3.0139	3.2208	3.4555	3.6799	3.8978	4.1112	4.5298	4.9421	5.3507	5.7568	6.1612	6.5642	6.9661	7.3671
h		330.65	1194.1	1219.1	1247.4	1274.3	1300.5	1326.1	1376.9	1427.6	1478.7	1530.5	1583.1	1636.5	1690.7	1745.7
s		0.5141	1.5695	1.5993	1.6313	1.6602	1.6867	1.7115	1.7573	1.7992	1.8383	1.8751	1.9099	1.9431	1.9748	2.0052

160 psia (363.55 °F)

	Sh	Sat. liq.	Sat. vap.	36.45	86.45	136.45	186.45	236.45	336.45	436.45	536.45	636.45	736.45	836.45	936.45	1036.45
v		0.01815	2.8336	3.0060	3.2288	3.4413	3.6469	3.8480	4.2420	4.6295	5.0132	5.3945	5.7741	6.1522	6.5293	6.9055
h		336.07	1195.1	1217.4	1246.0	1273.3	1299.6	1325.4	1376.4	1427.2	1478.4	1530.3	1582.9	1636.3	1690.5	1745.6
s		0.5206	1.5641	1.5906	1.6231	1.6522	1.6790	1.7039	1.7499	1.7919	1.8310	1.8678	1.9027	1.9359	1.9676	1.9980

180 psia (373.08 °F)

	Sh	Sat. liq.	Sat. vap.	26.92	76.92	126.92	176.92	226.92	326.92	426.92	526.92	626.92	726.92	826.92	926.92	1026.92
v		0.01827	2.5312	2.6474	2.8508	3.0433	3.2286	3.4093	3.7621	4.1084	4.4508	4.7907	5.1289	5.4657	5.8014	6.1363
h		346.19	1196.9	1213.8	1243.4	1271.2	1297.9	1324.0	1375.3	1426.3	1477.7	1529.7	1582.4	1635.9	1690.2	1745.3
s		0.5328	1.5543	1.5743	1.6078	1.6376	1.6647	1.6900	1.7362	1.7784	1.8176	1.8545	1.8894	1.9227	1.9545	1.9849

200 psia (381.80 °F)

	Sh	Sat. liq.	Sat. vap.	18.20	68.20	118.20	168.20	218.20	318.20	418.20	518.20	618.20	718.20	818.20	918.20	1018.20
v		0.01839	2.2873	2.3598	2.5480	2.7247	2.8939	3.0583	3.3783	3.6915	4.0008	4.3077	4.6128	4.9165	5.2191	5.5209
h		355.51	1198.3	1210.1	1240.6	1269.0	1296.2	1322.6	1374.3	1425.5	1477.0	1529.1	1581.9	1635.4	1689.8	1745.0
s		0.5438	1.5454	1.5593	1.5938	1.6242	1.6518	1.6773	1.7239	1.7663	1.8057	1.8426	1.8776	1.9109	1.9427	1.9732

220 psia (389.88 °F)

	Sh	Sat. liq.	Sat. vap.	10.12	60.12	110.12	160.12	210.12	310.12	410.12	510.12	610.12	710.12	810.12	910.12	1010.12
v		0.01850	2.0863	2.1240	2.2999	2.4638	2.6199	2.7710	3.0642	3.3504	3.6327	3.9125	4.1905	4.4671	4.7426	5.0173
h		364.17	1199.6	1206.3	1237.8	1266.9	1294.5	1321.2	1373.2	1424.7	1476.3	1528.5	1581.4	1635.0	1689.4	1744.7
s		0.5540	1.5374	1.5453	1.5808	1.6120	1.6400	1.6658	1.7128	1.7553	1.7948	1.8318	1.8668	1.9002	1.9320	1.9625

240 psia (397.39 °F)

	Sh	Sat. liq.	Sat. vap.	2.61	52.61	102.61	152.61	202.61	302.61	402.61	502.61	602.61	702.61	802.61	902.61	1002.61
v		0.01860	1.9177	1.9268	2.0928	2.2462	2.3915	2.5316	2.8024	3.0661	3.3259	3.5831	3.8385	4.0926	4.3456	4.5977
h		372.27	1200.6	1202.4	1234.9	1264.6	1292.7	1319.7	1372.1	1423.8	1475.6	1527.9	1580.9	1634.6	1689.1	1744.3
s		0.5634	1.5299	1.5320	1.5687	1.6006	1.6291	1.6552	1.7025	1.7452	1.7848	1.8219	1.8570	1.8904	1.9223	1.9528

260 psia (404.44 °F)

	Sh	Sat. liq.	Sat. vap.	45.56	95.56	145.56	195.56	295.56	395.56	495.56	595.56	695.56	795.56	895.56	995.56
v		0.01870	1.7742	1.9173	2.0619	2.1981	2.3289	2.5808	2.8256	3.0663	3.3044	3.5408	3.7758	4.0097	4.2427
h		379.90	1201.5	1231.9	1262.4	1290.9	1318.2	1371.1	1423.0	1474.9	1527.3	1580.4	1634.2	1688.7	1744.0
s		0.5722	1.5230	1.5573	1.5899	1.6189	1.6453	1.6930	1.7359	1.7756	1.8128	1.8480	1.8814	1.9133	1.9439

Table B.3 Superheated Steam (Continued)

Abs Press. Lb/Sq In. (Sat. Temp)		Sat. Water	Sat. Steam	Temperature – Degrees Fahrenheit													
				450	500	550	600	650	700	800	900	1000	1100	1200	1300	1400	1500
280 (411.07)	Sh			38.93	88.93	138.93	188.93	238.93	288.93	388.93	488.93	588.93	688.93	788.93	888.93	988.93	1088.93
	v	0.01880	1.6505	1.7665	1.9037	2.0322	2.1551	2.2748	2.3909	2.6194	2.8437	3.0655	3.2855	3.5042	3.7217	3.9384	4.1543
	h	387.12	1202.3	1228.8	1260.0	1289.1	1316.8	1343.4	1370.0	1422.1	1474.2	1526.8	1579.9	1633.8	1688.4	1743.7	1799.8
	s	0.5805	1.5166	1.5464	1.5798	1.6093	1.6361	1.6608	1.6841	1.7273	1.7671	1.8043	1.8395	1.8730	1.9050	1.9356	1.9649
300 (417.35)	Sh			32.65	82.65	132.65	182.65	232.65	282.65	382.65	482.65	582.65	682.65	782.65	882.65	982.65	1082.65
	v	0.01889	1.5427	1.6356	1.7665	1.8883	2.0044	2.1139	2.2263	2.4407	2.6509	2.8585	3.0643	3.2688	3.4721	3.6746	3.8764
	h	393.99	1202.9	1225.7	1257.7	1287.2	1315.2	1342.1	1368.9	1421.3	1473.6	1526.2	1579.4	1633.3	1688.0	1743.4	1799.6
	s	0.5882	1.5105	1.5361	1.5703	1.6003	1.6274	1.6526	1.6758	1.7192	1.7591	1.7964	1.8317	1.8652	1.8972	1.9278	1.9572
320 (423.31)	Sh			26.69	76.69	126.69	176.69	226.69	276.69	376.69	476.69	576.69	676.69	776.69	876.69	976.69	1076.69
	v	0.01899	1.4480	1.5207	1.6462	1.7623	1.8725	1.9670	2.0823	2.2843	2.4821	2.6774	2.8708	3.0628	3.2538	3.4438	3.6332
	h	400.53	1203.4	1222.5	1255.2	1285.3	1313.7	1340.8	1367.8	1420.5	1472.9	1525.6	1578.9	1632.9	1687.6	1743.1	1799.3
	s	0.5956	1.5048	1.5261	1.5612	1.5918	1.6192	1.6447	1.6680	1.7116	1.7516	1.7890	1.8243	1.8579	1.8899	1.9206	1.9500
340 (428.99)	Sh			21.01	71.01	121.01	171.01	221.01	271.01	371.01	471.01	571.01	671.01	771.01	871.01	971.01	1071.01
	v	0.01908	1.3640	1.4191	1.5399	1.6511	1.7561	1.8360	1.9552	2.1463	2.3333	2.5175	2.7000	2.8811	3.0611	3.2402	3.4186
	h	406.80	1203.8	1219.2	1252.8	1283.4	1312.2	1339.5	1366.7	1419.6	1472.2	1525.0	1578.4	1632.5	1687.3	1742.8	1799.0
	s	0.6026	1.4994	1.5165	1.5525	1.5836	1.6114	1.6373	1.6606	1.7044	1.7445	1.7820	1.8174	1.8510	1.8831	1.9138	1.9432
360 (434.41)	Sh			15.59	65.59	115.59	165.59	215.59	265.59	365.59	465.59	565.59	665.59	765.59	865.59	965.59	1065.59
	v	0.01917	1.2891	1.3285	1.4454	1.5521	1.6525	1.7175	1.8421	2.0237	2.2009	2.3755	2.5482	2.7196	2.8898	3.0592	3.2279
	h	412.81	1204.1	1215.8	1250.3	1281.5	1310.6	1338.1	1365.6	1418.7	1471.5	1524.4	1577.9	1632.1	1686.9	1742.5	1798.8
	s	0.6092	1.4943	1.5073	1.5441	1.5758	1.6040	1.6302	1.6536	1.6976	1.7379	1.7754	1.8109	1.8445	1.8766	1.9073	1.9368
380 (439.61)	Sh			10.39	60.39	110.39	160.39	210.39	260.39	360.39	460.39	560.39	660.39	760.39	860.39	960.39	1060.39
	v	0.01925	1.2218	1.2472	1.3606	1.4635	1.5598	1.6100	1.7410	1.9139	2.0825	2.2484	2.4124	2.5750	2.7366	2.8973	3.0572
	h	418.59	1204.4	1212.4	1247.7	1279.5	1309.0	1336.8	1364.5	1417.9	1470.8	1523.8	1577.4	1631.6	1686.5	1742.2	1798.5
	s	0.6156	1.4894	1.4982	1.5360	1.5683	1.5969	1.6235	1.6470	1.6911	1.7315	1.7692	1.8047	1.8384	1.8705	1.9012	1.9307
400 (444.60)	Sh			5.40	55.40	105.40	155.40	205.40	255.40	355.40	455.40	555.40	655.40	755.40	855.40	955.40	1055.40
	v	0.01934	1.1610	1.1738	1.2841	1.3836	1.4763	1.5646	1.6499	1.8151	1.9759	2.1339	2.2901	2.4450	2.5987	2.7515	2.9037
	h	424.17	1204.6	1208.8	1245.1	1277.5	1307.4	1335.9	1363.4	1417.0	1470.1	1523.3	1576.9	1631.2	1686.2	1741.9	1798.2
	s	0.6217	1.4847	1.4894	1.5282	1.5611	1.5901	1.6163	1.6406	1.6850	1.7255	1.7632	1.7988	1.8325	1.8647	1.8955	1.9250

Steam table — superheated steam (Sh = degrees of superheat, °F; v = specific volume; h = enthalpy; s = entropy). The first two data columns are the saturated-water and saturated-steam properties at each pressure.

Abs Press. lb/sq in. (Sat. Temp.)		Sat. Water	Sat. Steam													
440 (454.03)	Sh			45.97	95.97	145.97	195.97	245.97	345.97	445.97	545.97	645.97	745.97	845.97	945.97	1045.97
	v	0.01950	1.0554	1.1517	1.2454	1.3319	1.4138	1.4926	1.6445	1.7918	1.9363	2.0790	2.2203	2.3605	2.4998	2.6384
	h	434.77	1204.8	1239.7	1273.4	1304.2	1333.2	1361.1	1415.3	1468.7	1522.1	1575.9	1630.4	1685.5	1741.2	1797.7
	s	0.6332	1.4759	1.5132	1.5474	1.5772	1.6040	1.6286	1.6734	1.7142	1.7521	1.7878	1.8216	1.8538	1.8847	1.9143
480 (462.82)	Sh			37.18	87.18	137.18	187.18	237.18	337.18	437.18	537.18	637.18	737.18	837.18	937.18	1037.18
	v	0.01967	0.9668	1.0409	1.1300	1.2115	1.2881	1.3615	1.5023	1.6384	1.7716	1.9030	2.0330	2.1619	2.2900	2.4173
	h	444.75	1204.8	1234.1	1269.1	1300.8	1330.5	1358.8	1413.6	1467.3	1520.9	1574.9	1629.5	1684.7	1740.6	1797.2
	s	0.6439	1.4677	1.4990	1.5346	1.5652	1.5925	1.6176	1.6628	1.7038	1.7419	1.7777	1.8116	1.8439	1.8748	1.9045
520 (471.07)	Sh			28.93	78.93	128.93	178.93	228.93	328.93	428.93	528.93	628.93	728.93	828.93	928.93	1028.93
	v	0.01982	0.8914	0.9466	1.0321	1.1094	1.1816	1.2504	1.3819	1.5085	1.6323	1.7542	1.8746	1.9940	2.1125	2.2302
	h	454.18	1204.5	1228.3	1264.8	1297.4	1327.7	1356.5	1411.8	1465.9	1519.7	1573.9	1628.7	1684.0	1740.0	1796.7
	s	0.6540	1.4601	1.4853	1.5223	1.5539	1.5818	1.6072	1.6530	1.6943	1.7325	1.7684	1.8024	1.8348	1.8657	1.8954
560 (478.84)	Sh			21.16	71.16	121.16	171.16	221.16	321.16	421.16	521.16	621.16	721.16	821.16	921.16	1021.16
	v	0.01998	0.8264	0.8653	0.9479	1.0217	1.0902	1.1552	1.2787	1.3972	1.5129	1.6266	1.7388	1.8500	1.9603	2.0699
	h	463.14	1204.2	1222.2	1260.3	1293.9	1324.9	1354.2	1410.0	1464.4	1518.6	1572.9	1627.8	1683.3	1739.4	1796.1
	s	0.6634	1.4529	1.4720	1.5106	1.5431	1.5717	1.5975	1.6438	1.6853	1.7237	1.7598	1.7939	1.8263	1.8573	1.8870
600 (486.20)	Sh			13.80	63.80	113.80	163.80	213.80	313.80	413.80	513.80	613.80	713.80	813.80	913.80	1013.80
	v	0.02013	0.7697	0.7944	0.8746	0.9456	1.0109	1.0726	1.1892	1.3008	1.4093	1.5160	1.6211	1.7252	1.8284	1.9309
	h	471.70	1203.7	1215.9	1255.6	1290.3	1322.0	1351.8	1408.3	1463.0	1517.4	1571.9	1627.0	1682.6	1738.8	1795.6
	s	0.6723	1.4461	1.4590	1.4993	1.5329	1.5621	1.5884	1.6351	1.6769	1.7155	1.7517	1.7859	1.8184	1.8494	1.8792
700 (503.08)	Sh				46.92	96.92	146.92	196.92	296.92	396.92	496.92	596.92	696.92	796.92	896.92	996.92
	v	0.02050	0.6556		0.7271	0.7928	0.8520	0.9072	1.0102	1.1078	1.2023	1.2948	1.3858	1.4757	1.5647	1.6530
	h	491.60	1201.8		1243.4	1281.0	1314.6	1345.6	1403.7	1459.4	1514.4	1569.4	1624.8	1680.7	1737.2	1794.3
	s	0.6928	1.4304		1.4726	1.5090	1.5399	1.5673	1.6154	1.6580	1.6970	1.7335	1.7679	1.8006	1.8318	1.8617
800 (518.21)	Sh				31.79	81.79	131.79	181.79	281.79	381.79	481.79	581.79	681.79	781.79	881.79	981.79
	v	0.02087	0.5690		0.6151	0.6774	0.7323	0.7828	0.8759	0.9631	1.0470	1.1289	1.2093	1.2885	1.3669	1.4446
	h	509.81	1199.4		1230.1	1271.1	1306.8	1339.3	1399.1	1455.8	1511.4	1566.9	1622.7	1678.9	1735.7	1792.9
	s	0.7111	1.4163		1.4472	1.4869	1.5198	1.5484	1.5980	1.6413	1.6807	1.7175	1.7522	1.7851	1.8164	1.8464
900 (531.95)	Sh				18.05	68.05	118.05	168.05	268.05	368.05	468.05	568.05	668.05	768.05	868.05	968.05
	v	0.02123	0.5009		0.5263	0.5869	0.6388	0.6858	0.7713	0.8504	0.9262	0.9998	1.0720	1.1430	1.2131	1.2825
	h	526.70	1196.4		1215.5	1260.6	1298.6	1332.7	1394.4	1452.2	1508.5	1564.4	1620.6	1677.1	1734.1	1791.6
	s	0.7279	1.4032		1.4223	1.4659	1.5010	1.5311	1.5822	1.6263	1.6662	1.7033	1.7382	1.7713	1.8028	1.8329

Table B.3 Superheated Steam (Continued)

Abs Press Lb/Sq In (Sat. Temp)		Sat Water	Sat Steam	\multicolumn{14}{c}{Temperature Degrees Fahrenheit}													
				550	600	650	700	750	800	850	900	1000	1100	1200	1300	1400	1500
1000 (544.58)	Sh			5.42	55.42	105.42	155.42	205.42	255.42	305.42	355.42	455.42	555.42	655.42	755.42	855.42	955.42
	v	0.02159	0.4460	0.4535	0.5137	0.5636	0.6080	0.6489	0.6875	0.7245	0.7603	0.8295	0.8966	0.9622	1.0266	1.0901	1.1529
	h	542.55	1192.9	1199.3	1249.3	1290.1	1325.9	1358.7	1389.6	1419.4	1448.5	1505.4	1561.9	1618.4	1675.3	1732.5	1790.3
	s	0.7434	1.3910	1.3973	1.4457	1.4833	1.5149	1.5426	1.5677	1.5908	1.6126	1.6530	1.6905	1.7256	1.7589	1.7905	1.8207
1100 (556.28)	Sh				43.72	93.72	143.72	193.72	243.72	293.72	343.72	443.72	543.72	643.72	743.72	843.72	943.72
	v	0.02195	0.4006		0.4531	0.5017	0.5440	0.5826	0.6188	0.6533	0.6865	0.7505	0.8121	0.8723	0.9313	0.9894	1.0468
	h	557.55	1189.1		1237.3	1281.2	1318.8	1352.9	1384.7	1415.2	1444.7	1502.4	1559.4	1616.3	1673.5	1731.0	1789.0
	s	0.7578	1.3794		1.4259	1.4664	1.4996	1.5284	1.5542	1.5779	1.6000	1.6410	1.6787	1.7141	1.7475	1.7793	1.8097
1200 (567.19)	Sh				32.81	82.81	132.81	182.81	232.81	282.81	332.81	432.81	532.81	632.81	732.81	832.81	932.81
	v	0.02232	0.3624		0.4016	0.4497	0.4905	0.5273	0.5615	0.5939	0.6250	0.6845	0.7418	0.7974	0.8519	0.9055	0.9584
	h	571.85	1184.8		1224.2	1271.8	1311.5	1346.9	1379.7	1410.8	1440.9	1499.4	1556.9	1614.2	1671.6	1729.4	1787.6
	s	0.7714	1.3683		1.4061	1.4501	1.4851	1.5150	1.5415	1.5658	1.5883	1.6298	1.6679	1.7035	1.7371	1.7691	1.7996
1300 (577.42)	Sh				22.58	72.58	122.58	172.58	222.58	272.58	322.58	422.58	522.58	622.58	722.58	822.58	922.58
	v	0.02269	0.3299		0.3570	0.4052	0.4451	0.4804	0.5129	0.5436	0.5729	0.6287	0.6822	0.7341	0.7847	0.8345	0.8836
	h	585.58	1180.2		1209.9	1261.9	1303.9	1340.8	1374.6	1406.4	1437.1	1496.3	1554.3	1612.0	1669.8	1727.9	1786.3
	s	0.7843	1.3577		1.3860	1.4340	1.4711	1.5022	1.5296	1.5544	1.5773	1.6194	1.6578	1.6937	1.7275	1.7596	1.7902
1400 (587.07)	Sh				12.93	62.93	112.93	162.93	212.93	262.93	312.93	412.93	512.93	612.93	712.93	812.93	912.93
	v	0.02307	0.3018		0.3176	0.3667	0.4059	0.4400	0.4712	0.5004	0.5282	0.5809	0.6311	0.6798	0.7272	0.7737	0.8195
	h	598.83	1175.3		1194.1	1251.4	1296.1	1334.5	1369.3	1402.0	1433.2	1493.2	1551.8	1609.9	1668.0	1726.3	1785.0
	s	0.7966	1.3474		1.3652	1.4181	1.4575	1.4900	1.5182	1.5436	1.5670	1.6096	1.6484	1.6845	1.7185	1.7508	1.7815
1600 (604.87)	Sh					45.13	95.13	145.13	195.13	245.13	295.13	395.13	495.13	595.13	695.13	795.13	895.13
	v	0.02387	0.2555			0.3026	0.3415	0.3741	0.4032	0.4301	0.4555	0.5031	0.5482	0.5915	0.6336	0.6748	0.7153
	h	624.20	1164.5			1228.3	1279.4	1321.4	1358.5	1392.8	1425.2	1486.9	1546.6	1605.6	1664.3	1723.2	1782.3
	s	0.8199	1.3274			1.3861	1.4312	1.4667	1.4968	1.5235	1.5478	1.5916	1.6312	1.6678	1.7022	1.7347	1.7657
1800 (621.02)	Sh					28.98	78.98	128.98	178.98	228.98	278.98	378.98	478.98	578.98	678.98	778.98	878.98
	v	0.02472	0.2186			0.2505	0.2906	0.3223	0.3500	0.3752	0.3988	0.4426	0.4836	0.5229	0.5609	0.5980	0.6343
	h	648.49	1152.3			1201.2	1261.1	1307.4	1347.2	1383.3	1417.1	1480.6	1541.4	1601.2	1660.7	1720.1	1779.7
	s	0.8417	1.3079			1.3526	1.4054	1.4446	1.4768	1.5049	1.5302	1.5753	1.6156	1.6528	1.6876	1.7204	1.7516
2000 (635.80)	Sh					14.20	64.20	114.20	164.20	214.20	264.20	364.20	464.20	564.20	664.20	764.20	864.20
	v	0.02565	0.1883			0.2056	0.2488	0.2805	0.3072	0.3312	0.3534	0.3942	0.4320	0.4680	0.5027	0.5365	0.5695
	h	672.11	1138.3			1168.3	1240.9	1292.6	1335.4	1373.5	1408.7	1474.1	1536.2	1596.9	1657.0	1717.0	1771.1
	s	0.8625	1.2881			1.3154	1.3794	1.4231	1.4578	1.4874	1.5138	1.5603	1.6014	1.6391	1.6743	1.7075	1.7389

Steam Tables (continued) — Superheated steam at high pressures

2500 psia (668.11 °F)

	Sat. liquid	Sat. vapor														
Sh			31.89	81.89	131.89	181.89	231.89	281.89	331.89	381.89	431.89	481.89	531.89	631.89	731.89	831.89
v	0.02859	0.1307	0.1681	0.2032	0.2293	0.2514	0.2712	0.2896	0.3068	0.3232	0.3390	0.3543	0.3692	0.3980	0.4259	0.4529
h	731.71	1093.3	1176.7	1250.6	1303.4	1347.4	1386.7	1423.1	1457.5	1490.7	1522.9	1554.6	1585.9	1647.8	1709.2	1770.4
s	0.9139	1.2345	1.3076	1.3701	1.4129	1.4472	1.4766	1.5029	1.5269	1.5492	1.5703	1.5903	1.6094	1.6456	1.6796	1.7116

3000 psia (695.33 °F)

	Sat. liquid	Sat. vapor														
Sh			4.67	54.67	104.67	154.67	204.67	254.67	304.67	354.67	404.67	454.67	504.67	604.67	704.67	804.67
v	0.03428	0.0850	0.0982	0.1483	0.1759	0.1975	0.2161	0.2329	0.2484	0.2630	0.2770	0.2904	0.3033	0.3282	0.3522	0.3753
h	801.84	1020.3	1060.5	1197.9	1267.0	1319.0	1363.2	1403.1	1440.2	1475.4	1509.4	1542.4	1574.8	1638.5	1701.4	1763.8
s	0.9728	1.1619	1.1966	1.3131	1.3692	1.4097	1.4429	1.4717	1.4976	1.5213	1.5434	1.5642	1.5841	1.6214	1.6561	1.6888

3500 psia

	750	800	850	900	950	1000	1050	1100	1150	1200	1300	1400	1500
v	0.1048	0.1364	0.1583	0.1764	0.1922	0.2066	0.2200	0.2326	0.2447	0.2563	0.2784	0.2995	0.3198
h	1127.1	1224.6	1287.8	1338.2	1382.2	1422.2	1459.7	1495.5	1529.9	1563.6	1629.2	1693.6	1757.2
s	1.2450	1.3242	1.3734	1.4112	1.4430	1.4709	1.4962	1.5194	1.5412	1.5618	1.6002	1.6358	1.6691

4000 psia

	750	800	850	900	950	1000	1050	1100	1150	1200	1300	1400	1500
v	0.0631	0.1052	0.1284	0.1463	0.1616	0.1752	0.1877	0.1994	0.2105	0.2210	0.2411	0.2601	0.2783
h	1007.4	1174.3	1253.4	1316.6	1360.2	1403.6	1443.6	1481.3	1517.3	1552.2	1619.8	1685.7	1750.6
s	1.1396	1.2754	1.3371	1.3807	1.4158	1.4461	1.4730	1.4976	1.5203	1.5417	1.5812	1.6177	1.6516

5000 psia

	750	800	850	900	950	1000	1050	1100	1150	1200	1300	1400	1500
v	0.0338	0.0591	0.0855	0.1038	0.1185	0.1312	0.1425	0.1529	0.1626	0.1718	0.1890	0.2050	0.2203
h	854.9	1042.9	1173.6	1252.9	1313.5	1364.6	1410.2	1452.1	1491.5	1529.1	1600.9	1670.0	1737.4
s	1.0070	1.1593	1.2612	1.3207	1.3645	1.4001	1.4309	1.4582	1.4831	1.5061	1.5481	1.5863	1.6216

Sh = superheat, F

v = specific volume, ft^3/lb_m

h = enthalpy, Btu/lb_m

s = entropy, Btu/lb_m R

Source: Values reprinted, by permission, from *1967 ASME Steam Tables*, C. A. Meyer et al., 1968. Copyright © 1968 by the American Society of Mechanical Engineers.

Table B.4 Thermodynamic Properties of Air at Low Pressure

T (°K)	h (kJ/kg)	u (kJ/kg)	P_r	ν_r	ϕ
100	99.93	71.23	.02977	964.19	4.6004
120	119.97	85.52	.05626	612.27	4.7831
140	140.01	99.82	.09635	417.06	4.9376
160	160.05	114.12	.15357	299.05	5.0714
180	180.09	128.42	.23168	223.01	5.1894
200	200.13	142.72	.33468	171.52	5.2950
220	220.18	157.03	.46684	135.26	5.3905
240	240.22	171.34	.63263	108.89	5.4777
260	260.68	185.65	.8368	89.188	5.5580
280	280.35	199.98	1.0842	74.129	5.6323
300	300.43	214.32	1.3801	62.393	5.7016
320	320.53	228.68	1.7301	53.091	5.7665
340	340.66	243.07	2.1398	45.608	5.8275
360	360.81	257.48	2.6154	39.509	5.8851
380	381.01	271.94	3.1633	34.481	5.9397
400	401.25	286.43	3.7902	30.292	5.9916
420	421.54	300.98	4.5035	26.769	6.0411
440	441.88	315.59	5.3106	23.781	6.0884
460	462.28	330.25	6.2197	21.228	6.1338
480	482.76	344.98	7.2391	19.032	6.1773
500	503.30	359.79	8.378	17.130	6.2193
520	523.93	374.67	9.645	15.474	6.2597
540	544.63	389.63	11.052	14.025	6.2988
560	565.42	404.68	12.608	12.749	6.3366
580	586.29	419.82	14.324	11.623	6.3732
600	607.26	435.04	16.212	10.623	6.4087
620	628.32	450.36	18.284	9.733	6.4433
640	649.47	465.77	20.553	8.938	6.4768
660	670.72	481.28	23.033	8.2249	6.5095
680	692.07	496.89	25.736	7.5839	6.5414
700	713.51	512.59	28.679	7.0058	6.5725
720	735.05	528.39	31.876	6.4832	6.6028
740	756.68	544.28	35.344	6.0097	6.6324
760	778.41	560.27	39.098	5.5795	6.6614

Table B.4 Thermodynamic Properties of Air at Low Pressure (*Continued*)

T (°K)	h (kJ/kg)	u (kJ/kg)	P_r	ν_r	φ
800	822.15	592.53	47.535	4.8306	6.7175
840	866.26	625.16	57.336	4.2052	677.13
880	910.73	658.15	68.655	3.6791	6.8230
920	955.55	691.48	81.663	3.2336	6.8728
960	1000.69	725.14	96.54	2.8543	6.9209
1000	1046.16	759.13	113.48	2.5294	6.9673
1050	1103.41	802.03	137.86	2.1862	7.0231
1100	1161.11	845.38	166.21	1.8996	7.0768
1150	1219.23	889.14	198.99	1.6588	7.1285
1200	1277.73	933.29	236.69	1.4552	7.1783
1250	1336.60	977.81	279.83	1.2822	7.2263
1300	1395.81	1022.67	328.98	1.1342	7.2728
1350	1455.35	1067.86	384.74	1.00716	7.3177
1400	1515.18	1113.34	447.73	.89752	7.3612
1450	1517.30	1159.11	518.63	.80250	7.4034
1500	1635.68	1205.14	598.14	.71982	7.4444
1550	1696.32	1251.42	687.01	.64759	7.4841
1600	1757.19	1297.94	786.04	.58426	7.5228
1650	1818.28	1344.68	896.05	.52855	7.5604
1700	1817.58	1391.63	1017.9	.47937	7.5970
1750	1441.09	1438.78	1152.6	.43582	7.6326
1800	2002.78	1486.12	1300.9	.39714	7.6674
1850	2064.65	1533.65	1464.0	.36270	7.7013
1900	2126.70	1581.34	1643.0	.33194	7.7344
1950	2188.91	1629.20	1838.8	.30439	7.7667
2000	2251.28	1677.22	2052.6	.27967	7.7983
2100	2376.46	1733.70	2539.3	.23737	7.8594
2200	2502.20	1870.73	3113.3	.20283	7.9179
2300	2628.45	1968.27	3785.6	.17439	7.9740
2400	2755.17	2066.29	4568.1	.15080	8.0279
2500	2882.34	2164.76	5473.7	.13110	8.0798
2600	3009.91	2263.63	6516.1	.11453	8.1299
2800	3266.21	2462.52	9071.4	.08860	8.2248
3000	3523.87	2662.78	1236.4	.06965	8.3137

Source: Abridged with permission from Table 1 in *Gas Tables: International Version*, 2d ed., by Joseph H. Keenan, Jing Chao, and Joseph Kaye, John Wiley & Sons, New York, copyright © 1980.

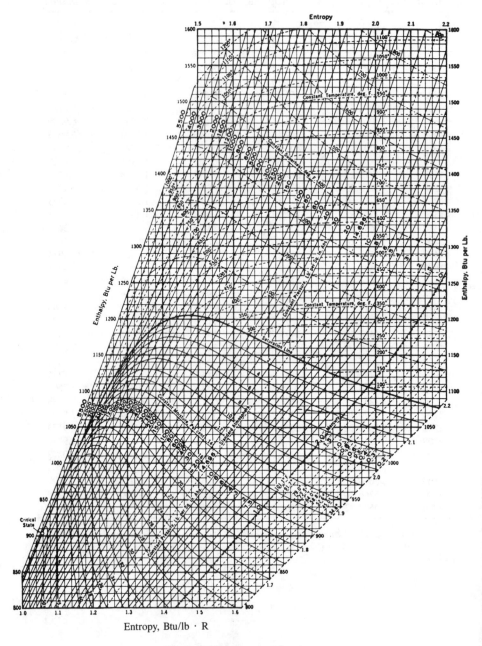

Figure B.1 Mollier chart for steam.

616

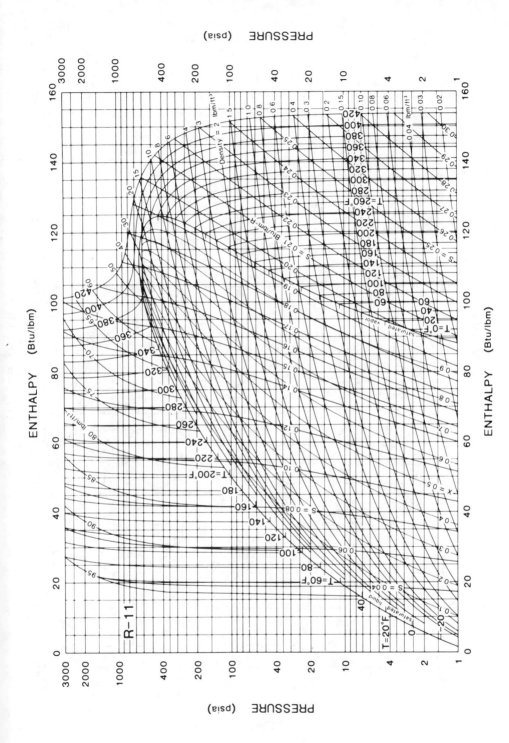

Figure B.2 Pressure-enthalpy diagram for Freon 11.

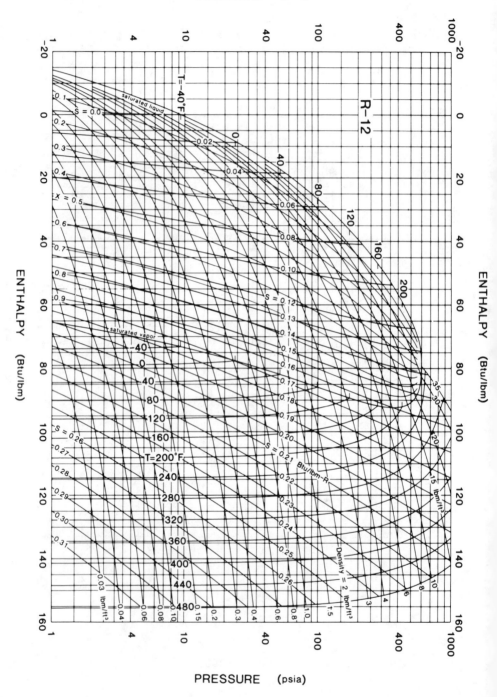

Figure B.3 Pressure-enthalpy diagram for Freon 12.

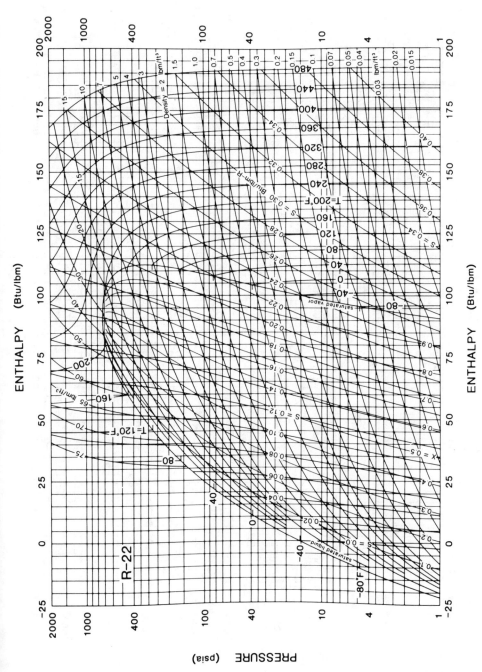

Figure B.4 Pressure-enthalpy diagram for Freon 22.

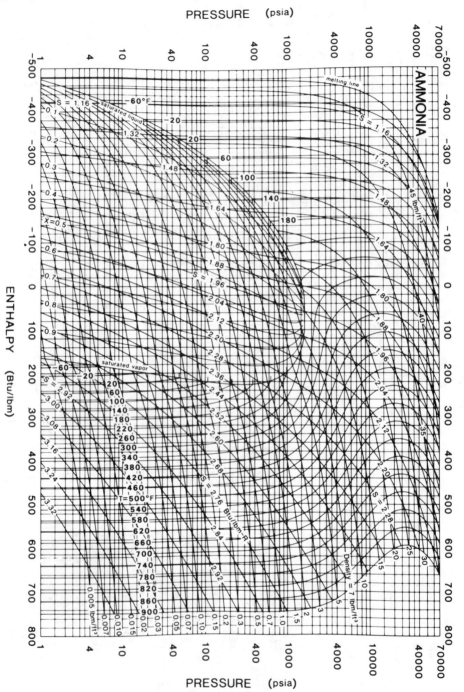

Figure B.5 Pressure-enthalpy diagram for ammonia.

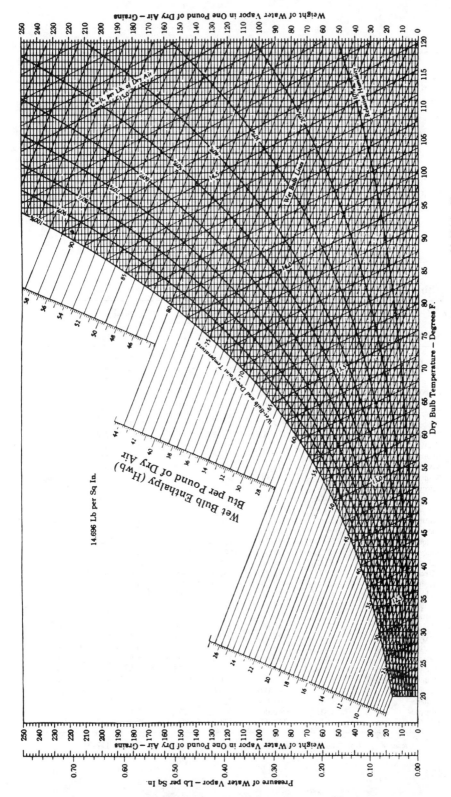

Figure B.6 Psychrometric chart for air-steam mixture; barometric pressure = 29.921 in. Hg.

THERMOCHEMICAL PROPERTIES

Table C.1 Critical Constants

Substance	Formula	Molecular weight	Temperature		Pressure		Volume (ft³/lb mol)
			°K	°R	atm	lb$_f$/in.²	
Ammonia	NH_3	17.03	405.5	729.8	111.3	1636	1.16
Argon	Ar	39.944	151	272	48.0	705	1.20
Bromine	Br_2	159.832	584	1052	102	1500	2.17
Carbon dioxide	CO_2	44.01	304.2	547.5	72.9	1071	1.51
Carbon monoxide	CO	28.01	133	240	64.5	507	1.49
Chlorine	Cl_2	70.914	417	751	76.1	1120	1.99
Deuterium (normal)	D_2	4.00	38.4	69.1	16.4	241	—
Helium	He	4.003	5.3	9.5	2.26	33.2	0.926
Helium	He	3.00	3.34	6.01	1.15	16.9	—
Hydrogen (normal)	H_2	2.016	33.3	59.9	12.8	188.1	1.04
Krypton	Kr	83.7	209.4	376.9	54.3	798	1.48
Neon	Ne	20.183	44.5	80.1	26.9	395	0.668
Nitrogen	N_2	28.016	126.2	227.1	33.5	492	1.44
Nitrous oxide	N_2O	44.02	309.7	557.1	71.7	1054	1.54
Oxygen	O_2	32.00	154.8	278.6	50.4	736	1.25
Sulfur dioxide	So_2	64.06	430.7	775.2	77.8	1143	1.95
Water	H_2O	18.016	647.4	1165.3	218.3	3204	0.09
Xenon	Xe	131.3	289.75	521.55	58.0	852	1.90
Benzene	C_6H_6	78.11	562	1012	48.6	714	4.17
n-Butane	C_4H_{10}	58.120	425.2	765.2	37.5	551	4.08
Carbon tetrachloride	CCl_4	153.81	556.4	1001.5	45.0	661	4.42
Chloroform	$CHCl_3$	119.39	536.6	965.8	54.0	794	3.85
Dichlorodifluoromethane	CCl_2F_2	120.92	384.7	692.4	39.6	582	3.49
Dichlorofluoromethane	$CHCl_2F$	102.93	451.7	813.0	51.0	749	3.16
Ethane	C_2H_6	30.068	305.5	549.8	48.2	708	2.37
Ethyl alcohol	C_2H_5OH	46.07	516.0	929.0	63.0	926.	2.68
Ethylene	C_2H_4	28.052	282.4	508.3	50.5	742	1.99
n-Hexane	C_6H_{14}	86.172	507.9	914.2	29.9	439	5.89
Methane	CH_4	16.012	191.1	343.9	45.8	673	1.59
Methyl alcohol	CH_3OH	32.04	513.2	923.7	78.5	1154	1.89
Methyl chloride	CH_3Cl	50.49	416.3	749.3	65.9	968	2.29
Propane	C_3H_8	44.094	370.0	665.9	42.0	617	3.20
Propene	C_3H_6	42.078	365.0	656.9	45.6	670	2.90
Propyne	C_3H_4	40.062	401	722	52.8	776	—
Trichlorofluoromethane	CCl_3F	137.38	471.2	848.1	43.2	635	3.97

Source: K. A. Kobe and R. E. Lynn, Jr., *Chemical Reviews,* vol. 52 (1953), pp. 117–236.

Table C.2 Zero-Pressure Properties of Gases

Gas	Chemical formula	Molecular weight	R (ft lb_f/lb_m R)	c_{p0} (Btu/lb_m R)	c_{v0} (Btu/lb_m R)	k^a
Air	—	28.97	53.34	0.240	0.171	1.400
Argon	Ar	39.94	38.66	0.1253	0.0756	1.667
Carbon dioxide	CO_2	44.01	35.10	0.203	0.158	1.285
Carbon monoxide	CO	28.01	55.16	0.249	0.178	1.399
Helium	He	4.003	386.0	1.25	0.753	1.667
Hydrogen	H_2	2.016	766.4	3.43	2.44	1.404
Methane	CH_4	16.04	96.35	0.532	0.403	1.32
Nitrogen	N_2	28.016	55.15	0.248	0.177	1.400
Oxygen	O_2	32.000	48.28	0.219	0.157	1.395
Steam	H_2O	18.016	85.76	0.445	0.335	1.329

$^a c_{p0}$, c_{v0}, and k are at 80 °F.

Table C.3 Enthalpy of Combustion of Some Hydrocarbons at 77 °F

Hydrocarbon	Formula	Liquid H_2O in products (negative of higher heating value)		Vapor H_2O in products (negative of lower heating value)	
		Liquid hydro-carbon, Btu/lb_m fuel	Gaseous hydro-carbon, Btu/lb_m fuel	Liquid hydro-carbon, Btu/lb_m fuel	Gaseous hydro-carbon, Btu/lb_m fuel
Paraffin Family					
Methane	CH_4		−23,861		−21,502
Ethane	C_2H_6		−22,304		−20,416
Propane	C_3H_8	−21,490	−21,649	−19,773	−19,929
Butane	C_4H_{10}	−21,134	−21,293	−19,506	−19,665
Pentane	C_5H_{12}	−20,914	−21,072	−19,340	−19,499
Hexane	C_6H_{14}	−20,772	−20,930	−19,233	−10,391
Heptane	C_7H_{16}	−20,668	−20,825	−19,457	−19,314
Octane	C_8H_{18}	−20,591	−20,747	−19,100	−19,256
Decane	$C_{10}H_{22}$	−20,484	−20,638	−19,020	−19,175
Dodecane	$C_{12}H_{26}$	−20,410	−20,564	−18,964	−19,118
Olefin Family					
Ethene	C_2H_4		−21,626		−20,276
Propene	C_3H_6		−21,033		−19,683
Butene	C_4H_8		−20,833		−19,483
Pentene	C_5H_{10}		−20,696		−19,346
Hexene	C_6H_{12}		−20,612		−19,262
Heptene	C_7H_{14}		−20,552		−19,202
Octene	C_8H_{16}		−20,507		−19,157
Nonene	C_9H_{18}		−20,472		−19,122
Decene	$C_{10}H_{20}$		−20,444		−19,094
Alkylbenzene Family					
Benzene	C_6H_6	−17,985	−18,172	−17,259	−17,446
Methylbenzene	C_7H_8	−18,247	−18,423	−17,424	−17,601
Ethylbenzene	C_8H_{10}	−18,488	−18,659	−17,596	−17,767
Propylbenzene	C_9H_{12}	−18,667	−18,832	−17,722	−17,887
Butylbenzene	$C_{10}H_{14}$	−18,809	−18,970	−17,823	−17,984

Table C.4 Enthalpy of Formation and Absolute Entropy of Various Substances at 77 °F and 1 Atmosphere Pressure

Substance	Formula	M	State	$\bar{h}_f^\circ$ Cal/gm mole	Btu/lb mole	$\bar{s}^\circ$ Cal/gm mole K Btu/lb mole R
Carbon monoxide[a]	CO	28.011	gas	−26,417	−47,551	47.214
Carbon dioxide[a]	CO_2	44.011	gas	−94,054	−169,297	51.072
Water[a,b]	H_2O	18.016	gas	−57,798	−104,036	45.106
Water[b]	H_2O	18.016	liq.	−68,317	−122,971	16.716
Methane[a]	CH_4	16.043	gas	−17,895	−32,211	44.490
Acetylene[a]	C_2H_2	26.038	gas	54,190	97,542	48.004
Ethylene[a]	C_2H_4	28.054	gas	12,496	22,493	52.447
Ethane[e]	C_2H_6	30.070	gas	−20,236	−36,425	54.85
Propane[e]	C_3H_8	44.097	gas	−24,820	−44,676	64.51
Butane[e]	C_4H_{10}	58.124	gas	−30,150	−54,270	74.12
Octane[e]	C_8H_{18}	114.23	gas	−49,820	−89,680	111.55
Octane[e]	C_8H_{18}	114.23	liq.	−59,740	−107,532	86.23
Carbon[a] (graphite)	C	12.011	solid	0	0	1.359

[a]From JANAF *Thermochemical Data,* The Dow Chemical Company, Thermal Laboratory, Midland, Mich.

[b]From *Circular 500,* National Bureau of Standards.

[c]From F. D. Rossim et al., *API Research Project 44.*

THERMOPHYSICAL PROPERTIES OF SOLIDS

Table D.1 Thermal Properties of Metals

Metal	Properties at 68°F				k (Btu/hr ft °F)									
	ρ (lb$_m$/ft³)	c_p (Btu/ lb$_m$ °F)	k (Btu/hr ft °F)	α (ft²/hr)	−148° −100°C	32°F 0°C	212°F 100°C	392°F 200°C	572°F 300°C	752°F 400°C	1112°F 600°C	1472°F 800°C	1832°F 1000°C	2192°F 1200°C
Aluminum														
Pure	169	0.214	132	3.665	134	132	132	132	132					
Al-Cu (Duralumin): 94–96 Al. 3–5 Cu. trace Mg	174	0.211	95	2.580	73	92	105	112						
Al-Mg (Hydronalium): 91–95 Al. 5–9 Mg	163	0.216	65	1.860	54	63	73	82						
Al-Si (Silumin): 87 Al, 13 Si	166	0.208	95	2.773	86	94	101	107						
Al-Si (Silumin, copper bearing): 86.5 Al, 12.5 Si, 1 Cu	166	0.207	79	2.311	69	79	83	88	93					
Al-Si (Alusil): 78–80 Al, 20–22 Si	614	0.204	93	2.762	83	91	97	101	103					
Al-Mg-Si: 97 Al, 1 Mg, 1 Si, 1 Mn	169	0.213	102	2.859	–	101	109	118						
Lead	710	0.031	20	0.924	21.3	20.3	19.3	18.2	17.2					
Iron														
Pure	493	0.108	42	0.785	50	42	39	36	32	28	23	21	20	21
Wrought iron (C < 0.05%)	490	0.11	34	0.634	–	34	33	30	28	26	21	19	19	19
Cast iron (C ≈ 4%)	454	0.10	30	0.666										
Steel (C$_{max}$ ≈ 1.5%)														

Carbon steel (C ≈														
0.5%)	489	0.111	31	0.570	—	32	30	28	26	24	20	17	17	18
1.0%	487	0.113	25	0.452	—	25	25	24	23	21	19	17	16	17
1.5%	484	0.116	21	0.376	—	21	21	21	20	19	18	16	16	17
Nickel steel (Ni ≈ 0%)	493	0.108	42	0.785										
10%	496	0.11	15	0.279										
20%	499	0.11	11	0.204										
30%	504	0.11	7	0.118										
40%	510	0.11	6	0.108										
50%	516	0.11	8	0.140										
60%	523	0.11	11	0.182										
70%	531	0.11	15	0.258										
80%	538	0.11	20	0.344										
90%	547	0.11	27	0.452										
100%	556	0.106	52	0.892										
Invar (Ni ≈ 36%)	508	0.11	6.2	0.108										
Chrome steel (Cr = 0%)	493	0.108	42	0.785	50	42	39	36	32	28	23	21	20	21
1%	491	0.11	35	0.645	—	36	32	30	27	24	21	19	19	11
2%	491	0.11	30	0.559	—	31	28	26	24	22	19	18	18	17
5%	489	0.11	23	0.430	—	23	22	21	21	19	17	17	17	
10%	486	0.11	18	0.344	—	18	18	18	17	17	16	16	17	
20%	480	0.11	13	0.258	—	13	13	13	13	14	14	15		
30%	476	0.11	11	0.204	—									
Cr-Ni (chrome-nickel)														
15 Cr, 10 Ni	491	0.11	11	0.204		9.4	10	10	11	11	13	15		
18 Cr, 8 Ni (V2A)	488	0.11	9.4	0.172	—								18	
20 Cr, 15 Ni	489	0.11	8.7	0.161										
25 Cr, 20 Ni	491	0.11	7.4	0.140										

Table D.1 Thermal Properties of Metals (Continued)

Metal	ρ (lbm/ft³)	cp (Btu/lbm °F)	k (Btu/hr ft °F)	α (ft²/hr)	−148° −100°C	32°F 0°C	212°F 100°C	392°F 200°C	572°F 300°C	752°F 400°C	1112°F 600°C	1472°F 800°C	1832°F 1000°C	2192°F 1200°C
Ni-Cr (nickel-chrome)														
80 Ni, 15 Cr	532	0.11	10	0.172										
60 Ni, 15 Cr	516	0.11	7.4	0.129										
40 Ni, 15 Cr	504	0.11	6.7	0.118										
20 Ni, 15 Cr	491	0.11	8.1	0.151	—	8.1	8.7	8.7	9.4	10	11	13		
Cr-Ni-Al: 6 Cr, 1.5 Al, 0.5 Si (Sicromal 8)	482	0.117	13	0.237										
24 Cr, 2.5 Al, 0.5 Si (Sicromal 12)	479	0.118	11	0.194										
Manganese steel (Ma = 0%)	493	0.118	42	0.784										
1%	491	0.11	29	0.538	—	22	21	21	21	20	19			
2%	491	0.11	22	0.376										
5%	490	0.11	13	0.247										
10%	487	0.11	10	0.194										
Tungsten steel (W = 0%)	493	0.108	42	0.785										
1%	494	0.107	38	0.720	—	36	34	31	28	26	21			
2%	497	0.106	36	0.677										
5%	504	0.104	31	0.591										
10%	519	0.100	28	0.527										
20%	551	0.093	25	0.484										
Silicon steel (Si = 0%)	493	0.108	42	0.785										
1%	485	0.11	24	0.451										
2%	479	0.11	18	0.344										
5%	463	0.11	11	0.215										

German silver: 62 Cu, 15 Ni, 22 Zn	538	0.094	14.4	0.290	11.1	—	18	23	26	28	32	36	39	40
Constantan: 60Cu, 40 Ni	557	0.098	13.1	0.237	12	—	12.8	15						
Magnesium														
Pure	109	0.242	99	3.762	103	99	97	94	91					
Mg-Al (electrolytic) 6-8% Al, 1-2% Zn	113	0.24	38	1.397	—	30	36	43	48					
Mg-Mn: 2% Mn	111	0.24	66	2.473	54	64	72	75						
Molybdenum	638	0.060	79	2.074	80	79	79							
Nickel														
Pure (99.9%)	556	0.1065	52	0.882	60	54	48	42	37	34				
Impure (99.2%)	556	0.106	40	0.677	—	40	37	34	32	30				
Ni-Cr: 90 Ni, 10 Cr	541	0.106	10	0.172	—	9.9	10.9	12.1	13.2	14.2	13.0			
	519	0.106	7.3	0.129	—	7.1	8.0	9.0	9.9	10.9				
Silver														
Purest	657	0.0559	242	6.601	242	241	240	238						
Pure (99.9%)	657	0.0559	235	6.418	242	237	240	216	209	208				
Tungsten	1208	0.0321	94	2.430	—	96	87	82	77	73		44		
Zinc, pure	446	0.0918	64.8	1.591	66	65	63	61	58	54	65			
Tin, pure	456	0.0541	37	1.505	43	38.1	34	33						
Copper														
Pure	559	0.0915	223	4.353	235	223	219	216	—	210	204			
Aluminum bronze:95 Cu, 5 Al	541	0.098	48	0.903										

Table D.1 Thermal Properties of Metals (*Continued*)

Metal	Properties at 68°F				k (Btu/hr ft °F)									
	ρ (lb$_m$/ft^3)	c_p (Btu/ lb$_m$ °F)	k (Btu/hr ft °F)	α (ft^2/hr)	−148° −100°C	32°F 0°C	212°F 100°C	392°F 200°C	572°F 300°C	752°F 400°C	1112°F 600°C	1472°F 800°C	1832°F 1000°C	2192°F 1200°C
Bronze: 75 Cu, 25 Sn	541	0.082	15	0.333										
Red brass: 85 Cu, 9 Sn, 6 Zn	544	0.092	35	0.699	—	34	41							
Brass: 70 Cu, 30 Zn	532	0.092	64	1.322	51	—	74	83	85	85				

Source: From E. R. G. Eckert and R. M. Drake, *Heat and Mass Transfer*; copyright 1959 McGraw-Hill; used with the permission of McGraw-Hill Book Company.

Table D.2 Thermal Properties of Some Nonmetals and Insulation

Substance	C_p (Btu/lb$_m$ °F)	ρ (lb$_m$/ft^3)	t (°F)	k (Btu/hr ft^2 °F)	α (ft^2/hr)
Structural					
Asphalt			68	0.43[a]	
Bakelite	0.38[b]	79.5[b]	68	0.134[b]	0.0044
Bricks					
Common	0.20[d]	100[d]	68	0.40[a]	0.02
Face		128[d]	68	0.76[a]	
Carborundum brick			1110	10.7[a]	
			2550	6.4[a]	
			392	1.34[a]	0.036
Chrome brick	0.20[d]	188[d]	1022	1.43[a]	0.038
			1652	1.15[a]	0.031
Diatomaceous earth			400	0.14[a]	
(fired)			1600	0.18[a]	
			932	0.60[a]	0.020
Fire clay brick (burnt	0.23[d]	128[d]	1472	0.62[a]	0.021
2426 °F)			2012	0.63[a]	0.021
			932	0.74[a]	0.022
Fire clay brick (burnt	0.23[d]	145[d]	1472	0.79[a]	0.024
2642 °F)			2012	0.81[a]	0.024
			392	0.58[a]	0.015
Fire clay brick (Missouri)	0.23[d]	165[f]	1112	0.85[a]	0.022
			2552	1.02[a]	0.027
			400	2.2[a]	
Magnesite	0.27[d]		1200	1.6[a]	
			2200	1.1[a]	
Cement, portland		94		0.17[a]	
Cement, mortar			75	0.67[a]	
Concrete	0.21[b]	119–144[b]	68	0.47–0.81[b]	0.019–0.027
Concrete, cinder			75	0.44[a]	
Glass, plate	0.2[b]	169[b]	68	0.44[b]	0.013
Glass, borosilicate		139[b]	86	0.63[b]	
Plaster, gympsum	0.2[d]	90[d]	70	0.28[a]	0.016
Plaster, metal lath			70	0.27[a]	
Plaster, wood lath			70	0.16[a]	
Stone					
Granite	0.195[d]	165[d]		1.0–2.3[a]	0.031–0.071
Limestone	0.217[d]	155[d]	210–570	0.73–0.77[a]	0.022–0.023
Marble	0.193[b]	156–169[b]	68	1.6[b]	0.054
Sandstone	0.17[b]	135–144[b]	68	0.94–1.2[b]	0.041–0.049
Wood, cross grain					
Balsa		8.8[a]	86	0.032[a]	
Cypress		29[d]	86	0.056[a]	
Fir	0.65[d]	26.0[b]	75	0.063[a]	0.0037
Oak	0.57[d]	39–30[b]	86	0.096[a]	0.0049
Yellow pine	0.67[d]	40[d]	75	0.085[a]	0.0032
White pine		27[d]	86	0.065[a]	
Wood, radial					
Oak	0.57[b]	38–30[b]	68	0.10–0.12[b]	0.0043 – 0.047
Fir	0.65[b]	26.0–26.3[b]	68	0.08[b]	0.048

Table D.2 Thermal Properties of Some Nonmetals and Insulation (*Continued*)

Substance	C_p (Btu/lb$_m$ °F)	ρ (lb$_m$/ft^3)	t (°F)	k (Btu/hr ft^2 °F)	α (ft^2/hr)
Insulating					
Asbestos		29.3[b]	$\{$ −328	0.043[b]	
			32	0.090[b]	
			32	0.087[b]	
			212	0.111[b]	
		36.0[b]	392	0.120[b]	
			752	0.129[b]	
		43.5[b]	$\{$ −328	0.09[b]	
			32	0.135[b]	
Asbestos cement				1.2[a]	
Asbestos cement board			68	0.43[a]	
Asbestos sheet			124	0.096[a]	
Asbestos felt (40 laminations per inch)			100	0.033[a]	
			300	0.040[a]	
			500	0.048[a]	
Asbestos felt (20 laminations per inch			100	0.045[a]	
			300	0.055[a]	
			500	0.065[a]	
Asbestos, corrugated (4 plies per inch)			100	0.05[a]	
			200	0.058[a]	
			300	0.069[a]	
Balsam wool		2.2[a]	90	0.023[a]	
Cardboard, corrugated				0.037[a]	
Celotex			90	0.028[a]	
Corkboard		10[b]	86	0.025[b]	
Cork, expanded scrap	0.45[b]	2.8–7.4[b]	68	0.021[b]	0.006–0.017
Cork, ground		9.4[b]	86	0.025[b]	
Insulating					
Diatomaceous earth (powdered)		10[e]	200	0.029[e]	
			400	0.038[e]	
			600	0.048[e]	
Diatomaceous earth (powdered)		14[e]	200	0.033[e]	
			400	0.039[e]	
			600	0.046[e]	
Diatomaceous earth (powdered)		18[e]	200	0.040[e]	
			400	0.045[e]	
			600	0.049[e]	
Felt, hair		8.2[c]	20	0.0237[c]	
			100	0.0269[c]	
			200	0.0310[c]	
Felt, hair		11.4[c]	20	0.0212[c]	
			100	0.0254[c]	
			200	0.0299[c]	
Felt, hair		12.8[c]	20	0.0233[c]	
			100	0.0262[c]	
			200	0.0295[c]	
Fiber insulating board		14.8[b]	70	0.028[b]	
			20	0.0217[c]	

Table D.2 Thermal Properties of Some Nonmetals and Insulation (*Continued*)

Substance	C_p (Btu/lb$_m$ °F)	ρ (lb$_m$/ft^3)	t (°F)	k (Btu/hr ft^2 °F)	α (ft^2/hr)
Glass wool		1.5[c]	100	0.0313[c]	
			200	0.0435[c]	
			20	0.0179[c]	
Glass wool		4.0[c]	100	0.0239[c]	
			200	0.0317[c]	
			20	0.0163[c]	
Glass wool		6.0[c]	100	0.0218[c]	
			200	0.0288[c]	
Kapok			86	0.020[a]	
			100	0.039[a]	
Magnesia, 85%		16.9[c]	200	0.041[a]	
			300	0.043[a]	
			400	0.046[a]	
Rock wool		4.0[c]	100	0.0224[c]	
			200	0.0317[c]	
			20	0.0171[c]	
Rock wool		8.0[c]	100	0.0228[c]	
			200	0.0299[c]	
			20	0.0183[c]	
Rock wool		12.0[c]	100	0.0226[c]	
			200	0.0281[c]	
Miscellaneous					
Aerogel, silica		8.5[b]	248	0.013[b]	
Clay	0.21[b]	91.0[b]	68	0.739[b]	0.039
Coal, anthracite	0.30[b]	75–94[b]	68	0.15[b]	0.005–0.006
Coal, powdered	0.31[b]	46[b]	86	0.067[b]	0.005
Cotton	0.31[b]	5[b]	68	0.034[b]	0.075
Earth, coarse	0.44[b]	128[b]	68	0.30[b]	0.0054
Ice	0.46[b]	57[b]	32	1.28[b]	0.048
Rubber, hard		74.8[b]	32	0.087[b]	
Sawdust			75	0.034[a]	
Silk	0.33[b]	3.6[b]	68	0.021[b]	0.017

Source: Adapted from (a) A. I. Brown and S. M. Marco, *Introduction to Heat Transfer*, 3rd ed., McGraw-Hill, New York, 1958; (b) E. R. G. Eckert, *Introduction to the Transfer of Heat and Mass*, McGraw-Hill, New York, 1950; (c) R. H. Heilman, *Industrial and Engineering Chemistry*, Vol. 28 (1936), p. 782; (d) L. S. Marks, *Mechanical Engineers' Handbook*, 6th ed., McGraw-Hill, New York, 1958; (e) R. Calvert, *Diatomaceous Earth*, Chemical Catalog Company, Inc., 1930; (f) H. F. Norton, *Journal of the American Ceramic Society*, Vol. 10 (1957), p. 30.

THERMOPHYSICAL PROPERTIES
OF SATURATED LIQUIDS

Table E Property Values of Liquids in Saturated State

T (°F)	ρ (lb/ft^3)	C_p (Btu/lb °F)	ν (ft^2/sec)	k (Btu/hr ft °F)	α (ft^2/hr)	Pr
			Water (H_2O)			
32	62.57	1.0074	1.925×10^{-5}	0.319	5.07×10^{-3}	13.6
68	62.46	0.9988	1.083	0.345	5.54	7.02
104	62.09	0.9980	0.708	0.363	5.86	4.34
140	61.52	0.9994	0.514	0.376	6.02	3.02
176	60.81	1.0023	0.392	0.386	6.34	2.22
212	59.97	1.0070	0.316	0.393	5.51	1.74
248	59.01	1.015	0.266	0.396	6.62	1.446
284	57.95	1.023	0.230	0.395	6.68	1.241
320	56.79	1.037	0.204	0.393	6.70	1.099
356	55.50	1.055	0.186	0.390	6.68	1.004
392	54.11	1.076	0.172	0.384	6.61	0.937
428	52.59	1.101	0.161	0.377	6.51	0.891
464	50.92	1.136	0.154	0.367	6.35	0.871
500	49.06	1.182	0.148	0.353	6.11	0.874
537	46.98	1.244	0.145	0.335	5.74	0.910
572	44.59	1.368	0.145	0.312	5.13	1.019

Table E Property Values of Liquids in Saturated State (*Continued*)

T (°F)	ρ (lb/ft^3)	C_p (Btu/lb °F)	ν (ft^2/sec)	k (Btu/hr ft °F)	α (ft^2/hr)	Pr
			Ammonia (NH$_3$)			
−58	43.93	1.066	0.468×10^{-5}	0.316	6.75×10^{-3}	2.60
−40	43.18	1.067	0.437	0.316	6.88	2.28
−22	42.41	1.069	0.417	0.317	6.98	2.15
−4	41.62	1.077	0.410	0.316	7.05	2.09
14	40.80	1.090	0.407	0.314	7.07	2.07
32	39.96	1.107	0.402	0.312	7.05	2.05
50	39.09	1.126	0.396	0.307	6.98	2.04
68	38.19	1.146	0.386	0.301	6.88	2.02
86	37.23	1.168	0.376	0.293	6.75	2.01
104	36.27	1.194	0.366	0.285	6.59	2.00
122	35.23	1.222	0.355	0.275	6.41	1.99
			Dichlorodifluoromethane, Freon-12 (CCl$_2$F$_2$)			
−58	96.56	0.2090	0.334×10^{-5}	0.039	1.94×10^{-3}	6.2
−40	94.81	0.2113	0.300	0.040	1.99	5.4
−22	92.99	0.2139	0.272	0.040	2.04	4.8
−4	91.18	0.2167	0.253	0.041	2.09	4.4
14	89.24	0.2198	0.238	0.042	2.13	4.0
32	87.24	0.2232	0.230	0.042	2.16	3.8
50	85.17	0.2268	0.219	0.042	2.17	3.6
68	83.04	0.2307	0.213	0.042	2.17	3.5
86	80.85	0.2349	0.209	0.041	2.17	3.5
104	78.48	0.2393	0.206	0.040	2.15	3.5
122	75.91	0.2440	0.204	0.039	2.11	3.5
			Glycerin [C$_3$H$_5$(OH)$_3$]			
32	79.66	0.540	0.0895	0.163	3.81×10^{-3}	84.7×10^3
50	79.29	0.554	0.323	0.164	3.74	31.0
68	78.91	0.570	0.0127	0.165	3.67	12.5
86	78.54	0.584	0.0054	0.165	3.60	5.38
104	78.16	0.600	0.0024	0.165	3.54	2.45
122	77.72	0.617	0.0016	0.166	3.46	1.63
			Ethylene Glycol [C$_2$H$_4$(OH)$_2$]			
32	70.59	0.548	61.92×10^{-5}	0.140	3.62×10^{-3}	615
68	69.71	0.569	20.64	0.144	3.64	204
104	68.76	0.591	9.35	0.148	3.64	93
140	67.90	0.612	5.11	0.150	3.61	51
176	67.27	0.633	3.21	0.151	3.57	32.4
212	66.08	0.655	2.18	0.152	3.52	22.4
			Engine Oil (Unused)			
32	56.13	0.429	0.0461	0.085	3.53×10^{-3}	47100
68	55.45	0.449	0.0097	0.084	3.38	10400
104	54.69	0.469	0.0026	0.083	3.23	2870
140	53.94	0.489	0.903×10^{-3}	0.081	3.10	1050
176	53.19	0.509	0.404	0.080	2.98	490

Table E Property Values of Liquids in Saturated State (*Continued*)

T (°F)	ρ (lb/ft^3)	C_p (Btu/lb °F)	ν (ft^2/sec)	k (Btu/hr ft °F)	α (ft^2/hr)	Pr

Engine Oil (Unused) (*Continued*)

T (°F)	ρ	C_p	ν	k	α	Pr
212	52.44	0.530	0.219	0.070	276	
248	51.75	0.551	0.133	0.078	2.75	175
284	51.00	0.572	0.086	0.077	2.66	116
320	50.31	0.593	0.060	0.076	2.57	84

Mercury (Hg)

32	850.78	0.0335	0.133×10^{-5}	4.74	166.6×10^{-3}	0.0288
68	847.71	0.0333	0.123	5.02	178.5	0.249
122	843.14	0.0331	0.112	5.43	194.6	0.0207
212	835.57	0.0328	0.0999	6.07	221.5	0.0162
302	828.06	0.0326	0.0918	6.64	246.2	0.0134
392	820.61	0.0375	0.0863	7.13	267.7	0.0116
482	813.16	0.0324	0.0823	7.55	287.0	0.0103
600	802	0.032	0.0724	8.10	316	0.0083

Source: From E. R. G. Eckert and R. M. Drake, *Heat and Mass Transfer,* 1959, McGraw-Hill, reprinted by permission.

THERMOPHYSICAL PROPERTIES OF GASES

Table F Property Values of Gases at Atmospheric Pressure

T ($^\circ$F)	ρ (lb/ft^3)	C_p (Btu/lb $^\circ$F)	μ (lb/sec ft)	k (Btu/hr ft $^\circ$F)	α (ft^2/hr)	Pr
			Air			
−280	0.2248	0.2452	0.4653×10^{-5}	0.005342	0.09691	0.770
−190	0.1478	0.2412	0.6910	0.007936	0.2226	0.753
−100	0.1104	0.2403	0.8903	0.01045	0.3939	0.739
−10	0.0882	0.2401	1.074	0.01287	0.5100	0.722
80	0.0735	0.2402	1.241	0.01516	0.8587	0.708
170	0.0623	0.2410	1.394	0.01735	1.156	0.697
260	0.0551	0.2422	1.536	0.01944	1.457	0.689
350	0.0489	0.2438	1.669	0.02142	1.636	0.683
440	0.0440	0.2459	1.795	0.02333	2.156	0.680
530	0.0401	0.2482	1.914	0.02519	2.531	0.680
620	0.0367	0.2520	2.028	0.02692	2.911	0.680
710	0.0339	0.2540	2.135	0.02862	3.324	0.682
800	0.0314	0.2568	2.239	0.03022	3.748	0.684
890	0.0294	0.2593	2.339	0.03183	4.175	0.686
980	0.0275	0.2622	2.436	0.03339	4.631	0.689
1070	0.0259	0.2650	2.530	0.03483	5.075	0.692
1160	0.0245	0.2678	2.620	0.03628	5.530	0.696
1250	0.0232	0.2704	2.703	0.03770	6.010	0.699
1340	0.0220	0.2727	2.790	0.03901	6.502	0.702
1520	0.0200	0.2772	2.955	0.04178	7.536	0.706
1700	0.0184	0.2815	3.109	0.04410	8.514	0.714
1880	0.0169	0.2860	3.258	0.04641	9.602	0.722
2060	0.0157	0.2900	3.398	0.04880	10.72	0.726

Table F Property Values of Gases at Atmospheric Pressure (*Continued*)

T (°F)	ρ (lb/ft^3)	C_p (Btu/lb °F)	μ (lb/sec ft)	k (Btu/hr ft °F)	α (ft^2/hr)	Pr
			Air (*Continued*)			
2240	0.0147	0.2939	3.533×10^{-5}	0.05098	11.80	0.734
2420	0.0138	0.2982	3.668	0.05348	12.88	0.741
2600	0.0130	0.3028	3.792	0.05550	14.00	0.749
2780	0.0123	0.3075	3.915	0.05750	15.09	0.759
2960	0.0116	0.3128	4.029	0.0591	16.40	0.767
3140	0.0110	0.3196	4.168	0.0612	17.41	0.783
3320	0.0105	0.3278	4.301	0.0632	18.36	0.803
3500	0.0100	0.3390	4.398	0.0646	19.05	0.831
3680	0.0096	0.3541	4.513	0.0663	19.61	0.863
3860	0.0091	0.3759	4.611	0.0681	19.92	0.916
4160	0.0087	0.4031	4.750	0.0709	20.21	0.972
			Hydrogen			
−406	0.05289	2.589	1.079×10^{-6}	0.0132	0.966	0.759
−370	0.03181	2.508	1.691	0.0209	0.262	0.721
−280	0.01534	2.682	2.830	0.0384	0.933	0.712
−190	0.01022	3.010	3.760	0.0567	1.84	0.718
−100	0.00766	3.234	4.578	0.0741	2.99	0.719
−10	0.00613	3.358	5.321	0.0902	4.38	0.713
80	0.00511	3.419	6.023	0.105	6.02	0.706
170	0.00438	3.448	6.689	0.119	7.87	0.697
260	0.00383	3.461	7.300	0.132	9.95	0.690
350	0.00341	3.463	7.915	0.145	12.26	0.682
440	0.00307	3.465	8.491	0.157	14.79	0.675
530	0.00279	3.471	9.055	0.169	17.50	0.668
620	0.00255	3.472	9.599	0.182	20.56	0.664
800	0.00218	3.481	10.68	0.203	26.75	0.659
980	0.00191	3.505	11.69	0.222	33.18	0.664
1160	0.00170	3.540	12.62	0.238	39.59	0.676
1340	0.00153	3.575	13.55	0.254	46.49	0.686
1520	0.00139	3.622	14.42	0.268	53.19	0.703
1700	0.00128	36.70	15.29	0.282	60.00	0.715
1880	0.00118	3.720	16.18	0.296	67.40	0.733
1940	0.00115	3.735	16.42	0.300	69.80	0.736
			Helium			
−456		1.242	5.66×10^{-7}	0.0061		
−400	0.0915	1.242	33.7	0.0204	0.1792	0.74
−200	0.211	1.242	84.3	0.0536	2.044	0.70
−100	0.0152	1.242	105.2	0.0680	3.599	0.694
0	0.0119	1.242	122.1	0.0784	5.299	0.70
200	0.00829	1.242	154.9	0.0977	9.490	0.71
400	0.00637	1.242	184.8	0.114	14.40	0.72
600	0.00517	1.242	209.2	0.130	20.21	0.72
800	0.00439	1.242	233.5	0.145	25.81	0.72
1000	0.00376	1.242	256.5	0.159	34.00	0.72
1200	0.0030	1.242	277.9	0.172	41.98	0.72

Table F Property Values of Gases at Atmospheric Pressure (*Continued*)

T (°F)	ρ (lb/ft^3)	C_p (Btu/lb °F)	μ (lb/sec ft)	k (Btu/hr ft °F)	α (ft^2/hr)	Pr
			Oxygen			
−280	0.2492	0.2264	5.220×10^{-6}	0.00522	0.09252	0.815
−190	0.1635	0.2192	7.721	0.00790	0.2204	0.773
−100	0.1221	0.2181	9.979	0.01054	0.3958	0.745
−10	0.0975	0.2187	12.01	0.01305	0.6120	0.725
80	0.0812	0.2198	13.86	0.01546	0.8662	0.709
170	0.0695	0.2219	15.56	0.01774	1.150	0.702
260	0.0609	0.2250	17.16	0.02000	1.460	0.695
350	0.0542	0.2285	18.66	0.02212	1.786	0.694
440	0.0487	0.2322	20.10	0.02411	2.132	0.697
530	0.0443	0.2360	21.48	0.02610	2.496	0.700
620	0.0406	0.2399	22.79	0.02792	2.867	0.704
			Nitrogen			
−280	0.2173	0.2561	4.611×10^{-6}	0.005460	0.09811	0.786
−100	0.1068	0.2491	8.700	0.01054	0.3962	0.747
80	0.0713	0.2486	11.99	0.01514	0.8542	0.713
260	0.0533	0.2498	14.77	0.01927	1.477	0.691
440	0.0426	0.2521	17.27	0.02302	2.143	0.684
620	0.0355	0.2569	19.56	0.02646	2.901	0.686
800	0.0308	0.2620	21.59	0.02960	3.668	0.691
980	0.0267	0.2681	23.41	0.03241	4.528	0.700
1160	0.0237	0.2738	25.19	0.03507	5.404	0.711
1340	0.0213	0.2789	26.88	0.03741	6.297	0.724
1520	0.0194	0.2832	28.41	0.03958	7.204	0.736
1700	0.0178	0.2875	29.90	0.04151	8.111	0.748
			Carbon Dioxide			
−64	0.1544	0.187	7.462×10^{-6}	0.006243	0.2294	0.818
−10	0.1352	0.192	0.460	0.007444	0.2868	0.793
80	0.1122	0.208	10.051	0.009575	0.4103	0.770
170	0.0959	0.215	11.561	0.01183	0.5738	0.755
260	0.0838	0.225	12.98	0.01422	0.7542	0.738
350	0.0744	0.234	14.34	0.01674	0.9615	0.721
440	0.0670	0.242	15.63	0.01937	1.195	0.702
530	0.0608	0.250	16.85	0.02208	1.453	0.685
620	0.0558	0.257	18.03	0.02491	1.737	0.668
			Carbon Monoxide			
−64	0.09699	0.2491	9.295×10^{-6}	0.01101	0.4557	0.758
−10	0.0525	0.2490	10.35	0.01239	0.5837	0.750
80	0.07109	0.2489	11.990	0.01459	0.8246	0.737
170	0.06082	0.2492	13.50	0.01666	1.099	0.728
260	0.05329	0.2504	14.91	0.01864	1.397	0.722
350	0.04735	0.2520	16.25	0.0252	1.720	0.718
440	0.04259	0.2540	17.51	0.02232	2.063	0.718
530	0.03872	0.2569	18.74	0.02405	2.418	0.721
620	0.03549	0.2598	19.89	0.02569	2.786	0.724

Table F Property Values of Gases at Atmospheric Pressure *(Continued)*

T (°F)	ρ (lb/ft^3)	C_p (Btu/lb °F)	μ (lb/sec ft)	k (Btu/hr ft °F)	α (ft^2/hr)	Pr
			Steam (H$_2$O Vapor)			
224	0.0366	0.294	8.54×10^{-6}	0.0142	0.789	0.1060
260	0.0346	0.481	9.03	0.0151	0.906	1.040
350	0.0306	0.473	10.25	0.0173	1.19	1.010
440	0.0275	0.474	11.45	0.0196	1.50	0.996
530	0.0250	0.477	12.66	0.0219	1.84	0.991
620	0.0228	0.484	13.89	0.0244	2.22	0.986
710	0.0211	0.491	15.10	0.0268	2.58	0.995
800	0.0196	0.498	16.30	0.0292	2.99	1.000
890	0.0183	0.506	17.50	0.0317	3.42	1.005
980	0.0171	0.514	18.72	0.0342	3.88	1.010
1070	0.0161	0.522	19.95	0.0368	4.38	1.019
			Ammonia (NH$_3$)			
−58	0.0239	0.525	4.875×10^{-6}	0.0099	0.796	0.93
32	0.0495	0.520	6.285	0.012	0.507	0.90
122	0.0405	0.520	7.415	0.0156	0.744	0.88
212	0.0349	0.534	8.659	0.0189	1.015	0.87
302	0.0308	0.553	9.859	0.0226	1.330	0.87
392	0.0275	0.572	11.08	0.0270	1.713	0.84

Source: From E. R. G. Eckert and R. M. Drake, *Heat and Mass Transfer,* 1959, McGraw-Hill, reprinted with permission.

APPENDIX
G

RADIATIVE PROPERTIES

Table G.1 Total Emissivity Data

Surface	°C	°F	ϵ
	Metals		
Aluminum			
Polished, 98% pure	200–600	400–1100	0.04–0.06
Commercial sheet	100	200	0.09
Rough plate	40	100	0.07
Heavily oxidized	100–550	200–1000	0.20–0.33
Antimony			
Polished	40–250	100–500	0.28–0.31
Bismuth			
Bright	100	200	0.34
Brass			
Highly polished	250	500	0.03
Polished	40	100	0.07
Dull plate	40–250	100–500	0.22
Oxidized	40–250	100–500	0.46–0.56
Chromiun			
Polished seet	40–550	100–1000	0.08–0.27
Cobalt			
Unoxidized	250–550	500–1000	0.13–0.23
Copper			
Highly polished electrolytic	100	200	0.02
Polished	40	100	0.04
Slightly polished	40	100	0.12
Polished, lightly tarnished	40	100	0.05
Dull	40	100	0.15
Black oxidized	40	100	0.76

Table G.1 Total Emissivity Data (*Continued*)

Surface	°C	°F	ϵ
Gold			
Pure, highly polished	100–600	200–1100	0.02–0.035
Inconel			
X, stably oxidized	230–900	450–1600	0.55–0.78
B, stably oxidized	230–1000	450–1750	0.32–0.55
X and B, polished	150–300	300–600	0.20
Iron and Steel			
Mild steel, polished	150–500	300–900	0.14–0.32
Steel, polished	40–250	100–500	0.07–0.10
Sheet steel, ground	1000	1700	0.55
Sheet steel, rolled	40	100	0.66
Sheet steel, strong rough oxide	40	100	0.80
Steel, oxidized at 110 °F	250	500	0.79
Cast iron, with skin	40	100	0.70–0.80
Cast iron, newly turned	40	100	0.44
Cast iron, polished	200	400	0.21
Cast iron, oxidized	40–250	100–500	0.57–0.66
Iron, red rusted	40	100	0.61
Iron, heavily rusted	40	100	0.85
Wrought iron, smooth	40	100	0.35
Wrought iron, dull oxidized	20–360	70–680	0.94
Stainless, polished	40	100	0.07–0.17
Stainless, after repated heating and cooling	230–930	450–1650	0.50–0.70
Lead			
Polished	40–250	100–500	0.05–0.08
Gray, oxidized	40	100	0.28
Oxidized at 390 °F	200	400	0.63
Oxidized at 1100 °F	40	100	0.63
Magnesium			
Polished	40–250	100–500	0.07–0.13
Manganin			
Bright rolled	100	200	0.05
Mercury			
Pure, clean	40–100	100–200	0.10–0.12
Molybdenum			
Polished	40–250	100–500	0.06–0.08
Polished	550–1100	1000–2000	0.11–0.18
Filament	550–2800	1000–5000	0.08–0.29
Monel			
After repeated heating and cooling	230–930	450–1650	0.45–0.70
Oxidized at 1100 °F	200–600	400–1100	0.41–0.46
Polished	40	100	0.17
Nickel			
Polished	40–250	100–500	0.05–0.07
Oxidized	40–250	100–500	0.35–0.49
Wire	250–1100	500–2000	0.10–0.19
Platinum			

Table G.1 Total Emissivity Data (*Continued*)

Surface	°C	°F	ϵ
Pure, polished plate	200–600	400–1100	0.05–0.10
Oxidized at 100°F	250–550	500–1000	0.07–0.11
Electrolytic	250–550	500–1000	0.06–0.10
Strip	550–1100	1000–2000	0.12–0.14
Filament	40–1100	100–2000	0.04–0.19
Wire	200–1370	400–2500	0.07–0.18
Silver			
Polished or deposited	40–550	100–1000	0.01–0.03
Oxidized	40–550	100–1000	0.02–0.04
German silver,[a]	250–550	500–1000	0.07–0.09
Tin			
Bright tinned iron	40	100	0.04–0.06
Bright	40	100	0.06
Polished sheet	100	200	0.05
Tungsten			
Filament	550–1100	1000–2000	0.11–0.16
Filament	2800	5000	0.39
Filament, aged	440–3300	100–6000	0.03–0.35
Polished	40–550	100–1000	0.04–0.08
Zinc			
Pure polished	40–250	100–500	0.02–0.03
Oxidized at 750°F	400	750	0.11
Galvanized, gray	40	100	0.28
Galvanized, fairly bright	40	100	0.23
Dull	40–250	100–500	0.21
Nonmetals			
Asbestos			
Board	40	100	0.96
Cement	40	100	0.96
Paper	40	100	0.93–0.95
Slate	40	100	0.97
Brick			
Red, rough	40	100	0.93
Silica	1000	1800	0.80–0.85
Fireclay	1000	1800	0.75
Ordinary refractory	1100	2000	0.59
Magnesite refractory	1000	1800	0.38
White refractory	1100	2000	0.29
Gray, glazed	1100	2000	0.75
Carbon			
Filament	1050–1420	1900–2600	0.53
Lampsoot	40	100	0.95
Clay			
Fired	100	200	0.91
Concrete			
Rough	40	100	0.94
Corundum			
Emery rough	100	200	0.86

[a]German silver is actually an alloy of copper, nickel, and zinc.

Table G.1 Total Emissivity Data (*Continued*)

Surface	°C	°F	ϵ
Glass			
Smooth	40	100	0.94
Quartz glass (2 mm)	250–550	500–1000	0.96–0.66
Pyrex	250–550	500–1000	0.94–0.75
Gypsum	40	100	0.80–0.90
Ice			
Smooth	0	32	0.97
Rough crystals	0	32	0.99
Hoarfrost	−18	0	0.99
Limestone	40–250	100–500	0.95–0.83
Marble			
Light gray, polished	40	100	0.93
White	40	100	0.95
Mica	40	100	0.75
Paints			
Aluminum, various ages and compositions	100	200	0.27–0.62
Black glass	40	100	0.90
Black lacquer	40	100	0.80–0.93
White paint	40	100	0.89–0.97
White lacquer	40	100	0.80–0.95
Various oil paints	40	100	0.92–0.96
Red lead	100	200	0.93
Paper			
White	40	100	0.95
Writing paper	40	100	0.98
Any color	40	100	0.92–0.94
Roofing	40	100	0.91
Plaster			
Lime, rough	40–250	100–500	0.92
Porcelain			
Glazed	40	100	0.93
Quartz	40–550	100–1000	0.89–0.58
Rubber			
Hard	40	100	0.94
Soft, gray rough	40	100	0.86
Sandstone	40–250	100–250	0.83–0.90
Snow	(−12) − (−6)	10–20	0.82
Water			
0.11 mm or more thick	40	100	0.96
Wood			
Oak, planed	40	100	0.90
Walnut, sanded	40	100	0.83
Spruce, sanded	40	100	0.82
Beech	40	100	0.94
Planed	40	100	0.78
Various	40	100	0.80–0.90
Sawdust	40	100	0.75

Source: From E. M. Sparrow and R. D. Cess, *Radiation Heat Transfer,* rev. ed., copyright © 1970 by Wadsworth Publishing Company, Inc., reprinted by permission of the authors.

ELECTRICAL CHARACTERISTICS
AND PROPERTIES

Table H.1 Cable Impedances

Approximate 60-cycle resistance, reactance, and impedance of copper conductor cable per 100 feet[a]

Three-conductor cable

AWG or MCM	In magnetic duct[b]						In nonmagnetic duct[c]					
	600 volts and 5 kV nonshielded			5 kV shielded and 15 kV			600 volts and 5 kV nonshielded			5 kV shielded and 15 kV		
	R	X	Z	R	X	Z	R	X	Z	R	X	Z
8	.811	.0577	.813	.811	.0658	.814	.811	.0503	.812	.811	.0574	.813
8 (solid)	.786	.0577	.788	.786	.0658	.789	.786	.0503	.787	.786	.0574	.788
6	.510	.0525	.513	.510	.0610	.514	.510	.0457	.512	.510	.0531	.513
6 (solid)	.496	.0525	.499	.496	.0610	.500	.496	.0457	.498	.496	.0531	.499
4	.321	.0483	.325	.321	.0568	.326	.321	.0422	.324	.321	.0495	.325
4 (solid)	.312	.0483	.316	.312	.0508	.317	.312	.4022	.315	.312	.0495	.316
2	.202	.0448	.207	.202	.0524	.209	.202	.0390	.206	.202	.0457	.207
1	.160	.0436	.166	.160	.0516	.168	.160	.0380	.164	.160	.0450	.166
1/0	.128	.0414	.135	.128	.0486	.137	.127	.0360	.132	.128	.0423	.135
2/0	.102	.0407	.110	.103	.0482	.114	.101	.0355	.107	.102	.0420	.110
3/0	.0805	.0397	.0898	.0814	.0463	.0936	.0766	.0346	.0841	.0805	.0403	.090
4/0	.0640	.0381	.0745	.0650	.0446	.0788	.0633	.0332	.0715	.0640	.0389	.0749
250	.0552	.0379	.0670	.0557	.0436	.0707	.0541	.0330	.0634	.0547	.0380	.0666
300	.0464	.0377	.0598	.0473	.0431	.0640	.0451	.0329	.0559	.0460	.0376	.0596
350	.0378	.0373	.0539	.0386	.0427	.0576	.0368	.0328	.0492	.0375	.0375	.0530
400	.0356	.0371	.0514	.0362	.0415	.0551	.0342	.0327	.0475	.0348	.0366	.0505
450	.0322	.0361	.0484	.0328	.0404	.0520	.0304	.0320	.0441	.0312	.0359	.0476
500	.0294	.0349	.0456	.0300	.0394	.0495	.0276	.0311	.0416	.0284	.0351	.0453
600	.0257	.0343	.0429	.0264	.0382	.0464	.0237	.0309	.0389	.0246	.0344	.0422
750	.0216	.0326	.0391	.0223	.0364	.0427	.0197	.0297	.0355	.0203	.0332	.0389

Three single-conductor cable

8	.811	.0754	.814	.811	.0860	.816	.811	.0603	.813	.813	.0688	.814
8 (solid)	.786	.0754	.790	.786	.0860	.791	.786	.0603	.788	.786	.0688	.789
6	.510	.0685	.615	.510	.0796	.516	.510	.0548	.513	.510	.0636	.514
6 (solid)	.496	.0685	.501	.496	.0796	.502	.496	.0548	.499	.496	.0636	.500
4	.321	.0632	.327	.321	.0742	.329	.321	.0506	.325	.321	.0594	.326
4 (solid)	.312	.0632	.318	.312	.0742	.321	.312	.0506	.316	.312	.0594	.318
2	.202	.0585	.210	.202	.0685	.214	.202	.0467	.207	.202	.0547	.209
1	.160	.0570	.170	.160	.0675	.174	.160	.0456	.166	.160	.0540	.169
1/0	.128	.0540	.139	.128	.0635	.143	.127	.0432	.134	.128	.0507	.138
2/0	.102	.0533	.115	.103	.0630	.121	.101	.0426	.110	.102	.0504	.114
3/0	.0805	.0519	.0958	.0814	.0605	.101	.0766	.0415	.0871	.0805	.0484	.0939
4/0	.0640	.0497	.0810	.0650	.0583	.0929	.0633	.0398	.0748	.0640	.0466	.0792
250	.0552	.0495	.0742	.0557	.0570	.0797	.0541	.0396	.0670	.0547	.0456	.0712
300	.0464	.0493	.0677	.0473	.0564	.0736	.0451	.0394	.0599	.0460	.0451	.0644
350	.0378	.0491	.0617	.0386	.0562	.0681	.0368	.0393	.0536	.0375	.0450	.0586
400	.0356	.0490	.0606	.0362	.0548	.0657	.0342	.0392	.0520	.0348	.0438	.0559
450	.0322	.0480	.0578	.0328	.0538	.0630	.0304	.0384	.0490	.0312	.0430	.0531
500	.0294	.0466	.0551	.0300	.0526	.0505	.0276	.0373	.0464	.0284	.0421	.0508
600	.0257	.0463	.0530	.0264	.0516	.0580	.0237	.0371	.0440	.0246	.0412	.0479
750	.0216	.0445	.0495	.0223	.0497	.0545	.0194	.0356	.0405	.0203	.0396	.0445

[a] Resistance based on tinned copper at 60 cycles per 1000 feet at 75°C. Reactance of 600-V and 5-kV nonshielded cable based on varnished-cambric insulation. Reactance of 5-kV shielded and 15-kV cable based on neoprene insulation.

[b] Also applies to steel interlocked armor used on 3/c cables.

[c] Also applies to aluminum interlocked armor used on 3/c cables

Source: Westinghouse Construction Specifications, Catalogue 55-000, 3d ed., 1975. Reprinted with permission.

Table H.2 Overhead Line Impedances[a]

Phase conductor wire size (cm)	Strands	Positive and negative sequence impedance components[b,c]		
		$R_1 = R_2$	$X_1 = X_2$	$Z_1 = Z_2$
Copper conductors				
500,000	19	.0246	.1195	.1216
450,000	19	.0273	.1206	.1252
400,000	19	.0307	.1220	.1258
350,000	19	.0348	.1235	.1284
300,000	19	.0407	.1254	.1318
250,000	19	.0487	.1275	.1364
4/0	19	.0574	.1294	.1415
3/0	12	.0723	.1309	.1494
2/0	7	.0911	.1360	.1640
1/0	7	.1150	.1386	.1799
1	7	.1449	.1449	.2027
2	7	.1809	.1434	.2301
3	3	.2280	.1460	.2708
4	1	.2847	.1506	.3220
6	1	.4527	.1559	.4792
8	1	.7197	.1612	.7405
Bare all-aluminum conductors				
795,000	37	.0248	.1138	.1165
750,000	37	.0263	.1146	.1174
715,500	37	.0277	.1150	.1184
700,000	61	.0282	.1152	.1186
636,000	37	.0309	.1163	.1199
600,000	61	.0328	.1169	.1216
556,500	37	.0352	.1180	.1233
500,000	37	.0392	.1189	.1250
477,000	37	.0411	.1195	.1263
450,000	37	.0436	.1203	.1278
400,000	37	.0498	.1214	.1309
397,500	19	.0492	.1220	.1316
350,000	37	.0557	.1231	.1347
336,400	37	.580	.1237	.1366
300,000	37	.0650	.1252	.1407
266,800	37	.0731	.1265	.1460
250,000	37	.0778	.1271	.1489
4/0	19	.0920	.1284	.1580
3/0	19	.1159	.1311	.1744
2/0	19	.1466	.1347	.1989
1/0	19	.1845	.1377	.2301
1	7	.2330	.1413	.2731
2	7	.2934	.1428	.3263
3	7	.3701	.1466	.3981
4	7	.4661	.1492	.4886
6	7	.7424	.1547	.7576

Table H.2 Overhead Line Impedances[a] (*Continued*)

Phase conductor wire size (cm)	Strands	Positive and negative sequence impedance components[b,c]		
		$R_1 = R_2$	$X_1 = X_2$	$Z_1 = Z_2$
ACSR conductors				
795,000	26	.0244	.1108	.1138
715,000	26	.0273	.1119	.1153
666,600	54	.0303	.1133	.1170
636,000	26	.0307	.1133	.1172
605,000	26	.0326	.1138	.1188
556,500	26	.0352	.1148	.1203
500,000	30	.0390	.1150	.1214
477,000	26	.0409	.1167	.1239
397,500	26	.0491	.1188	.1284
336,400	26	.5080	.1206	.1341
300,000	26	.0648	.1220	.1379
266,800	26	.0729	.1233	.1430
4/0	6	.1121	.1453	.1833
3/0	6	.1369	.1528	.2055
2/0	6	.1695	.1566	.2311
1/0	6	.2121	.1595	.2655
1	6	.2614	.1612	.3078
2	6	.3201	.1612	.3570
3	6	.3920	.1604	.4233
4	6	.4867	.1600	.5133
6	6	.7538	.1627	.7689

[a]Conductor impedances for an assumed geometric mean spacing of 4.69 ft and a conductor temperature of 50 °C. (See note below.) Impedances are in ohms per 1000 ft.

[b]For geometric mean spacing of 4.0 ft subtract 0.0034 from $X_1 = X_2$ and solve for $Z_1 = Z_2$. For geometric mean spacing of 3.5 ft subtract 0.0064 from $X_1 = X_2$ and solve for $Z_1 = Z_2$. For geometric mean spacing of 3.0 ft subtract 0.0100 from $X_1 = X_2$ and solve for $Z_1 = Z_2$. For geometric mean spacing of 5.0 ft add 0.0017 to $X_1 = X_2$ and solve for $Z_1 = Z_2$. $Z = \sqrt{R^2 + X^2}$.

[c]For temperatures other than 50 °C, convert resistances according to: $R_2/R_1 = (T + t_2)/(T + t_1)$, where R_1 is resistance of the conductor at temperature t_1 and R_2 is resistance of the conductor at temperature t_2. The constant T is approximately 234.5 for copper and 228.0 for aluminum.

Source: Distribution System Protection Manual, Bulletin 71022, McGraw-Edison Company, Power Systems Division. Reprinted with permission.

Table H.3 Transformer Losses

Typical values of no-load loss and total loss (%) for distribution transformers of standardized design (at 75°C)

	Voltage rating of primary winding																	
	2.4 kV		4.8 kV		7.2 kV		12 kV		24.9 Grd Y/ 14.4 kV		23 kV		34.5 kV		46 kV		69 kV	
kVA rating	No-load loss	Total loss	No-load loss	Total loss	No-load loss	Total loss	No-load loss	Total loss	No-load loss	Total loss	No-load loss	Total loss	No-load loss	Total loss	No-load loss	Total loss	No-load loss	Total loss
Single phase																		
5	.72	2.60	.77	2.84	.86	3.00	.90	3.24	1.02	3.82	—	—	—	—	—	—	—	—
10	.57	2.00	.57	2.20	.67	2.50	.68	2.60	.84	3.10	—	—	—	—	—	—	—	—
15	.52	1.95	.51	2.08	.60	2.30	.60	2.30	.70	2.80	—	—	—	—	—	—	—	—
25	.43	1.74	.43	1.88	.52	2.02	.52	2.04	.60	2.48	.80	2.92	.88	3.18	—	—	—	—
50	.35	1.58	.35	1.68	.43	1.77	.45	1.78	.58	2.06	.64	2.45	.72	2.54	.84	2.70	—	—
100	.32	1.45	.32	1.51	.37	1.54	.40	1.55	—	—	.52	1.97	.57	2.10	.66	2.23	.83	2.35
333	.34	1.42	.34	1.46	.34	1.36	.34	1.42	—	—	.37	1.41	.40	1.50	.44	1.59	.47	1.64
500	.29	1.25	.29	1.30	.29	1.25	.29	1.27	—	—	.33	1.27	.35	1.34	.38	1.38	.41	1.44
Three phase																		
9	1.00	2.94	1.00	3.17	1.00	3.47	—	—	—	—	—	—	—	—	—	—	—	—
15	1.80	2.66	.80	2.94	.96	3.14	1.04	3.46	—	—	—	—	—	—	—	—	—	—
30	.62	2.27	.62	2.46	.75	2.65	.79	2.84	—	—	—	—	—	—	—	—	—	—
75	.46	1.96	.46	2.08	.59	2.14	.63	2.20	—	—	—	—	—	—	—	—	—	—
150	.37	1.58	.37	1.72	.52	1.86	.54	1.92	—	—	.71	2.33	—	—	—	—	—	—
300	.36	1.63	.36	1.69	.46	1.77	.48	1.78	—	—	.58	1.93	.64	2.02	.75	2.15	—	—
500	.36	1.52	.36	1.54	.44	1.57	.45	1.57	—	—	.51	1.74	.55	1.75	.62	1.91	.74	1.96

Typical values for no-load loss and total loss (%) for network transformers (liquid-filled) of standardized design

Losses

kVA rating loss	5.0 kV		8.66 kV		15 kV		25 kV		34.5 kV	
	No-load loss	Total loss	No-load loss	Total loss	No-load loss	Total loss	No-load loss	Total loss	No-load loss	Total loss
300	.417	1.23	.417	1.23	.417	1.23	—	—	—	—
500	.354	1.16	.354	1.16	.354	1.16	.390	1.26	.436	1.38
750	.318	1.16	.318	1.16	.318	1.16	.354	1.17	.387	1.25
1000	.300[a]	0.99[a]	.300	0.99	.300	0.99	.330	1.08	.365	1.15
1500[a]	—	—	.273	0.98	.273	0.98	.097	1.03	.307	1.09
2000[a]	—	—	.260	0.93	.260	0.93	.275	0.97	.295	1.00
2500[a]	—	—	.238	0.92	.238	0.92	.248	0.96	.256	0.99

[a]Not available at 216/120 V. For temperatures different from 75°C, see the note in Appendix Table 2.

Source: Distribution Systems, Electric Utility Engineering Reference Book, vol. 3, Westinghouse Corporation. Reprinted with permission.

Table H.4a Transformer Impedances: Standard Reactances and Impedances for Ratings 500 kVA and Below (for 60-Cycle Transformers)

	Rated-voltage class (kV)							
	2.5		15		25		69	
Single-phase kVA rating[a]	Average reactance (%)	Average impedance (%)	Average reactance (%)	Average impedance (%)	Average reactance (%)	Average impedance (%)	Average reactance (%)	Average impedance (%)
3	1.1	2.2	0.8	2.8				
10	1.5	2.2	1.3	2.4	4.4	5.2		
25	2.0	2.5	1.7	2.3	4.8	5.2		
50	2.1	2.4	2.1	2.5	4.9	5.2	6.3	6.5
100	3.1	3.3	2.9	3.2	5.0	5.2	6.3	6.5
500	4.7	4.8	4.9	5.0	5.1	5.2	6.4	6.5

[a]For three-phase transformers use 1/3 of the three-phase kVA rating, and enter table with rated line-to-line voltages.

Table H.4b Transformer Impedances: Standard Range in Impedances for Two-Winding Power Transformers Rated at 55 °C Rise (Both 25- and 60-Cycle Transformers)

High-voltage winding insulation class (kV)	Low-voltage winding insulation class (kV)	Class OA OW OA/FA[a] OA/FA/FOA[a]		Class FOA FOW	
		Min.	Max.	Min.	Max.
15	15	4.5	7.0	6.75	10.5
25	15	5.5	8.0	8.25	12.0
34.5	15	6.0	8.0	9.0	12.0
	25	6.5	9.0	12.0	
46	25	6.5	9.0	9.75	13.5
	34.5	7.0	10.0	10.5	15.0
69	34.5	7.0	10.0	10.5	15.0
	46.	8.0	11.0	12.0	16.5
92	34.5	7.5	10.5	11.25	15.75
	69	8.5	12.5	12.75	18.75
115	34.5	8.0	12.0	12.0	18.0
	69	9.0	14.0	13.5	21.0
	92	10.0	15.0	15.0	23.25
138	34.5	8.5	13.0	12.75	19.5
	69	9.5	15.0	14.25	22.5
	115	10.5	17.0	15.75	25.5
161	46	9.5	15.0	13.5	21.0
	92	10.5	16.0	15.75	24.0
	138	11.5	18.0	17.25	27.0
196	46	10	15.0	15.0	22.5
	92	11.5	17.0	17.25	25.5
	161	12.5	19.0	18.75	28.5
230	46	11.0	16.0	16.5	24.0
	92	12.5	18.0	18.75	27.0
	161	14.0	20.0	21.0	30.0

[a]The impedances are expressed in percent on the self-cooled rating of OA/FA and OA/FA/FOA. Definition of transformer classes: OA—oil-immersed, self cooled; OW—oil-immersed, water-cooled; OA/FA—oil-immersed, self-cooled/forced-air-cooled; OA/FA/FOA—oil-immersed, self-cooled/forced-air-cooled/forced oil cooled; FOA—oil-immersed, forced-oil-cooled with forced air cooler; FOW—oil-immersed, forced-oil-cooled with water cooler.

Note: The through impedance of a two-winding autotransformer can be estimated knowing rated circuit voltages, by multiplying impedance obtained from this table by the factor $(HV - LV)/HV$.

Table H.4c Transformer Impedances: Typical Impedances of Distribution Transformers of Standardized Design

	Voltage rating of primary winding																
	2.4 kV		4.8 kV		7.2 kV		12 kV		24.9/14.4 Grd Y		23 kV		34.5 kV		46 kV		69 kV	
kVA rating	%R	%Z	%R	%Z	%R	%Z	%R	%Z	%R	%Z	%R	%Z	%R	%Z	%R	%Z	%R	%Z
Single phase																		
5	1.9	2.3	2.1	2.4	2.1	2.6	2.3	2.7	2.8	4.0	—	—	—	—	—	—	—	—
10	1.4	1.9	1.6	2.0	1.9	2.3	1.9	2.1	2.3	3.0	—	—	—	—	—	—	—	—
15	1.4	1.8	1.6	1.9	1.7	2.1	1.7	2.0	2.1	2.6	—	—	—	—	—	—	—	—
25	1.3	1.8	1.5	1.8	1.6	2.2	1.5	1.9	1.9	2.0	2.0	5.2	2.2	5.2	—	—	—	—
50	1.2	2.1	1.3	2.2	1.3	2.2	1.3	2.3	1.9	1.7	1.7	5.2	1.7	5.2	1.8	5.7	—	—
100	1.1	2.0	1.2	1.9	1.2	2.0	1.2	2.2	—	—	1.4	5.2	1.5	5.2	1.5	5.7	1.4	6.5
333	1.1	4.8	1.1	4.8	1.0	5.0	1.0	5.0	—	—	1.0	5.2	1.1	5.2	1.1	5.7	1.1	6.5
500	1.0	4.8	1.0	4.8	1.0	5.0	1.0	5.0	—	—	0.9	5.2	1.0	5.2	1.0	5.7	1.0	6.5
Three phase																		
9	2.0	2.4	2.2	2.4	2.5	2.5	—	—	—	—	—	—	—	—	—	—	—	—
15	1.9	2.5	2.1	2.5	2.2	2.6	2.4	2.8	—	—	—	—	—	—	—	—	—	—
30	1.6	2.4	1.8	2.5	1.9	2.6	2.1	3.1	—	—	—	—	—	—	—	—	—	—
75	1.5	3.2	1.6	3.2	1.6	2.9	1.6	3.3	—	—	—	—	—	—	—	—	—	—
150	1.2	4.2	1.4	4.3	1.3	3.5	1.4	4.3	—	—	1.6	5.2	—	—	—	—	—	—
300	1.3	4.8	1.3	4.8	1.3	5.0	1.3	5.0	—	—	1.3	5.2	1.4	5.2	1.4	5.7	—	—
500	1.2	4.8	1.2	4.8	1.1	5.0	1.1	5.0	—	—	1.2	5.2	1.2	5.2	1.3	5.7	1.2	6.5

Source: Transmission and Distribution, Westinghouse Corporation, Distribution Systems, Electric Utility Engineering Reference Book, vol. 3, Westinghouse Corporation. Reprinted with permission.

STANDARD PIPE AND TUBING SIZES

Table I.1 Standard Pipe Sizes

Nominal pipe size (in.)	Outside diameter (in.)	Schedule no.	Wall thickness (in.)	Inside diameter (in.)	Cross-sectional area metal (in.^{2})	Inside sectional area (ft^2)
1/8	0.405	40	0.068	0.269	0.072	0.00040
		80	0.095	0.215	0.093	0.00025
1/4	0.540	40	0.088	0.364	0.125	0.00072
		80	0.119	0.302	0.157	0.00050
3/8	0.675	40	0.091	0.493	0.167	0.00133
		80	0.126	0.423	0.217	0.00098
1/2	0.840	40	0.109	0.622	0.250	0.00211
		80	0.147	0.546	0.320	0.00163
		160	0.187	0.466	0.384	0.00118
3/4	1.050	40	0.113	0.824	0.333	0.00371
		80	0.154	0.742	0.433	0.00300
		160	0.218	0.614	0.570	0.00206
1	1.315	40	0.133	1.049	0.494	0.00600
		80	0.179	0.957	0.639	0.00499
		160	0.250	0.815	0.837	0.00362
1 1/2	1.900	40	0.145	1.610	0.799	0.01414
		80	0.200	1.500	1.068	0.01225
		160	0.281	1.338	1.429	0.00976
2	2.375	40	0.154	2.067	1.075	0.02330
		80	0.218	1.939	1.477	0.02050
		160	0.343	1.689	2.190	0.01556
2 1/2	2.875	40	0.203	2.469	1.704	0.03322
		80	0.276	2.323	2.254	0.02942
		160	0.375	2.125	2.945	0.02463
3	3.500	40	0.216	3.068	2.228	0.05130
		80	0.300	2.900	3.016	0.04587
		160	0.437	2.626	4.205	0.03761
4	4.500	40	0.237	4.026	3.173	0.08840
		80	0.337	3.826	4.407	0.07986
		120	0.437	3.626	5.578	0.07170
		160	0.531	3.438	6.621	0.06447
5	5.563	40	0.258	5.047	4.304	0.1390
		80	0.375	4.813	6.112	0.1263
		120	0.500	4.563	7.953	0.1136
		160	0.625	4.313	9.696	0.1015
6	6.625	40	0.280	6.065	5.584	0.2006
		80	0.432	5.761	8.405	0.1810
		120	0.562	5.501	10.71	0.1650
		160	0.718	5.189	13.32	0.1469

Table I.1 Standard Pipe Sizes (*Continued*)

Nominal pipe size (in.)	Outside diameter (in.)	Schedule no.	Wall thickness (in.)	Inside diameter (in.)	Cross-sectional area metal (in.^{2})	Inside sectional area (ft^2)
8	8.6.25	20	0.250	8.125	6.570	0.3601
		30	0.277	8.071	7.260	0.3553
		40	0.322	7.981	8.396	0.3474
		60	0.406	7.813	10.48	0.3329
		80	0.500	7.625	12.76	0.3171
		100	0.593	7.439	14.96	0.3018
		120	0.718	7.189	17.84	0.2819
		140	0.812	7.001	19.93	0.2673
		160	0.906	6.813	21.97	0.2532
10	10.75	20	0.250	10.250	8.24	0.5731
		30	0.307	10.136	10.07	0.5603
		40	0.365	10.020	11.90	0.5475
		60	0.500	9.750	16.10	0.5158
		80	0.593	9.564	18.92	0.4989
		100	0.718	9.314	22.63	0.4732
		120	0.843	9.064	26.24	0.4481
		140	1.000	8.750	30.63	0.4176
		160	1.125	8.500	34.02	0.3941
12	12.75	20	0.250	12.250	9.82	0.8185
		30	0.330	12.090	12.87	0.7972
		40	0.406	11.938	15.77	0.7773
		60	0.562	11.626	21.52	0.7372
		80	0.687	11.376	26.03	0.7058
		100	0.843	11.064	31.53	0.6677
		120	1.000	10.750	36.91	0.6303
		140	1.125	10.500	41.08	0.6013
		160	1.312	10.126	47.14	0.5592

Table I.2 Standard Tubing Gages

Outside diameter (in.)	Wall thickness B.W.G. and stubs' gage	(in.)	Inside diameter (in.)	Cross-sectional area (ft²)	Inside sectional area (ft²)
1/2	12	0.109	0.282	0.1338	0.000433
	14	0.083	0.334	0.1087	0.000608
	16	0.065	0.370	0.0888	0.000747
	18	0.049	0.402	0.0694	0.000882
	20	0.035	0.430	0.0511	0.001009
3/4	12	0.109	0.532	0.2195	0.00154
	13	0.095	0.560	0.1955	0.00171
	14	0.083	0.584	0.1739	0.00186
	15	0.072	0.606	0.1534	0.00200
	16	0.065	0.620	0.1398	0.00210
	17	0.058	0.634	0.1261	0.00219
	18	0.049	0.652	0.1079	0.00232
1	12	0.109	0.782	0.3051	0.00334
	13	0.095	0.810	0.2701	0.00358
	14	0.083	0.834	0.2391	0.00379
	15	0.072	0.856	0.2099	0.00400
	16	0.065	0.870	0.1909	0.00413
	17	0.058	0.884	0.1716	0.00426
	18	0.049	0.902	0.1463	0.00444
1 1/4	12	0.109	1.032	0.3907	0.00581
	13	0.095	1.060	0.3447	0.00613
	14	0.083	1.084	0.3042	0.00641
	15	0.072	1.106	0.2665	0.00677
	16	0.065	1.120	0.2419	0.00684
	17	0.058	1.134	0.2172	0.00701
	18	0.049	1.152	0.1848	0.00724
1 1/2	12	0.109	1.282	0.4763	0.00896
	13	0.095	1.310	0.4193	0.00936
	14	0.083	1.334	0.3694	0.00971
	15	0.072	1.358	0.3187	0.0100
	16	0.065	1.370	0.2930	0.0102
	17	0.058	1.384	0.2627	0.0107
	18	0.049	1.402	0.2234	0.0109
1 3/4	10	0.134	1.482	0.6803	0.0120
	11	0.120	1.510	0.6145	0.0124
	12	0.109	1.532	0.5620	0.0128
	13	0.095	1.560	0.4939	0.0133
	14	0.083	1.584	0.4346	0.0137
	15	0.072	1.606	0.3796	0.0141
	16	0.065	1.620	0.3441	0.0143
2	10	0.134	1.732	0.7855	0.0164
	11	0.120	1.760	0.7084	0.0169
	12	0.109	1.782	0.6475	0.0173
	13	0.095	1.810	0.5686	0.0179
	14	0.083	1.834	0.4998	0.0183
	15	0.072	1.856	0.4359	0.0188
	16	0.065	1.870	0.3951	0.0191

INDEX